T0132607

More Concise Algebraic Topology

CHICAGO LECTURES IN MATHEMATICS SERIES
Editors: Spencer J. Bloch, Peter Constantin, Benson Farb,
Norman R. Lebovitz, Carlos Kenig, and J. P. May

Other *Chicago Lectures in Mathematics* titles available from the University of Chicago Press

Simplical Objects in Algebraic Topology, by J. Peter May (1967, 1993)

Fields and Rings, Second Edition, by Irving Kaplansky (1969, 1972)

Lie Algebras and Locally Compact Groups, by Irving Kaplansky (1971)

Several Complex Variables, by Raghavan Narasimhan (1971)

Torsion-Free Modules, by Eben Matlis (1973)

Stable Homotopy and Generalised Homology, by J. F. Adams (1974)

Rings with Involution, by I.N. Herstein (1976)

Theory of Unitary Group Representation, by George V. Mackey (1976)

Commutative Semigroup Rings, by Robert Gilmer (1984)

Infinite-Dimensional Optimization and Convexity, by Ivar Ekeland and Thomas Turnbull (1983)

Navier-Stokes Equations, by Peter Constantin and Ciprian Foias (1988)

Essential Results of Functional Analysis, by Robert J. Zimmer (1990)

Fuchsian Groups, by Svetlana Katok (1992)

Unstable Modules over the Steenrod Algebra and Sullivan's Fixed Point Set Conjecture,
    by Lionel Schwartz (1994)

Topological Classification of Stratified Spaces, by Shmuel Weinberger (1994)

Lectures on Exceptional Lie Groups, by J. F. Adams (1996)

Geometry of Nonpositively Curved Manifolds, by Patrick B. Eberlein (1996)

*Dimension Theory in Dynamical Systems: Contemporary Views
    and Applications*, by Yakov B. Pesin (1997)

Harmonic Analysis and Partial Differential Equations: Essays in Honor of Alberto Calderón,
    edited by Michael Christ, Carlos Kenig, and Cora Sadosky (1999)

A Concise Course in Algebraic Topology, by J. P. May (1999)

Topics in Geometric Group Theory, by Pierre de la Harpe (2000)

Exterior Differential Systems and Euler-Lagrange Partial Differential Equations,
    by Robert Bryant, Phillip Griffiths, and Daniel Grossman (2003)

Ratner's Theorems on Unipotent Flows, by Dave Witte Morris (2005)

Geometry, Rigidity, and Group Actions, edited by Benson Farb and David Fisher (2011)

Groups of Circle Diffeomorphisms, by Andrés Navas (2011)

More Concise Algebraic Topology

LOCALIZATION, COMPLETION, AND MODEL CATEGORIES

J. P. May and K. Ponto

THE UNIVERSITY OF CHICAGO PRESS · CHICAGO AND LONDON

J. P. May is professor of mathematics at the University of Chicago; he is author or coauthor of many papers and books, including *Simplicial Objects in Algebraic Topology* and *A Concise Course in Algebraic Topology* in this series.

K. Ponto is assistant professor of mathematics at the University of Kentucky; she is an algebraic topologist whose research has focused on topological fixed point theory.

The University of Chicago Press, Chicago 60637
The University of Chicago Press, Ltd., London
© 2012 by The University of Chicago
All rights reserved. Published 2012.
Printed in the United States of America
20 19 18 17 16 15 14 13   2 3 4 5

ISBN-13: 978-0-226-51178-8 (cloth)
ISBN-10: 0-226-51178-2 (cloth)

Library of Congress Cataloging-in-Publication Data

May, J. P.
  More concise algebraic topology : localization, completion, and model categories / J. P. May and K. Ponto.
      p.   cm. — (Chicago lectures in mathematics series)
   ISBN-13: 978-0-226-51178-8 (hardcover : alk. paper)
   ISBN-10: 0-226-51178-2 (hardcover : alk. paper)
  1. Algebraic topology.   I. Ponto, Kate.   II. Title.   III. Series: Chicago lectures in mathematics.
   QA612.M388   2012
   514'.2—dc22   2011012400

⊗ This paper meets the requirements of ANSI/NISO Z39.48-1992 (Permanence of Paper).

Contents

PART 4 || *An Introduction to Model Category Theory*

Introduction

There is general agreement on the rudiments of algebraic topology, the things that every mathematician should know. This material might include the fundamental group; covering spaces; ordinary homology and cohomology in its singular, cellular, axiomatic, and represented versions; higher homotopy groups and the Hurewicz theorem; basic homotopy theory including fibrations and cofibrations; Poincaré duality for manifolds and for manifolds with boundary. The rudiments should also include a reasonable amount of categorical language and at least enough homological algebra for the universal coefficient and Künneth theorems. This material is treated in such recent books as [3, 34, 36, 59, 93] and many earlier ones. What next? Possibly K-theory, which is treated in [3] and, briefly, [93], and some idea of cobordism theory [36, 93]. None of the most recent texts goes much beyond the material just mentioned, all of which dates at latest from the early 1960s. Regrettably, only one of these texts, [34], includes anything about spectral sequences, but [79, 98, 123] help make up for that.

The subject of algebraic topology is very young. Despite many precursors and earlier results, firm foundations only date from the landmark book of Eilenberg and Steenrod [45], which appeared in 1952. It is not an exaggeration to say that even the most recent published texts do not go beyond the first decade or so of the serious study of the subject. For that reason, people outside the field very often know little or nothing about some of its fundamental branches that have been developed over the past half century. A partial list of areas a student should learn is given in the suggestions for further reading of [93], and a helpful guide to further development of the subject (with few proofs) has been given by Selick [123].

It seems to us that the disparity between the lack of accessibility of the published sources and the fundamental importance of the material is nowhere greater than in the theory of localization and completion of topological spaces.[1] It makes little more sense to consider modern algebraic topology without localization and completion of spaces than it does to consider modern algebra without localization and completion of rings. These tools have been in common use ever since they were introduced in the early 1970s. Many papers in algebraic topology start with the blanket assumption that all spaces are to be localized or completed at a given prime p. Readers of such papers are expected to know what this means. Experts know that these constructions can be found in such basic 1970s references as [21, 62, 133]. However, the standard approaches favored by the experts are not easily accessible to the novices, especially in the case of completions. In fact, these notions can and should be introduced at a much more elementary level. The notion of completion is particularly important because it relates directly to mod p cohomology, which is the invariant that algebraic topologists most frequently compute.

In the first half of this book, we set out the basic theory of localization and completion of nilpotent spaces. We give the most elementary treatment we know, making no use of simplicial techniques or model categories. We assume only a little more than a first course in algebraic topology, such as can be found in [3, 34, 36, 59, 93]. We require and provide more information about some standard topics, such as fibration and cofibration sequences, Postnikov towers, and homotopy limits and colimits, than appears in those books, but this is fundamental material of independent interest. The only other preliminary that we require and that cannot be found in most of the books cited above is the Serre spectral sequence. There are several accessible sources for that, such as [34, 79, 98, 123], but to help make this book more self-contained, we give a concise primer on spectral sequences in Chapter 24; it is taken from 1960s notes of the first author and makes no claim to originality.

The second half of the book is quite different and consists of two parts that can be read independently of each other and of the first half. While written with algebraic topologists in mind, both parts should be of more general interest. They are devoted to topics in homotopical algebra and in pure algebra that are needed by all algebraic topologists and many others. By far the longer of these parts is an introduction to model category theory. This material can easily be overemphasized, to the detriment of concrete results and the nuances needed to prove them. For example, its use would in no way simplify anything in the first half of the book. However, its use allows us to complete the first half by giving a conceptual construction and characterization of localizations and completions of general, not necessarily nilpotent, spaces. More

---

1. These topics are not mentioned in [3, 36, 59].

fundamentally, model category theory has become the central organizational principle of homotopical algebra, a subject that embraces algebraic topology, homological algebra, and much modern algebraic geometry. Anybody interested in any of these fields needs to know model category theory. It plays a role in homotopical algebra analogous to the role played by category theory in mathematics. It gives a common language for the subject that greatly facilitates comparisons, and it allows common proofs of seemingly disparate results.

The short last part of the book is something of a bonus track, in that it is peripheral to the main thrust of the book. It develops the basic theory of bialgebras and Hopf algebras. Its main point is a redevelopment of the structure theory of Hopf algebras, due originally to Milnor and Moore [104] but with an addendum from [85]. Hopf algebras are used in several places in the first half of the book, and they are fundamental to the algebra of algebraic topology.

We say a bit about how our treatments of these topics developed and how they are organized. The starting point for our exposition of localization and completion comes from unpublished lecture notes of the first author that date from sometime in the early 1970s. That exposition attempted a synthesis in which localization and completion were treated as special cases of a more general elementary construction. The synthesis did not work well because it obscured essential differences. Those notes were reworked to a more accessible form by the second author and then polished to publishable form by the authors working together. There are some new results, but we make little claim to originality. Most of the results and many of the proofs are largely the same as in one or another of Bousfield and Kan [21], Sullivan [133], and Hilton, Mislin, and Roitberg [62].

However, a central feature of the subject is the fracture theorems for the passage back and forth between local and global information. It is here that the treatments of localization and completion differ most from each other. The relevant material has been reworked from scratch, and the treatment in the first author's 1970s notes has largely been jettisoned. In fact, the literature in this area requires considerable clarification, and we are especially concerned to give coherent accounts of the most general and accurate versions of the fracture theorems for nilpotent spaces. These results were not fully understood at the time the primary sources [21, 62, 133] were written, and there are seriously incorrect statements in some of the important early papers. Moreover, generalizations of the versions of these results that appear in the primary sources were proven after they were written and can only be found in relatively obscure papers that are known just to a few experts. We have introduced several new ideas that we think clarify the theorems, and we have proven some results that are essential for full generality and that we could not find anywhere in the literature.

The first half of the book is divided into three parts: preliminaries, localizations, and completions. The reader may want to skim the first part, referring back to it as needed. Many of the preliminaries are essential for the later parts, but mastery of their details is not needed on a first reading. Specific indications of material that can be skipped are given in the introductions to the first four chapters. The first chapter is about cofibrations, fibrations, and actions by the fundamental group. The second is about elementary homotopy colimits and homotopy limits and $\lim^1$ exact sequences. This both sets the stage for later work and rounds out material that was omitted from [93] but that all algebraic topologists should know. The third chapter deals with nilpotent spaces and their approximation by Postnikov towers, giving a more thorough treatment of the latter than can be found in existing expository texts. This is the most essential preliminary to our treatment of localizations and completions.

The fourth chapter shows how to prove that various groups and spaces are nilpotent and is more technical; although it is logically placed, the reader may want to return to it later. The reader might be put off by nilpotent spaces and groups at a first reading. After all, the vast majority of applications involve simply connected or, more generally, simple spaces. However, the proofs of some of the fracture theorems make heavy use of connected components of function spaces $F(X, Y)$. Even when X and Y are simply connected CW complexes, these spaces are rarely simple, but they are nilpotent when X is finite. Moreover, nilpotent spaces provide exactly the right level of generality for an elementary exposition, and the techniques used to prove results for nilpotent spaces are not very different from those used for simple spaces.

We say just a bit about the literature for spaces that are not nilpotent and about alternative constructions. There are several constructions of localizations and completions of general spaces that agree when restricted to nilpotent spaces. The most important of these is Bousfield localization, which we shall construct model theoretically. These more general constructions are still not well understood calculationally, and knowledge of them does not seem helpful for understanding the most calculationally important properties of localization and completion, such as their homotopical behavior and the fracture theorems.

We construct localizations of abelian groups, nilpotent groups, and nilpotent spaces in Chapter 5, and we construct completions of abelian groups, nilpotent groups, and nilpotent spaces in the parallel Chapter 10. We characterize localizations and describe their behavior under standard topological constructions in Chapter 6, and we do the same for completions in the parallel Chapter 11. We prove the fracture theorems for localizations in Chapters 7 and 8 and the fracture theorems for completions in the parallel Chapters 12 and 13. In many cases, the same results, with a few words changed, are considered in the same order in the cases of localization and completion. This is intentional, and it allows us to explain and emphasize both similarities and differences. As we have already indicated, although these chapters are parallel, it substantially clarifies the constructions and results not to subsume both under a single general construction. We give a few results about rationalization of spaces in Chapter 9.

We say a little here about our general philosophy and methodology, which goes back to a paper of the first author [91] on "The dual Whitehead theorems". As is explained there and will be repeated here, we can dualize the proof of the first theorem below (as given for example in [93]) to prove the second.

THEOREM 0.0.1. *A weak homotopy equivalence* $e : Y \longrightarrow Z$ *between CW complexes is a homotopy equivalence.*

THEOREM 0.0.2. *An integral homology isomorphism* $e : Y \longrightarrow Z$ *between simple spaces is a weak homotopy equivalence.*

The argument is based on the dualization of cell complexes to cocell complexes, of which Postnikov towers are examples, and of the Homotopy Extension and Lifting Property (HELP) to co-HELP. Once this dualization is understood, it becomes almost transparent how one can construct and study the localizations and completions of nilpotent spaces simply by inductively localizing or completing their Postnikov towers one cocell at a time. Our treatment of localization and completion is characterized by a systematic use of cocellular techniques dual to familiar cellular techniques. In the case of localization, but not of completion, there is a dual cellular treatment applicable to simply connected spaces.

We turn now to our treatment of model categories, and we first try to answer an obvious question. There are several excellent introductory sources for model category theory [43, 54, 65, 66, 113]. Why add another one? One reason is that, for historical reasons, the literature of model category theory focuses overwhelmingly on a simplicial point of view, and especially on model categories enriched in simplicial sets. There is nothing wrong with that point of view, but it obscures essential features that are present in the classical contexts of algebraic topology and homological algebra and that are not present in the simplicial context. Another reason is that we feel that some of the emphasis in the existing literature focuses on technicalities at the expense of the essential conceptual simplicity of the ideas.

We present the basic general theory of model categories and their associated homotopy categories in Chapter 14. For conceptual clarity, we offer a slight reformulation of the original definition of a model category that focuses on weak factorization systems (WFSs): a model category consists of a subcategory of weak equivalences together with a pair of related WFSs. This point of view separates out the main constituents of the definition in a way that we find illuminating. We discuss compactly generated and cofibrantly generated model categories in Chapter 15; we shall describe the difference shortly. We also describe proper model categories in Chapter 15. We give essential categorical perspectives in Chapter 16.

To make the general theory flow smoothly, we have deferred examples to the parallel Chapters 17 and 18. On a first reading, the reader may want to skip directly from Chapter 14 to Chapters 17 and 18. These chapters treat model structures on categories of spaces and on categories of chain complexes in parallel. In both, there are three intertwined model structures, which we call the h-model structure, the q-model structure, and the m-model structure.

The h stands for homotopy equivalence or Hurewicz. The weak equivalences are the homotopy equivalences, and the cofibrations and fibrations are defined by the HEP (Homotopy Extension Property) and the CHP (Covering Homotopy Property). Such fibrations were first introduced by Hurewicz. The q stands for Quillen or quasi-isomorphism. The weak equivalences are the weak equivalences of spaces or the quasi-isomorphisms of chain complexes. The fibrations are the Serre fibrations of spaces or the epimorphisms of chain complexes, and the cofibrations in both cases are the retracts of cell complexes.

The m stands for mixed, and the m-model structures, due to Cole [33], combine the good features of the h- and q-model structures. The weak equivalences are the q-equivalences and the fibrations are the h-fibrations. The m-cofibrant objects are the spaces of the homotopy types of CW complexes or the chain complexes of the homotopy types of complexes of projective modules (at least in the bounded below case). We argue that classical algebraic topology, over at least the mathematical lifetime of the first author, has implicitly worked in the m-model structure. For example, the first part of this book implicitly works there. Modern approaches to classical homological algebra work similarly. We believe that this trichotomy of model category structures, and especially the precise analogy between these structures in topology and algebra, gives the best possible material for an introduction to model category theory. We reiterate that these features are not present in the simplicial world, which in any case is less familiar to those just starting out.

The m-model structure can be viewed conceptually as a colocalization model structure, which we rename a resolution model structure. In our examples, it codifies CW approximation of spaces or projective resolutions of modules, where these are explicitly understood as up to homotopy constructions. Colocalization is dual to Bousfield localization, and this brings us to another reason for our introduction to model category theory, namely a perceived need for as simple and accessible an approach to Bousfield localization as possible. This is such a centrally important tool in modern algebraic topology (and algebraic geometry) that every student should see it. We give a geodesic development that emphasizes the conceptual idea and uses as little special language as possible.

In particular, we make no use of simplicial theory and minimal use of cofibrantly generated model categories, which were developed historically as a codification of the methods Bousfield introduced in his original construction of localizations [16]. An idiosyncratic feature of our presentation of model category theory is that we emphasize a dichotomy between cofibrantly generated model categories and compactly generated model categories. The small object argument for constructing the WFSs in these model structures is presented in general, but in the most basic examples it can be applied using only cell complexes of the familiar form colim X_n, without use of cardinals bigger than ω. Transfinite techniques are essential to the theory of localization, but we feel that the literature focuses on them to an inordinate extent. Disentangling the optional from the essential use of such methods leads to a more user-friendly introduction to model category theory.

Although we make no use of it, we do describe the standard model structure on simplicial sets. In the literature, the proof of the model axioms is unpleasantly lengthy. We sketch a new proof, due to Bousfield and the first author, that is shorter and focuses more on basic simplicial constructions and less on the intertwining of simplicial and topological methods.

The last part of the book, on bialgebras and Hopf algebras, is again largely based on unpublished notes of the first author that date from the 1970s. Since we hope our treatment has something to offer to algebraists as well as topologists, one introductory remark is obligatory. In algebraic topology, algebras are always graded and often connected, meaning that they are zero in negative degrees and the ground field in degree zero. Under this assumption, bialgebras automatically have antipodes, so that there is no distinction between Hopf algebras and bialgebras. For this reason, and for historical reasons, algebraic topologists generally use the term "Hopf algebra" for both notions, but we will be careful about the distinction.

Chapter 20 gives the basic theory as used in all subjects and Chapter 21 gives features that are particularly relevant to the use of Hopf algebras in algebraic topology, together with quick applications to cobordism and K-theory. Chapters 22 and 23 give the structure theory for Hopf algebras in characteristic zero and in positive characteristic, respectively. The essential organizing principle of these two chapters is that all of the main theorems on the structure of connected Hopf algebras can be derived from the Poincaré-Birkhoff-Witt theorem on the structure of Lie algebras (in characteristic zero) and of restricted Lie algebras (in positive characteristic). The point is that passage to associated graded algebras from the augmentation ideal filtration gives a primitively generated Hopf algebra, and such Hopf algebras are universal enveloping Hopf algebras of Lie algebras or of restricted Lie algebras. This point of view is due to [104], but we will be a little more explicit.

Our point of view derives from the first author's thesis, in which the cited filtration was used to construct a spectral sequence for the computation of the cohomology of Hopf algebras starting from the cohomology of (restricted) Lie algebras [84], and from his short paper [85]. This point of view allows us to simplify the proofs of some of the results in [104], and it shows that the structure theorems are more widely applicable than seems to be known.

Precisely, the structure theorems apply to ungraded bialgebras and, more generally, to nonconnected graded bialgebras whenever the augmentation ideal filtration is complete. We emphasize this fact in view of the current interest in more general Hopf algebras, especially the quantum groups. A cocommutative Hopf algebra (over a field) is a group in the cartesian monoidal category of coalgebras, the point being that the tensor product of cocommutative coalgebras is their categorical product. In algebraic topology, the Hopf algebras that arise naturally are either commutative or cocommutative. By duality, one may as well focus on the cocommutative case. It was a fundamental insight of Drinfel'd that dropping cocommutativity allows very interesting examples with "quantized" deviation from cocommutativity. These are the "quantum groups". Because we are writing from the point of view of algebraic topology, we shall not say anything about them here, but the structure theorems are written with a view to possible applications beyond algebraic topology.

Sample applications of the theory of Hopf algebras within algebraic topology are given in several places in the book, and they pervade the subject as a whole. In Chapter 9, the structure theory for rational Hopf algebras is used to describe the category of rational H-spaces and to explain how this information

is used to study H-spaces in general. In Chapter 22, we explain the Hopf algebra proof of Thom's calculation of the real cobordism ring and describe how
the method applies to other unoriented cobordism theories. We also give the
elementary calculational proof of complex Bott periodicity. Of course, there is
much more to be said here. Our goal is to highlight for the beginner important
sample results that show how directly the general algebraic theorems relate to
the concrete topological applications.

Some conventions and notations

This book is perhaps best viewed as a sequel to [93], although we have tried to make it reasonably self-contained. Aside from use of the Serre spectral sequence, we assume no topological preliminaries that are not to be found in [93], and we redo most of the algebra that we use.

To keep things familiar, elementary, and free of irrelevant pathology, we work throughout the first half in the category $\mathscr{U}$ of compactly generated spaces (see [93, Ch. 5]). It is by now a standard convention in algebraic topology that spaces mean compactly generated spaces, and we adopt that convention. While most results will not require this, we implicitly restrict to spaces of the homotopy types of CW complexes whenever we talk about passage to homotopy. This allows us to define the homotopy category $\mathrm{Ho}\mathscr{U}$ simply by identifying homotopic maps; it is equivalent to the homotopy category of all spaces in $\mathscr{U}$, not necessarily CW homotopy types, that is formed by formally inverting the weak homotopy equivalences.

We nearly always work with based spaces. To avoid pathology, we assume once and for all that basepoints are nondegenerate, meaning that the inclusion $\ast \longrightarrow X$ is a cofibration (see [93, p. 56]). We write $\mathscr{T}$ for the category of nondegenerately based compactly generated spaces, that is, nondegenerately based spaces in $\mathscr{U}$.[1] Again, whenever we talk about passage to homotopy, we implicitly restrict to spaces of the based homotopy type of CW complexes. This allows us to define the homotopy category $\mathrm{Ho}\mathscr{T}$ by identifying maps that are homotopic in the based sense, that is, through homotopies h such that each h_t is a based map. The category $\mathscr{T}$, and its restriction to CW homotopy types, has been the preferred working place of algebraic topologists for very many

---

1. This conflicts with [93], where $\mathscr{T}$ was defined to be the category of based spaces in $\mathscr{U}$ (denoted $\mathscr{U}_\ast$ here). That choice had the result that the "nondegenerately based" hypothesis reappears with monotonous regularity in [93].

years; for example, the first author has worked explicitly in this category ever since he wrote [87], around forty years ago.

We ask the reader to accept these conventions and not to quibble if we do not repeat these standing assumptions in all of our statements of results. The conventions mean that, when passing to homotopy categories, we implicitly approximate all spaces by weakly homotopy equivalent CW complexes, as we can do by [93, §10.5]. In particular, when we use Postnikov towers and pass to limits, which are not of the homotopy types of CW complexes, we shall implicitly approximate them by CW complexes. We shall be a little more explicit about this in Chapters 1 and 2, but we shall take such CW approximation for granted in later chapters.

The expert reader will want a model theoretic justification for working in $\mathscr{T}$. First, as we explain in §17.1, the category $\mathscr{U}_*$ of based spaces in $\mathscr{U}$ inherits an h-model, or Hurewicz model, structure from $\mathscr{U}$. In that model structure, all objects are fibrant and the cofibrant objects are precisely the spaces in $\mathscr{T}$. Second, as we explain in §17.4, $\mathscr{U}_*$ also inherits an m-model, or mixed model, structure from $\mathscr{U}$. In that model structure, all objects are again fibrant and the cofibrant objects are precisely the spaces in $\mathscr{T}$ that have the homotopy types of based CW complexes. Cofibrant approximation is precisely approximation of spaces by weakly homotopy equivalent CW complexes in $\mathscr{T}$. This means that working in $\mathscr{T}$ and implicitly approximating spaces by CW complexes is part of the standard model-theoretic way of doing homotopy theory. The novice will learn later in the book how very natural this language is, but it plays no role in the first half. We believe that to appreciate model category theory, the reader should first have seen some serious homotopical algebra, such as the material in the first half of this book.

It is convenient to fix some notations that we shall use throughout.

NOTATIONS 0.0.3. We fix some notations concerning based spaces.

(i) Spaces are assumed to be path connected unless explicitly stated otherwise, and we use the word connected to mean path connected from now on. We also assume that all given spaces X have universal covers, denoted $\tilde{X}$.

(ii) For based spaces X and Y, let $[X, Y]$ denote the set of maps $X \longrightarrow Y$ in Ho$\mathscr{T}$; equivalently, after CW approximation of X if necessary, it is the set of based homotopy classes of based maps $X \longrightarrow Y$.

(iii) Let $F(X, Y)$ denote the space of based maps $X \longrightarrow Y$. It has a canonical basepoint, namely the trivial map. We write $F(X, Y)_f$ for the component of a map f and give it the basepoint f. When using these notations, we can allow Y to be a general space, but to have the right weak homotopy type we must insist that X has the homotopy type of a CW complex.

(iv) The smash product $X \wedge Y$ of based spaces X and Y is the quotient of the product $X \times Y$ by the wedge (or one-point union) $X \vee Y$. We have adjunction homeomorphisms

$$F(X \wedge Y, Z) \cong F(X, F(Y, Z))$$

and consequent bijections

$$[X \wedge Y, Z] \cong [X, F(Y, Z)].$$

(v) For an unbased space K, let K_+ denote the union of K and a disjoint basepoint. The based cylinder $X \wedge (I_+)$ is obtained from $X \times I$ by collapsing the line through the basepoint of X to a point. Similarly, we have the based cocylinder $F(I_+, Y)$. It is the space of unbased maps $I \longrightarrow Y$ based at the constant map to the basepoint. These specify the domain and, in adjoint form, the codomain of based homotopies, that is, homotopies that are given by based maps $h_t \colon X \longrightarrow Y$ for $t \in I$.

We also fix some algebraic notations and point out right away some ways that algebraic topologists think differently than algebraic geometers and others about even very basic algebra.

NOTATIONS 0.0.4. Let T be a fixed set of primes and p a single prime.

(i) Let $\mathbb{Z}_T$ denote the ring of integers localized at T, that is, the subring of $\mathbb{Q}$ consisting of rationals expressible as fractions k/ℓ, where ℓ is a product of primes *not* in T. We let $\mathbb{Z}[T^{-1}]$ denote the subring of fractions k/ℓ, where ℓ is a product of primes in T. In particular, $\mathbb{Z}[p^{-1}]$ has only p inverted. Let $\mathbb{Z}_{(p)}$ denote the ring of integers localized at the prime ideal (p) or, equivalently, at the singleton set $\{p\}$.

(ii) Let $\mathbb{Z}_p$ denote the ring of p-adic integers. Illogically, but to avoid conflict of notation, we write $\hat{\mathbb{Z}}_T$ for the product over $p \in T$ of the rings $\mathbb{Z}_p$. We then write $\hat{\mathbb{Q}}_T$ for the ring $\hat{\mathbb{Z}}_T \otimes \mathbb{Q}$; when $T = \{p\}$, this is the ring of p-adic numbers.

(iii) Let $\mathbb{F}_p$ denote the field with p elements and $\mathbb{F}_T$ denote the product over $p \in T$ of the fields $\mathbb{F}_p$. Let $\mathbb{Z}/n$ denote the quotient group $\mathbb{Z}/n\mathbb{Z}$. We sometimes consider the ring structure on $\mathbb{Z}/n$, and then $\mathbb{Z}/p = \mathbb{F}_p$ for a prime p.

(iv) We write $A_{(p)}$ and $\hat{A}_p$ for the localization at p and the p-adic completion of an abelian group A. Thus $\hat{\mathbb{Z}}_p$ is the underlying abelian group of the ring $\mathbb{Z}_p$.

(v) We write A_T and $\hat{A}_T$ for the localization and completion of A at T; the latter is the product over $p \in T$ of the $\hat{A}_p$.

(vi) Let $\mathscr{A}b$ denote the category of abelian groups. We sometimes ignore the maps and use the notation $\mathscr{A}b$ for the collection of all abelian groups. More generally, $\mathscr{A}$ will denote any collection of abelian groups that contains 0.

(vii) We often write $\otimes$, Hom, Tor, and Ext for $\otimes_{\mathbb{Z}}$, $\mathrm{Hom}_{\mathbb{Z}}$, $\mathrm{Tor}_1^{\mathbb{Z}}$, and $\mathrm{Ext}_{\mathbb{Z}}^1$. We assume familiarity with these functors.

WARNING 0.0.5. We warn the reader that algebraic notations in the literature of algebraic topology have drifted over time and are quite inconsistent. The reader may find $\mathbb{Z}_p$ used for either our $\mathbb{Z}_{(p)}$ or for our $\mathbb{F}_p$; the latter choice is used ubiquitously in the "early" literature, including most of the first author's papers. In fact, regrettably, we must warn the reader that $\mathbb{Z}_p$ means $\mathbb{F}_p$ in the book [93]. The p-adic integers only began to be used in algebraic topology in the 1970s, and old habits die hard. In both the algebraic and topological literature, the ring $\mathbb{Z}_p$ is sometimes denoted $\hat{\mathbb{Z}}_p$; we would prefer that notation as a matter of logic, but the notation $\mathbb{Z}_p$ has by now become quite standard.

WARNING 0.0.6 (CONVENTIONS ON GRADED ALGEBRAIC STRUCTURES). We think of homology and cohomology as graded abelian groups. For most algebraists, a graded abelian group A is the direct sum over degrees of its homogeneous subgroups A_n, or, with cohomological grading, A^n. In algebraic topology, unless explicitly stated otherwise, when some such notation as H^{**} is often used, graded abelian groups mean sequences of abelian groups A_n. That is, algebraic topologists do not usually allow the addition of elements of different degrees. To see just how much difference this makes, consider a Laurent series algebra $k[x, x^{-1}]$ over a field k, where x has positive even degree. To an algebraic topologist, this is a perfectly good graded field: every nonzero element is a unit. To an algebraist, it is not. This is not an esoteric difference. With $k = \mathbb{F}_p$, such graded fields appear naturally in algebraic topology as the coefficients of certain generalized cohomology theories, called Morava K-theories, and their homological algebra works exactly as for any other field, a fact that has real calculational applications.

The tensor product $A \otimes B$ of graded abelian groups is specified by

$$(A \otimes B)_n = \sum_{p+q=n} A_p \otimes B_q.$$

In categorical language, the category $\mathscr{A}b_*$ of graded abelian groups is a symmetric monoidal category under $\otimes$, meaning that $\otimes$ is unital (with unit $\mathbb{Z}$ concentrated in degree 0), associative, and commutative up to coherent natural isomorphisms. Here again, there is a difference of conventions. For an algebraic topologist, the commutativity isomorphism $\gamma : A \otimes B \longrightarrow B \otimes A$ is specified by

$$\gamma(a \otimes b) = (-1)^{pq} b \otimes a$$

where $\deg(a) = p$ and $\deg(b) = q$. A graded k-algebra with product ϕ is commutative if $\phi \circ \gamma = \phi$; elementwise, this means that $ab = (-1)^{pq} ba$. In the algebraic literature, such an algebra is said to be graded commutative or sometimes even supercommutative, but in algebraic topology this notion of commutativity is and always has been the default (at least since the early 1960s). Again, this is not an esoteric difference. To an algebraic topologist, a polynomial algebra $k[x]$ where x has odd degree is not a commutative k-algebra unless k has characteristic 2. The homology $H_*(\Omega S^n; k)$, n even, is an example of such a noncommutative algebra.

The algebraist must keep these conventions in mind when reading the material about Hopf algebras in this book. To focus on commutativity in the algebraist's sense, one can double the degrees of all elements and so eliminate the appearance of odd degree elements.

Acknowledgments

Many people have helped us with this book in a variety of ways. Several generations of the senior author's students have had input during its very long gestation. Notes in the 1970s by Zig Fiedorowicz influenced our treatment of nilpotent spaces and Postnikov towers, and Lemma 3.4.2 is due to him. The mixed or resolution model structures that play a central role in our treatment of model structures on spaces and chain complexes are due to Mike Cole. Part of the treatment of model structures comes from the book [97] with Johann Sigurdsson. Notes and a 2009 paper [118] by Emily Riehl influenced our treatment of the basic definitions in model category theory. She and Mike Shulman, as a postdoc, made an especially thorough reading of the model category theory part and found many mistakes and infelicities. Mohammed Abouzaid, Anna Marie Bohmann, Rolf Hoyer, and John Lind read and commented on several parts of the book. Mona Merling went through Part 1 with a meticulous eye to excesses of concision. Many other students over the past thirty plus years have also had input.

We have many thanks to offer others. John Rognes texed §21.4 for his own use in 1996. Kathryn Hess suggested that we include a treatment of model category theory, which we had not originally intended, hence she is responsible for the existence of that part of the book. Bill Dwyer helped us with the fracture theorems. We owe an especially big debt to Pete Bousfield. He gave us many insights and several proofs that appear in the chapters on the fracture theorems and in the sections on Bousfield localization. Moreover, our treatment of the model structure on the category of simplicial sets arose from correspondence with him and is primarily his work.

We also thank an anonymous reviewer for both complimentary words and cogent criticism.

January 1, 2010

PART 1

Preliminaries: Basic homotopy theory and nilpotent spaces

1

COFIBRATIONS AND FIBRATIONS

We shall make constant use of the theory of fibration and cofibration sequences, and this chapter can be viewed as a continuation of the basic theory of such sequences as developed in [93, Ch. 6–8]. We urge the reader to review that material, although we shall recall most of the basic definitions as we go along. The material here leads naturally to such more advanced topics as model category theory [65, 66, 97], which we turn to later, and triangulated categories [94, 111, 138]. However, we prefer to work within the more elementary foundations of [93] in the first half of this book. We concentrate primarily on just what we will use later, but we round out the general theory with several related results that are of fundamental importance throughout algebraic topology. The technical proofs in §3 and the details of §4 and §5 should not detain the reader on a first reading.

1.1. Relations between cofibrations and fibrations

Remember that we are working in the category $\mathcal{T}$ of nondegenerately based compactly generated spaces. Although the following folklore result was known long ago, it is now viewed as part of Quillen model category theory, and its importance can best be understood in that context. For the moment, we view it as merely a convenient technical starting point.

LEMMA 1.1.1. *Suppose that i is a cofibration and p is a fibration in the following diagram of based spaces, in which $p \circ g = f \circ i$.*

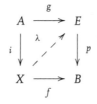

If either i or p is a homotopy equivalence, then there exists a map λ such that the diagram commutes.

This result is a strengthened implication of the definitions of cofibrations and fibrations. As in [93, p. 41], reinterpreted in the based context, a map i is a (based) cofibration if there is a lift λ in all such diagrams in which p is the map $p_0 \colon F(I_+, Y) \longrightarrow Y$ given by evaluation at 0 for some space Y. This is a restatement of the homotopy extension property, or HEP. Dually, as in [93, p. 47], a map p is a (based) fibration if there is a lift λ in all such diagrams in which i is the inclusion $i_0 \colon Y \longrightarrow Y \wedge I_+$ of the base of the cylinder. This is the covering homotopy property, or CHP. These are often called Hurewicz cofibrations and fibrations to distinguish them from other kinds of cofibrations and fibrations (in particular Serre fibrations) that also appear in model structures on spaces.

The unbased version of Lemma 1.1.1 is proven in Proposition 17.1.4, using no intermediate theory, and the reader is invited to skip there to see it. The based version follows, but rather technically, using Lemmas 1.3.3 and 1.3.4 below. The deduction is explained model theoretically in Corollary 17.1.2 and Remark 17.1.3.

One can think of model category theory as, in part, a codification of the notion of duality, called Eckmann-Hilton duality, that is displayed in the definitions of cofibrations and fibrations and in Lemma 1.1.1. We shall be making concrete rather than abstract use of such duality for now, but it pervades our point of view throughout. We leave the following dual pair of observations as exercises. Their proofs are direct from the definitions of pushouts and cofibrations and of pullbacks and fibrations. In the first, the closed inclusion hypothesis serves to ensure that we do not leave the category of compactly generated spaces [93, p. 38].

EXERCISE 1.1.2. Suppose given a commutative diagram

$$
\begin{array}{ccccc}
Y & \xleftarrow{\ f\ } & X & \xrightarrow{\ i\ } & Z \\
\beta \downarrow & & \| & & \downarrow \xi \\
Y' & \xleftarrow{\ f'\ } & X & \xrightarrow{\ i'\ } & Z'
\end{array}
$$

in which i and i' are closed inclusions and β and ξ are cofibrations. Prove that the induced map of pushouts

$$Y \cup_X Z \longrightarrow Y' \cup_X Z'$$

is a cofibration. Exhibit an example to show that the conclusion does not hold for a more general diagram of the same shape with the equality $X = X$ replaced by a cofibration $X \longrightarrow X'$. (Hint: interchange i' and $=$ in the diagram.)

EXERCISE 1.1.3. Suppose given a commutative diagram

$$
\begin{array}{ccc}
Y & \xrightarrow{f} & X & \xleftarrow{p} & Z \\
\downarrow{\beta} & & \| & & \downarrow{\xi} \\
Y' & \xrightarrow{f'} & X & \xleftarrow{p'} & Z'
\end{array}
$$

in which β and ξ are fibrations. Prove that the induced map of pullbacks

$$Y \times_X Z \longrightarrow Y' \times_X Z'$$

is a fibration. Again, the conclusion does not hold for a more general diagram of the same shape with the equality $X = X$ replaced by a fibration $X \longrightarrow X'$.

We shall often use the following pair of results about function spaces. The first illustrates how to use the defining lifting properties to construct new cofibrations and fibrations from given ones.

LEMMA 1.1.4. *Let $i \colon A \longrightarrow X$ be a cofibration and Y be a space. Then the induced map $i^* \colon F(X, Y) \longrightarrow F(A, Y)$ is a fibration and the fiber over the basepoint is $F(X/A, Y)$.*

PROOF. To show that $i^* \colon F(X, Y) \longrightarrow F(A, Y)$ is a fibration it is enough to show that there is a lift in any commutative square

$$
\begin{array}{ccc}
Z & \xrightarrow{f} & F(X, Y) \\
\downarrow{i_0} & \nearrow & \downarrow{i^*} \\
Z \wedge I_+ & \xrightarrow{h} & F(A, Y).
\end{array}
$$

By adjunction, we obtain the following diagram from that just given.

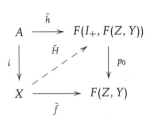

Here $\tilde{h}(a)(t)(z) = h(z,t)(a)$ and $\tilde{f}(x)(z) = f(z)(x)$ where $a \in A$, $z \in Z$, $x \in X$, and $t \in I$. Since $i\colon A \longrightarrow X$ is a cofibration there exists a lift $\tilde{H}$. The map $H\colon Z \wedge I_+ \longrightarrow F(X,Y)$ specified by $H(z,t)(x) = \tilde{H}(x)(t)(z)$ for $x \in X$, $z \in Z$, and $t \in I$ gives a lift in the original diagram. Therefore $i^*\colon F(X,Y) \longrightarrow F(A,Y)$ is a fibration. The basepoint of $F(A,Y)$ sends A to the basepoint of Y, and its inverse image in $F(X,Y)$ consists of those maps $X \longrightarrow Y$ that send A to the basepoint. These are the maps that factor through X/A, that is, the elements of $F(X/A, Y)$. $\qquad\square$

For the second, we recall the following standard definitions from [93, pp. 57, 59]. They will be used repeatedly throughout the book. By Lemmas 1.3.3 and 1.3.4 below, our assumption that basepoints are nondegenerate ensures that the terms "cofibration" and "fibration" in the following definition can be understood in either the based or the unbased sense.

DEFINITION 1.1.5. Let $f\colon X \longrightarrow Y$ be a (based) map. The homotopy cofiber Cf of f is the pushout $Y \cup_f CX$ of f and $i_0\colon X \longrightarrow CX$. Here the cone CX is $X \wedge I$, where I is given the basepoint 1. Since i_0 is a cofibration, so is its pushout $i\colon Y \longrightarrow Cf$ [93, p. 42]. The homotopy fiber Ff of f is the pullback $X \times_f PY$ of f and $p_1\colon PY \longrightarrow Y$. Here the path space PY is $F(I,Y)$, where I is given the basepoint 0; thus it consists of paths that start at the basepoint of Y. Since p_1 is a fibration (by Lemma 1.1.4), so is its pullback $\pi\colon Ff \longrightarrow X$ [93, p. 47].

We generally abbreviate "homotopy cofiber" to "cofiber". This is unambiguous since the word "cofiber" has no preassigned meaning. When $f\colon X \longrightarrow Y$ is a cofibration, the cofiber is canonically equivalent to the quotient Y/X. We also generally abbreviate "homotopy fiber" to "fiber". Here there is ambiguity when the given based map is a fibration, in which case the actual fiber $f^{-1}(*)$ and the homotopy fiber are canonically equivalent. By abuse, we then use whichever term seems more convenient.

LEMMA 1.1.6. *Let $f: X \longrightarrow Y$ be a map and Z be a space. Then the homotopy fiber Ff^* of the induced map of function spaces $f^*: F(Y, Z) \longrightarrow F(X, Z)$ is homeomorphic to $F(Cf, Z)$, where Cf is the homotopy cofiber of f.*

PROOF. The fiber Ff^* is $F(Y, Z) \times_{F(X,Z)} PF(X, Z)$. Clearly $PF(X, Z)$ is homeomorphic to $F(CX, Z)$. Technically, in view of the convention that I has basepoint 0 when defining P and 1 when defining C, we must use the homeomorphism $I \longrightarrow I$ that sends t to $1 - t$ to see this. Since the functor $F(-, Z)$ converts pushouts to pullbacks, the conclusion follows. □

1.2. The fill-in and Verdier lemmas

In formal terms, the results of this section describe the homotopy category $\mathrm{Ho}\mathscr{T}$ as a "pretriangulated category". However, we are more interested in describing precisely what is true before passage to the homotopy category, since some easy but little known details of that will ease our later work.

The following dual pair of "fill-in lemmas" will be at the heart of our theories of localization and completion. They play an important role throughout homotopy theory. They are usually stated entirely in terms of homotopy commutative diagrams, but the greater precision that we describe will be helpful.

LEMMA 1.2.1. *Consider the following diagram, in which the left square commutes up to homotopy and the rows are canonical cofiber sequences.*

$$
\begin{array}{ccccccc}
X & \xrightarrow{f} & Y & \xrightarrow{i} & Cf & \xrightarrow{\pi} & \Sigma X \\
\downarrow{\alpha} & & \downarrow{\beta} & & \downarrow{\gamma} & & \downarrow{\Sigma\alpha} \\
X' & \xrightarrow{f'} & Y' & \xrightarrow{i} & Cf' & \xrightarrow{\pi} & \Sigma X'
\end{array}
$$

There exists a map γ such that the middle square commutes and the right square commutes up to homotopy. If the left square commutes strictly, then there is a unique $\gamma = C(\alpha, \beta)$ such that both right squares commute, and then the cofiber sequence construction gives a functor from the category of maps and commutative squares to the category of sequences of spaces and commutative ladders between them.

PROOF. Recall again that $Cf = Y \cup_X CX$, where the pushout is defined with respect to $f: X \longrightarrow Y$ and the inclusion $i_0: X \longrightarrow CX$ of the base of the cone.

Let $h\colon X \times I \longrightarrow Y'$ be a (based) homotopy from $\beta \circ f$ to $f' \circ \alpha$. Define $\gamma(y) = \beta(y)$ for $y \in Y \subset Cf$, as required for commutativity of the middle square, and define

$$\gamma(x, t) = \begin{cases} h(x, 2t) & \text{if } 0 \le t \le 1/2 \\ (\alpha(x), 2t - 1) & \text{if } 1/2 \le t \le 1 \end{cases}$$

for $(x, t) \in CX$. The homotopy commutativity of the right square is easily checked. When the left square commutes, we can and must redefine γ on CX by $\gamma(x, t) = (\alpha(x), t)$ to make the right square commute. For functoriality, we have in mind the infinite sequence of spaces extending to the right, as displayed in [93, p. 57], and then the functoriality is clear. $\qquad\square$

Exercise 1.1.2 gives the following addendum, which applies to the comparison of cofiber sequences in which the left hand squares display composite maps.

ADDENDUM 1.2.2. If $X = X'$, α is the identity map, the left square commutes, and β is a cofibration, then the canonical map $\gamma\colon Cf \longrightarrow Cf'$ is a cofibration.

It is an essential feature of Lemma 1.2.1 that, when the left square only commutes up to homotopy, the homotopy class of γ depends on the choice of the homotopy and is not uniquely determined.

The dual result admits a precisely dual proof, where now the functoriality statement refers to the infinite sequence of spaces extending to the left, as displayed in [93, p. 59]. Recall that $Ff = X \times_Y PY$, where the pullback is defined with respect to $f\colon X \longrightarrow Y$ and the end-point evaluation $p_1\colon PY \longrightarrow Y$.

LEMMA 1.2.3. *Consider the following diagram, in which the right square commutes up to homotopy and the rows are canonical fiber sequences.*

$$
\begin{array}{ccccccc}
\Omega Y & \xrightarrow{\iota} & Ff & \xrightarrow{\pi} & X & \xrightarrow{f} & Y \\
\Omega\alpha \downarrow & & \downarrow \gamma & & \downarrow \beta & & \downarrow \alpha \\
\Omega Y' & \xrightarrow{\iota} & Ff' & \xrightarrow{\pi} & X' & \xrightarrow{f'} & Y'
\end{array}
$$

There exists a map γ such that the middle square commutes and the left square commutes up to homotopy. If the right square commutes strictly, then there is a unique $\gamma = F(\alpha, \beta)$ such that both left squares commute, hence the fiber sequence construction gives a functor from the category of maps and commutative squares to the category of sequences of spaces and commutative ladders between them.

ADDENDUM 1.2.4. *If $Y = Y'$, α is the identity map, the right square commutes, and β is a fibration, then the canonical map $\gamma \colon Ff \longrightarrow Ff'$ is a fibration.*

The addenda above deal with composites, and we have a dual pair of "Verdier lemmas" that encode the relationship between composition and cofiber and fiber sequences. We shall not make formal use of them, but every reader should see them since they are precursors of the basic defining property, Verdier's axiom, in the theory of triangulated categories [94, 138].[1]

LEMMA 1.2.5. *Let h be homotopic to $g \circ f$ in the following braid of cofiber sequences and let $j'' = \Sigma i(f) \circ \pi(g)$. There are maps j and j' such that the diagram commutes up to homotopy, and there is a homotopy equivalence $\xi \colon Cg \longrightarrow Cj$ such that $\xi \circ j' \simeq i(j)$ and $j'' = \pi(j) \circ \xi$.*

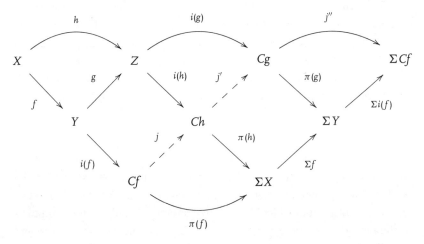

The square and triangle to the left of j and j' commute; if $h = g \circ f$, then there are unique maps j and j' such that the triangle and square to the right of j and j' commute.

---

1. In [138], diagrams like these are written as "octagons", with identity maps inserted. For this reason, Verdier's axiom is often referred to in the literature of triangulated categories as the "octahedral" axiom. In this form, the axiom is often viewed as mysterious and obscure. Lemmas 1.2.1 and 1.2.3 are precursors of another axiom used in the usual definition of triangulated categories, but that axiom is shown to be redundant in [94].

PROOF. Let $H\colon g \circ f \simeq h$. The maps j and j' are obtained by application of Lemma 1.2.1 to H regarded as a homotopy $g \circ f \simeq h \circ \mathrm{id}_X$ and the reverse of H regarded as a homotopy $\mathrm{id}_Z \circ h \simeq g \circ f$. The square and triangle to the left of j and j' are center squares of fill-in diagrams and the triangle and square to the right of j and j' are right squares of fill-in diagrams. The diagram commutes when $h = g \circ f$ and j and j' are taken to be

$$j = g \cup \mathrm{id}\colon Y \cup_f CX \longrightarrow Z \cup_h CX \text{ and } j' = \mathrm{id} \cup Cf\colon Z \cup_h CX \longrightarrow Z \cup_g CY,$$

as in the last part of Lemma 1.2.1. Define ξ to be the inclusion

$$Cg = Z \cup_g CY \longrightarrow (Z \cup_h CX) \cup_j C(Y \cup_f CX) = Ch \cup_j CCf = Cj$$

induced by $i(h)\colon Z \longrightarrow Ch$ and the map $CY \longrightarrow C(Cf)$ obtained by applying the cone functor to $i(f)\colon Y \longrightarrow Cf$. Then $j'' = \pi(j) \circ \xi$ since $\pi(g)$ collapses Z to a point, $\pi(j)$ collapses $Ch = Z \cup_h CX$ to a point, and both maps induce $\Sigma i(f)$ on $Cg/Z = \Sigma Y$. Using mapping cylinders and noting that j and j' are obtained by passage to quotients from maps $j\colon Mf \longrightarrow Mg$ and $j'\colon Mg \longrightarrow Mh$, we see by a diagram chase that ξ is an equivalence in general if it is so when $h = g \circ f$. In this case, we claim that there is a deformation retraction $r\colon Cj \longrightarrow Cg$ so that $r \circ \xi = \mathrm{id}$ and $r \circ i(j) = j'$. This means that there is a homotopy $k\colon Cj \times I \longrightarrow Cj$ relative to Cg from the identity to a map into Cg. In effect, looking at the explicit description of Cj, k deforms CCX to $CX \subset CY$. The details are fussy and left to the reader, but the intuition becomes clear from the observation that the quotient space $Cj/\xi(Cg)$ is homeomorphic to the contractible space $C\Sigma X$. □

REMARK 1.2.6. There is a reinterpretation that makes the intuition still clearer and leads to an alternative proof. We can use mapping cylinders as in [93, p. 43] to change the spaces and maps in our given diagram so as to obtain a homotopy equivalent diagram in which f and g are cofibrations and h is the composite cofibration $g \circ f$. As in [93, p. 58], the cofibers of f, g, and h are then equivalent to Y/X, Z/Y, and Z/X, respectively, and the equivalence ξ just becomes the evident homeomorphism $Z/Y \cong (Z/X)/(Y/X)$.

LEMMA 1.2.7. Let f be homotopic to $h \circ g$ in the following braid of fiber sequences and let $j'' = \iota(g) \circ \Omega p(h)$. There are maps j and j' such that the diagram commutes up to homotopy, and there is a homotopy equivalence $\xi\colon Fj \longrightarrow Fg$ such that $j' \circ \xi \simeq p(j)$ and $j'' \simeq \xi \circ \iota(j)$.

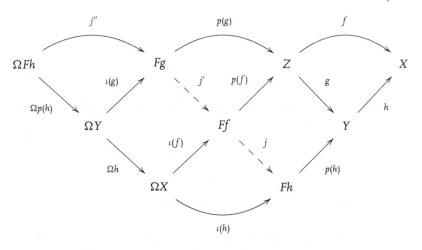

If $f = h \circ g$, then there are unique maps j and j' such that the diagram commutes, and then ξ can be so chosen that $j' \circ \xi = p(j)$ and $j'' = \xi \circ \iota(j)$.

1.3. Based and free cofibrations and fibrations

So far, we have been working in the category $\mathcal{T}$ of based spaces, and we usually continue to do so. However, we often must allow the basepoint to vary, and we sometimes need to work without basepoints. Homotopies between maps of unbased spaces, or homotopies between based maps that are not required to satisfy $h_t(*) = *$, are often called free homotopies. In this section and the next, we are concerned with the relationship between based homotopy theory and free homotopy theory.

Much that we have done in the previous two sections works just as well in the category $\mathcal{U}$ of unbased spaces as in the category $\mathcal{T}$. For example, using unreduced cones and suspensions, cofiber sequences work the same way in the two categories. However, the definition of the homotopy fiber Ff of a map $f: X \longrightarrow Y$ requires the choice of a basepoint to define the path space PY. We have both free and based notions of cofibrations and fibrations, and results such as Lemma 1.1.1 apply to both. It is important to keep track of which notion is meant when interpreting homotopical results. For example, we understand free cofibrations and free fibrations in the following useful result. It is the key to our approach to the Serre spectral sequence, and we shall have other uses for it. Recall that a homotopy $h: X \times I \longrightarrow X$ is said to be a deformation if h_0 is the identity map of X.

LEMMA 1.3.1. *Let $p\colon E \longrightarrow B$ be a fibration and $i\colon A \longrightarrow B$ a cofibration. Then the inclusion $D = p^{-1}(A) \longrightarrow E$ is a cofibration.*

PROOF. As in [93, p. 43], we can choose a deformation h of B and a map $u\colon B \longrightarrow I$ that represent (B, A) as an NDR-pair. By the CHP, we can find a deformation H of E that covers h, $p \circ H = h \circ (p \times \mathrm{id})$. Define a new deformation J of E by

$$J(x, t) = \begin{cases} H(x, t) & \text{if } t \le u(p(x)) \\ H(x, u(p(x))) & \text{if } t \ge u(p(x)). \end{cases}$$

Then J and $u \circ p$ represent (E, D) as an NDR-pair. $\qquad\square$

One of the many motivations for our standing assumption that basepoints are nondegenerate is that it ensures that based maps are cofibrations or fibrations in the free sense if and only if they are cofibrations or fibrations in the based sense. For fibrations, this is implied by the case $(X, A) = (Y, *)$ of a useful analogue of Lemma 1.1.1 called the Covering Homotopy and Extension Property.

LEMMA 1.3.2 **(CHEP).** *Let $i\colon A \longrightarrow X$ be a free cofibration and $p\colon E \longrightarrow B$ be a free fibration. Let $j\colon Mi = X \times \{0\} \cup A \times I \longrightarrow X \times I$ be the inclusion of the free mapping cylinder Mi in the cylinder $X \times I$. For any commutative square*

there is a homotopy H that makes the diagram commute.

PROOF. Here $h\colon A \times I \longrightarrow E$ is a homotopy of the restriction of $f\colon X \longrightarrow E$ to A, and $\bar{h}$ is a homotopy of pf whose restriction to A is covered by h. The conclusion is a special case of the free version of Lemma 1.1.1 proven in Proposition 17.1.4 since j is a cofibration and a homotopy equivalence by [93, p. 43]. $\square$

LEMMA 1.3.3. *Let $p\colon E \longrightarrow B$ be a map between based spaces. If p is a based fibration, then p is a free fibration. If p is a free fibration and Y is nondegenerately based, then p satisfies the based CHP with respect to homotopies $Y \wedge (I_+) \longrightarrow B$.*

PROOF. For the first statement, we apply the based CHP to based homotopies $(Y \times I)_+ \cong Y_+ \wedge I_+ \longrightarrow B$ to obtain the free CHP. For the second statement, we must obtain lifts in diagrams of based spaces

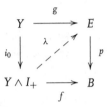

when Y is nondegenerately based, and these are supplied by the case $(X, A) = (Y, *)$ of the CHEP, with h constant at the basepoint of E. $\square$

The following result, like the previous one, was stated without proof in [93, pp. 56, 59]. Since its proof is not obvious (as several readers of [93] have complained), we give it in detail. Unfortunately, the argument is unpleasantly technical.[2]

LEMMA 1.3.4. *Let* $i: A \longrightarrow X$ *be a map between based spaces. If* i *is a free cofibration, then* i *is a based cofibration. If* A *and* X *are nondegenerately based and* i *is a based cofibration, then* i *is a free cofibration.*

PROOF. The first statement is clear: since the basepoint is in A, free lifts in Lemma 1.1.1 are necessarily based when the given maps are based. Thus assume that i is a based cofibration and A and X are nondegenerately based. The problem here is that the maps in a given test diagram for the HEP (as in [93, p. 41]) need not preserve basepoints, and the based HEP only gives information when they do. One might try to deform the unbased data into new based data to which the based HEP applies, but we shall instead check a slight variant of the NDR-pair criterion for i to be a free cofibration [93, p. 43].

Just as for free cofibrations, the fact that we are working with compactly generated spaces ensures that i is a closed inclusion since the based mapping cylinder Mi is a retract of $X \wedge I_+$. It suffices to prove that (X, A) is an NDR-pair. This means that there is a map $u: X \longrightarrow I$ such that $u^{-1}(0) = A$ and a deformation h of X relative to A, so that $h(x, 0) = x$ and $h(a, t) = a$, such that $h(x, 1) \in A$ if $u(x) < 1$. Inspection of the proof of the theorem on [93, p. 43] shows that if we start with a free cofibration i, then we obtain a

2. The proofs of Lemmas 1.3.3 and 1.3.4 are due to Strøm [131, p. 14] and [132, p. 440].

pair (h, u) with the stronger property that $h(x, t) \in A$ if $u(x) < t$. It follows that the characterization theorem for free cofibrations remains true if we redefine NDR-pairs by requiring this stronger condition, and we use the stronger condition throughout the proof that follows. We proceed in three steps.

Step (i). Let $w \colon X \longrightarrow I$ be any map such that $w^{-1}(0) = *$. Then there is a deformation $k' \colon X \times I \longrightarrow X$ relative to A and a map $w' \colon X \longrightarrow I$ such that $w'(x) \leq w(x)$, $A = (w')^{-1}(0)$, and

1.3.5
$$k'(x, t) \in A \quad \text{if} \quad w'(x) < \min(t, w(x)).$$

PROOF. Let $M^+ i = X \cup_A (A \times I) \subset X \times I$ be the free mapping cylinder of i, where X is identified with $X \times \{0\}$. Define

$$M(w) = X \cup \{(a, t) | t \leq w(a)\} \subset M^+ i.$$

The basepoint of $A \subset X$ gives $M(w)$ a basepoint. The reduced mapping cylinder Mi is obtained from $M^+ i$ by collapsing the line $\{*\} \times I$ to a point. Define a based map $f \colon Mi \longrightarrow M(w)$ by $f(x) = x$ for $x \in X$ and $f(a, t) = (a, \min(t, w(a)))$. Since i is a based cofibration, f extends to a based map $g \colon X \wedge I_+ \longrightarrow M(w)$. Define

$$k'(x, t) = \pi_1 g(x, t) \quad \text{and} \quad w'(x) = \sup\{\min(t, w(x)) - \pi_2 g(x, t) \mid t \in I\}$$

where π_1 and π_2 are the projections from $M(w)$ to X and I. Clearly $k'(x, 0) = x$, $k'(a, t) = a$ for $a \in A$, $w'(x) \leq w(x)$, and $w'(a) = 0$ for $a \in A$, so that $A \subset (w')^{-1}(0)$. To see that this inclusion is an equality, suppose that $w'(x) = 0$. Then $\min(t, w(x)) \leq \pi_2 g(x, t)$ for all $t \in I$. If $k'(x, t) \notin A$ for any t, then $\pi_2 g(x, t) = 0$, so that $t = 0$ or $w(x) = 0$. If $w(x) = 0$, then $x = *$ and $k'(x, t) = * \in A$. We conclude that if $t > 0$, then $k'(x, t) \in A$. Since A is closed in X, it follows that $k'(x, 0) = x$ is also in A. Finally, (1.3.5) holds since $k'(x, t) \notin A$ implies $\pi_2 g(x, t) = 0$ and thus $w'(x) \geq \min(t, w(x))$. $\qquad \square$

Step (ii). There is a representation (ℓ, z) of $(X, *)$ as an NDR-pair such that

$$\ell(A \times I) \subset A.$$

PROOF. Let (k_X, w_X) and (k_A, w_A) represent $(X, *)$ and $(A, *)$ as NDR-pairs, where $k_X(x, t) = *$ if $w_X(x) < t$ and $k_A(a, t) = *$ if $w_A(a) < t$. Since i is a based cofibration, we may regard k_A as taking values in X and extend it to a based map

$$\tilde{k}_A \colon X \wedge I_+ \longrightarrow X$$

such that $\tilde{k}_A(x, 0) = x$. Construct (k', w') from w_X as in step (i) and extend w_A to $\tilde{w}_A \colon X \longrightarrow I$ by

$$\tilde{w}_A(x) = \left(1 - \frac{w'(x)}{w_X(x)}\right) w_A(k'(x, 1)) + w'(x).$$

We interpret this as $w'(x)$ when $w'(x) = w_X(x)$; in particular, since $w_X(*) = 0$, $\tilde{w}_A(*) = 0$. The definition makes sense since $k'(x, 1) \in A$ if $w'(x) < w_X(x)$, by (1.3.5). We claim that $\tilde{w}_A^{-1}(0) = *$. Thus suppose that $\tilde{w}_A(x) = 0$. Then $w'(x) = 0$, so that $x \in A$, and this implies that $k'(x, 1) = x$ and $w_A(x) = 0$, so that $x = *$. The required pair (ℓ, z) is now defined by

$$\ell(x, t) = \begin{cases} \tilde{k}_A(x, t/\tilde{w}_A(x)) & \text{if } t \leq \tilde{w}_A(x) \\ k_X(\tilde{k}_A(x, 1), t - \tilde{w}_A(x)) & \text{if } t > \tilde{w}_A(x) \end{cases}$$

and

$$z(x) = \min(1, \tilde{w}_A(x) + w_X(\tilde{k}_A(x, 1))).$$

To see that $\ell(A \times I) \subset A$, recall that $\tilde{k}_A$ and $\tilde{w}_A$ extend k_A and w_A and let $x \in A$. Clearly $\tilde{k}_A(x, t) \in A$ for all t. If $t > w_A(x)$, then $w_A(x) < 1$ and thus $k_A(x, 1) = *$. Therefore $k_X(k_A(x, 1), t - w_A(x)) = * \in A$. Note that $\ell(x, t) = *$ if $z(x) < t$. $\square$

Step (iii). Completion of the proof. Construct $h' \colon X \times I \longrightarrow X$ and $z' \colon X \longrightarrow I$ by applying step (i) to $z \colon X \longrightarrow I$. Then define

$$h(x, t) = \begin{cases} \ell(h'(x, t), \min(t, z'(x)/z(x))) & \text{if } x \neq * \\ * & \text{if } x = * \end{cases}$$

and

$$u(x) = z'(x) - z(x) + \sup\{z(h'(x, t)) \mid t \in I\}.$$

Then (h, u) represents (X, A) as an NDR-pair (in the strong sense). $\square$

1.4. Actions of fundamental groups on homotopy classes of maps

For based spaces X and Y, let $[X, Y]$ denote the set of based homotopy classes of based maps $X \longrightarrow Y$. For unbased spaces X and Y, let $[X, Y]_{\text{free}}$ denote the set of free homotopy classes of maps $X \longrightarrow Y$. If we choose basepoints for X and Y and use them to define $[X, Y]$, then we obtain a function $[X, Y] \longrightarrow [X, Y]_{\text{free}}$ by forgetting the basepoints. There is a classical description of this function in terms of group actions, and we describe that and related material on group actions in homotopy theory in this section.

We later consider varying basepoints in a given space, and we want them all to be nondegenerate. Therefore, we tacitly restrict attention to spaces that have that property. This holds for locally contractible spaces, such as CW complexes.

DEFINITION 1.4.1. Let X and Y be spaces. As usual, we assume that they are connected and nondegenerately based. Let $i: * \longrightarrow X$ be the inclusion of the basepoint and $p: Y \longrightarrow *$ be the trivial map. A loop α based at $* \in Y$ and a based map $f: X \longrightarrow Y$ give a map $f \cup \alpha: Mi = X \cup I \longrightarrow Y$. Applying Lemma 1.3.2 to i, p, and $f \cup \alpha$, we obtain a homotopy $h: X \times I \longrightarrow Y$ such that $h_0 = f$ and $h(*, t) = \alpha(t)$. Another use of Lemma 1.3.2 shows that the (based) homotopy class of h_1 depends only on the path class $[\alpha]$ and the homotopy class $[f]$. With the usual conventions on composition of loops [93, p. 6], the definition $[\alpha][f] = [h_1]$ gives a left action of $\pi_1(Y, *)$ on the set $[X, Y]$ with orbit set $[X, Y]/\pi_1(Y, *)$. It is clear that two based maps that are in the same orbit under the action of $\pi_1(Y, *)$ are freely homotopic.

LEMMA 1.4.2. *The induced function $[X, Y]/\pi_1(Y) \longrightarrow [X, Y]_{\text{free}}$ is a bijection.*

PROOF. By HEP, any map $X \longrightarrow Y$ is freely homotopic to a based map. If two based maps f and g are freely homotopic via a homotopy h, then the restriction of h to $\{*\} \times I$ gives a loop α, and $[g] = [\alpha][f]$ by the definition of the action. $\square$

An H-space is a based space Y with a product $\mu: Y \times Y \longrightarrow Y$, written $x \cdot y$ or by juxtaposition, whose basepoint is a two-sided unit up to based homotopy. That is, the maps $y \mapsto * \cdot y$ and $y \mapsto y \cdot *$ are both homotopic to the identity map. Equivalently, the composite of the inclusion $Y \vee Y \longrightarrow Y \times Y$ and the product is homotopic to the fold map $\nabla: Y \vee Y \longrightarrow Y$, which restricts to the identity map on each wedge summand Y. Using our standing assumption that basepoints are nondegenerate, we see that the given product is homotopic to a product for which the basepoint of Y is a strict unit. Therefore, we may as well assume henceforward that H-spaces have strict units.

PROPOSITION 1.4.3. *For an H-space Y, the action of $\pi_1(Y, *)$ on $[X, Y]$ is trivial and therefore $[X, Y] \cong [X, Y]_{\text{free}}$.*

PROOF. For a map $f: X \longrightarrow Y$ and a loop α based at $* \in Y$, the homotopy $h(x, t) = \alpha(t) \cdot f(x)$ satisfies $h_1 = f$. Using this choice of homotopy in Definition 1.4.1, as we may, we see that $[\alpha][f] = [f]$. $\square$

The following definition hides some elementary verifications that we leave to the reader.

DEFINITION 1.4.4. Take $X = S^n$ in Definition 1.4.1. The definition then specializes to define an action of the group $\pi_1(Y, *)$ on the group $\pi_n(Y, *)$. When $n = 1$, this is the conjugation action of $\pi_1(Y, *)$ on itself. A (connected) space Y is simple if $\pi_1(Y, *)$ is abelian and acts trivially on $\pi_n(Y, *)$ for all $n \geq 2$.

COROLLARY 1.4.5. *Any H-space is a simple space.*

In the rest of this section, we revert to the based context and consider extra structure on the long exact sequences of sets of homotopy classes of (based) maps that are induced by cofiber and fiber sequences. These long exact sequences are displayed, for example, in [93, pp. 57, 59]. As observed there, the sets $[X, Y]$ are groups if X is a suspension or Y is a loop space and are abelian groups if X is a double suspension or Y is a double loop space. However, there is additional structure at the ends of these sequences that will play a role in our work. Thus, for a based map $f : X \longrightarrow Y$ and a based space Z, consider the exact sequence of pointed sets induced by the canonical cofiber sequence of f:

$$[\Sigma Y, Z] \xrightarrow{(\Sigma f)^*} [\Sigma X, Z] \xrightarrow{\pi^*} [Cf, Z] \xrightarrow{i^*} [Y, Z] \xrightarrow{f^*} [X, Z].$$

LEMMA 1.4.6. *The following statements hold.*

(i) *The group $[\Sigma X, Z]$ acts from the right on the set $[Cf, Z]$.*

(ii) *$\pi^* : [\Sigma X, Z] \longrightarrow [Cf, Z]$ is a map of right $[\Sigma X, Z]$-sets.*

(iii) *$\pi^*(x) = \pi^*(x')$ if and only if $x = (\Sigma f)^*(y) \cdot x'$ for some $y \in [\Sigma Y, Z]$.*

(iv) *$i^*(z) = i^*(z')$ if and only if $z = z' \cdot x$ for some $x \in [\Sigma X, Z]$.*

(v) *The image of $[\Sigma^2 X, Z]$ in $[\Sigma Cf, Z]$ is a central subgroup.*

(In (iii) and (iv), we have used the notation $\cdot$ to indicate the action.)

PROOF. In (i), the action is induced by applying the contravariant functor $[-, Z]$ to the "coaction" $Cf \longrightarrow Cf \vee \Sigma X$ of ΣX on Cf that is specified by pinching $X \times \{1/2\} \subset Cf$ to a point (and of course linearly expanding the half intervals of the resulting wedge summands homeomorphic to Cf and ΣX to full intervals). Then (ii) is clear since the quotient map $Cf \longrightarrow \Sigma X$ commutes with the pinch map. To show (iii), observe that since $[\Sigma X, Z]$ is a group, (ii) implies that $\pi^*(x) = \pi^*(x')$ if and only if $\pi^*(x \cdot (x')^{-1}) = *$. By exactness, this holds if and only if $x \cdot (x')^{-1} = (\Sigma f)^*(y)$ and thus $x = (\Sigma f)^*(y) \cdot x'$ for some $y \in [\Sigma Y, Z]$. For (iv), if $i^*(z) = i^*(z')$, then the HEP for the cofibration i implies that the homotopy classes z and z' can be represented by maps c and c' from Cf to Z that restrict to the same map on Y. Then $z = z' \cdot x$, where x is represented by the map $\Sigma X \longrightarrow Z$ that is obtained by regarding ΣX as the union of upper and lower cones on X, using $c'|_{CX}$ with cone coordinate reversed on the lower cone, and using $c|_{CX}$ on the upper cone.

Finally, for (v), let $G = [\Sigma Cf, Z]$ and let $H \subset G$ be the image of $[\Sigma^2 X, Z]$. The suspension of the pinch map $Cf \longrightarrow Cf \vee \Sigma X$ gives a right action $*$ of H on G, a priori different from the product in G. We may obtain the product, gg' say, on G from the pinch map defined using the suspension coordinate of ΣCf. Then the usual proof of the commutativity of $[\Sigma^2 X, Z]$ applies to show that $hg = gh$ for $h \in H$ and $g \in G$. In detail, if $1 \in H$ is the identity element, then, for $g, g' \in G$ and $h, h' \in H$,

$$1 * h = h, \quad g * 1 = g, \quad \text{and} \quad (g * h)(g' * h') = (gg') * (hh').$$

Therefore

$$gh = (g * 1)(1 * h) = (g1) * (1h) = g * h$$
$$= (1g) * (h1) = (1 * h)(g * 1) = hg. \qquad \square$$

Dually, consider the exact sequence of pointed sets induced by the canonical fiber sequence of $f : X \longrightarrow Y$:

$$[Z, \Omega X] \xrightarrow{(\Omega f)_*} [Z, \Omega Y] \xrightarrow{\iota_*} [Z, Ff] \xrightarrow{p_*} [Z, X] \xrightarrow{f_*} [Z, Y].$$

This is of greatest interest when $Z = S^0$. Since $[-, -]$ refers to based homotopy classes, $[S^0, X] = \pi_0(X)$ and the sequence becomes

$$\pi_1(X) \xrightarrow{f_*} \pi_1(Y) \xrightarrow{\iota_*} \pi_0(Ff) \xrightarrow{p_*} \pi_0(X) \xrightarrow{f_*} \pi_0(Y).$$

LEMMA 1.4.7. *The following statements hold.*

(i) *The group $[Z, \Omega Y]$ acts from the right on the set $[Z, Ff]$.*
(ii) *$\iota_* : [Z, \Omega Y] \longrightarrow [Z, Ff]$ is a map of right $[Z, \Omega Y]$-sets.*
(iii) *$\iota_*(y) = \iota_*(y')$ if and only if $y = (\Omega f)_*(x) \cdot y'$ for some $x \in [Z, \Omega X]$.*
(iv) *$p_*(z) = p_*(z')$ if and only if $z = z' \cdot y$ for some $y \in [Z, \Omega Y]$.*
(v) *The image of $[Z, \Omega^2 Y]$ in $[Z, \Omega Ff]$ is a central subgroup.*

1.5. Actions of fundamental groups in fibration sequences

It is more usual and more convenient to think of Lemma 1.4.7 in terms of fibrations. That is, instead of starting with an arbitrary map f, in this section we start with a fibration $p : E \longrightarrow B$. We give some perhaps well-known (but hard to find) results about fundamental group actions in fibrations.

Observe that a fibration $p\colon E \longrightarrow B$ need not be surjective,[3] but either every point or no point of each component of B is in the image of E. For nontriviality, we assume that p is surjective. We may as well assume that B is connected, since otherwise we could restrict attention to the components of E over a chosen component of B. If we were given a general surjective map $p\colon E \longrightarrow B$, we would replace E by its mapping path fibration Np (which depends on a choice of basepoint in B) to apply the results to follow. That is, we can think of applying the results to the fibration $Np \longrightarrow B$ as applying them to the original map p "up to homotopy".

Although we have been working in the based context, we now think of p as a free fibration and let the basepoint $b \in B$ vary. We choose a point $e \in E$ such that $p(e) = b$ and we let $F_b = p^{-1}(b)$. Also let E_e be the component of E that contains e and F_e be the component of F_b that contains e. We view these as based spaces with basepoint e (as recorded in the notation), retaining our standing assumption that basepoints are nondegenerate. As in [93, p. 64], we then have the exact sequence

1.5.1

$$\cdots \longrightarrow \pi_n(F_e, e) \xrightarrow{\iota_*} \pi_n(E_e, e) \xrightarrow{p_*} \pi_n(B, b) \xrightarrow{\partial} \pi_{n-1}(F_e, e) \longrightarrow$$

$$\cdots \longrightarrow \pi_1(E_e, e) \xrightarrow{p_*} \pi_1(B, b) \xrightarrow{\partial} \pi_0(F_b, e) \xrightarrow{\iota_*} \pi_0(E, e).$$

Notice that we have replaced E and F_b by E_e and F_e in the higher homotopy groups. Since the higher homotopy groups only depend on the component of the basepoint this doesn't change the exact sequence.

By [93, p. 52], there is a functor $\lambda = \lambda_{(B,p)}\colon \Pi B \longrightarrow \mathrm{Ho}\mathscr{U}$ that sends a point b of the fundamental groupoid ΠB to the fiber F_b. This specializes to give a group homomorphism $\pi_1(B, b) \longrightarrow \pi_0(\mathrm{Aut}(F_b))$. Here $\mathrm{Aut}(F_b)$ is the topological monoid of (unbased) homotopy equivalences of F_b. We think of $\pi_1(B, b)$ as acting "up to homotopy" on the space F_b, meaning that an element $\beta \in \pi_1(B, b)$ determines a well-defined homotopy class of homotopy equivalences $F_b \longrightarrow F_b$. (If p is a covering space, the action is by homeomorphisms, as in [93, p. 29].) As we shall use later, it follows that $\pi_1(B)$ acts on the homology and cohomology groups of F_b.

Observe that we can apply this to the path space fibration $\Omega Y \to PY \to Y$ of a based space Y. We thus obtain an action of $\pi_1(Y, *)$ on ΩY. Since ΩY is simple, by Proposition 1.4.3, $[\Omega Y, \Omega Y] \cong [\Omega Y, \Omega Y]_{\mathrm{free}}$ and we can

---

3. In [93, p. 47], fibrations were incorrectly required to be surjective maps.

view $\pi_1(Y, *)$ as acting through basepoint-preserving homotopy equivalences of the fiber ΩY. There results another action of $\pi_1(Y, *)$ on $\pi_n(Y, *)$. We leave it to the reader to check that this action agrees with that defined in Definition 1.4.4.

We will need a more elaborate variant of the functor $\lambda = \lambda_{(B,p)} : \Pi B \rightarrow \text{Ho}\mathcal{U}$. Here, instead of starting with paths in B, we start with paths in E and work with components of the fibers and the total space, regarded as based spaces. Thus let $\alpha : I \longrightarrow E$ be a path from e to e'. Let $b = p(e)$ and $b' = p(e')$, and let $\beta = p \circ \alpha$ be the resulting path from b to b'. Consider the following diagram, in which $F = F_e$ and $\iota : F \longrightarrow E$ is the inclusion.

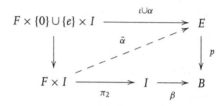

The CHEP, Lemma 1.3.2, gives a homotopy $\tilde{\alpha}$ that makes the diagram commute. At the end of this homotopy, we have a map $\tilde{\alpha}_1 : F_e \longrightarrow F_{e'}$ such that $\tilde{\alpha}_1(e) = e'$. By a slight variant of the argument of [93, p. 51], which again uses the CHEP, the based homotopy class of maps $\tilde{\alpha}_1 : (F_e, e) \longrightarrow (F_{e'}, e')$ such that $\tilde{\alpha}_1(e) = e'$ that are obtained in this way depends only on the path class $[\alpha]$. That is, homotopic paths $e \longrightarrow e'$ give homotopic maps $\tilde{\alpha}_1$ for any choices of lifts $\tilde{\alpha}$. We define $\lambda[\alpha] = [\tilde{\alpha}_1]$. These arguments prove the first statement of the following result.

THEOREM 1.5.2. *There is a functor $\lambda = \lambda_{(E,p)} : \Pi E \longrightarrow \text{Ho}\mathcal{T}$ that assigns the (based) component F_e of the fiber F_b to a point $e \in E$ with $p(e) = b$. The functor λ restricts to give a homomorphism $\pi_1(E, e) \longrightarrow \pi_0(\text{Aut}(F_b))$ and thus an action of $\pi_1(E, e)$ on $\pi_n(F_b, e)$. The following diagram commutes up to the natural transformation $U\lambda_{(E,p)} \longrightarrow \lambda_{(B,p)}p_*$ given by the inclusions $F_e \longrightarrow F_b$.*

$$
\begin{array}{ccc}
\Pi E & \xrightarrow{\;\;p_*\;\;} & \Pi B \\
{\scriptstyle \lambda_{(E,p)}} \downarrow & & \downarrow {\scriptstyle \lambda_{(B,p)}} \\
\text{Ho}\mathcal{T} & \xrightarrow[\;\;U\;\;]{} & \text{Ho}\mathcal{U}
\end{array}
$$

Here U is the functor obtained by forgetting basepoints.

PROOF. It remains to prove the last statement.

By definition, for a path $\beta: b \longrightarrow b'$, $\lambda_{(B,p)}[\beta] = [\tilde{\beta}_1]$, where $\tilde{\beta}$ is a homotopy that makes the following diagram commute.

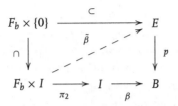

If $b = p(e)$, we may restrict F_b to its components F_e. Letting $e' = \tilde{\beta}(e, 1)$, $\tilde{\beta}$ restricts on $\{e\} \times I$ to a path $\alpha: e \longrightarrow e'$. Then $\tilde{\beta}_1$ is a homotopy of the sort used to define $\lambda_{(E,p)}[\alpha]$. Turning the argument around, if we start with a given path α and define $\beta = p \circ \alpha$, then the map $\tilde{\alpha}$ used in the specification of $\lambda_{E,p}[\alpha] = [\tilde{\alpha}_1]$ serves as a choice for the restriction to $F_e \times I$ of a lift $\tilde{\beta}$ in the diagram above. This says that the restriction of $\lambda_{(B,p)}$ to the component F_e is $\lambda_{E,p}[\alpha]$, which is the claimed naturality statement. $\square$

REMARK 1.5.3. In the argument just given, if β is a loop there need be no choice of $\tilde{\beta}$ such that $\tilde{\beta}_1(e) = e$ unless $\pi_1: \pi_1(E_e, e) \longrightarrow \pi_1(B, b)$ is surjective. Similarly, unless $\pi_1: \pi_1(E_e, e) \longrightarrow \pi_1(B, b)$ is injective, loops α in E_e can give non-homotopic action maps $[\alpha]: F_e \longrightarrow F_e$ even though they have the same image under p_*.

The naturality and homotopy invariance statements proven for $\lambda_{(B,p)}$ in [93, pp. 52–53] apply with obvious changes of statement to $\lambda_{(E,p)}$. We write r for the trivial fibration $r: Y \longrightarrow *$ for any space Y. When $p = r$ in the construction above, the resulting action of $\pi_1(Y, *)$ on $\pi_n(Y, *)$ agrees with that discussed earlier. In effect, in the based context, taking $X = Y$ in Definition 1.4.1 we see that we used this construction to give our original definition of fundamental group actions. The more concrete construction in Definition 1.4.4 is then obtained via maps $S^n \longrightarrow Y$.

PROPOSITION 1.5.4. *The long exact sequence (1.5.1), ending at $\pi_1(B, b)$, is an exact sequence of $\pi_1(E, e)$-groups and therefore of $\pi_1(F, e)$-groups. In more detail, the following statements hold.*

(i) *For $j \in \pi_1(F_b, e)$ and $x \in \pi_n(F_b, e)$, $jx = \iota_*(j)x$.*

(ii) *For $g \in \pi_1(E, e)$ and $z \in \pi_n(B, b)$, $gz = p_*(g)z$.*

(iii) *For $g \in \pi_1(E, e)$ and $x \in \pi_n(F_b, e)$, $\iota_*(gx) = g\iota_*(x)$.*

(iv) For $g \in \pi_1(E, e)$ and $y \in \pi_n(E, e)$, $p_*(gx) = gp_*(x)$.

(v) For $g \in \pi_1(E, e)$ and $z \in \pi_n(B, b)$, $\partial(gz) = g\partial(z)$.

PROOF. We define the actions of $\pi_1(E, e)$ on $\pi_n(F_b, e)$, $\pi_n(E, e)$ and $\pi_n(B, b)$ to be those given by $\lambda_{(E,p)}$, $\lambda_{(E,r)}$ and $\lambda_{(B,r)} \circ p_*$, respectively. We let $\pi_1(F, e)$ act on these groups by pullback along $\iota_*: \pi_1(F_e, e) \longrightarrow \pi_1(E, e)$. By (i) and inspection, this implies that its actions are given by $\lambda_{(F_e,r)}$, $\lambda_{(E,r)} \circ \iota_*$, and the trivial action, respectively. The maps in the exact sequence are maps of $\pi_1(E, e)$-groups by (iii), (iv), and (v).

By restricting the construction of $\lambda_{(E,p)}$ to loops $\alpha: I \longrightarrow F_e$, we see that $\lambda_{(F_b,r)}[\alpha] = \lambda_{(E,p)}[\iota \circ \alpha]$. This implies part (i), and part (ii) is immediate from the definition of the action of $\pi_1(E, e)$ on $\pi_1(B, b)$. Part (iv) holds by the naturality of the action of π_1 on π_n, which applies to any map, not just a fibration such as p.

To prove (iii), we use the notations of the diagram on the previous page. Since $\iota: F \longrightarrow E$ is a cofibration, by Lemma 1.3.1, we can apply the CHEP to the diagram

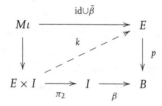

to obtain a deformation $k: E \times I \longrightarrow E$. Since $\tilde{\beta}_1: F_b \longrightarrow F_{b'} \subset E$ represents $\iota_* \circ \lambda_{(E,p)}([\alpha])$ and this map is $k_1 \circ \iota$, it also represents $\lambda_{(E,r)} \circ \iota$. This implies (iii).

To prove (v), recall from [93, p. 64] that $\partial: \pi_n(B, b) \longrightarrow \pi_{n-1}(F, e)$ is the composite of the inverse of the isomorphism $p_*: \pi_n(E, F, e) \longrightarrow \pi_n(B, b)$ and the boundary map of the pair (E, F), which is obtained by restricting representative maps $(D^n, S^{n-1}) \longrightarrow (E, F)$ to $S^{n-1} \longrightarrow F$. Let α be a loop at e in E that represents g and let $p(e) = b$ and $p \circ \alpha = \beta$. Let $h: B \times I \longrightarrow B$ be a deformation $(h_0 = \mathrm{id})$ such that $h(b, t) = \beta(t)$. As in Definition 1.4.1, for a based map $f: X \longrightarrow B$, $[\beta][f] = [h_1]$. To describe this action in terms of the pair (E, F), use the CHEP to obtain lifts j and k in the following diagrams.

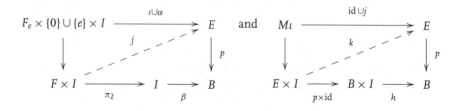

Then $(k, j): (E, F) \times I \longrightarrow (E, F)$ is a relative homotopy that restricts to j and covers h. Precomposing with a map $\tilde{z}: (D^n, S^{n-1}) \longrightarrow (E, F)$ that represents $z \in \pi_1(B, b)$, the composite $h_1 \tilde{z}$ represents $[\beta]z$, which is $gz = p_*(g)z$. Its restriction to S^{n-1} represents $\partial(gz)$ and, by the first of the above pair of diagrams, it also represents $g\partial(z)$. $\qquad \square$

REMARK 1.5.5. Consider (1.5.1). Lemma 1.4.7 implies that $\pi_0(F_b, e)$ is a right $\pi_1(B, b)$-set and $\partial: \pi_1(B, b) \longrightarrow \pi_0(F_b, e)$ is a map of $\pi_1(B, b)$-sets and therefore of $\pi_1(E, e)$-sets. However, we are especially interested in the next step to the left. Assume that the image of $p_*: \pi_1(E, e) \longrightarrow \pi_1(B, b)$ is a normal subgroup. Then coker p_* is a group contained in the based set $\pi_0(F_b, e)$ of components of F_b with base component F_e (later sometimes denoted $[e]$). We denote this group by $\tilde{\pi}_0(F_b, e)$ and have the exact sequence

$$\pi_1(E, e) \xrightarrow{\;p_*\;} \pi_1(B, b) \xrightarrow{\;\partial\;} \tilde{\pi}_0(F_b, e) \xrightarrow{\;i_*\;} *.$$

It is an exact sequence of $\pi_1(E, e)$-groups and thus of $\pi_1(F_b, e)$-groups, where $\pi_1(F_b, e)$ acts trivially on the last two groups.

2

HOMOTOPY COLIMITS AND HOMOTOPY LIMITS; $\lim^1$

The material of this chapter is again of general interest. We describe the most basic homotopy colimits and limits, with focus on their precise algebraic behavior. In particular, we develop the dual homotopical $\lim^1$ exact sequences. We shall not go into the general theory of homotopy colimits, but the material here can serve as an introduction to such more advanced sources as [42, 65, 128]. In this book, homotopy limits, and especially homotopy pullbacks, will play a central role in the fracture theorems of Chapters 8 and 13.

In §3 and §4, we describe the algebraic properties of the functor $\lim^1$ and give a concrete topological example where nontrivial $\lim^1$ terms appear. In §5 and §6, we give some observations about the homology of filtered colimits and sequential limits and advertise a kind of universal coefficient theorem for profinite abelian groups. While §1 and §2 are vital to all of our work, the later sections play a more peripheral role and need only be skimmed on a first reading.

2.1. Some basic homotopy colimits

Intuitively, homotopy colimits are constructed from ordinary categorical colimits by gluing in cylinders. These give domains for homotopies that allow us to replace equalities between maps that appear in the specification of ordinary colimits by homotopies between maps. There is always a natural map from a homotopy colimit to the corresponding ordinary colimit, and in some but not all cases there is a convenient criterion for determining whether or not that natural map is a homotopy equivalence. Since homotopies between given maps are not unique, not even up to homotopy, homotopy colimits give weak colimits in the homotopy category in the sense that they satisfy the existence but not the uniqueness property of ordinary colimits. We shall spell out the relevant algebraic property quite precisely for homotopy pushouts (or double mapping cylinders), homotopy coequalizers (or mapping tori), and sequential homotopy colimits (or telescopes). We record the analogous results for the

24

constructions Eckmann-Hilton dual to these in the next section. We work in
the based context, but the unbased analogues should be clear.

DEFINITION 2.1.1. The homotopy pushout (or double mapping cylinder)
$M(f, g)$ of a pair of maps $f \colon A \longrightarrow X$ and $g \colon A \longrightarrow Y$ is the pushout written
in alternative notations as

$$(X \vee Y) \cup_{A \vee A} A \wedge I_+ \quad \text{or} \quad X \cup_f (A \wedge I_+) \cup_g Y.$$

It is the pushout defined with respect to $f \vee g \colon A \vee A \longrightarrow X \vee Y$ and the
cofibration $(i_0, i_1) \colon A \vee A \longrightarrow A \wedge I_+$.

Explicitly, with the alternative notation, we start with $X \vee (A \wedge I_+) \vee Y$ and
then identify $(a, 0)$ with $f(a)$ and $(a, 1)$ with $g(a)$. In comparision with the
ordinary pushout $X \cup_A Y$, we are replacing A by the cylinder $A \wedge I_+$. Except
that the line $* \times I$ through the basepoint should be collapsed to a point, the
following picture should give the idea.

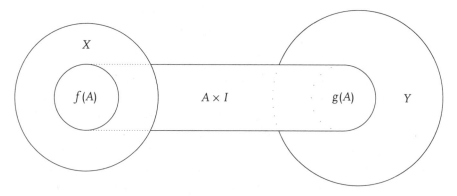

PROPOSITION 2.1.2. *For any space Z, the natural map of pointed sets*

$$[M(f, g), Z] \longrightarrow [X, Z] \times_{[A, Z]} [Y, Z]$$

*is a surjection. Its kernel is isomorphic to the set of orbits of $[\Sigma A, Z]$ under the right
action of the group $[\Sigma X, Z] \times [\Sigma Y, Z]$ specified by*

$$a * (x, y) = (\Sigma f)^*(x)^{-1} \cdot a \cdot (\Sigma g)^*(y)$$

for $a \in [\Sigma A, Z]$, $x \in [\Sigma X, Z]$, and $y \in [\Sigma Y, Z]$.

PROOF. The pullback in the statement is the set of homotopy classes $([\alpha], [\beta])$
in $[X, Z] \times [Y, Z]$ such that $[\alpha]$ and $[\beta]$ have the same image in $[A, Z]$, which
means that $\alpha \circ f$ is homotopic to $\beta \circ g$.

We have an evident cofibration $j\colon X \vee Y \longrightarrow M(f,g)$. As in [93, p. 57], it gives rise to an exact sequence of pointed sets

$$[\Sigma X, Z] \times [\Sigma Y, Z] \xrightarrow{\pi^*} [Cj, Z] \xrightarrow{i^*} [M(f,g), Z] \xrightarrow{j^*} [X, Z] \times [Y, Z].$$

If $\alpha\colon X \longrightarrow Z$ and $\beta\colon Y \longrightarrow Z$ are such that $\alpha f \simeq \beta g$, then any choice of homotopy $A \wedge I_+ \longrightarrow Z$ determines a map $\gamma\colon M(f,g) \longrightarrow Z$ that restricts to α and β on X and Y. Thus j^* induces a surjection onto the pullback in the statement of the result. That is, for homotopy classes $[\alpha]$ and $[\beta]$ such that $f^*[\alpha] = g^*[\beta]$, there is a homotopy class $[\gamma]$, not uniquely determined, such that $j^*[\gamma] = ([\alpha], [\beta])$. This is what we mean by saying that the homotopy colimit $M(f,g)$ is a weak pushout of f and g in $\text{Ho}\mathscr{T}$.

By Lemma 1.4.6(iv), $\ker j^* = \operatorname{im} i^*$ is the set of orbits of $[\Sigma A, Z]$ under the action of $[\Sigma X, Z] \times [\Sigma Y, Z]$ specified there. Since j is a cofibration, the canonical quotient map $\psi\colon Cj \longrightarrow \Sigma A$ is an equivalence ([93, p. 58]). A homotopy inverse ξ to ψ can be specified by

$$\xi(a,t) = \begin{cases} (f(a), 1-3t) \in CX & \text{if } 0 \le t \le 1/3 \\ (a, 3t-1) \in A \wedge I_+ & \text{if } 1/3 \le t \le 2/3 \\ (g(a), 3t-2) \in CY & \text{if } 2/3 \le t \le 1. \end{cases}$$

The pinch map on Cj used to define the action in Lemma 1.4.6 pinches the equators $X \times \{1/2\}$ and $Y \times \{1/2\}$ of CX and CY to the basepoint, so mapping Cj to $\Sigma X \vee Cj \vee \Sigma Y$. If we first apply ξ and then this pinch map on Cj, we obtain the same result as if we first apply the pinch map $\Sigma A \longrightarrow \Sigma A \vee \Sigma A \vee \Sigma A$ that pinches $A \times \{1/6\}$ and $A \times \{5/6\}$ to a point and then apply ξ. Up to homotopy, ξ restricts on the three copies of ΣA to $-\Sigma f$, the identity, and Σg. Therefore the action defined in Lemma 1.4.6 agrees with the action specified in the statement. $\square$

The following result is often called the "gluing lemma".

LEMMA 2.1.3. *Assume given a commutative diagram*

in which f and f′ are cofibrations. If α, β, and γ are homotopy equivalences, then so is their pushout

$$X \cup_A Y \longrightarrow X' \cup_{A'} Y'.$$

PROOF. One can prove this directly by expanding on arguments about cofiber equivalence given in [93, §4.5]. However, that would be fairly lengthy and digressive. Instead we take the opportunity to advertise the model category theory that appears later in the book, specifically in §15.4. The notion of a left proper model category is specified in Definition 15.4.1, and it is proven in Proposition 15.4.4 that a model category is left proper if and only if the conclusion of the gluing lemma holds. This applies to a very large class of categories in which one can do homotopy theory. By Theorem 17.1.1 and Corollary 17.1.2, it applies in particular to the category of spaces and the category of based spaces. □

COROLLARY 2.1.4. *If f is a cofibration and g is any map, then the natural quotient map $M(f,g) \longrightarrow X \cup_A Y$ is a homotopy equivalence.*

PROOF. An explicit proof of the unbased version is given in [93, p. 78], but the conclusion is also a direct consequence of the previous result. To see that, let Mf be the mapping cylinder of f and observe that $M(f,g)$ is the pushout of the canonical cofibration $A \longrightarrow Mf$ and $g: A \longrightarrow Y$. Taking α and γ to be identity maps and β to be the canonical homotopy equivalence $Mf \longrightarrow X$, the previous result applies. □

Observe that the fold map $\nabla: X \vee X \longrightarrow X$ is conceptually dual to the diagonal map $\Delta: X \longrightarrow X \times X$.

DEFINITION 2.1.5. The homotopy coequalizer (or mapping torus) $T(f,f')$ of a pair of maps $f,f': X \longrightarrow Y$ is the homotopy pushout of $(f,f'): X \vee X \longrightarrow Y$ and $\nabla: X \vee X \longrightarrow X$.

We have written the definition in a way that mimics the construction of coequalizers from pushouts and coproducts in any category (see below). However, unravelling the definition, we see that $(X \wedge X) \wedge I_+$ can be identified with $(X \wedge I_+) \vee (X \wedge I_+)$. The identification along ∇ has the effect of gluing these two cylinders into a single cylinder of twice the length. Therefore $T(f,f')$ is homeomorphic to the quotient of $Y \vee (X \wedge I_+)$ obtained by identifying $(x, 0)$

with $f(x)$ and $(x, 1)$ with $f'(x)$. This gives the source of the alternative name. We urge the reader to draw a picture.

The categorical coequalizer of f and f' can be constructed as $Y/\sim$, where $f(x) \sim f'(x)$ for x in X. The required universal property ([93, p. 16]) clearly holds. Equivalently, the coequalizer is the pushout of $(f, f'): X \vee X \longrightarrow Y$ and $\nabla: X \vee X \longrightarrow X$. In principle, the map (f, f') might be a cofibration, but that almost never happens in practice. Since cofibrations must be inclusions, it can only happen if the intersection of the images of f and f' is just the base-point of Y, and that does not generally hold. The gluing lemma applies to categorical coequalizers when it does hold, but there is no convenient generally applicable analogue of the gluing lemma. For that reason, there is no convenient general criterion that ensures that the natural map from the homotopy coequalizer to the categorical coequalizer is a homotopy equivalence.

The equalizer $E(\alpha, \beta)$ of functions $\alpha, \beta: S \longrightarrow U$ is $\{s | \alpha(s) = \beta(s)\} \subset S$, as we see by checking the universal property ([93, p. 16]). Equivalently, it is the pullback of $(\alpha, \beta): S \longrightarrow U \times U$ and $\Delta: U \longrightarrow U \times U$.

PROPOSITION 2.1.6. *For any space Z, the natural map of pointed sets*

$$[T(f, f'), Z] \longrightarrow E(f^*, f'^*)$$

is a surjection, where $f^*, f'^*: [Y, Z] \longrightarrow [X, Z]$. Its kernel is isomorphic to the set of orbits of $[\Sigma X, Z]$ under the right action of the group $[\Sigma Y, Z]$ specified by

$$x * y = (\Sigma f)^*(y)^{-1} \cdot x \cdot (\Sigma f')^*(y)$$

for $x \in [\Sigma X, Z]$ and $y \in [\Sigma Y, Z]$.

PROOF. Abbreviate $G = [\Sigma X, Z]$ and $H = [\Sigma Y, Z]$ and let $\theta = (\Sigma f)^*$ and $\theta' = (\Sigma f')^*$. By Proposition 2.1.2 and the definition of $T(f, f')$,

$$[T(f, f'), Z] \longrightarrow E(f^*, f'^*)$$

is a surjection with kernel the set of orbits of $G \times G$ under the right action of $H \times G$ specified by

$$(w, x) * (y, z) = ((\theta y)^{-1}, (\theta' y))^{-1}(w, x)(z, z) = ((\theta y)^{-1} wz, (\theta' y)^{-1} xz)$$

for $w, x, z \in G$ and $y \in H$. Define $\mu: G \times G \longrightarrow G$ by $\mu(w, x) = wx^{-1}$ and define $\nu: H \times G \longrightarrow H$ by $\nu(y, z) = y$. With H acting on G as in the statement, μ is ν-equivariant and induces a bijection on orbits by an easy algebraic verification. $\qquad \square$

DEFINITION 2.1.7. The homotopy colimit (or mapping telescope) tel X_i of a sequence of maps $f_i \colon X_i \longrightarrow X_{i+1}$ is the homotopy coequalizer of the identity map of $Y = \vee_i X_i$ and $\vee_i f_i \colon Y \longrightarrow Y$. It is homeomorphic to the union of mapping cylinders described in [93, p. 113], which gives the more usual description.

The definition of $\lim^1 G_i$ for an inverse sequence of abelian groups is given, for example, in [93, p. 146]. It generalizes to give a definition for not necessarily abelian groups. However, the result is only a set in general, not a group.

DEFINITION 2.1.8. Let $\gamma_i \colon G_{i+1} \longrightarrow G_i$, $i \geq 0$, be homomorphisms of groups. Define a right action of the group $G = \times_i G_i$ on the set $S = \times_i G_i$ by

$$(s_i) * (g_i) = (g_i^{-1} s_i \gamma_i(g_{i+1})).$$

The set of orbits of S under this action is called $\lim^1 G_i$. Observe that $\lim G_i$ is the set of elements of G that fix the element $(1) \in S$ whose coordinates are the identity elements of the G_i. Equivalently, $\lim G_i$ is the equalizer of the identity map of S and $\times_i \gamma_i \colon S \longrightarrow S$.

PROPOSITION 2.1.9. *For any space Z, the natural map of pointed sets*

$$[\operatorname{tel} X_i, Z] \longrightarrow \lim[X_i, Z]$$

is a surjection with kernel isomorphic to $\lim^1[\Sigma X_i, Z]$.

PROOF. With our definition of tel X_i as a homotopy coequalizer, this is immediate from Proposition 2.1.6. $\qquad \square$

The following "ladder lemma" is analogous to the gluing lemma above.

LEMMA 2.1.10. *Assume given a commutative diagram*

$$
\begin{array}{ccccccccc}
X_0 & \xrightarrow{f_0} & X_1 & \longrightarrow & \cdots & \longrightarrow & X_i & \xrightarrow{f_i} & X_{i+1} & \longrightarrow & \cdots \\
\downarrow{\scriptstyle \alpha_0} & & \downarrow{\scriptstyle \alpha_1} & & & & \downarrow{\scriptstyle \alpha_i} & & \downarrow{\scriptstyle \alpha_{i+1}} & & \\
X_0' & \xrightarrow{f_0'} & X_1' & \longrightarrow & \cdots & \longrightarrow & X_i' & \xrightarrow{f_i'} & X_{i+1}' & \longrightarrow & \cdots
\end{array}
$$

in which the f_i and f_i' are cofibrations. If the maps α_i are homotopy equivalences, then so is their colimit

$$\operatorname{colim} X_i \longrightarrow \operatorname{colim} X_i'.$$

PROOF. This follows inductively from the proposition about cofiber homotopy equivalence given in [93, p. 44]. □

COROLLARY 2.1.11. If the maps $f_i\colon X_i \longrightarrow X_{i+1}$ are cofibrations, then the natural quotient map tel $X_i \longrightarrow \operatorname{colim} X_i$ is a homotopy equivalence.

PROOF. The map in question identifies a point (x, t) in the cylinder $X_i \wedge I_+$ with the point $(f_i(x), 0)$ in the base of the next cylinder $X_{i+1} \wedge I_+$. We may describe tel X_i as the colimit of a sequence of partial telescopes Y_i, each of which comes with a deformation retraction $\alpha_i\colon Y_i \longrightarrow X_i$. These partial telescopes give a ladder to which the previous result applies. □

This applies in particular to the inclusions of skeleta of a CW complex X, and we have the following important definition.

DEFINITION 2.1.12. Let X be a (based) CW complex with n-skeleton X^n. A map $f\colon X \longrightarrow Z$ is called a phantom map if the restriction of f to X^n is null homotopic for all n.

WARNING 2.1.13. This is the original use of the term "phantom map", but the name is also used in some, but by no means all, of the more recent literature for maps $f\colon X \longrightarrow Z$ such that $f \circ g$ is null homotopic for all maps $g\colon W \longrightarrow X$, where W is a finite CW complex. One might differentiate by renaming the original notion "skeletally phantom" or renaming the new notion "finitely phantom". Skeletally phantom implies finitely phantom since any g as above factors through X^n for some n. Of course, the two notions agree when X has finite skeleta.

The name comes from the fact that, with either definition, a phantom map $f\colon X \longrightarrow Z$ induces the zero map on all homotopy, homology, and cohomology groups, since these invariants depend only on skeleta, and in fact only on composites $f \circ g$, where g has finite domain. With the original definition, Corollary 2.1.11 and Proposition 2.1.9 give the following identification of the phantom homotopy classes.

COROLLARY 2.1.14. *The set of homotopy classes of phantom maps* $X \longrightarrow Z$ *can be identified with* $\lim^1 [\Sigma X^n, Z]$.

PROOF. By Corollary 2.1.11, the quotient map tel $X^n \longrightarrow X$ is a homotopy equivalence, hence the conclusion follows from Proposition 2.1.9. $\qquad\square$

In §2.3, we specify a simple algebraic condition, called the Mittag-Leffler condition, on an inverse sequence $\{G_i, \gamma_i\}$ that ensures that $\lim^1 G_i$ is a single point. In the situation that occurs most often in topology, the G_i are countable, and then Theorem 2.3.3 below shows that $\lim^1 G_i$ is uncountable if the Mittag-Leffler condition fails and a certain normality of subgroups condition holds. In particular, the cited result has the following consequence.

COROLLARY 2.1.15. *If the skeleta* X^n *are finite, the homotopy groups* $\pi_q(Z)$ *are countable, the groups* $[\Sigma X^n, Z]$ *are abelian, and the Mittag-Leffer condition fails for the inverse system* $\{[\Sigma X^n, Z]\}$, *then* $\lim^1 [\Sigma X^n, Z]$ *is an uncountable divisible abelian group.*

As a concrete example, this result applies to show that $[\mathbb{C}P^\infty, S^3]$ contains an uncountable divisible subgroup. This is a result due to Gray [55] that we shall explain in §2.4. Here the groups $[\Sigma\mathbb{C}P^n, S^3]$ are abelian since S^3 is a topological group. Although it is hard to specify phantom maps concretely on the point-set level, their existence is not an exotic phenomenon.

2.2. Some basic homotopy limits

We gave the definitions and results in the previous section in a form that makes their dualization to definitions and results about homotopy limits as transparent as possible. We leave the details of proofs to the interested reader. These constructions will be used later in the proofs of fracture theorems for localizations and completions.

DEFINITION 2.2.1. The homotopy pullback (or double mapping path fibration) $N(f, g)$ of a pair of maps $f \colon X \longrightarrow A$ and $g \colon Y \longrightarrow A$ is the pullback written in alternative notations as

$$(X \times Y) \times_{A \times A} F(I_+, A) \quad \text{or} \quad X \times_f F(I_+, A) \times_g Y.$$

It is the pullback defined with respect to $f \times g \colon X \times Y \longrightarrow A \times A$ and the fibration $(p_0, p_1) \colon F(I_+, A) \longrightarrow A \times A$.

Explicitly, with the alternative notation, $N(f,g)$ is the subspace of $X \times F(I_+, A) \times Y$ that consists of those points (x, ω, y) such that $\omega(0) = f(x)$ and $\omega(1) = g(y)$. Here $(p_0, p_1): F(I_+, A) \longrightarrow A \times A$ is a fibration by application of Lemma 1.1.4 to the cofibration $(i_0, i_1): A \vee A \longrightarrow A \wedge I_+$. Its fiber over the basepoint is ΩA. Observe that pullback of this fibration along $f \times g: X \times Y \longrightarrow A \times A$ gives a fibration $(p_0, p_1): N(f,g) \longrightarrow X \times Y$ with fiber ΩA over the basepoint.

PROPOSITION 2.2.2. *For any space Z, the natural map of pointed sets*

$$[Z, N(f,g)] \longrightarrow [Z, X] \times_{[Z,A]} [Z, Y]$$

is a surjection. Its kernel is isomorphic to the set of orbits of $[Z, \Omega A]$ under the right action of the group $[Z, \Omega X] \times [Z, \Omega Y]$ specified by

$$a * (x, y) = (\Omega f)_*(x)^{-1} \cdot a \cdot (\Omega g)_*(y)$$

for $a \in [Z, \Omega A]$, $x \in [Z, \Omega X]$, and $y \in [Z, \Omega Y]$.

The following consequence, which uses the fibration $(p_0, p_1): N(f,g) \longrightarrow X \times Y$ and the exact sequence of [93, p. 64], will be especially important in our study of fracture theorems.

COROLLARY 2.2.3. *Let $f: X \longrightarrow A$ and $g: Y \longrightarrow A$ be maps between connected spaces. There is a long exact sequence*

$$\cdots \longrightarrow \pi_{n+1}(A) \longrightarrow \pi_n N(f,g) \xrightarrow{(p_0, p_1)_*} \pi_n(X) \times \pi_n(Y) \xrightarrow{f_* - g_*} \pi_n(A)$$

$$\longrightarrow \cdots \longrightarrow \pi_1(A) \longrightarrow \pi_0 N(f,g) \longrightarrow *.$$

The space $N(f,g)$ is connected if and only if every element of $\pi_1(A)$ is the product of an element of $f_*\pi_1(X)$ and an element of $g_*\pi_1(Y)$. For $n \geq 1$, the natural map

$$\pi_n(N(f,g)) \longrightarrow \pi_n(X) \times_{\pi_n(A)} \pi_n(Y)$$

is an isomorphism if and only if every element of the abelian group $\pi_{n+1}(A)$ is the sum of an element of $f_*\pi_{n+1}(X)$ and an element of $g_*\pi_{n+1}(Y)$.

If $n = 1$, the inaccurate notation $f_* - g_*$ means the pointwise product $f_* g_*^{-1}$; the additive notation is appropriate when $n > 1$. In both cases, the condition on homotopy groups says that the relevant homomorphism $f_* - g_*$ is an epimorphism.

LEMMA 2.2.4. *Assume given a commutative diagram*

$$
\begin{array}{ccccc}
X & \xrightarrow{\ f\ } & A & \xleftarrow{\ g\ } & Y \\
{\scriptstyle\beta}\downarrow & & {\scriptstyle\alpha}\downarrow & & \downarrow{\scriptstyle\gamma} \\
X' & \xrightarrow[\ f'\ ]{} & A' & \xleftarrow[\ g'\ ]{} & Y'
\end{array}
$$

in which f and f' are fibrations. If α, β, and γ are homotopy equivalences, then so is their pullback

$$ X \times_A Y \longrightarrow X' \times_{A'} Y'. $$

PROOF. The model category theory that we give later is self-dual in the very strong sense that results for a model category, when applied to its opposite category, give dual conclusions. This lemma is an illustrative example. It is the model theoretic dual of Lemma 2.1.3, and the proof of Lemma 2.1.3 that we outlined above dualizes in this sense. The notion of a right proper model category is specified in Definition 15.4.1, and by Proposition 15.4.4, a model category is right proper if and only if the conclusion of this "cogluing lemma" holds. By Theorem 17.1.1 and Corollary 17.1.2, the category of spaces and the category of based spaces are both left and right proper, so the conclusions of both Lemma 2.1.3 and this lemma hold in both categories. □

COROLLARY 2.2.5. *If f is a fibration and g is any map, then the natural injection $X \times_A Y \longrightarrow N(f,g)$ is a homotopy equivalence.*

DEFINITION 2.2.6. *The homotopy equalizer (or double fiber) $F(f,f')$ of a pair of maps $f, f' : X \longrightarrow Y$ is the homotopy pullback of $(f,f') : X \longrightarrow Y \times Y$ and $\Delta : Y \longrightarrow Y \times Y$.*

Again, we have written the definition so as to mimic the categorical construction of equalizers from pullbacks and products. Unraveling the definition, we find that the space $F(f,f')$ is homeomorphic to the pullback of the natural fibration $N(f,f') \longrightarrow X \times X$ along $\Delta : X \longrightarrow X \times X$.

PROPOSITION 2.2.7. *For any space Z, the natural map of pointed sets*

$$ [Z, F(f,f')] \longrightarrow E(f_*,f'_*) $$

is a surjection, where $f_*, f'_* : [Z,X] \longrightarrow [Z,Y]$. Its kernel is isomorphic to the set of orbits of $[Z, \Omega Y]$ under the right action of the group $[Z, \Omega X]$ specified by

$$y * x = (\Omega f)_*(x)^{-1} \cdot y \cdot (\Omega f')_*(x)$$

for $x \in [Z, \Omega X]$ and $y \in [Z, \Omega Y]$.

DEFINITION 2.2.8. The homotopy limit (or mapping microscope)[1] mic X_i of a sequence of maps $f_i \colon X_{i+1} \longrightarrow X_i$ is the homotopy equalizer of the identity map of $Y = \times_i X_i$ and $\times_i f_i \colon Y \longrightarrow Y$. It is homeomorphic to the limit of the sequence of partial microscopes Y_n defined dually to the partial telescopes in [93, p. 113].

Explicitly, let $\pi_0 \colon Y_0 = F(I_+, X_0) \longrightarrow X_0$ be p_1, evaluation at 1. Assume inductively that $\pi_n \colon Y_n \longrightarrow X_n$ has been defined. Define Y_{n+1} to be the pullback displayed in the right square of the following diagram. Its triangle defines $\pi_{n+1} \colon Y_{n+1} \longrightarrow X_{n+1}$, and the limit of the maps $Y_{n+1} \longrightarrow Y_n$ is homeomorphic to mic X_i.

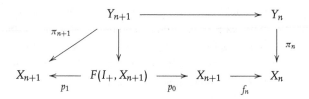

PROPOSITION 2.2.9. *For any space Z, the natural map of pointed sets*

$$[Z, \text{mic } X_i] \longrightarrow \lim[Z, X_i]$$

is a surjection with kernel isomorphic to $\lim^1[Z, \Omega X_i]$. In particular, there are natural short exact sequences

$$0 \longrightarrow \lim{}^1 \pi_{n+1}(X_i) \longrightarrow \pi_n(\text{mic } X_i) \longrightarrow \lim \pi_n(X_i) \longrightarrow 0.$$

LEMMA 2.2.10. *Assume given a commutative diagram*

$$
\begin{array}{ccccccccc}
\cdots & \longrightarrow & X_{i+1} & \overset{f_i}{\longrightarrow} & X_i & \longrightarrow & \cdots & \longrightarrow & X_1 & \overset{f_0}{\longrightarrow} & X_0 \\
& & \downarrow{\alpha_{i+1}} & & \downarrow{\alpha_i} & & & & \downarrow{\alpha_1} & & \downarrow{\alpha_0} \\
\cdots & \longrightarrow & X_{i+1} & \underset{f_i}{\longrightarrow} & X_i & \longrightarrow & \cdots & \longrightarrow & X_1 & \underset{f_0}{\longrightarrow} & X_0
\end{array}
$$

---

1. This joke name goes back to 1970s notes of the first author.

in which the f_i and f_i' are fibrations. If the α_i are homotopy equivalences, then so is their limit

$$\lim X_i \longrightarrow \lim X_i'.$$

PROOF. This follows inductively from the proposition about fiber homotopy equivalence given in [93, p. 50]. □

PROPOSITION 2.2.11. If the maps $f_i \colon X_{i+1} \longrightarrow X_i$ are fibrations, then the natural injection $\lim X_i \longrightarrow \text{mic}\, X_i$ is a homotopy equivalence.

2.3. Algebraic properties of $\lim^1$

The $\lim^1$ terms that appear in Propositions 2.1.9 and 2.2.9 are an essential, but inconvenient, part of algebraic topology. In practice, they are of little significance in most concrete applications, the principle reason being that they generally vanish on passage either to rationalization or to completion at any prime p, as we shall see in §6.8 and §11.6. We give some algebraic feel for this construction here, but only the Mittag-Leffler condition for the vanishing of $\lim^1 G_i$ will be relevant to our later work.

We consider a sequence of homomorphisms $\gamma_i \colon G_{i+1} \longrightarrow G_i$. For $j > i$, let $\gamma_i^j = \gamma_i \gamma_{i+1} \cdots \gamma_{j-1} \colon G_j \longrightarrow G_i$ and let $G_i^j = \text{im}\,\gamma_i^j$. We say that the sequence $\{G_i, \gamma_i\}$ satisfies the Mittag-Leffler condition if for each i there exists $j(i)$ such that $G_i^k = G_i^{j(i)}$ for all $k > j(i)$. That is, these sequences of images eventually stabilize. For example, this condition clearly holds if each γ_i is an epimorphism or if each G_i is a finite group.

The following results collect the basic properties of $\lim^1 G_i$. The main conclusion is that either the Mittag-Leffler condition holds and $\lim^1 G_i = 0$ or, under further hypotheses that usually hold in the situations encountered in algebraic topology, the Mittag-Leffler condition fails and $\lim^1 G_i$ is uncountable.

PROPOSITION 2.3.1. A short exact sequence

$$\{1\} \longrightarrow \{G_i', \gamma_i'\} \longrightarrow \{G_i, \gamma_i\} \longrightarrow \{G_i'', \gamma_i''\} \longrightarrow \{1\}$$

of towers of groups gives rise to a natural exact sequence of pointed sets

$$1 \to \lim G_i' \to \lim G_i \to \lim G_i'' \overset{\delta}{\to} \lim^1 G_i' \to \lim^1 G_i \to \lim^1 G_i'' \to 1.$$

Moreover, the group $\lim G_i''$ *acts from the right on the set* $\lim^1 G_i'$, δ *is a map of right* $\lim G_i''$*-sets, and two elements of* $\lim^1 G_i'$ *map to the same element of* $\lim^1 G_i$ *if and only if they lie in the same orbit.*

PROOF. The identity elements of groups are considered as their basepoints, and sequences of identity elements give the basepoints necessary to the statement. By an exact sequence of pointed sets, we just mean that the image in each term is the set of elements that map to the basepoint in the next term. We use square brackets to denote orbits and we regard G_i' as a subgroup of G_i. We define the right action of $\lim G_i''$ on $\lim^1 G_i'$ by letting

$$[s_i'](g_i'') = [g_i^{-1} s_i' \gamma_i (g_{i+1})],$$

where (g_i'') is a point in $\lim G_i''$. Here $g_i \in G_i$ is any element that maps to $g_i'' \in G_i''$. Define $g_i' = g_i^{-1} s_i' \gamma_i (g_{i+1})$. Then g_i' is in G_i' because the equality $g_i'' = \gamma_i''(g_{i+1}'')$ implies that g_i' maps to 1 in G_i''. Define δ by $\delta(g'') = [1]g''$, where [1] denotes the orbit of the sequence of identity elements of the G_i'. With these definitions, the verification is straightforward, if laborious. $\square$

LEMMA 2.3.2. *For any strictly increasing sequence* $\{j(i)\}$, *the diagram*

$$\cdots \longrightarrow G_{i+1}^{j(i+1)} \xrightarrow{\gamma_i} G_i^{j(i)} \longrightarrow \cdots \longrightarrow G_1^{j(1)} \xrightarrow{\gamma_0} G_0^{j(0)}$$
$$\downarrow\cap \qquad\qquad \downarrow\cap \qquad\qquad\qquad \downarrow\cap \qquad\qquad \downarrow\cap$$
$$\cdots \longrightarrow G_{i+1} \xrightarrow{\gamma_i} G_i \longrightarrow \cdots \longrightarrow G_1 \xrightarrow{\gamma_0} G_0$$

induces an isomorphism $\lim G_i^{j(i)} \longrightarrow \lim G_i$ *and a surjection* $\lim^1 G_i^{j(i)} \longrightarrow \lim^1 G_i$. *The latter function is a bijection if* $G_i^{j(i)}$ *is a normal subgroup of* G_i *for each* i.

PROOF. Since the sequence $\{j(i)\}$ is strictly increasing, the isomorphism of lim groups holds by cofinality. If (g_i) is in $\lim G_i$, then g_j maps to g_i for each $j > i$. In particular $g_{j(i)}$ maps to g_i. This shows that the map of lim groups is surjective, and injectivity is clear. To see the surjectivity on $\lim^1$, let $(t_i) \in \times_i G_i$ and $(g_i) \in \times_i G_i$. The definition of the right action given in Definition 2.1.8 shows that $(t_i) = (s_i)(g_i)$, where

$$s_i = g_i t_i \gamma_i (g_{i+1})^{-1} \in G_i.$$

If we choose

$$g_i = \gamma_i^{j(i)}(t_{j(i)})^{-1}\gamma_i^{j(i)-1}(t_{j(i)-1})^{-1}\cdots\gamma_i^{i+1}(t_{i+1})^{-1}t_i^{-1},$$

then we find that $s_i = \gamma_i^{j(i)+1}(t_{j(i)+1})\cdots\gamma_i^{j(i+1)}(t_{j(i+1)})$ is in $G_i^{j(i)}$. This displays the orbit $[t_i] \in \lim^1 G_i$ as an element in the image of $\lim^1 G_i^{j(i)}$. If $G_i^{j(i)}$ is normal in G_i for each i, then $\lim G_i/G_i^{j(i)}$ is defined, and it is the trivial group since another cofinality argument shows that an element other than (e_i) would have to come from a nontrivial element of $\lim G_i/\lim G_i^{j(i)}$. Therefore $\lim^1 G_i^{j(i)} \to \lim^1 G_i$ is an injection by the exact sequence of Proposition 2.3.1. $\square$

THEOREM 2.3.3. $\lim^1 G_i$ satisfies the following properties.

(i) If $\{G_i, \gamma_i\}$ satisfies the Mittag-Leffler condition, then $\lim^1 G_i$ is trivial.

(ii) If G_i^j is a normal subgroup of G_i for each $j > i$ and each G_i is countable, then either $\{G_i, \gamma_i\}$ satisfies the Mittag-Leffler condition or $\lim^1 G_i$ is uncountable.

(iii) If each G_i is a finitely generated abelian group, then $\lim^1 G_i$ is a divisible abelian group.

PROOF. For (i), we may assume that $G_i^k = G_i^{j(i)}$ for $k > j(i)$, where the $j(i)$ form a strictly increasing sequence. Let $(t_i) \in \times_i G_i^{j(i)}$. There are elements $h_i \in G_i^{j(i)}$ such that $t_i = h_i^{-1}\gamma_i(h_{i+1})$. To see this, let $h_0 = 1$ and assume inductively that h_a has been constructed for $a \leq i$. Then $h_i t_i$ is in $G_i^{j(i)} = G_i^{j(i+1)}$, say $h_i t_i = \gamma_i^{j(i+1)}(g)$. Let $h_{i+1} = \gamma_{i+1}^{j(i+1)}(g)$. Then $h_{i+1} \in G_{i+1}^{j(i+1)}$ and $h_i t_i = \gamma_i(h_{i+1})$, as required. This implies that $(t_i) = (1)(h_i)$, so that the orbit set $\lim^1 G_i^{j(i)}$ contains only the element $[1]$. By the surjectivity result of the previous lemma, this implies that $\lim^1 G_i$ also contains only the single element $[1]$.

For (ii), fix $i \geq 0$ and, letting j vary, consider the diagram

$$
\begin{array}{ccccccccc}
\cdots & \longrightarrow & G_i^{i+j+1} & \longrightarrow & G_i^{i+j} & \longrightarrow & \cdots & \longrightarrow & G_i^{i+2} & \longrightarrow & G_i^{i+1} \\
& & \downarrow & & \downarrow & & & & \downarrow & & \downarrow \\
\cdots & = & G_i & = & G_i & = & \cdots & = & G_i & = & G_i \\
& & \downarrow & & \downarrow & & & & \downarrow & & \downarrow \\
\cdots & \longrightarrow & H_i^{j+1} & \longrightarrow & H_i^j & \longrightarrow & \cdots & \longrightarrow & H_1^2 & \longrightarrow & H_i^1.
\end{array}
$$

Here $H_i^j = G_i/G_i^{i+j}$. By Proposition 2.3.1 and the fact that $\lim^1$ is trivial on constant systems, there results an exact sequence

$$* \longrightarrow \lim G_i^{i+j} \longrightarrow G_i \longrightarrow \lim H_i^j \longrightarrow \lim^1 G_i^{i+j} \longrightarrow *.$$

Applying Proposition 2.3.1 to the exact sequence

$$\{1\} \longrightarrow \{\ker \gamma_i^j\} \longrightarrow \{G_j\} \longrightarrow \{G_i^{i+j}\} \longrightarrow \{1\},$$

we see that we also have a surjection $\lim^1 G_j \longrightarrow \lim^1 G_i^{i+j}$. If $\lim H_i^j$ is uncountable, then so are $\lim^1 G_i^{i+j}$ (since G_i is assumed to be countable) and $\lim^1 G_j$. Here the index runs over $j \geq i$ for our fixed i, but $\lim^1 G_j$ is clearly independent of i. The cardinality of $\lim H_i^j$ is the product over $j \geq 1$ of the cardinalities of the kernels G_i^{i+j}/G_i^{i+j+1} of the epimorphisms $H_i^{j+1} \longrightarrow H_i^j$. Therefore $\lim H_i^j$ is countable if and only if G_i^{i+j}/G_i^{i+j+1} has only one element for all but finitely many values of j, and the latter assertion (for all i) is clearly equivalent to the Mittag-Leffler condition. This proves (ii).

For (iii), let $n > 1$ and apply Proposition 2.3.1 to the spliced short exact sequences

Since G_i/nG_i is finite, $\lim^1 G_i/nG_i = 0$ by (i). Therefore, by the triangles in the diagram, multiplication by n on $\lim^1 G_i$ is the composite epimorphism

$$\lim^1 G_i \longrightarrow \lim^1 nG_i \longrightarrow \lim^1 G_i. \qquad \square$$

2.4. An example of nonvanishing $\lim^1$ terms

As promised, we here give an example due to Gray [55] that shows how easy it can be to prove that the Mittag-Leffler condition fails.

LEMMA 2.4.1. Let X be a based CW complex such that $\pi_i(X) = 0$ for $i < q$ and $\pi_q(X) = \mathbb{Z}$, where $q \geq 2$ is even. Assume that X is a CW complex with q-skeleton S^q

and give ΣX the induced CW structure with $(\Sigma X)^n = \Sigma X^n$. Let $f\colon S^{q+1} \longrightarrow \Sigma X$ generate $\pi_{q+1}(\Sigma X) = \mathbb{Z}$. Suppose $g \circ f$ has degree zero for every map $g\colon \Sigma X \longrightarrow S^{q+1}$. Then the sequence of groups $[\Sigma X^n, S^{q+1}]$ does not satisfy the Mittag-Leffler condition.

PROOF. For $n \geq q$, we have a cofiber sequence

$$J_n \xrightarrow{\ j_n\ } X^n \xrightarrow{\ k_n\ } X^{n+1},$$

where J_n is a wedge of n-spheres, j_n is given by attaching maps, and k_n is the inclusion. Suspending, these induce exact sequences

$$[\Sigma J_n, S^{q+1}] \xleftarrow{\ (\Sigma j_n)^*\ } [\Sigma X^n, S^{q+1}] \xleftarrow{\ (\Sigma k_n)^*\ } [\Sigma X^{n+1}, S^{q+1}].$$

Since the functor $[-, Z]$ converts finite wedges to finite products and thus to finite direct sums when it takes values in abelian groups, $[\Sigma J_n, S^{q+1}]$ is a direct sum of homotopy groups $\pi_{n+1}(S^{q+1})$. As we shall recall in §6.7, since q is even, $\pi_{n+1}(S^{q+1})$ is finite for all $n > q$.

Since $X^q = S^q$ and $\pi_q(X) = \mathbb{Z}$, j_q must be null homotopic; if not, its homotopy class would be an element of $\pi_q(X^q)$ that would map to zero in $\pi_q(X)$. With its cell structure induced from that of X, the $q+1$-skeleton of ΣX^q is S^{q+1}. Let $g_0\colon \Sigma X^q \longrightarrow S^{q+1}$ be the identity. Since j_q is null homotopic, so is Σj_q. Therefore $(\Sigma j_q)^*[g_0] = 0$ and $[g_0] = (\Sigma k_q)^*[g_1]$ for some $[g_1] \in [\Sigma X^{q+1}, S^{q+1}]$. Inductively, starting with $m_1 = 1$, for $n \geq 1$ we can choose positive integers m_n and maps $g_n\colon \Sigma X^{q+n} \longrightarrow S^{q+1}$ such that $(\Sigma k_{q+n-1})^*(g_n) = m_n g_{n-1}$. Indeed, for $n \geq 2$, $(\Sigma j_{q+n-1})^*([g_{n-1}])$ is an element of a finite group, so is annihilated by some m_n, and then $m_n[g_{n-1}]$ is in the image of $(\Sigma k_{q+n-1})^*$.

Now suppose that the Mittag-Leffler condition holds. This means that, for n_0 large enough,

$$\mathrm{im}[\Sigma X^{q+n}, S^{q+1}] \longrightarrow [\Sigma X^q, S^{q+1}] = \mathrm{im}[\Sigma X^{q+n_0}, S^{q+1}] \longrightarrow [\Sigma X^q, S^{q+1}]$$

for all $n > n_0$. Then we can take $m_n = 1$ for $n > n_0$. By the surjectivity part of the $\lim^1$ exact sequence, there is a map $g\colon \Sigma X \longrightarrow S^{q+1}$ whose restriction to ΣX^{q+n} is homotopic to g_n for each n. Thinking of $f\colon S^{q+1} \longrightarrow \Sigma X$ as the composite of the identity map on the $q+1$-skeleton composed with inclusions of skeleta and thinking of g as the colimit of its restrictions to skeleta, we see that the composite $g \circ f$ has degree $m_1 \cdots m_{n_0}$. This contradicts the assumption that $g \circ f$ has degree zero. $\qquad\square$

LEMMA 2.4.2. *With $q = 2$, $\mathbb{C}P^\infty$ satisfies the hypotheses of Lemma 2.4.1.*

PROOF. With the notations of Lemma 2.4.1, we must show that $g \circ f$ has degree zero for any map $g\colon \Sigma\mathbb{C}P^\infty \longrightarrow S^3$. Suppose that $g \circ f$ has degree $m \neq 0$. Up to sign, $g^*(i) = mx$ in cohomology with any ring R of coefficients, where $i \in H^3(S^3)$ is the fundamental class and $x \in H^3(\mathbb{C}P^\infty)$ is the generator. Taking $R = \mathbb{F}_p$ where p is prime to m, we see that g^* is an isomorphism in degree 3. But the first Steenrod operation P^1 (see, e.g., [86, 130]) satisfies $P^1 x \neq 0$, which by naturality contradicts $P^1(i) = 0$. $\qquad\square$

COROLLARY 2.4.3. *There are uncountably many phantom maps $\mathbb{C}P^\infty \longrightarrow S^3$.*

We have just seen how easy it is to prove that phantom maps exist, but it is very far from obvious how to write them down in any explicit form.

2.5. The homology of colimits and limits

It was observed in [93, p. 113] that homology commutes with sequential colimits of inclusions. The same holds more generally for suitably well-behaved filtered colimits, which are defined to be colimits of diagrams defined on filtered categories.

DEFINITION 2.5.1. A small category $\mathscr{D}$ is filtered if

(i) For any two objects d and d', there is an object e that admits morphisms $d \longrightarrow e$ and $d' \longrightarrow e$.

(ii) For any two morphisms $\alpha, \beta\colon d \longrightarrow e$, there is a morphism $\gamma\colon e \longrightarrow f$ such that $\gamma\alpha = \gamma\beta$.

This definition suffices for many applications. However, we insert the following more general definitions [11, p. 268] since they will later play a significant role in model category theory. The reader may ignore the generality now, but it will be helpful later to have seen an elementary example of how these definitions are used before seeing such arguments in model category theory. The union of a finite set of finite sets is a finite set, and we recall that regular cardinals are defined to be those with the precisely analogous property.

DEFINITION 2.5.2. A cardinal is an ordinal that is minimal among those of the same cardinality. A cardinal λ is regular if for every set I of cardinality less than λ and every set $\{S_i | i \in I\}$ of sets S_i, each of cardinality less than λ, the cardinality of the union of the S_i is less than λ.

DEFINITION 2.5.3. Let λ be a regular cardinal. A small category $\mathscr{D}$ is λ-filtered if

(i) For any set of objects $\{d_i | i \in I\}$ indexed by a set I of cardinality less than λ, there is an object e that admits morphisms $d_i \longrightarrow e$ for all $i \in I$.
(ii) For any set of morphisms $\{\alpha_i : d \longrightarrow e | i \in I\}$ indexed by a set I of cardinality less than λ, there is a morphism $\gamma : e \longrightarrow f$ such that $\gamma\alpha_i = \gamma\alpha_j$ for all indices i and j in I.

In Definition 2.5.1, we restricted to the ordinal with two elements, but by finite induction we see that our original definition of a filtered category is actually the same notion as an ω-filtered category. As with any category, a filtered category $\mathscr{D}$ may or may not have colimits and a functor, or diagram, defined on $\mathscr{D}$ may or may not preserve them. The precise meaning of the assumptions on colimits in the following result will become clear in the proof. By a sequential colimit, we just mean a colimit indexed on the nonnegative integers, viewed as a category whose only nonidentity maps are $m \longrightarrow n$ for $m < n$.

PROPOSITION 2.5.4. Let X be the colimit of a diagram $X_* : \mathscr{D} \longrightarrow \mathscr{U}$ of closed inclusions of spaces, where $\mathscr{D}$ is λ-filtered for some regular cardinal λ. If $\lambda > \omega$, assume in addition that $\mathscr{D}$ has sequential colimits and that X_* preserves them. Let K be a compact space. Then any map $f : K \longrightarrow X$ factors through some X_d.

PROOF. The colimit X has the topology of the union, so that a subspace is closed if and only if it intersects each X_d in a closed subset, and each X_d is a closed subspace of X. Since we are working with compactly generated spaces, $f(K)$ is a closed compact subspace of X [93, p. 37]. Assume that f does not factor through any X_d. Starting with any object d_0, we can choose a sequence of objects d_n and maps $\alpha_n : d_{n-1} \longrightarrow d_n$ in $\mathscr{D}$ and a sequence of elements k_n in K such that $f(k_n)$ is in the complement of the image of $X_{d_{n-1}}$ in X_{d_n}. Indeed, let $n \geq 1$ and suppose that d_i, α_i, and k_i have been chosen for $0 < i < n$. There is an element k_n such that $f(k_n)$ is not in $X_{d_{n-1}}$. There must be some object d'_{n-1} such that $f(k_n)$ is in $X_{d'_{n-1}}$. There is an object d_n that admits maps $\alpha_n : d_{n-1} \longrightarrow d_n$ and $\alpha'_n : d'_{n-1} \longrightarrow d_n$. Then $f(k_n)$ is in the image of the inclusion $X_{d'_{n-1}} \longrightarrow X_{d_n}$ induced by α'_n but is not in the image of the inclusion $X_{d_{n-1}} \longrightarrow X_{d_n}$ induced by α_n.

We may view the countable ordered set $\{d_n\}$ as a subcategory of $\mathscr{D}$. If $\lambda = \omega$, then $\{d_n\}$ is cofinal in $\mathscr{D}$ and X can be identified with $\operatorname{colim} X_{d_n}$; for notational convenience, we then write $X_e = X$ (with no e in mind). If $\lambda > \omega$, then, by assumption, $\{d_n\}$ has a colimit $e \in \mathscr{D}$ and $X_e = \operatorname{colim} X_{d_n}$. Since X_e is one of the spaces in our colimit system, our hypotheses imply that X_e is a closed subspace of X. Therefore, in both cases $f(K) \cap X_e$ is a closed subspace of the compact space $f(K)$ and is again compact. Since compactly generated spaces are T_1 (points are closed), the set $S_0 = \{f(k_n)\}$ and each of its subsets $S_m = \{f(k_{m+i})|i \geq 0\}$, $m \geq 0$, is closed in $f(K) \cap X_e$. Any finite subset of the set $\{S_m\}$ has nonempty intersection, but the intersection of all of the S_m is empty. Since $f(K) \cap X_e$ is compact, this is a contradiction.[2] $\qquad\square$

EXAMPLE 2.5.5. The hypothesis on sequential colimits is essential. For a counterexample without it, let $X = [0, 1] \subset \mathbb{R}$, and observe that X is the colimit of its countable closed subsets X_d, partially ordered under inclusion. This gives an $\aleph_0$-filtered indexing category. Obviously the identity map of X does not factor through any X_d. For $n \geq 0$, let $X_n = \{0\} \cup \{1/i | 1 \leq i \leq n\}$. As a set, the colimit of the X_n is $X_\infty = \{0\} \cup \{1/i | i \geq 1\}$. Topologized as a subspace of X, X_∞ is a countable closed subspace of X, so it qualifies as the colimit of $\{X_n\}$ in our indexing category of countable closed subspaces. However, its colimit topology is discrete, so our colimit hypothesis fails. (Observe too that if we redefine the X_n without including $\{0\}$, then X_∞ is not closed in X, showing that our indexing category does not have all sequential colimits.)[3]

COROLLARY 2.5.6. *For X_* and X as in Proposition 2.5.4,*

$$H_*(X) \cong \operatorname{colim}_{d \in D} H_*(X_d),$$

where homology is taken with coefficients in any abelian group. Similarly, when X_* takes values in $\mathscr{T}$, $\pi_*(X) \cong \operatorname{colim}_{d \in D} \pi_*(X_d)$.

PROOF. We can compute homology with singular chains. Since the simplex Δ_n is compact, any singular simplex $f : \Delta_n \longrightarrow X$ factors through some X_d. Similarly, for the second statement, any based map from S^n or $S^n \wedge I_+$ to X factors through some X_d. $\qquad\square$

---

2. This pleasant argument is an elaboration of a lemma of Dold and Thom [37, Hilfsatz 2.14].

3. This example is due to Rolf Hoyer.

We need a kind of dual result, but only for sequential limits of based spaces. The following standard observation will be needed in the proof. Recall the notion of a q-equivalence from [93, p. 67].

LEMMA 2.5.7. *A q-equivalence $f : X \longrightarrow Y$ induces isomorphisms on homology and cohomology groups in dimensions less than q.*

PROOF. Using mapping cylinders, we can replace f by a cofibration [93, p. 42]. Since weak equivalences induce isomorphisms on homology and cohomology, relative CW approximation [93, p. 76] and cellular approximation of maps [93, p. 74] show that we can replace X and Y by CW complexes with the same q-skeleton and can replace f by a cellular map that is the identity on the q-skeleton. Since in dimensions less than q the cellular chains and thus the homology and cohomology groups of a CW complex depend only on its q-skeleton, the conclusion follows. □

DEFINITION 2.5.8. A tower (or inverse sequence) of spaces $f_n : X_{n+1} \longrightarrow X_n$ is convergent if for each q, there is an n_q such that the canonical map $X \longrightarrow X_n$ is a q-equivalence for all $n \geq n_q$.

PROPOSITION 2.5.9. *Let $X = \lim X_n$, where $\{X_n\}$ is a convergent tower of fibrations. Then the canonical maps induce isomorphisms*

$$\pi_*(X) \cong \lim \pi_*(X_n), \quad H_*(X) \cong \lim H_*(X_n), \quad \text{and} \quad H^*(X) \cong \operatorname{colim} H^*(X_n).$$

PROOF. We may replace X by mic X_n. The inverse systems of homotopy groups satisfy the Mittag-Leffler condition, so that the $\lim^1$ error terms are trivial, and the isomorphism on homotopy groups follows. The isomorphism on homology and cohomology groups is immediate from Lemma 2.5.7. □

2.6. A profinite universal coefficient theorem

In this brief and digressive algebraic section, we advertise an observation about cohomology with coefficients in a profinite abelian group. We view it as a kind of universal coefficient theorem for such groups. For present purposes, we understand a profinite abelian group B to be the filtered limit of a diagram $\{B_d\}$ of finite abelian groups. Here filtered limits are defined in evident analogy with filtered colimits. They are limits of diagrams that are indexed on the opposite category $\mathscr{D}^{op}$ of a filtered category $\mathscr{D}$, as specified in Definition 2.5.1.

THEOREM 2.6.1. *Let X be a chain complex of free abelian groups and let $B = \lim B_d$ be a profinite abelian group. Then the natural homomorphism*

$$H^*(X; B) \longrightarrow \lim H^*(X; B_d)$$

is an isomorphism, and each $H^q(X; B)$ is profinite.

PROOF. Let A be any abelian group. We claim first that

$$\text{Hom}\,(A, B) \cong \lim \text{Hom}\,(A, B_d)$$

and the derived functors $\lim^n \text{Hom}\,(A, B_d)$ are zero for all $n > 0$. It would be altogether too digressive to develop the theory of derived functors here, and we shall content ourselves by pointing out where the required arguments can be found. When A is finitely generated, so that each $\text{Hom}\,(A, B_d)$ is finite, the claim follows from Roos [119, Prop. 1] or, more explicitly, Jensen [73, Prop. 1.1]. In the general case, write A as the filtered colimit of its finitely generated subgroups A_i, where i runs through an indexing set $\mathscr{I}$. By Roos [119, Thm. 3], there is a spectral sequence that converges from

$$E_2^{p,q} = \lim_d{}^p \lim_i{}^q \text{Hom}\,(A_i, B_d)$$

to the derived functors $\lim^n$ of the system of groups $\{\text{Hom}\,(A_i, B_d)\}$ indexed on $\mathscr{I} \times \mathscr{D}^{\mathrm{op}}$. Since all of these groups are finite, these $\lim^n$ groups are zero for $n > 0$, by a generalized version of the Mittag-Leffler criterion [119, cor. to Prop. 2], and the zeroth group is $\text{Hom}\,(A, B)$. Since $E_2^{p,q} = 0$ for $q > 0$, $E_2 = E_\infty$ and the groups $E_2^{p,0}$ must be zero for $p > 0$. This implies the claim.

We claim next that

$$\text{Ext}\,(A, B) \cong \lim \text{Ext}\,(A, B_d).$$

Write A as a quotient F/F' of free abelian groups and break the exact sequence

$$0 \longrightarrow \text{Hom}\,(A, B_d) \longrightarrow \text{Hom}\,(F, B_d) \longrightarrow \text{Hom}\,(F', B_d) \longrightarrow \text{Ext}\,(A, B_d) \longrightarrow 0$$

of diagrams into two short exact sequences in the evident way. There result two long exact sequences of $\lim^n$ groups, and the vanishing of $\lim^n$ on the Hom systems therefore implies both that $\lim^n (\text{Ext}\,(A, B_d)) = 0$ for $n > 0$ and that the displayed exact sequence remains exact on passage to limits. Our claim follows by use of the five lemma.

Taking A to be a homology group $H_q(X)$ and applying the universal coefficient theorem to the calculation of $H^*(X; B)$ and the $H^*(X; B_d)$, we now see by the five lemma that

$$H^*(X; B) \cong \lim H^*(X; B_d).$$

Finally, to see that each $H^q(X; B)$ is profinite, write $X = \lim X_j$, where X_j runs through those subcomplexes of X such that each $H_q(X_j)$ is finitely generated; write $\mathscr{J}$ for the resulting set of indices j. Then each $H^q(X_j; B_d)$ is a finite abelian group and $H_q(X) = \operatorname{colim} H_q(X_j)$. The arguments just given demonstrate that

$$H^*(X; B) \cong \lim H^*(X_j; B_d),$$

where the limit is taken over $\mathscr{J} \times \mathscr{D}^{\mathrm{op}}$. $\qquad\square$

3

NILPOTENT SPACES AND POSTNIKOV TOWERS

In this chapter, we define nilpotent spaces and Postnikov towers and explain the relationship between them. We are especially interested in restrictions of these notions that are specified in terms of some preassigned collection $\mathscr{A}$ of abelian groups, and we assume once and for all that the zero group is in any such chosen collection. We define $\mathscr{A}$-nilpotent spaces and Postnikov $\mathscr{A}$-towers, and we prove that any $\mathscr{A}$-nilpotent space is weakly equivalent to a Postnikov $\mathscr{A}$-tower. The role of the collection $\mathscr{A}$ is to allow us to develop results about spaces built up from a particular kind of abelian group (T-local, T-complete, etc.) in a uniform manner.

As we discussed in the Introduction, nilpotent spaces give a comfortable level of generality for the definition of localizations and completions. The theory is not much more complicated than it is for simple spaces, and nilpotency is needed for the fracture theorems.

The material of this chapter is fundamental to the philosophy of the entire book. We expect most readers to be reasonably comfortable with CW complexes but to be much less comfortable with Postnikov towers, which they may well have never seen or seen only superficially. We want the reader to come away from this chapter with a feeling that these are such closely dual notions that there is really no reason to be more comfortable with one than the other. We also want the reader to come away with the idea that cohomology classes, elements of $\tilde{H}^n(X; \pi)$, are interchangable with (based) homotopy classes of maps, elements of $[X, K(\pi, n)]$. This is not just a matter of theory but rather a powerful concrete tool for working with these elements to prove theorems.

3.1. $\mathscr{A}$-nilpotent groups and spaces

A group is nilpotent if it has a central series that terminates after finitely many steps. It is equivalent that either its lower central series or its upper central series terminates after finitely many steps [56, p. 151]. We simultaneously generalize this definition in two ways. First, the successive quotients in the lower central series are abelian groups. One direction of generalization is to require normal sequences whose successive quotients satisfy more restrictive conditions. For example, we might require them to be $\mathbb{Z}_T$-modules, where $\mathbb{Z}_T$ is the localization of $\mathbb{Z}$ at a set of primes T. The other direction is to start with an action of a second group on our given group, rather than to restrict attention to the group acting on itself by conjugation as is implicit in the usual notion of nilpotency.

Let $\mathcal{A}$ be a collection of abelian groups (containing 0). The main example is the collection $\mathcal{A}b$ of all abelian groups or, more generally, the collection $\mathcal{A}_R$ of modules over a commutative ring R. The rings $R = \mathbb{Z}_T$ and $R = \mathbb{Z}_p$ are of particular interest.[1] While $\mathcal{A}_R$ is an abelian category and we will often need that structure to prove things we want, we shall also encounter examples of interest where we really do only have a collection of abelian groups. We could regard such a collection $\mathcal{A}$ as a full subcategory of $\mathcal{A}b$ but, in the absence of kernels and cokernels in $\mathcal{A}$, that is not a useful point of view.

Call a group G a π-group if it has a (left) action of the group π as automorphisms of G. This means that we are given a homomorphism from π to the automorphism group of G. We are thinking of π as $\pi_1(X)$ and G as $\pi_n(X)$ for a space X. Allowing general non-abelian groups G unifies the cases $n = 1$ and $n > 1$.

DEFINITION 3.1.1. Let G be a π-group. A finite normal series

$$\{1\} = G_q \subset G_{q-1} \subset \cdots \subset G_0 = G$$

of subgroups of G is said to be an $\mathcal{A}$-central π-series if

(i) G_{j-1}/G_j is in $\mathcal{A}$ and is a central subgroup of G/G_j.
(ii) G_j is a π-subgroup of G and π acts trivially on G_{j-1}/G_j.

If such a sequence exists, the action of π on G is said to be nilpotent, and the π-group G is said to be $\mathcal{A}$-nilpotent of nilpotency class at most q; the nilpotency class of G is the smallest q for which such a sequence exists.

NOTATION 3.1.2. We abbreviate notation by saying that an $\mathcal{A}b$-nilpotent π-group is a nilpotent π-group and that an $\mathcal{A}_R$-nilpotent π-group is an

---

1. We warn the knowledgeable reader that, in contrast to the theory in [21], we really do mean the p-adic integers $\mathbb{Z}_p$ and not the field $\mathbb{F}_p$ here.

R-nilpotent π-group. When it is clearly understood that a given group π is acting on G, we sometimes just say that G is $\mathscr{A}$-nilpotent, leaving π understood.

Of course, we can separate out our two generalizations of the notion of nilpotency. Ignoring π, a group G is said to be $\mathscr{A}$-nilpotent if its action on itself by inner automorphisms, $x \cdot g = xgx^{-1}$, is $\mathscr{A}$-nilpotent. When $\mathscr{A} = \mathscr{A}b$, this is the standard notion of nilpotency. On the other hand, when G is abelian, a π-group is just a module over the group ring $\mathbb{Z}[\pi]$. The purpose of unifying the notions is to unify proofs of the results we need, such as the following one. We will use it often.

LEMMA 3.1.3. *Let* $1 \longrightarrow G' \xrightarrow{\phi} G \xrightarrow{\psi} G'' \longrightarrow 1$ *be an exact sequence of* π*-groups. If the extension is central and* G' *and* G'' *are* $\mathscr{A}$*-nilpotent* π*-groups, then* G *is an* $\mathscr{A}$*-nilpotent* π*-group. Conversely, if* $\mathscr{A}$ *is closed under passage to subgroups and quotient groups and* G *is an* $\mathscr{A}$*-nilpotent* π*-group, then* G' *and* G'' *are* $\mathscr{A}$*-nilpotent* π*-groups.*

PROOF. For the first statement, let

$$1 = G'_p \subset G'_{p-1} \subset \cdots \subset G'_0 = G'$$

and

$$1 = G''_q \subset G''_{q-1} \subset \cdots \subset G''_0 = G''$$

be $\mathscr{A}$-central π-series. Then the sequence of inclusions

$$1 = \phi(G'_p) \subset \phi(G'_{p-1}) \subset \cdots \subset \phi(G'_0)$$
$$= \psi^{-1}(G''_q) \subset \psi^{-1}(G''_{q-1}) \subset \cdots \subset \psi^{-1}(G''_0) = G$$

is an $\mathscr{A}$-central π-series for G'. The centrality assumption, which in particular implies that G' is abelian, is essential to the conclusion.

Conversely, suppose that G is $\mathscr{A}$-nilpotent. Then there is an $\mathscr{A}$-central π-series

$$1 = G_q \subset G_{q-1} \subset \cdots \subset G_0 = G.$$

We may identify $\phi(G')$ with G' and G'' with G/G'. Define subgroups $G'_i = G_i \cap G'$ of G'. Since $\mathscr{A}$ is closed under passage to subgroups, this gives a finite $\mathscr{A}$-central π-series for G'. The quotient groups G_i/G'_i are isomorphic to the quotient groups $(G_i \cdot G')/G'$, and these are subgroups of G''. Since $\mathscr{A}$ is closed under passage to quotient groups as well as subgroups, G_i'' is an $\mathscr{A}$-central π-series for G''. □

We use the action of $\pi_1 X$ on the groups $\pi_n X$ from Definition 1.4.4 and the definition of an $\mathscr{A}$-nilpotent $\pi_1 X$-group to define the notion of an $\mathscr{A}$-nilpotent space. Observe that we are discarding unnecessary generality above, since now either $\pi = \pi_1 X$ is acting on itself by conjugation or π is acting on the abelian group $\pi_n X$ for $n \geq 2$.

DEFINITION 3.1.4. A connected based space X is said to be $\mathscr{A}$-nilpotent if $\pi_n X$ is an $\mathscr{A}$-nilpotent $\pi_1 X$-group for each $n \geq 1$. This means that $\pi_1(X)$ is $\mathscr{A}$-nilpotent and acts nilpotently on $\pi_n(X)$ for $n \geq 2$. When $\mathscr{A} = \mathscr{A}b$ we say that X is nilpotent. When $\mathscr{A} = \mathscr{A}_R$, we say that X is R-nilpotent.

Recall that a connected space X is simple if $\pi_1 X$ is abelian and acts trivially on $\pi_n X$. Clearly simple spaces and, in particular, simply connected spaces, are nilpotent. Connected H-spaces are simple and are therefore nilpotent. While it might seem preferable to restrict attention to simple or simply connected spaces, nilpotent spaces have significantly better closure properties under various operations. For an important example already mentioned, we shall see in Theorem 6.3.2 that if X is a finite CW complex, Y is a nilpotent space, and $f \colon X \longrightarrow Y$ is any map, then the component $F(X, Y)_f$ of f in $F(X, Y)$ is nilpotent. This space is generally not simple even when X and Y are simply connected.

3.2. Nilpotent spaces and Postnikov towers

We defined $\mathscr{A}$-nilpotent spaces in the previous section. The definition depends only on the homotopy groups of X. We need a structural characterization that allows us to work concretely with such spaces. We briefly recall two well-known results that we generalize before going into this. The classical result about Postnikov towers reads as follows.

THEOREM 3.2.1. *A connected space X is simple if and only if it admits a Postnikov tower of principal fibrations.*

We recall what this means. We can always construct maps $\alpha_n \colon X \longrightarrow X_n$ such that α_n induces an isomorphism on π_i for $i \leq n$ and $\pi_i X_n = 0$ for

$i > n$ just by attaching cells inductively to kill the homotopy groups of X in dimension greater than n. When X is simple, and only then, we can arrange further that X_{n+1} is the homotopy fiber of a "k-invariant"

$$k^{n+2}: X_n \longrightarrow K(\pi_{n+1}X, n+2).$$

This is what it means for X to have a "Postnikov tower of principal fibrations". The name comes from the fact that X_{n+1} is then the pullback along k^{n+2} of the path space fibration over $K(\pi_{n+1}X, n+2)$. Of course, the fiber of the resulting map $p_{n+1}: X_{n+1} \longrightarrow X_n$ is an Eilenberg-MacLane space $K(\pi_{n+1}X, n+1)$.

The quickest construction is perhaps the one outlined on [93, p. 179]. Proceeding inductively, the idea is to check that $\pi_i C\alpha_n$ is zero if $i \leq n+1$ and is $\pi_{n+1}X$ if $i = n+2$. One then constructs k^{n+2} by killing the higher homotopy groups of the cofiber $C\alpha_n$ and defines X_{n+1} to be the fiber of k^{n+2}. However, the proof there is not complete since the check requires a slightly strengthened version of homotopy excision or the relative Hurewicz theorem, neither of which were proven in [93].

We shall give a complete proof of a more general result that gives an analogous characterization of $\mathscr{A}$-nilpotent spaces. In the special case of ordinary nilpotent spaces, it is usually stated as follows. We say that a Postnikov tower admits a principal refinement if for each n, $p_{n+1}: X_{n+1} \longrightarrow X_n$ can be factored as a composite

$$X_{n+1} = Y_{r_n} \xrightarrow{q_{r_n}} Y_{r_n-1} \xrightarrow{q_{r_n-1}} \cdots Y_1 \xrightarrow{q_1} Y_0 = X_n$$

where, for $1 \leq i \leq r_n$, q_i is the pullback of the path space fibration over $K(G_i, n+2)$ along a map $k_i : Y_{i-1} \to K(G_i, n+2)$. The fiber of q_i is the Eilenberg-MacLane space $K(G_i, n+1)$, where the G_i are abelian groups.

THEOREM 3.2.2. *A connected space X is nilpotent if and only if the Postnikov tower of X admits a principal refinement.*

3.3. Cocellular spaces and the dual Whitehead theorem

As a preliminary, we explain cocellular spaces and the dual Whitehead theorems in this section. The arguments here were first given in [91], which is a short but leisurely expository paper. We recall the definitions of the cocellular constructions that we shall use from that source and refer to it for some easily supplied details that are best left to the reader as pleasant exercises. Again, $\mathscr{A}$ is any collection of abelian groups with $0 \in \mathscr{A}$.

DEFINITION 3.3.1. Let $\mathscr{K}$ be any collection of spaces that contains $*$ and is closed under loops. A $\mathscr{K}$-tower is a based space X together with a sequence of based maps $k_n \colon X_n \longrightarrow K_n$, $n \geq 0$, such that

(i) $X_0 = *$,
(ii) K_n is a product of spaces in $\mathscr{K}$,
(iii) $X_{n+1} = Fk_n$, and
(iv) X is the limit of the X_n.

An $\mathscr{A}$-tower is a $\mathscr{K}\mathscr{A}$-tower, where $\mathscr{K}\mathscr{A}$ is the collection of Eilenberg-Mac Lane spaces $K(A, m)$ such that $A \in \mathscr{A}$ and $m \geq 0$.

Thus X_{n+1} is the pullback in the following map of fiber sequences.

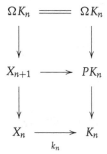

We think of the maps $X_{n+1} \longrightarrow X_n$ as giving a decreasing filtration of X, and of course the fiber over the basepoint of this map is ΩK_n. That is dual to thinking of the inclusions of skeleta $X^n \longrightarrow X^{n+1}$ of a CW complex X as giving it an increasing filtration, and of course the quotient space X^{n+1}/X^n is a wedge of suspensions ΣS^n.

REMARK 3.3.2. The collection of $\mathscr{K}$-towers has very general closure properties. Since right adjoints, such as $P(-)$ or, more generally, $F(X, -)$, preserve all categorical limits and since limits, such as pullbacks and sequential limits, commute with other limits, we find easily that products, pullbacks, and sequential limits of $\mathscr{K}$-towers are again $\mathscr{K}$-towers. The more restrictive collections of Postnikov $\mathscr{A}$-towers that we shall introduce shortly have weaker closure properties; compare Lemma 3.5.2 below. For this reason, it is sometimes more convenient to work with $\mathscr{K}\mathscr{A}$-towers than with Postnikov $\mathscr{A}$-towers.

We focus on $\mathscr{K}\mathscr{A}$-towers in what follows, and we assume that all given $K(A, n)$'s are of the homotopy types of CW complexes. It follows that the X_n

also have the homotopy types of CW complexes (see for example [103, 121]), but it does not follow and is not true that X has the homotopy type of a CW complex. The composite of the canonical composite $X \longrightarrow X_{n+1} \longrightarrow PK_n$ and the projection of PK_n onto one of its factors $PK(A, m)$ is called a cocell. The composite of k_n and one of the projections $K_n \longrightarrow K(A, m)$ is called a coattaching map. Note that the dimensions m that occur in cocells for a given n are allowed to vary.

There is a precisely dual definition of a based cell complex. We recall it in full generality for comparison, but we shall only use a very special case.

DEFINITION 3.3.3. Let $\mathscr{J}$ be a collection of based spaces that contains $*$ and is closed under suspension. A $\mathscr{J}$-cell complex is a based space X together with a sequence of based maps $j_n : J_n \longrightarrow X_n$, $n \geq 0$, such that

  (i) $X_0 = *$,
 (ii) J_n is a wedge of spaces in $\mathscr{J}$,
(iii) $X_{n+1} = Cj_n$, and
 (iv) X is the colimit of the X_n.

Thus X_{n+1} is the pushout in the following map of cofiber sequences.

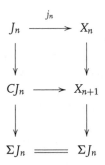

The restriction of the composite $Cj_n \longrightarrow X_{n+1} \longrightarrow X$ to a wedge summand CJ, $J \in \mathscr{J}$, is called a cell and the restriction of j_n to a wedge summand J is called an attaching map.

The case to focus on is $\mathscr{J}A = \{\Sigma^n A | n \geq 0\}$ for a fixed space A. For example, since $\mathscr{J}S^0 = \{S^n | n \geq 0\}$, $\mathscr{J}S^0$-cell complexes are the same as based cell complexes with based attaching maps. All connected spaces have approximations by $\mathscr{J}S^0$-cell complexes [93, p. 75]. In general, since attaching maps defined on S^m for $m > 0$ land in the component of the basepoint, the non-basepoint components of $\mathscr{J}S^0$-cell complexes are discrete. It is more sensible to consider

$\mathcal{J}S^1$-cell complexes, which have a single vertex and model connected spaces without extra discrete components. Similarly, all simply connected spaces have approximations by $\mathcal{J}S^2$-cell complexes [93, p. 85]. Many of the standard arguments for CW complexes that are given, for example, in [93, Ch. 10] work just as well for $\mathcal{J}$-cell complexes in general. They are described in that generality in [91]. We note parenthetically that $\mathcal{J}A$-cell complexes have been studied in many later papers, such as [28, 40], where they are called A-cellular spaces.

Based CW complexes with based attaching maps are the same as $\mathcal{J}S^0$-cell complexes in which cells are attached only to cells of lower dimension. In the context here, such an X has two filtrations, the one given by the spaces X_n in Definition 3.3.3, which tells at what stage cells are attached, and the skeletal filtration, in which X^n denotes the union of the cells of dimension at most n. In practice, when X is connected, we can arrange that the two filtrations coincide. However, in other mathematical contexts, it is the cellular filtration $\{X_n\}$ that matters. In particular, model category theory focuses on cell complexes rather than CW complexes, which in fact play no role in that theory. It often applies to categories in which cell complexes can be defined just as in Definition 3.3.3, but there is no useful notion of a CW complex because the cellular approximation theorem [93, p. 74] fails. We leave the following parenthetical observation as an exercise.

EXERCISE 3.3.4. Let X be an n-dimensional based connected CW complex and π be an abelian group. Then the reduced cellular cochains of X with coefficients in π are given by $\tilde{C}^q(X; \pi) \cong \pi_{n-q}F(X^q/X^{q-1}, K(\pi, n))$. The differentials are induced by the topological boundary maps $X^q/X^{q-1} \longrightarrow \Sigma X^{q-1}/X^{q-2}$ that are defined in [93, p. 96]; compare [93, pp. 117, 147].

Analogously, the Postnikov towers of Theorems 3.2.1 and 3.2.2 are special kinds of $\mathcal{A}b$-towers. The generality of our cellular and cocellular definitions helps us to give simple dual proofs of results about them. For a start, the following definition is dual to the definition of a subcomplex.

DEFINITION 3.3.5. A map $p: Z \longrightarrow B$ is said to be a projection onto a quotient tower if Z and B are $\mathcal{A}$-towers, p is the limit of maps $Z_n \longrightarrow B_n$, and the composite of p and each cocell $B \longrightarrow PK(A, m)$ of B is a cocell of Z (for the same n).

DEFINITION 3.3.6. A map $\xi: X \longrightarrow Y$ is an $\mathcal{A}$-cohomology isomorphism if $\xi^*: H^*(Y; A) \longrightarrow H^*(X; A)$ is an isomorphism for all $A \in \mathcal{A}$.

A word-by-word dualization of the proof of the homotopy extension and lifting property [93, p. 73] gives the following result. The essential idea is just to apply the representability of cohomology,

$$\tilde{H}^n(X; A) \cong [X, K(A, n)],$$

which is dual to the representability of homotopy groups,

$$\pi_n X = [S^n, X],$$

and induct up the cocellular filtration of an $\mathscr{A}$-tower. The reader is urged to carry out the details herself, but they can be found in [91, 4*]. As a hint, one starts by formulating and proving the dual of the based version of the lemma on [93, p. 68]. The proof of the cited lemma simplifies considerably in the based case, and the proof of its dual is correspondingly easy.

THEOREM 3.3.7 (COHELP). *Let B be a quotient tower of an $\mathscr{A}$-tower Z and let $\xi: X \longrightarrow Y$ be an $\mathscr{A}$-cohomology isomorphism. If $p_1 \circ h = g \circ \xi$ and $p_0 \circ h = p \circ f$ in the following diagram, then there exist $\tilde{g}$ and $\tilde{h}$ that make the diagram commute.*

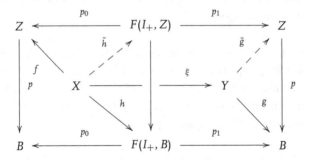

The following result, which is [91, Thm. 6$^{\sharp}$], gives a generalization of Theorem 0.0.2. Remember our standing assumption that all given spaces are of the homotopy types of CW complexes. As we have noted, our towers $Z = \lim Z_n$ are rarely of the homotopy types of CW complexes; it is for this reason that weak homotopy type rather than homotopy type appears in the following statement.

THEOREM 3.3.8 (DUAL WHITEHEAD (FIRST FORM)). *The following statements are equivalent for a map $\xi: X \longrightarrow Y$ between connected spaces X and Y of the weak homotopy types of $\mathscr{A}$-towers.*

(i) ξ *is an isomorphism in* $\text{Ho}\mathcal{T}$.

(ii) $\xi_*\colon \pi_*(X) \longrightarrow \pi_*(Y)$ *is an isomorphism.*

(iii) $\xi^*\colon H^*(Y; A) \longrightarrow H^*(X; A)$ *is an isomorphism for all* $A \in \mathcal{A}$.

(iv) $\xi^*\colon [Y, Z] \longrightarrow [X, Z]$ *is a bijection for all* $\mathcal{A}$*-towers* Z.

If $\mathcal{A}$ *is the collection of modules over a commutative ring* R, *then the following statement is also equivalent to those above.*

(v) $\xi_*\colon H_*(X; R) \longrightarrow H_*(Y; R)$ *is an isomorphism.*

SKETCH PROOF. The equivalence of (i) and (ii) is immediate from the definition of $\text{Ho}\mathcal{T}$, and (ii) implies (iii) and (v) by the weak equivalence axiom for cohomology and homology. To see that (v) implies (iii) for a general ring R, we must use the "universal coefficient spectral sequence", but we shall only apply this when R is a PID, so that the ordinary universal coefficient theorem applies. The crux of the matter is the implication (iii) implies (iv), and this is restated separated in the following result. The final implication (iv) implies (i) is formal. Taking $Z = X$ in (iv), we obtain a map $\xi^{-1}\colon Y \longrightarrow X$ such that $\xi^{-1} \circ \xi \simeq \text{id}$. Taking $Z = Y$, we see that $\xi \circ \xi^{-1} \simeq \text{id}$ since $\xi^*[\xi \circ \xi^{-1}] = \xi^*[\text{id}]$ in $[Y, Y]$. □

THEOREM 3.3.9 (DUAL WHITEHEAD (SECOND FORM)). *If* $\xi\colon X \longrightarrow Y$ *is an* $\mathcal{A}$*-cohomology isomorphism between connected spaces and* Z *is an* $\mathcal{A}$*-tower, then* $\xi^*\colon [Y, Z] \longrightarrow [X, Z]$ *is a bijection.*

SKETCH PROOF. This is where the force of dualizing familiar cellular arguments really kicks in. In view of our standing hypothesis that given spaces such as X and Y have the homotopy types of CW complexes, we may interpret $[X, Z]$ as the set of homotopy classes of maps $X \longrightarrow Z$. Now, as observed in [91, Thm. $5^\sharp$], the conclusion follows directly from coHELP in exactly the same way that the dual result on cell complexes [93, p. 73] follows directly from HELP. The surjectivity of ξ^* results by application of coHELP to the quotient tower $Z \longrightarrow *$. Just as the cofibration $(i_0, i_1)\colon (\partial I)_+ \longrightarrow I_+ \wedge X$ is the inclusion of a subcomplex when X is a based CW complex, so the fibration $(p_0, p_1)\colon F(I_+, Z) \longrightarrow F(\partial I_+, Z)$ is the projection of a quotient tower when Z is an $\mathcal{A}$-tower (compare Remark 3.3.2). Application of coHELP to this quotient tower implies the injectivity of ξ^*. □

The equivalence between (iii) and (v) in Theorem 3.3.8 leads us to the following fundamental definition and easy observation. Despite its simplicity, the observation is philosophically important to our treatment of localization and completion (and does not seem to be mentioned in the literature). For the moment, we drop all hypotheses on our given spaces.

DEFINITION 3.3.10. Let R be a commutative ring and $f: X \longrightarrow Y$ be a map.

(i) f is an R-homology isomorphism if $f_*: H_*(X; R) \longrightarrow H_*(Y; R)$ is an isomorphism.
(ii) f is an R-cohomology isomorphism if $f^*: H^*(Y; M) \longrightarrow H^*(X; M)$ is an isomorphism for all R-modules M.

In contrast to Theorem 3.3.8, the following result has no $\mathscr{A}$-tower hypothesis.

PROPOSITION 3.3.11. *Let R be a PID. Then $f: X \longrightarrow Y$ is an R-homology isomorphism if and only if it is an R-cohomology isomorphism.*

PROOF. The forward implication is immediate from the universal coefficient theorem (e.g., [93, p. 132]). For the converse, it suffices to show that the reduced homology of the cofiber of f is zero. If Z is a chain complex of free R-modules such that $H^*(Z; M) = 0$ for all R-modules M, then the universal coefficient theorem implies that $\operatorname{Hom}(H_n(Z), M) = 0$ for all n and all R-modules M. Taking $M = H_n(Z)$, $\operatorname{Hom}(H_n(Z), H_n(Z)) = 0$, so the identity map of $H_n(Z)$ is zero and $H_n(Z) = 0$. Applying this observation to the reduced chains of the cofiber of f, we see that the homology of the cofiber of f is zero. $\qquad\square$

3.4. Fibrations with fiber an Eilenberg-MacLane space

The following key result will make clear exactly where actions of the fundamental group and nilpotency of group actions enter into the theory of Postnikov towers.[2] For the novice in algebraic topology, we shall go very slowly through the following proof since it gives our first application of the Serre spectral sequence and a very explicit example of how one uses the representability of cohomology,

3.4.1 $$\tilde{H}^n(X; A) \cong [X, K(A, n)],$$

to obtain homotopical information. We regard $K(A, n)$ as a name for any space whose only nonvanishing homotopy group is $\pi_n(K(A, n)) = A$. With

---

2. The first author learned this result and its relevance from Zig Fiedorowicz in the 1970s.

our standing CW homotopy type hypothesis, any two such spaces are homotopy equivalent. We shall make use of the fact that addition in the cohomology group on the left is induced by the loop space multiplication on $K(A, n) = \Omega K(A, n+1)$ on the right; the proof of this fact is an essential feature of the verification that cohomology is representable in [93, §22.2]. We shall also make use of the fact that application of π_n induces a bijection from the homotopy classes of maps $K(A, n) \longrightarrow K(A, n)$ to $\mathrm{Hom}\,(A, A)$. One way to see that is to quote the Hurewicz and universal coefficient theorems [93, pp. 116, 132].

LEMMA 3.4.2. *Let $f: X \longrightarrow Y$ be a map of connected based spaces whose (homotopy) fiber Ff is an Eilenberg-Mac Lane space $K(A, n)$ for some abelian group A and $n \geq 1$. Then the following statements are equivalent.*

(i) *There is a map $k: Y \longrightarrow K(A, n+1)$ and an equivalence $\xi: X \longrightarrow Fk$ such that the following diagram commutes, where π is the canonical fibration with (actual) fiber $K(A, n) = \Omega K(A, n+1)$.*

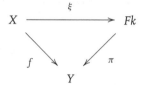

(ii) *There is a map $k: Y \longrightarrow K(A, n+1)$ and an equivalence $\lambda: Nf \longrightarrow Fk$ such that the following diagram commutes, where π is as in (i) and $\rho: Nf \to Y$ is the canonical fibration with (actual) fiber Ff.*

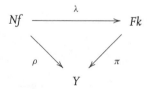

(iii) *The group $\pi_1(Y)$ acts trivially on the space $Ff = K(A, n)$.*
(iv) *The group $\pi_1(Y)$ acts trivially on the group $A = \pi_n(Ff)$.*

PROOF. In (ii), Nf is the mapping path fibration of f, as defined in [93, pp. 48, 59]. We first elaborate on the implications of (i). Consider the following diagram, in which ι and π are used generically for canonical maps in fiber sequences, as specified in [93, p. 59].

$$
\begin{array}{ccccccc}
\Omega Y & \xrightarrow{\iota} & Ff & \xrightarrow{\pi} & X & \xrightarrow{f} & Y \\
\Big\| & & \Big\downarrow{\chi} & & \Big\downarrow{\xi} & & \Big\| \\
\Omega Y & \xrightarrow[-\Omega k]{} & \Omega K(A, n+1) & \xrightarrow{\iota} & Fk & \xrightarrow{\pi} & Y & \xrightarrow{k} & K(A, n+1)
\end{array}
$$

In the bottom row, the actual fiber $\Omega K(A, n+1)$ of the fibration π is canonically equivalent to its homotopy fiber $F\pi$. The dotted arrow χ making the two left squares commute up to homotopy comes from Lemma 1.2.3, using that the first four terms of the bottom row are equivalent to the fiber sequence generated by the map π [93, p. 59]. Since ξ is an equivalence, a comparison of long exact sequences of homotopy groups (given by [93, p. 59]) shows that χ is a weak equivalence and therefore, by our standing assumption that all given spaces have the homotopy types of CW complexes, an equivalence. Therefore the diagram displays equivalences showing that the sequence

$$
K(A, n) \xrightarrow{\pi} X \xrightarrow{f} Y \xrightarrow{k} K(A, n+1)
$$

is equivalent to the fiber sequence generated by the map k.

(i) implies (ii). Consider the following diagram.

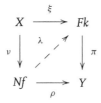

Here $\rho \circ \nu$ is the canonical factorization of f as the composite of a cofibration and homotopy equivalence ν and a fibration ρ with fiber Ff, as in [93, pp. 48, 59][3]. Since ν is an equivalence and a cofibration and π is a fibration,

---

3. The proof that ν is a cofibration is missing from [93, p. 48]. It is easily supplied by verification of the NDR-pair condition of [93, p. 43], using the deformation h on [93, p. 48] and the map $u: Nf \longrightarrow I$ defined by letting $u(x, \omega)$ be the supremum of $\{1 - t | \omega(s) = f(x)$ for $0 \le s \le t\}$, where $\omega: I \longrightarrow Y$ is such that $f(x) = \omega(0)$.

Lemma 1.1.1 gives a lift λ that makes the diagram commute. Since ξ and ν are equivalences, so is λ. Although not needed here, λ is a fiber homotopy equivalence over Y by [93, p. 50].

(ii) implies (iii). The action of $\pi_1(Y)$ on Ff is obtained by pulling back the action of $\pi_1 K(A, n+1)$ on $K(A, n)$ along $k_*\colon \pi_1(Y) \longrightarrow \pi_1 K(A, n+1)$. Since $K(A, n+1)$ is simply connected, the action is trivial in the sense that each element of $\pi_1(Y)$ acts up to homotopy as the identity map of Ff.

(iii) implies (iv). This holds trivially since the action of $\pi_1(Y)$ on $\pi_n(Ff)$ is induced from the action of $\pi_1(Y)$ on Ff by passage to homotopy groups.

(iv) implies (iii). This holds since homotopy classes of maps $Ff \longrightarrow Ff$ correspond bijectively to homomorphisms $A \longrightarrow A$.

(iii) implies (i). This is the crux of the matter. Write $Ff = K(A, n)$. We shall construct a commutative diagram

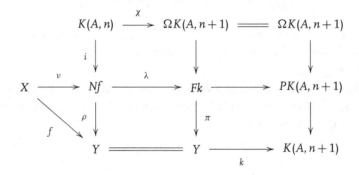

in which the three columns display fibrations (the bottom vertical arrows) and the inclusions of their fibers (the top vertical arrows); the map χ and therefore also the map λ are equivalences. Then $\xi = \lambda \circ \nu\colon X \longrightarrow Fk$ will be an equivalence such that $\pi \circ \xi = f$, proving (i).

The fibration in the right column is the path space fibration, and we are given the fibration in the left column. The lower left triangle commutes by the definition of Nf. We must first construct k. This is where the Serre spectral sequence enters, and we will summarize everything we need in Chapter 24. Taking coefficients in A, the cohomology Serre spectral sequence for the left fibration ρ converges to $H^*(Nf; A) \cong H^*(X; A)$. Our assumption (iii) implies that the local coefficient system that enters into the calculation of the E_2 term is trivial and therefore

$$E_2^{p,q} = H^p(Y; H^q(K(A, n); A)).$$

In fact, we shall only need to use the triviality of the local coefficients when $q = n$ is the Hurewicz dimension, where it is transparent. Indeed, the Hurewicz isomorphism preserves the action of $\pi_1(Y)$, so that $\pi_1(Y)$ acts trivially on $H_n(K(A, n); \mathbb{Z})$ and therefore, by the universal coefficient theorem, on $H^n(K(A, n); A)$.

Clearly, $E_2^{p,q} = 0$ for $0 < q < n$. By the universal coefficient theorem,

$$E_2^{0,n} = H^n(K(A, n); A) \cong \mathrm{Hom}\,(A, A).$$

We let ι_n denote the fundamental class, which is given by the identity homomorphism of A. Similarly we let ι_{n+1} be the fundamental class in $H^{n+1}(K(A, n+1); A)$. The differentials $d_r(\iota_n)$ land in zero groups for $2 \leq r \leq n$, and we have the transgression differential

$$\tau(\iota_n) = d_{n+1}(\iota_n) = j \in H^{n+1}(Y; A) = E_{n+1}^{n+1,0}.$$

The class j is represented by a map $k \colon Y \longrightarrow K(A, n+1)$, so that $k^*(\iota_{n+1}) = j$. By definition, the fiber Fk is the pullback displayed in the the lower right square of our diagram.

We claim that $k \circ \rho$ is null homotopic. Indeed, a map $Nf \longrightarrow K(A, n+1)$ is null homotopic if and only if it represents 0 in $H^{n+1}(Nf; A)$, and the cohomology class represented by $k \circ \rho$ is

$$\rho^* k^*(\iota_{n+1}) = \rho^*(j) = \rho^* d_{n+1}(\iota_n) = 0.$$

The last equality holds since ρ^* can be identified with the edge homomorphism

$$H^{n+1}(Y; A) \cong E_2^{n+1,0} = E_{n+1}^{n+1,0} \longrightarrow E_{n+1}^{n+1,0}/d_{n+1} E_{n+1}^{0,n}$$

$$= E_\infty^{n+1,0} \subset H^{n+1}(Nf; A).$$

Choose a homotopy $h \colon Nf \times I \longrightarrow K(A, n+1)$ from the trivial map to $k \circ \rho$. Then h determines the map $\lambda' \colon Nf \longrightarrow Fk$ specified by $\lambda'(z) = (\rho(z), h(z))$ for $z \in Nf$, where $h(z)(t) = h(z, t)$. This makes sense since $Fk = \{(y, \omega)\}$ where $y \in Y$, $\omega \in PK(A, n+1)$, and $k(y) = \omega(1)$. Observe for later use that $h(z) \in \Omega K(A, n+1)$ when $z \in i(Ff) = \rho^{-1}(*)$.

Clearly $\pi \circ \lambda' = \rho$ since π is induced by projection on the first coordinate. Thus λ' restricts to a map $\chi' \colon K(A, n) = Ff \longrightarrow \Omega K(A, n+1)$ on fibers. This gives a diagram of the sort displayed at the start of the proof that (iv) implies (i), but with primes on the left-hand horizontal maps. Here χ' need not be an equivalence, but we claim that we can correct λ' and χ' to new maps λ and χ such that χ and therefore λ are equivalences. In the Serre spectral sequence of the path space fibration on the right, $d_{n+1}(\iota_n) = \iota_{n+1}$. By naturality, comparing the left column to the right column in our diagram, we have

$$d_{n+1}((\chi')^*(\iota_n)) = k^*(d_{n+1}(\iota_n)) = k^*(\iota_{n+1}) = j = d_{n+1}(\iota_n),$$

where the last d_{n+1} is that of the spectral sequence of π. Therefore d_{n+1} for the spectral sequence of ρ satisfies

$$d_{n+1}(\iota_n - (\chi')^*(\iota_n)) = 0$$

so that we have the cycle

$$\iota_n - (\chi')^*(\iota_n) \in E_{n+2}^{0,n} = E_\infty^{0,n} = H^n(Nf;A)/F^1 H^n(Nf;A).$$

Choose a representative $\ell: Nf \longrightarrow \Omega K(A,n+1) \simeq K(A,n)$ for a cohomology class that represents $\iota_n - (\xi')^*(io_n)$ in this quotient group. For any choice of ℓ, the restriction $\ell \circ i$ to $K(A,n) = Ff$ represents $\iota_n - (\chi')^*(\iota_n)$. Now define

$$\lambda(z) = (\rho(z), h(z) \cdot \ell(z)),$$

where the dot denotes concatenation of the path $h(z)$ with the loop $\ell(z)$. We again have $\pi\lambda = \rho$, and we thus have an induced map $\chi: K(A,n) = Ff \to \Omega K(A,n+1)$ on fibers. For $z \in Ff$, $\chi(z) = \chi'(z) \cdot \ell i(z)$. Since loop multiplication on $\Omega K(A,n+1)$ induces addition on cohomology classes,

$$\chi^*(\iota_n) = (\chi')^*(\iota_n) + i^*\ell^*(\iota_n) = (\chi')^*(\iota_n) + \iota_n - (\chi')^*(\iota_n) = \iota_n.$$

Therefore $\chi \in [K(A,n), \Omega K(A,n+1)] \cong H^n(K(A,n);A)$ corresponds to the identity map $A \longrightarrow A$. This proves that χ and therefore λ are equivalences. $\square$

3.5. Postnikov $\mathscr{A}$-towers

Just as cell complexes are too general for convenience, suggesting restriction to CW complexes, so $\mathscr{A}$-towers are too general for convenience, suggesting restriction to Postnikov $\mathscr{A}$-towers.

DEFINITION 3.5.1. A Postnikov $\mathscr{A}$-tower is an $\mathscr{A}$-tower $X = \lim X_i$ (see Definition 3.3.1) such that each K_i is a $K(A_i, n_i + 1)$ with $A_i \in \mathscr{A}$, $n_{i+1} \geq n_i \geq 1$, and only finitely many $n_i = n$ for each $n \geq 1$. A map $\psi : X \longrightarrow Y$ between Postnikov $\mathscr{A}$-towers is cocellular if it is the limit of maps $\psi_i : X_i \longrightarrow Y_i$. For example, a projection onto a quotient tower of a Postnikov $\mathscr{A}$-tower X is a cocellular map.

A Postnikov $\mathscr{A}$-tower X is connected since $X_0 = *$ and the K_i are simply connected. By the long exact sequences of the fibrations appearing in the definition of an $\mathscr{A}$-tower, the homotopy groups of X are built up in order, with each homotopy group built up in finitely many stages. We shall be more precise about this shortly. Note that a product of Eilenberg-Mac Lane spaces $\Pi_j K(A_j, j)$ is an Eilenberg-Mac Lane space $K(\Pi_j A_j, j)$. When $\mathscr{A}$ is closed under products, this makes it especially reasonable to use a single cocell at each stage of the filtration.

We shall make little formal use of the following result, but we urge the reader to supply the proofs. They are precisely dual to the proofs of familiar results about CW complexes that are given in [93, pp. 72–73].[4]

LEMMA 3.5.2. *Let X and Y be Postnikov $\mathscr{A}$-towers, let W be a quotient tower of X, and let $\psi : Y \longrightarrow W$ be a cocellular map.*

  (i) *$X \times Y$ is a Postnikov $\mathscr{A}$-tower with one cocell for each cocell of X and each cocell of Y.*
 (ii) *$Y \times_W X$ is a Postnikov $\mathscr{A}$-tower with one cocell for each cocell of X that does not factor through a cocell of W.*
(iii) *If X is simply connected, ΩX is a Postnikov $\mathscr{A}$-tower whose coattaching maps are the loops of the coattaching maps of X.*

Recall that we assume that all given spaces have the homotopy types of CW complexes, although limits constructed out of such spaces, such as $\mathscr{A}$-towers, will not have this property. Recall too that we are working in $\text{Ho}\mathscr{T}$, where spaces are implicitly replaced by CW approximations or, equivalently, where all weak equivalences are formally inverted. In view of this framework, the results of this section show that we can freely replace $\mathscr{A}$-nilpotent spaces and maps between them by weakly equivalent Postnikov $\mathscr{A}$-towers and cocellular maps between them.

---

4. For (iii), the precise dual states that the suspension of a CW complex X is a CW complex whose attaching maps are the suspensions of those of X. That requires based attaching maps as in Definition 3.3.3.

DEFINITION 3.5.3. An $\mathscr{A}$-cocellular approximation of a space X is a weak equivalence from X to a Postnikov $\mathscr{A}$-tower.

The definition should be viewed as giving a kind of dual to CW approximation.[5] Just as CW approximation is the basis for the cellular construction of the homology and cohomology of general spaces, cocellular approximation is the basis for the cocellular construction of localizations and completions of general nilpotent spaces.

A Postnikov $\mathscr{A}$-tower is obtained from a sequence of maps of fiber sequences

$$
\begin{array}{ccc}
K(A_i, n_i) & = & K(A_i, n_i) \\
\downarrow & & \downarrow \\
X_{i+1} & \longrightarrow & PK(A_i, n_i+1) \\
\downarrow & & \downarrow \\
X_i & \xrightarrow{\ k_i\ } & K(A_i, n_i+1).
\end{array}
$$

The left column gives an exact sequence of homotopy groups (central extension)

$$
0 \longrightarrow A_i \longrightarrow \pi_{n_i} X_{i+1} \longrightarrow \pi_{n_i} X_i \longrightarrow 0.
$$

Since $X_{i+1} \longrightarrow X_i$ on the left is the fibration induced by pullback along k_i from the path space fibration on the right, we see by using the naturality of the group actions of Proposition 1.5.4 with respect to maps of fibration sequences that $\pi_1(X)$ is nilpotent and acts trivially on the A_i that enter into the computation of $\pi_n(X)$ for $n > 1$. Therefore, using Lemma 3.1.3, we see that any Postnikov $\mathscr{A}$-tower is an $\mathscr{A}$-nilpotent space. The following result gives a converse to this statement.

THEOREM 3.5.4. *Let X be an $\mathscr{A}$-nilpotent space.*

(i) *There is a Postnikov $\mathscr{A}$-tower $P(X)$ and a weak equivalence $\xi_X \colon X \longrightarrow P(X)$; that is, ξ is a cocellular approximation of X.*

(ii) *If $\psi \colon X \longrightarrow X'$ is a map of $\mathscr{A}$-nilpotent spaces, then there is a cocellular map $P(\psi) \colon P(X) \longrightarrow P(X')$ such that $P(\psi) \circ \xi_X$ is homotopic to $\xi_{X'} \circ \psi$.*

---

5. It is similar to fibrant approximation, which is dual to cofibrant approximation in model category theory.

PROOF. The idea is to use any given $\mathscr{A}$-central $\pi_1 X$-series for the groups $\pi_n X$ to construct spaces X_i in a Postnikov $\mathscr{A}$-tower such that X is weakly equivalent to $\lim X_i$. Taking $\mathscr{A} = \mathscr{A}b$, it is clear that Theorems 3.2.1 and 3.2.2 are special cases. For the first, a simple space is a nilpotent space such that each homotopy group is built in a single step. For the second, the notion of a principal refinement of a Postnikov tower in the classical sense, with each homotopy group built up in a single step, is just a reformulation of our notion of a Postnikov $\mathscr{A}b$-tower.

Thus, to prove (i), assume given $\mathscr{A}$-central $\pi_1 X$-series

$$1 = G_{n,r_n} \subset \cdots \subset G_{n,0} = \pi_n X$$

for $n \geq 1$. Let $A_{n,j} = G_{n,j}/G_{n,j+1}$ for $0 \leq j < r_n$, so that $A_{n,j} \in \mathscr{A}$ and $\pi_1 X$ acts trivially on $A_{n,j}$. Using these groups, we define spaces $Y_{n,j}$ and maps

$$\tau_{n,j}\colon X \longrightarrow Y_{n,j}$$

such that

(i) $\tau_{n,j}$ induces an isomorphism $\pi_q X \longrightarrow \pi_q Y_{n,j}$ for $q < n$;
(ii) $\pi_n Y_{n,j} = \pi_n X/G_{n,j}$ and the map $\pi_n X \longrightarrow \pi_n Y_{n,j}$ induced by $\tau_{n,j}$ is the epimorphism $\pi_n X \longrightarrow \pi_n X/G_{n,j}$; and
(iii) $\pi_q Y_{n,j} = 0$ for $q > n$.

The spaces $Y_{n,j}$ and maps $X \longrightarrow Y_{n,j}$ are constructed by attaching $(n+1)$-cells to X to kill the subgroup $G_{n,j}$ of $\pi_n(X)$, using maps $S^n \longrightarrow X$ that represent generators of $G_{n,j}$ as attaching maps, and then attaching higher-dimensional cells to kill the homotopy groups in dimensions greater than n. To start work and to implement the transition from finishing work on the nth homotopy group to starting work on the $(n+1)$st, we set $Y_{1,0} = *$ and $Y_{n+1,0} = Y_{n,r_n}$. The maps $\tau_{n,j}$ are just the inclusion maps. From the constructions of $Y_{n,j}$ and $Y_{n,j+1}$ we have the solid arrow maps in the diagram

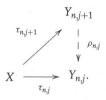

We construct a map $\rho_{n,j}$ that makes the diagram commute directly from the definition. We urge the knowledgeable reader to resist the temptation to

reformulate the argument in terms of obstruction theory. Let $Y_s \longrightarrow Y_{s+1}$ be the sth stage of the construction of the relative cell complex $\tau_{n,j+1} \colon X \longrightarrow Y_{n,j+1}$, so that Y_{s+1} is constructed from Y_s by attaching $(s+1)$-cells. Starting with $\tau_{n,j} \colon X \longrightarrow Y_{n,j}$, assume that we have constructed $\rho_s \colon Y_s \longrightarrow Y_{n,j}$ such that the following diagram commutes.

Since $G_{n,j+1} \subset G_{n,j} \subset \pi_n X$ and $\pi_q Y_{n,j} = 0$ for $q > n$, the composite of ρ_s and any attaching map $S^n \longrightarrow Y_s$ used in the construction of Y_{s+1} from Y_s represents an element of a homotopy groups that has already been killed. Thus this composite is null homotopic and therefore extends over the cone CS^s. Using these extensions to map the attached cells, we extend ρ_s to $\rho_{s+1} \colon Y_{s+1} \longrightarrow Y_{n,j}$. Passing to colimits over s we obtain the desired map $\rho_{n,j}$. Clearly $\rho_{n,j}$ induces an isomorphism on all homotopy groups except the nth, where it induces the quotient homomorphism $\pi_n(X)/G_{n,j+1} \longrightarrow \pi_n(G)/G_{n,j}$. Since $A_{n,j}$ is the kernel of this homomorphism, the only nonzero homotopy group of the fiber $F\rho_{n,j}$ is the nth, which is $A_{n,j}$. Moreover, by the naturality of fundamental group actions, we see that $\pi_1 Y_{n,j}$ acts trivially on this group.

We now correct this construction to obtain commutative diagrams

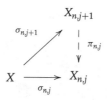

with the same behavior on homotopy groups together with "k-invariants"

$$k_{n,j} \colon X_{n,j} \longrightarrow K(A_{n,j}, n+1)$$

such that $X_{n,j+1} = Fk_{n,j}$ and $\pi_{n,j}$ is the canonical fibration, where $X_{1,0} = *$ and $X_{n+1,0} = X_{n,r_n}$. In fact, we construct equivalences $\chi_{n,j} \colon Y_{n,j} \longrightarrow X_{n,j}$ and commutative diagrams

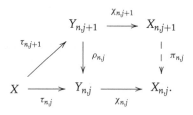

Assuming that we have already constructed the equivalence $\chi_{n,j}$, we simply apply Lemma 3.4.2 to the composite $f = \chi_{n,j}\rho_{n,j} \colon Y_{n,j+1} \longrightarrow X_{n,j}$. Since $Ff = K(A_{n,j}, n)$ and $\pi_1 X_{n,j}$ acts trivially on $A_{n,j}$, part (i) of that result gives the required k-invariant $k_{n,j}$ together with an equivalence

$$\chi_{n,j+1} \colon Y_{n,j+1} \longrightarrow X_{n,j+1} = Fk_{n,j}$$

making the diagram just displayed commute.

We define $\sigma_{n,j} = \chi_{n,j} \circ \tau_{n,j} \colon X \longrightarrow X_{n,j}$. We then define $P(X) = \lim X_{n,j}$ and let $\xi \colon X \longrightarrow P(X)$ be the map obtained by passage to limits from the maps $\sigma_{n,j}$. Then ξ is a weak equivalence since $\sigma_{n,j}$ induces an isomorphism on homotopy groups in degrees less than n, and n is increasing.

For the naturality statement, we are free to refine given $\mathscr{A}$-central $\pi_1 X$ and $\pi_1 X'$ series for $\pi_n X$ and $\pi_n X'$ by repeating terms, and we may therefore assume that the $\mathscr{A}$-central $\pi_1 X'$ series $1 = G'_{n,r} \subset \cdots \subset G'_{n,0} = \pi_n X'$ for $n \geq 1$ satisfies $\psi_*(G_{n,j}) \subset G'_{n,j}$. Let $A'_{n,j} = G'_{n,j}/G'_{n,j+1}$ and perform all of the constructions above with X replaced by X'.

We claim first that there exist maps $\theta_{n,j} \colon Y_{n,j} \longrightarrow Y'_{n,j}$ such that the following diagram commutes.

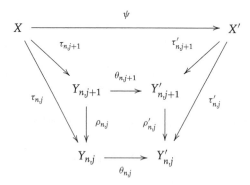

For this, assuming that $\theta_{n,j}$ has been constructed, we look inductively at the stages $Y_s \longrightarrow Y_{s+1}$ and $Y'_s \longrightarrow Y'_{s+1}$ of the cellular constructions of $Y_{n,j+1}$ and

$Y_{n,j+1}$ from X and X'. We assume that $\theta_s \colon Y_s \longrightarrow Y'_s$ has been constructed in the following diagram.

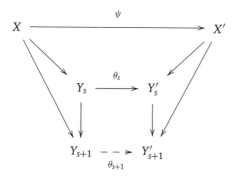

The composites of attaching maps $S^q \longrightarrow Y_s$ with the composite $Y_s \longrightarrow Y'_{s+1}$ in the diagram are null homotopic, either because of our hypothesis that $\psi_*(G_{n,j}) \subset G'_{n,j}$ or by the vanishing of homotopy groups, hence we can construct θ_{s+1} by using homotopies to extend over cells.

By induction, suppose that we have constructed maps $\psi_{n,j}$ and homotopies $h_{n,j} \colon \chi'_{n,j} \circ \theta_{n,j} \simeq \psi_{n,j} \circ \chi_{n,j}$ as displayed in the following diagram.

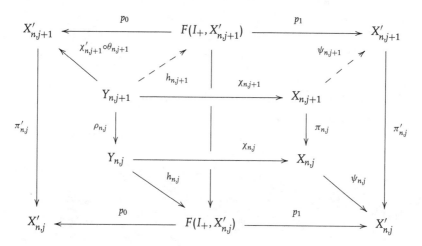

Here the inductive hypothesis implies that the solid arrow portion of the diagram commutes. Therefore we can apply Theorem 3.3.7 to obtain a map $\psi_{n,j+1}$ such that $\pi'_{n,j} \circ \psi_{n,j+1} = \psi_{n,j} \circ \pi_{n,j}$ and a homotopy

$$h_{n,j+1} \colon \chi'_{n,j+1} \circ \theta_{n,j+1} \simeq \psi_{n,j+1} \circ \chi_{n,j+1}.$$

To start the induction, note that $X_{1,0}$ and $X'_{1,0}$ are both $*$ and $h_{1,0}$ and $\psi_{1,0}$ are both constant maps.

Define $P(\psi) = \lim \psi_{n,j}\colon P(X) \longrightarrow P(X')$. In the following diagram all squares and triangles that are not made homotopy commutative by $h_{n,j}$ and $h_{n,j+1}$ are actually commutative. Therefore the maps $h_{n,j}$ determine a homotopy from $\xi_{X'} \circ \psi$ to $P(\psi) \circ \xi_X$ on passage to limits.

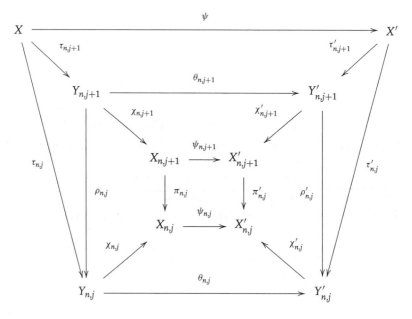

This completes the proof of the naturality statement. $\qquad\qquad\qquad\square$

4

DETECTING NILPOTENT GROUPS AND SPACES

We collect together a number of technical preliminaries that will be needed in our treatment of localizations and completions and are best treated separately from either. The reader is advised to skip this chapter on a first reading. The essential point is to detect when groups and spaces are $\mathscr{A}$-nilpotent. When discussing naturality, we require that $\mathscr{A}$ be a category rather than just a collection of abelian groups. For some results, we must assume the much stronger hypothesis that $\mathscr{A}$ is an abelian category, and we then change notation from $\mathscr{A}$ to $\mathscr{C}$ for emphasis.

4.1. Nilpotent actions and cohomology

We give a few easy results about identifying nilpotent group actions here. We are only interested in group actions on abelian groups, and we agree to call an abelian π-group a π-module. It is sometimes convenient to think in terms of group rings. An action of π on an abelian group A is equivalent to an action of the group ring $\mathbb{Z}[\pi]$ on A. Let I denote the augmentation ideal of $\mathbb{Z}[\pi]$, namely the kernel of the augmentation $\varepsilon \colon \mathbb{Z}[\pi] \longrightarrow \mathbb{Z}$ specified by $\varepsilon(g) = 1$ for $g \in \pi$. Thus I is generated by the elements $g - 1$ and I^q is generated by the products $(g_1 - 1) \cdots (g_q - 1)$. It follows that A is π-nilpotent of class q if and only if $I^q A = 0$ but $I^{q-1} A \neq 0$.

The following observation gives a simple criterion for when a nilpotent π-module is a $\mathscr{C}$-nilpotent π-module, assuming that $\mathscr{C}$ is an abelian category with infinite direct sums. Since we have only quite specific examples in mind, we shall not recall the formal definition of abelian categories; see, for example [79, p. 198]. Informally, they are additive subcategories of $\mathscr{A}b$ that are closed under finite direct sums, kernels, and cokernels; more accurately, they come with faithful and exact forgetful functors to $\mathscr{A}b$. The examples to keep in mind are the categories of modules over a commutative ring, thought of as abelian groups by neglect of the module structure.

LEMMA 4.1.1. *Let $\mathscr{C}$ be an abelian category with infinite direct sums. Let $C \in \mathscr{C}$ be a nilpotent π-module such that $x \colon C \longrightarrow C$ is a morphism in $\mathscr{C}$ for each $x \in \pi$. Then C is a $\mathscr{C}$-nilpotent π-module.*

PROOF. Here $C \in \mathscr{C}$, but the action of π on C need not be trivial. The lemma says that if the action of each element of π is a morphism in $\mathscr{C}$, then we can find a π-series with π-trivial subquotients such that the subquotients lie in $\mathscr{C}$. Clearly the quotient C/IC is π-trivial, and it is in $\mathscr{C}$ because it is a cokernel of a morphism in $\mathscr{C}$, namely the morphism $\oplus_{x \in \pi} C \longrightarrow C$ whose restriction to the xth summand C is the map $x - 1 \colon C \longrightarrow C$. Our assumption on the action ensures that this map is in $\mathscr{C}$. Now IC is in $\mathscr{C}$ since it is the kernel of the quotient map $C \longrightarrow C/IC$, which is in $\mathscr{C}$. Let $C_i = I^i C$ for $i \geq 0$. Inductively, each C_i satisfies the original hypotheses on C and each inclusion $C_{i+1} \subset C_i$ is a map in $\mathscr{C}$ since it is the kernel of the quotient map $I^i C \longrightarrow I^i C / I^{i+1} C$ in $\mathscr{C}$. The descending π-series $\{C_i\}$ satisfies $C_n = 0$ for some n since C is π-nilpotent. $\square$

In two examples, we will encounter full abelian subcategories of $\mathscr{A}b$, full meaning that all maps of abelian groups between objects in the category are again in the category. The first plays a role in the theory of localizations and the second plays a role in the theory of completions. In these cases, we conclude that all nilpotent π-modules in $\mathscr{C}$ are necessarily $\mathscr{C}$-nilpotent π-modules.

LEMMA 4.1.2. *Let T be a set of primes. Any homomorphism of abelian groups between T-local abelian groups is a homomorphism of $\mathbb{Z}_T$-modules. Thus the category of $\mathbb{Z}_T$-modules is a full abelian subcategory of $\mathscr{A}b$.*

LEMMA 4.1.3. *Let T be a nonempty set of primes and let $\mathbb{F}_T = \times_{p \in T} \mathbb{F}_p$. Observe that $\mathbb{F}_T$-modules are products over $p \in T$ of vector spaces over $\mathbb{F}_p$. Any homomorphism of abelian groups between $\mathbb{F}_T$-modules is a map of $\mathbb{F}_T$-modules. Thus the category of $\mathbb{F}_T$-modules is a full abelian subcategory of $\mathscr{A}b$.*

LEMMA 4.1.4. *If a group π acts nilpotently on each term of a complex of π-modules, then the induced action on its homology is again nilpotent.*

PROOF. This is a consequence of Lemma 3.1.3, using the exact sequences

$$0 \longrightarrow Z_n(C) \longrightarrow C_n \longrightarrow B_{n-1}(C) \longrightarrow 0$$

$$0 \longrightarrow B_n(C) \longrightarrow Z_n(C) \longrightarrow H_n(C) \longrightarrow 0$$

relating the cycles $Z_n(C)$, boundaries $B_n(C)$, and homology groups of a chain complex C. $\square$

For the following lemma, we assume that the reader has seen the definition

$$H^*(\pi; B) = \text{Ext}^*_{\mathbb{Z}[\pi]}(\mathbb{Z}, B)$$

of the cohomology of the group π with coefficients in a π-module B. Here π acts trivially on $\mathbb{Z}$; more formally, $\mathbb{Z}[\pi]$ acts through $\varepsilon \colon \mathbb{Z}[\pi] \longrightarrow \mathbb{Z}$. We shall later also use homology,

$$H_*(\pi; B) = \text{Tor}^{\mathbb{Z}[\pi]}_*(\mathbb{Z}, B).$$

We urge the reader to do the exercises of [93, pp. 127, 141–142], which use universal covers to show that there are natural isomorphisms

4.1.5 $H_*(\pi; B) \cong H_*(K(\pi, 1); B)$ and $H^*(\pi; B) \cong H^*(K(\pi, 1); B)$

when π acts trivially on B. The latter will be used in conjunction with the representability of cohomology. In degree zero, the groups in (4.1.5) are B, and it makes sense to remove them by defining

$$\widetilde{H}_*(\pi; B) = \widetilde{H}_*(K(\pi, 1); B) \quad and \quad \widetilde{H}^*(\pi; B) = \widetilde{H}^*(K(\pi, 1); B).$$

LEMMA 4.1.6. *Let $\mathcal{A}$ be a collection of abelian groups. Let $\xi \colon \pi \longrightarrow \pi'$ be a homomorphism of groups such that $\xi^* \colon H^*(\pi'; A) \longrightarrow H^*(\pi; A)$ is an isomorphism for all $A \in \mathcal{A}$, where π and π' act trivially on A. Then $\xi^* \colon H^*(\pi'; B) \longrightarrow H^*(\pi; B)$ is an isomorphism for all $\mathcal{A}$-nilpotent π'-modules B, where π acts on B through ξ.*

PROOF. Since B is an $\mathcal{A}$-nilpotent π'-module, there is a chain of π'-modules

$$0 = B_q \subset B_{q-1} \subset \cdots \subset B_0 = B$$

such that $A_i = B_i/B_{i+1} \in \mathcal{A}$ and π' acts trivially on A_i. The short exact sequence of π' groups

$$0 \longrightarrow B_{i+1} \longrightarrow B_i \longrightarrow A_i \longrightarrow 0$$

induces a pair of long exact sequences in cohomology and a map between them.

$$
\begin{array}{ccccccc}
H^n(\pi'; B_{i+1}) & \longrightarrow & H^n(\pi'; B_i) & \longrightarrow & H^n(\pi'; A_i) & \longrightarrow & H^{n+1}(\pi'; B_{i+1}) \\
\downarrow & & \downarrow & & \downarrow & & \downarrow \\
H^n(\pi; B_{i+1}) & \longrightarrow & H^n(\pi; B_i) & \longrightarrow & H^n(\pi; A_i) & \longrightarrow & H^{n+1}(\pi; B_{i+1})
\end{array}
$$

By induction and the five lemma, this is an isomorphism of long exact sequences for all i, and the conclusion follows. □

4.2. Universal covers of nilpotent spaces

Let X be a nilpotent space. By Theorem 3.5.4, we may assume that X is a
Postnikov tower, which for the moment we denote by $\lim Y_i$. After finitely many
stages of the tower, we reach $Y_j = K(\pi_1 X, 1)$, and then the higher stages of the
tower build up the higher homotopy groups. In this section, it is not relevant
that $\pi_1 X$ is nilpotent, only that it acts nilpotently on the higher homotopy
groups. We therefore define a modified tower by setting $X_0 = *$, $X_1 = Y_j$,
and $X_i = Y_{j+i-1}$ for $i \geq 2$. We still have $X = \lim X_i$, but now the fiber of the
fibration $\pi_i \colon X_{i+1} \longrightarrow X_i$ is $K(A_i, n_i)$ for some abelian group A_i with trivial
action by $\pi_1 X$ and some $n_i \geq 2$. Although our interest is in nilpotent spaces,
we can now drop the assumption that $\pi_1(X)$ is nilpotent and work with a tower
of the form just specified. We note that we really do want to work with the
actual limit here, only later applying CW approximation. This ensures that not
only the maps π_i for $i \geq 1$ but also the projections $\sigma_i \colon X \longrightarrow X_i$ for $i \geq 1$ are
fibrations. These fibrations all induce isomorphisms on π_1.

Since $X_1 = K(\pi_1 X, 1)$, we can take $\tilde{X}$ to be the fiber of $\sigma_1 \colon X \longrightarrow X_1$.[1]
Of course, $\tilde{X}$ is simply connected rather than just nilpotent, hence it has an
ordinary Postnikov tower in which each homotopy group is built up in a single
step. However, it is useful to construct a refined Postnikov tower for $\tilde{X}$ from
our modified Postnikov tower $\lim X_i$. To this end, let $\tilde{X}_1 = *$ and, for $i > 1$, let
$\tilde{X}_i$ be the fiber of $\sigma_i \colon X_i \longrightarrow X_1$. Since σ_i induces an isomorphism on π_1, $\tilde{X}_i$ is
a universal cover of X_i. By Lemma 1.2.3, since the right-hand squares in the
following diagram commute (not just up to homotopy), there are dotted arrow
maps $\tilde{\sigma}_{i+1}$ and $\tilde{\pi}_i$ that make the diagram commute.

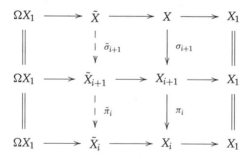

---

1. We assumed in Notations 0.0.3 that all spaces have universal covers, but not all nilpotent
spaces are semi-locally simply connected; we understand $\tilde{X}$ to mean the space just defined, even
though, strictly speaking, it need only be a fibration over X, not a cover.

By the functoriality statement in Lemma 1.2.3, $\tilde{\pi}_i \circ \tilde{\sigma}_{i+1} = \tilde{\sigma}_i$. Moreover, by Addendum 1.2.4, the maps $\tilde{\sigma}_i$ and $\tilde{\pi}_i$ are fibrations. Clearly, in degrees greater than 1, the long exact sequence of homotopy groups of the second column maps isomorphically to the long exact sequence of homotopy groups of the third column. By passage to limits, the maps $\tilde{\sigma}_i$ induce a map

$$\xi : \tilde{X} \longrightarrow \lim \tilde{X}_i,$$

and this map induces an isomorphism on homotopy groups and is thus a weak equivalence. It is reasonable to think of $\lim \tilde{X}_i$ as $P(\tilde{X})$. It is a refined Postnikov tower of $\tilde{X}$ that, in effect, uses the Postnikov tower of X to interpolate fibrations into the ordinary Postnikov tower of $\tilde{X}$ in such a way that the homotopy groups $\pi_n(\tilde{X})$ for $n \geq 2$ are built up in exactly the same way that the original Postnikov tower of X builds up the isomorphic homotopy groups $\pi_n(X)$.

PROPOSITION 4.2.1. *Let $\mathscr{C}$ be an abelian category with infinite direct sums, let $C \in \mathscr{C}$, and assume that the cohomology functors $H^q(-; C)$ on chain complexes of free abelian groups take values in $\mathscr{C}$ (as holds when $\mathscr{C}$ is the category of modules over a commutative ring). Let X be a nilpotent space and let $\tilde{X}$ be its universal cover. Then $\pi_1 X$ acts $\mathscr{C}$-nilpotently on each $H^q(\tilde{X}; C)$.*

PROOF. The action of $\pi_1 X$ on $H^q(\tilde{X}; C)$ is induced from the action of $\pi_1 X$ on $\tilde{X}$ by passage to singular chains $C_*(\tilde{X})$, then to singular cochains $\mathrm{Hom}\,(C_*(\tilde{X}), C)$, and finally to cohomology. Since the cohomology functor $H^q(-; C)$ takes values in $\mathscr{C}$, the action of $\pi_1 X$ on $H^q(\tilde{X}; C)$ is through morphisms in $\mathscr{C}$. Thus, by Lemma 4.1.1, it suffices to show that $\pi_1 X$ acts nilpotently on $H^q(\tilde{X}; C)$.

Since $\pi_1 X = \pi_1 X_i$ acts trivially on A_i, Lemma 3.4.2 implies that it acts trivially on the space $K(A_i, n_i)$ and therefore acts trivially on $H^*(K(A_i, n_i); A)$ for any abelian group A. We claim that $\pi_1 X$ acts trivially on $H^*(\tilde{X}_i; C)$. This is clear when $i = 1$ since $\tilde{X}_1 = *$, and we proceed by induction on i. We have the following commutative diagram whose rows and columns are fiber sequences.

The Serre spectral sequence of the top row of the diagram has E_2 term

$$H^p(\tilde{X}_i; H^q(K(A_i, n_i); C)).$$

By the induction hypothesis, we see that $\pi_1 X$ acts nilpotently on the E_2 term. By Lemma 4.1.4, it therefore acts nilpotently on each E_r and thus on E_∞. By Lemma 3.1.3, we conclude that it acts nilpotently on $H^*(\tilde{X}_{i+1}; C)$. □

4.3. $\mathscr{A}$-maps of $\mathscr{A}$-nilpotent groups and spaces

Now return to the definitions of §3.1. Whenever one defines objects, one should define maps between them. Since it is inconvenient to focus on canonical choices of $\mathscr{A}$-central π-series, we adopt an ad hoc definition that depends on choices and ignore the question of how to make a well-defined category using the maps that we define. The topology of cocellular maps of Postnikov $\mathscr{A}$-towers is tailor-made to mesh with this algebraic definition. We assume that $\mathscr{A}$ is a category and not just a collection of abelian groups. Thus not all homomorphisms of abelian groups between groups in $\mathscr{A}$ need be morphisms of $\mathscr{A}$.

DEFINITION 4.3.1. Let G and H be $\mathscr{A}$-nilpotent π-groups. We say that a morphism $f: G \longrightarrow H$ of π-groups is an $\mathscr{A}$-morphism if there exist $\mathscr{A}$-central π-series

$$1 = G_{q+1} \subset G_q \subset \cdots \subset G_1 = G$$

and

$$1 = H_{q+1} \subset H_q \subset \cdots \subset H_1 = H$$

such that $f(G_j) \subset H_j$ and each induced homomorphism $G_j/G_{j+1} \to H_j/H_{j+1}$ is a map in the category $\mathscr{A}$.

DEFINITION 4.3.2. We say that a map $f: X \longrightarrow Y$ of $\mathscr{A}$-nilpotent spaces is an $\mathscr{A}$-map if each $f_*: \pi_n X \longrightarrow \pi_n Y$ is an $\mathscr{A}$-morphism of $\pi_1 X$-groups, where $\pi_1 X$ acts on $\pi_n Y$ through $f_*: \pi_1 X \longrightarrow \pi_1 Y$.

NOTATION 4.3.3. We refer to R-morphisms or R-maps in these definitions when $\mathscr{A} = \mathscr{A}_R$ is the category of modules over a commutative ring R. We refer to fR-maps when R is Noetherian and $\mathscr{A}$ is the category of finitely generated R-modules.

Again, the cases $R = \mathbb{Z}_T$ and $R = \mathbb{Z}_p$ are especially important to us. However, they behave quite differently since a map of abelian groups between

R-modules is automatically a map of R-modules when R is the localization $\mathbb{Z}_T$ but not when R is the completion $\mathbb{Z}_p$.

In the latter case, the following results will be used to get around the inconvenient kernel and quotient group hypotheses in Lemma 3.1.3. These results require our given category $\mathscr{A}$ to be abelian, and we again change notation from $\mathscr{A}$ to $\mathscr{C}$ to emphasize this change of assumptions.

LEMMA 4.3.4. *Let $\mathscr{C}$ be an abelian category. The kernel and, if the image is normal, the cokernel of a $\mathscr{C}$-morphism $f : G \longrightarrow H$ between $\mathscr{C}$-nilpotent π-groups are again $\mathscr{C}$-nilpotent π-groups.*

PROOF. Assume that we are given $\mathscr{C}$-central π-series as described and displayed in Definition 4.3.1. We use the language of additive relations $A \longrightarrow B$, where A and B are abelian groups. While these can be viewed as just homomorphisms from a subgroup of A to a quotient group of B, the elementary formal theory in [79, II§6] is needed to make effective use of them. This works for abelian categories, but the reader should think of the category of modules over a commutative ring.

Our given category $\mathscr{C}$ is abelian, and the abelian groups and homomorphisms in the following argument are all in $\mathscr{C}$. The map f determines an additive relation

$$G_j/G_{j+1} \longrightarrow H_{j+r}/H_{j+r+1}.$$

It sends an element $x \in G_j$ such that $f(x) \in H_{j+r}$ to the coset of $f(x)$. We can construct a singly graded spectral sequence from these additive relations. Its E^0-term is

$$E_j^0 = G_j/G_{j+1} \oplus H_j/H_{j+1}$$

with differential given by $d_j^0(g, h) = (0, f(g))$. The kernel of d_j^0 is

$$\{(g, h) | f(g) \in H_{j+1}\} \subset G_j/G_{j+1} \oplus H_j/H_{j+1}.$$

The image of d_j^0 is

$$\{(0, f(g)) | g \in G_j\} \subset G_j/G_{j+1} \oplus H_j/H_{j+1}.$$

The differential $d_j^1 : E_j^1 \longrightarrow E_{j+1}^1$ is given by $d_j^1(g, h) = (0, f(g))$. Since d_j^1 is defined on the kernel of d_j^0, the image is contained in E_{j+1}^1. The map defined on the kernel descends to a well-defined map on the homology E_j^1. Each further differential is given by the same formula. The kernel of d_j^r is

$$\{(g, h)|f(g) \in H_{j+r+1}\} \subset G_j/G_{j+1} \oplus H_j/H_{j+1}$$

and the image of d^r_{j-r} is

$$\{(0, f(g'))|g' \in G_{j-r}/G_{j-r+1}\} \subset G_j/G_{j+1} \oplus H_j/H_{j+1}.$$

For $r > q$, the length of our given central series, we see that

$$E^\infty_j = E^r_j = \{(g, h)|f(g) = 0\}/\{(0, f(g'))|g' \in G\} \subset G_j/G_{j+1} \oplus H_j/H_{j+1}.$$

This can be rewritten as

$$E^\infty_j = [(G_j \cap \ker f)/(G_{j+1} \cap \ker f)] \oplus [\operatorname{im}(H_j \to \operatorname{coker} f)/\operatorname{im}(H_{j+1} \to \operatorname{coker} f)].$$

The required $\mathscr{C}$-central π-series for $\ker f$ and $\operatorname{coker} f$ are given by $G_j \cap \ker f$ and $\operatorname{im}(H_j \longrightarrow \operatorname{coker} f)$. $\qquad\square$

LEMMA 4.3.5. *Let $\mathscr{C}$ be an abelian category. Assume given a long exact sequence of π-modules*

$$\cdots \longrightarrow B_{n+1} \xrightarrow{g_{n+1}} C_{n+1} \xrightarrow{h_{n+1}} A_n \xrightarrow{f_n} B_n \xrightarrow{g_n} C_n \longrightarrow \cdots$$

in which the g_n are $\mathscr{C}$-morphisms between $\mathscr{C}$-nilpotent π-modules B_n and C_n. Then the A_n are $\mathscr{C}$-nilpotent π-modules. The analogues with hypotheses on the maps f_n or h_n and conclusion for the π-modules C_n or B_n are also true.

PROOF. By Lemma 4.3.4, $\ker g_n$ and $\operatorname{coker} g_{n+1}$ are $\mathscr{C}$-nilpotent π-modules, hence $\operatorname{coker} h_{n+1}$ and $\ker f_n$ are $\mathscr{C}$-nilpotent π-modules. Let

$$0 = D_l \subset D_{l-1} \subset \cdots \subset D_0 = \operatorname{coker} h_{n+1}$$

$$0 = E_m \subset E_{m-1} \subset \cdots \subset E_0 = \ker f_n$$

be $\mathscr{A}$-central π-series. By Lemma 3.1.3 and the short exact sequence

$$0 \longrightarrow \ker f_n \longrightarrow A_n \longrightarrow \operatorname{coker} h_{n+1} \longrightarrow 0,$$

the $\{E_i\}$ and the inverse images of the $\{D_j\}$ give an $\mathscr{A}$-central π-series for A_n. $\qquad\square$

4.4. Nilpotency and fibrations

As in §1.5, let $p: E \longrightarrow B$ be a surjective fibration. We allow basepoints to vary. We record three results that relate the nilpotency of the components of the spaces $F_b = p^{-1}(b)$, E, and B. We may as well assume that B is connected, since otherwise we could restrict to the components of E that map to a given component of B. For each $e \in E$ with $p(e) = b$, let E_e be the component of E that contains e, let F_e be the component of F_b that contains e, and let $\iota \colon F_e \longrightarrow E_e$ be the inclusion.

PROPOSITION 4.4.1. *Assume that E_e is nilpotent.*

(i) *F_e is nilpotent and the $\pi_1(F_e, e)$-nilpotency class of $\pi_n(F_b, e)$, $n \geq 1$, is at most one greater than the $\pi_1(E_e, e)$-nilpotency class of $\pi_n(E_e, e)$.*

(ii) *If B is also nilpotent, then $\pi_1(E_e, e)$ acts nilpotently on F_e.*

PROOF. We agree to write F and E for the components F_e and E_e and to omit basepoints from all notations. We use the action of $\pi_1(E)$ on the long exact homotopy sequence (1.5.1) that is made explicit in Proposition 1.5.4. Thus we have the long exact sequence

$$\cdots \longrightarrow \pi_{n+1}(B) \xrightarrow{\partial_{n+1}} \pi_n(F) \xrightarrow{\iota_n} \pi_n(E) \xrightarrow{p_n} \pi_n(B) \longrightarrow \cdots$$

of $\pi_1(E)$-groups and homomorphisms of $\pi_1(E)$-groups.

For (i), we focus on the case $n \geq 2$; the proof when $n = 1$ is similar, using that the image of ∂_2 in $\pi_1(F)$ is a central subgroup by Lemma 1.4.7(v). Write I_E and I_F for the augmentation ideals of $\pi_1(E)$ and $\pi_1(F)$. Assume that $\pi_n(E)$ is $\pi_1(E)$-nilpotent of class q, so that $I_E^q \mathbb{Z}[\pi_n(E)] = 0$. Let $h \in I_F^q$ and $x \in \mathbb{Z}[\pi_n(F)]$. Then $\iota_*(h) = 0$ and thus $\iota_*(hx) = \iota_*(h)\iota_*(x) = 0$, so there exists $z \in \mathbb{Z}[\pi_{n+1}(B)]$ such that $\partial(z) = hx$. For $g \in \pi_1(F)$, $\iota_*(g)z = p_*\iota_*(g)z = z$ by the definition of the action of $\pi_1(E)$ on $\pi_{n+1}(B)$ and the fact that $p_*\iota_*$ is the trivial homomorphism. Thus $(\iota_*(g) - 1)z = 0$. But, since ∂ is a map of $\pi_1(E)$-modules,

$$\partial(\iota_*(g) - 1)z = (\iota_*(g) - 1)\partial(z) = (\iota_*(g) - 1)hx = (g - 1)hx,$$

where Proposition 1.5.4(i) gives the last equality. This shows that $I_F^{q+1}\mathbb{Z}[\pi_n(F)] = 0$.

For (ii), we use that our long exact sequence breaks into short exact sequences.

$$1 \longrightarrow \ker(p_n) \longrightarrow \pi_n(E) \longrightarrow \pi_n(E)/\ker(p_n) \longrightarrow 1$$

$$1 \longrightarrow \operatorname{im}(p_n) \longrightarrow \pi_n(B) \longrightarrow \operatorname{coker}(p_n) \longrightarrow 1$$

$$1 \longrightarrow \operatorname{coker}(p_{n+1}) \longrightarrow \pi_n(F) \longrightarrow \ker(p_n) \longrightarrow 1$$

Lemma 3.1.3 implies that $\ker(p_n)$ and $\operatorname{coker}(p_n)$ are nilpotent $\pi_1(E)$-groups, and Lemma 1.4.7(v), adapted in the evident way to replace the homotopy fiber there with the actual fiber here, shows that the last extension is central when $n = 1$. These observations and Lemma 3.1.3 imply that $\pi_n(F)$ is a nilpotent $\pi_1(E)$-group. Observe that since $\pi_1(F)$ acts through $\pi_1(E)$ on $\pi_n(F)$, by Proposition 1.5.4(i), this gives another proof that F is nilpotent. $\square$

Returning to the notations of the previous section, we let $\mathscr{C}$ be an abelian category in the following two results. We continue to work in the unbased setting.

PROPOSITION 4.4.2. *If all components of B and E are $\mathscr{C}$-nilpotent and each restriction of p to a component of E is a $\mathscr{C}$-map onto a component of B, then each component of each fiber of p is $\mathscr{C}$-nilpotent. If, further, the image of $p_*: \pi_1(E, e) \longrightarrow \pi_1(B, b)$ is normal, then the cokernel $\tilde{\pi}_0(F_b, e)$ of p_* is a $\mathscr{C}$-nilpotent group.*

PROOF. Here we use Proposition 1.5.4 to regard (1.5.1) as a long exact sequence both of $\pi_1(E, e)$-groups and, by pullback along ι_*, of $\pi_1(F, e)$-groups. For $n \geq 2$, $\pi_n(F, e)$ is a $\mathscr{C}$-nilpotent $\pi_1(F, e)$-group by Lemma 4.3.5. Lemmas 3.1.3 and 4.3.4 imply that $\pi_1(F, e)$ and $\tilde{\pi}_0(F_b, e)$ are $\mathscr{C}$-nilpotent; compare Remark 1.5.5. $\square$

PROPOSITION 4.4.3. *If $f: A \longrightarrow B$ and the fibration $p: E \longrightarrow B$ are $\mathscr{C}$-maps between $\mathscr{C}$-nilpotent spaces, then the component of the basepoint of $A \times_B E$ is a $\mathscr{C}$-nilpotent space.*

PROOF. Let X denote the cited basepoint component. Omitting basepoints from the notations for homotopy groups, Corollaries 2.2.3 and 2.2.5 give an exact sequence of homotopy groups ending with

$$\cdots \longrightarrow \pi_2(A) \times \pi_2(E) \xrightarrow{f_* - p_*} \pi_2(B) \xrightarrow{\partial} \pi_1(X) \longrightarrow \pi_1(A) \times_{\pi_1(B)} \pi_1(E) \longrightarrow 1.$$

By Proposition 1.5.4, this is a long exact sequence of $\pi_1(X)$-groups. Since f and p are $\mathscr{C}$-maps between $\mathscr{C}$-nilpotent spaces, $f_* - p_*$ is a $\mathscr{C}$-map and, as we see from Lemma 4.3.4 or 4.3.5, the action of $\pi_1(X)$ on $\pi_n(X)$ is $\mathscr{C}$-nilpotent for $n \geq 2$. The image of ∂ is a central subgroup of $\pi_1(X)$, by Lemma 1.4.7, and $\pi_1(A) \times_{\pi_1(B)} \pi_1(E)$ is a kernel of a $\mathscr{C}$-morphism, hence we also conclude by Lemma 4.3.4 that $\pi_1(X)$ is a $\mathscr{C}$-nilpotent $\pi_1(X)$-group. A similar argument would work starting with the long exact sequence of the fibration $A \times_B E \to A$ with fiber $F = p^{-1}(*)$, but here the possible nonconnectivity of F would complicate the argument for $\pi_1(X)$. $\square$

4.5. Nilpotent spaces and finite type conditions

The spaces that one usually encounters in algebraic topology satisfy finiteness conditions. For CW complexes X, it is standard (going back at least to Wall's 1965 paper [140]) to say that X is of finite type if its skeleta are finite, that is, if it has finitely many n-cells for each n. More generally, a space X is said to be of finite type if it is weakly equivalent to a CW complex of finite type. This is a topological specification, and it is the meaning of "finite type" that we adopt.

Unfortunately, it has more recently become almost as standard to say that a space X is of finite type if its integral homology groups $H_i(X; \mathbb{Z})$ are finitely generated for each i. This is an algebraic specification, and we say that X is homologically of finite type. Some authors instead ask that the homotopy groups $\pi_i(X)$ be finitely generated for all i. Our main goal in this section is to prove that these three conditions are equivalent when X is nilpotent.

This will later help us determine what can be said about localizations and completions of nilpotent spaces of finite type. This is particularly important in the case of completion, whose behavior on homotopy groups is much simpler in the finite-type case than it is in general. We introduce the following notations in anticipation of consideration of localizations and completions.

NOTATION 4.5.1. For a Noetherian ring R, let $f\mathscr{A}_R$ denote the abelian category of finitely generated R-modules. To abbreviate notation, we say that an $f\mathscr{A}_R$-nilpotent π-group is fR-nilpotent. Similarly, we say that an $f\mathscr{A}_R$-nilpotent space is fR-nilpotent. We speak of f-nilpotent groups and spaces when $R = \mathbb{Z}$.

Thus an f-nilpotent space is a nilpotent space X such that $\pi_1(X)$ is f-nilpotent and acts f-nilpotently on $\pi_i(X)$ for $i \geq 2$. This is a statement about subquotients of these groups, but it turns out to be equivalent to the statement that these groups themselves, or the homology groups $H_i(X; \mathbb{Z})$, are finitely generated. These equivalences are not at all obvious, and we have not found a complete proof in the literature. We shall sketch the proof of the following theorem, which is well-known when X is simple. It is beyond our scope to give full details in the general case, but we shall give the essential ideas.

THEOREM 4.5.2. Let X be a nilpotent space. Then the following statements are equivalent.

(i) X is weakly equivalent to a CW complex with finite skeleta.
(ii) X is f-nilpotent.

(iii) $\pi_i(X)$ *is finitely generated for each* $i \geq 1$.
(iv) $\pi_1(X)$ *and* $H_i(\tilde{X}; \mathbb{Z})$ *for* $i \geq 2$ *are finitely generated.*
(v) $H_i(X; \mathbb{Z})$ *is finitely generated for each* $i \geq 1$.

The implication (i) implies (v) is general, requiring no hypotheses on X. We restate it in terms of our definitions of spaces being of finite type.

LEMMA 4.5.3. *If* X *is of finite type, then* X *is homologically of finite type.*

PROOF. The cellular chains of a CW complex X with finite skeleta are finitely generated in each degree, and the homology groups of a chain complex that is finitely generated in each degree are finitely generated. □

In outline, the rest of the proof goes as follows. We first explain a classical result of Serre that implies a generalization of the equivalence of (iii) and (iv). We then outline a purely group-theoretic proof that (ii) and (iii) are equivalent. Using information from that proof and a spectral sequence argument, we see next that (v) implies (iv). Finally, we observe that Wall's classical characterization of spaces of finite type [140] applies to show that (iv) implies (i).

Serre proved the following result using the Serre spectral sequence and what are now called Serre classes of abelian groups [125]. Their introduction was a crucial precursor to the theory of localization. While our argument follows his original one, tom Dieck [36, 20.6] has shown that his results can actually be proven without any use of spectral sequences.

THEOREM 4.5.4 (SERRE). *Let* X *be simply connected. Then all* $\pi_i(X)$ *are finitely generated if and only if all* $H_i(X; \mathbb{Z})$ *are finitely generated.*

Since $\pi_i(X) = \pi_i(\tilde{X})$ for $i \geq 2$, application of this result to the universal cover of a space X gives the promised generalization of the equivalence of (iii) and (iv) in Theorem 4.5.2.

COROLLARY 4.5.5. *For any (connected) space* X, *not necessarily nilpotent,* $\pi_i(X)$ *is finitely generated for each* $i \geq 1$ *if and only if* $\pi_1(X)$ *and* $H_i(\tilde{X}; \mathbb{Z})$ *for each* $i \geq 2$ *are finitely generated.*

REMARK 4.5.6. One might ask instead of (iv) that $H_i(\tilde{X}; \mathbb{Z})$ for $i \geq 2$ be finitely generated over $\mathbb{Z}[\pi_1(X)]$, rather than over $\mathbb{Z}$. These conditions are equivalent when $\pi_1(X)$ is finitely generated and nilpotent, but they are not equivalent in general and the previous corollary requires the stronger hypothesis.

The proof of Theorem 4.5.4 begins with the following result. We take all homology groups to have coefficients $\mathbb{Z}$ in the rest of this section.

THEOREM 4.5.7. *If X is simply connected, then all $H_i(X)$ are finitely generated if and only if all $H_i(\Omega X)$ are finitely generated.*

PROOF. The Serre spectral sequence of the path space fibration over X satisfies $E^2_{p,q} = H_p(X; H_q(\Omega X))$ and converges to $H_*(PX)$, which is zero except for $H_0(PX) = \mathbb{Z}$. Using the universal coefficient theorem in homology,

$$H_p(X; H_q(\Omega X)) \cong (H_p(X) \otimes H_q(\Omega X)) \oplus \mathrm{Tor}\,(H_{p-1}(\Omega X), H_q(\Omega X)).$$

In particular, $E^2_{p,0} = H_p(X)$ and $E^2_{0,q} = H_q(\Omega X)$. From here, inductive arguments that Serre codified in [125] give the conclusion. □

COROLLARY 4.5.8. *If π is a finitely generated abelian group and $n \geq 1$, then each $H_i(K(\pi, n))$ is finitely generated.*

PROOF. Either by a standard first calculation in the homology of groups [24, p. 35] or by direct topological construction of a model for $K(\pi, 1)$ as a CW complex with finitely many cells in each dimension (e.g., [93, p. 126]), the result is true when $n = 1$. Since $\Omega K(\pi, n + 1) = K(\pi, n)$, the conclusion follows from the theorem by induction on n. □

PROOF OF THEOREM 4.5.4. Suppose that the homotopy groups of X are finitely generated. The Postnikov tower of X gives fibrations

$$K(\pi_n(X), n) \longrightarrow X_n \longrightarrow X_{n-1}.$$

These have Serre spectral sequences that converge from the groups

$$E^2_{p,q} = H_p(X_{n-1}; H_q(K(\pi_n(X), n)))$$

in total degree $p + q$ to $H_*(X_n)$. By Corollary 4.5.8 and induction on n, we may assume that each $H_i(X_{n-1})$ is finitely generated and deduce that each $H_i(X_n)$ is finitely generated. Since this holds for all n, $H_i(X)$ is finitely generated.

Conversely, suppose that the homology groups of X are finitely generated. Define $q_n \colon X\langle n \rangle \longrightarrow X$ to be the fiber of the fibration $X \longrightarrow X_{n-1}$ given by a Postnikov tower of X. By the long exact sequence of homotopy groups, we see that $\pi_i(X\langle n \rangle) = 0$ for $i \leq n-1$ and $q_{n_*} \colon \pi_i(X\langle n \rangle) \longrightarrow \pi_i(X)$ is an isomorphism for $i \geq n$. A map $X\langle n \rangle \longrightarrow X$ with these properties is said to be an $(n-1)$-connected cover of X, and then $\pi_n(X) \cong H_n(X\langle n \rangle)$ by the Hurewicz theorem. Since X is simply connected, we can write $X = X\langle 2 \rangle$, and then $\pi_2(X) \cong H_2(X)$ is finitely generated. For $n \geq 3$, we can apply Lemma 1.2.3 to the diagram

$$
\begin{array}{ccccccc}
\Omega X_n & \longrightarrow & X\langle n \rangle & \xrightarrow{\ q_n\ } & X & \longrightarrow & X_n \\
\downarrow & & \vdots \, {\scriptstyle r_n} & & \| & & \downarrow {\scriptstyle \alpha} \\
\Omega X_{n-1} & \longrightarrow & X\langle n-1 \rangle & \xrightarrow[\ q_{n-1}\ ]{} & X & \longrightarrow & X_{n-1}
\end{array}
$$

to obtain a map r_n that makes the diagram commute up to homotopy. By the resulting map of exact sequences of homotopy groups, we see that the homotopy fiber of r_n must be a space $K(\pi_{n-1}(X), n-1)$. Applying the Serre spectral sequences of the fibration sequences

$$
K(\pi_{n-1}(X), n-1) \longrightarrow X\langle n \rangle \longrightarrow X\langle n-1 \rangle
$$

inductively, starting with $X = X\langle 2 \rangle$, we see that the homology groups of $X\langle n \rangle$ are finitely generated and therefore the $\pi_n(X) \cong H_n(X\langle n \rangle)$ are finitely generated. $\qquad\square$

The following result shows that (ii) and (iii) of Theorem 4.5.2 are equivalent. The proof uses a little more group theory than we wish to present in detail.

PROPOSITION 4.5.9. *Let G be a nilpotent group. Then the following statements are equivalent.*

(i) *G is f-nilpotent.*
(ii) *$G/[G, G]$ is finitely generated.*
(iii) *G is finitely generated.*
(iv) *Every subgroup of G is finitely generated.*

Moreover, when these conditions hold, the group ring $\mathbb{Z}[G]$ is (left and right) Noetherian and the group G is finitely presentable.

PROOF. Any central series of G gives an exact sequence

$$
1 \longrightarrow G' \xrightarrow{\ \subset\ } G \xrightarrow{\ \pi\ } G/G' \longrightarrow 1
$$

where G' is nilpotent of lower class than G and G/G' is abelian. Using the lower central series of G, we may as well take G' to be the commutator subgroup $[G, G]$. If G is f-nilpotent, then certainly G/G' is finitely generated, so that (i) implies (ii). Suppose that (ii) holds. Choose elements h_i of G such that the elements $\pi(h_i)$ generate G/G' and let H be the subgroup of G generated by the h_i. For $g \in G$, there is an element $h \in H$ such that $\pi(g) = \pi(h)$, and then $g = gh^{-1}h$ is an element of $G'H$. By [56, Cor. 10.3.3], this implies that $H = G$ and therefore G is finitely generated, showing that (ii) implies (iii).

The proof that (iii) implies (iv) is given in [56, p. 426]. The proof goes in two steps. First, if G is finitely generated, then G is supersolvable, meaning that it has a finite normal series with cyclic subquotients. Second, every subgroup of a supersolvable group is supersolvable and is therefore finitely generated. Finally, (iv) implies (i) since if every subgroup of G is finitely generated, then G/G' and G' are finitely generated. By induction on the nilpotency class, we can assume that G' is f-nilpotent, the abelian case being clear, and then the displayed short exact sequence shows that G is also f-nilpotent.

The Noetherian statement should look very plausible in view of (iv), but we refer the reader to [57, Thm. 1] for the proof. It applies more generally to polycyclic groups, which can be characterized as the solvable groups all of whose subgroups are finitely generated. The statement that G is finitely presentable is an exercise in Bourbaki [15, Ex. 17b, p. 163]. The essential point is that if N and H are finitely presentable and G is an extension of N by H, then G is finitely presentable. □

Returning to the topological context of Theorem 4.5.2, assume that X is nilpotent and let $\pi = \pi_1(X)$.

SKETCH PROOF THAT (V) IMPLIES (IV). Since $H_1(X; \mathbb{Z}) \cong \pi/[\pi, \pi]$, the implication (ii) implies (i) of Proposition 4.5.9 shows that π is finitely generated. There are several choices of spectral sequences that can be used to show that $H_i(\tilde{X}; \mathbb{Z})$ is finitely generated for $i \geq 2$. One can use the Serre spectral sequence obtained from the fiber sequence $\tilde{X} \longrightarrow X \longrightarrow K(\pi, 1)$, which converges from $H_*(\pi; H_*(\tilde{X}))$ to $H_*(X)$, using a backward induction from the assumption that π and the $H_i(X)$ are finitely generated. A more direct argument uses the Eilenberg-Moore spectral sequence which, in its homological version and ignoring details of grading, converges from the homology groups $E^2_{*,*} = H_*(\pi; H_*(X))$ to $H_*(\tilde{X})$. The spectral sequence converges by a result of Dwyer [41], the essential point being that π acts nilpotently on $H_*(\tilde{X})$, by the homology analogue of Proposition 4.2.1. Using that the group ring $\mathbb{Z}[\pi]$ is

Noetherian, one can check that the groups of the E_2-term are finitely generated and deduce that the groups $H_i(\tilde{X})$ are finitely generated. Alternatively, one can use the lower central series spectral sequence for the calculation of the groups $\pi_i(X)$, as explained (with comparable brevity) in [21, p. 153]. $\qquad\square$

SKETCH PROOF THAT (IV) IMPLIES (I). If X is simply connected and each $\pi_i(X)$ is finitely generated, the result is easy. There is a standard construction of a weak equivalence $\Gamma X \longrightarrow X$ from a CW complex ΓX to X given, for example, in [93, p. 75]. Using minimal sets of generators for the homotopy groups and kernels of maps of homotopy groups that appear in the construction, we see that we only need to attach a finite number of cells at each stage. A slight variant of the construction makes the argument a little clearer. Instead of attaching cylinders in the proof there, we can attach cells to kill the generators of the kernel of $\pi_i(X_i) \longrightarrow \pi_i(X)$ in the inductive argument. These kernels are finitely generated at each stage, so we only need to attach finitely many cells to kill them. Wall [140] refines this argument to deal with a nontrivial fundamental group π. In [140, Thm. A], he gives necessary and sufficient conditions for X to be of finite type. In [140, Thm. B], he shows that these conditions are satisfied if $\mathbb{Z}[\pi]$ is Noetherian, π is finitely presented, and each $H_i(\tilde{X})$ is finitely generated over $\mathbb{Z}[\pi]$. We have the first two conditions by Proposition 4.5.9, and the last condition certainly holds if each $H_i(\tilde{X})$ is finitely generated as an abelian group. $\qquad\square$

PART 2

Localizations of spaces at sets of primes

5

LOCALIZATIONS OF NILPOTENT GROUPS
AND SPACES

We develop localization at T for abelian groups, nilpotent groups, and nilpotent spaces. Of course, localization of abelian groups is elementary and direct. However, following Bousfield and Kan [21], we first construct localizations of spaces and then use them to construct localizations of nilpotent groups topologically rather than algebraically. A purely algebraic treatment is given by Hilton, Mislin, and Roitberg [60, 62]. We discuss localizations of abelian groups in §5.1, the essential point being to determine the behavior of localization on homology. We define localizations of spaces and show how to localize the Eilenberg-Mac Lane spaces of abelian groups in §5.2. We construct localizations of nilpotent spaces by induction up their Postnikov towers in §5.3. We specialize to obtain localizations of nilpotent groups in §5.4. We discuss their general algebraic properties in §5.5, and we discuss finiteness conditions in §5.6, leading up to a characterization of $f\mathbb{Z}_T$-nilpotent groups. The reader may wish to skip the last two sections on a first reading. Their main purpose is to develop algebra needed later to prove the fracture theorems in full generality.

Recall our notational conventions from the Introduction. In particular, T is a fixed set of primes, possibly empty, throughout this chapter and the next. Maps ϕ will always denote localizations.

5.1. Localizations of abelian groups

Recall that an abelian group B is said to be T-local if it admits a structure of $\mathbb{Z}_T$-module, necessarily unique. It is equivalent that the multiplication map $q: B \longrightarrow B$ is an isomorphism for all primes q not in T. We have the following easy observation.

LEMMA 5.1.1. *Let*

$$0 \longrightarrow A' \longrightarrow A \longrightarrow A'' \longrightarrow 0$$

be a short exact sequence of abelian groups. If any two of A', A, and A" are T-local, then so is the third.

The localization at T of an abelian group A is a map $\phi\colon A \longrightarrow A_T$ to a T-local abelian group A_T that is universal among such maps. This means that any homomorphism $f\colon A \longrightarrow B$, where B is T-local, factors uniquely through ϕ. That is, there is a unique homomorphism $\tilde{f}$ that makes the following diagram commute.

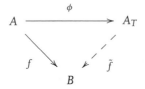

We can define ϕ explicitly by setting $A_T = A \otimes \mathbb{Z}_T$ and letting $\phi(a) = a \otimes 1$. Clearly A is T-local if and only if ϕ is an isomorphism. Since $\mathbb{Z}_T$ is a torsion-free abelian group, it is a flat $\mathbb{Z}$-module. We record the following important consequence.

LEMMA 5.1.2. *Localization is an exact functor from abelian groups to $\mathbb{Z}_T$-modules.*

We shall focus on cohomology when defining localizations of spaces. This is natural when thinking about Postnikov towers and the dual Whitehead theorem, and it leads to efficient proofs. In preparation for this, we describe the homological behavior of localization of abelian groups in some detail.

THEOREM 5.1.3. *The induced map*

$$\phi_*\colon H_*(A; \mathbb{Z}_T) \longrightarrow H_*(A_T; \mathbb{Z}_T)$$

is an isomorphism for all abelian groups A. If B is T-local, then the homomorphism

$$\tilde{H}_*(B; \mathbb{Z}) \longrightarrow \tilde{H}_*(B; \mathbb{Z}_T)$$

induced by the homomorphism $\mathbb{Z} \longrightarrow \mathbb{Z}_T$ is an isomorphism and thus $\tilde{H}_*(B; \mathbb{Z})$ is T-local in every degree.

PROOF. There are several ways to see this result, and the reader is urged to use her favorite. In view of (4.1.5), one is free to carry out the proof using algebra, topology, or a combination of the two. Any module over a PID R is the filtered colimit of its finitely generated submodules, and any finitely generated

R-module is a finite direct sum of cyclic R-modules (R-modules with a single generator). We apply this with $R = \mathbb{Z}$ and $R = \mathbb{Z}_T$. In these cases, the finite cyclic modules can be taken to be of prime power order, using only primes in T in the case of $\mathbb{Z}_T$, and the infinite cyclic modules are isomorphic to $\mathbb{Z}$ or to $\mathbb{Z}_T$.

The localization functor commutes with colimits since it is a left adjoint, and the homology of a filtered colimit of abelian groups is the colimit of their homologies. To see this topologically, for example, one can use the standard simplicial construction of classifying spaces [93, p. 126] to give a construction of $K(A, 1)$'s that commutes with filtered colimits, and one can then use that homology commutes with filtered colimits, by Proposition 2.5.4. Finite sums of abelian groups are finite products, and as we have already noted it is clear that a product $K(A, 1) \times K(A', 1)$ is a $K(A \times A', 1)$. By the Künneth theorem (e.g., [93, p. 130]), the conclusions of the theorem hold for a finite direct sum if they hold for each of the summands. This reduces the problem to the case of cyclic R-modules.

One can check the cyclic case directly, but one can decrease the number of checks needed by using the Lyndon-Hochschild-Serre (LHS) spectral sequence of Proposition 24.5.3. That spectral sequence allows one to deduce the result for cyclic groups of prime power order inductively from the result for cyclic groups of prime order.

Thus suppose first that A is cyclic of prime order q. Since $\mathbb{Z}_T$ is T-local, so are both the source and target homology groups. We focus on the reduced homology groups since $K(\pi, 1)$'s are connected and the zeroth homology group with coefficients in R is always R. If $q = 2$, $K(A, 1) = \mathbb{R}P^\infty$ and if q is odd, $K(A, 1)$ is the analogous lens space S^∞/A. In both cases, we know the integral homology explicitly (e.g., by an exercise in [93, p. 103]). The nonzero reduced homology groups are all cyclic of order q, that is, copies of A. Taking coefficents in $\mathbb{Z}_T$ these groups are zero if $q \notin T$ and A if $q \in T$, and of course $A_T = 0$ if $q \notin T$ and $A_T = A$ if $q \in T$. Thus the conclusions hold in the finite cyclic case.

Finally, consider $A = \mathbb{Z}$, so that $A_T = \mathbb{Z}_T$. The circle S^1 is a $K(\mathbb{Z}, 1)$. Our first example of a localized space is S^1_T, which not surprisingly turns out to be $K(\mathbb{Z}_T, 1)$. A_T can be constructed as the colimit of copies of $\mathbb{Z}$ together with the maps induced by multiplication by the primes not in T. For example, if we order the primes q_i not in T by size and define r_n inductively by $r_1 = q_1$ and $r_n = r_{n-1}q_1 \cdots q_n = q_1^n \cdots q_n$, then $\mathbb{Z}_T$ is the colimit over n of the maps $r_n \colon \mathbb{Z} \longrightarrow \mathbb{Z}$. We can realize these maps on $\pi_1(S^1)$ by using the r_n^{th} power map $S^1 \longrightarrow S^1$. Using the telescope construction [93, p. 113] to convert these multiplication maps into inclusions and passing to colimits, we obtain a space $K(\mathbb{Z}_T, 1)$; the van Kampen theorem gives that the colimit has fundamental group $\mathbb{Z}_T$, and the higher homotopy groups are zero because the image of a map from S^n into the colimit is contained in a finite stage of the telescope, and each such finite stage is equivalent to S^1. The commutation of homology with colimits gives that the only nonzero reduced integral homology group of $K(\mathbb{Z}_T, 1)$ is its first, which is $\mathbb{Z}_T$.

An alternative proof in this case uses the LHS spectral sequence of the quotient group $\mathbb{Z}_T/\mathbb{Z}$. Groups such as this will play an important role in completion theory and will be discussed in Chapter 10. The spectral sequence has the form

$$E^2_{p,q} = H_p(\mathbb{Z}_T/\mathbb{Z}; H_q(\mathbb{Z}; \mathbb{Z}_T)) \implies H_{p+q}(\mathbb{Z}_T; \mathbb{Z}_T).$$

The group $\mathbb{Z}_T/\mathbb{Z}$ is local away from T, hence the terms with $p > 0$ are zero, and the spectral sequence collapses to the edge isomorphism

$$\phi_* : H_*(\mathbb{Z}; \mathbb{Z}_T) \longrightarrow H_*(\mathbb{Z}_T; \mathbb{Z}_T). \qquad \square$$

COROLLARY 5.1.4. *The induced map*

$$\phi^* : H^*(A_T; B) \longrightarrow H^*(A; B)$$

is an isomorphism for all $\mathbb{Z}_T$-modules B.

On H^1, by the representability of cohomology and the topological interpretation (4.1.5) of the cohomology of groups, this says that

$$\phi^* : [K(A_T, 1), K(B, 1)] \longrightarrow [K(A, 1), K(B, 1)]$$

is an isomorphism. On passage to fundamental groups, this recovers the defining universal property of localization.

In fact, as we shall use heavily in §5.4, for any groups G and H, not necessarily abelian, passage to fundamental groups induces a bijection

5.1.5 $$[K(G, 1), K(H, 1)] \cong \mathrm{Hom}(G, H).$$

(This is an exercise in [93, p. 119]). One way to see this is to observe that the classifying space functor from groups to Eilenberg-Mac Lane spaces (e.g., [93, p. 126]) gives an inverse bijection to π_1, but it can also be verified directly from the elementary construction of $K(G, 1)$'s that is obtained by realizing π_1 as the fundamental group of a space X [93, p. 35] and then killing the higher homotopy groups of X.

5.2. The definition of localizations of spaces

Recall that we take all spaces to be path connected. We have the following three basic definitions. Recall Definition 3.3.10 and Proposition 3.3.11.

DEFINITION 5.2.1. A map $\xi: X \longrightarrow Y$ is a $\mathbb{Z}_T$-equivalence if the induced map $\xi_*: H_*(X; \mathbb{Z}_T) \longrightarrow H_*(Y; \mathbb{Z}_T)$ is an isomorphism or, equivalently, if the induced map $\xi^*: H^*(Y; B) \longrightarrow H^*(X; B)$ is an isomorphism for all $\mathbb{Z}_T$-modules B.

DEFINITION 5.2.2. A space Z is T-local if $\xi^*: [Y, Z] \longrightarrow [X, Z]$ is a bijection for all $\mathbb{Z}_T$-equivalences $\xi: X \longrightarrow Y$.

Diagrammatically, this says that for any map $f: X \longrightarrow Z$, there is a map $\tilde{f}$, unique up to homotopy, that makes the following diagram commute up to homotopy.

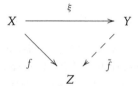

DEFINITION 5.2.3. A map $\phi: X \longrightarrow X_T$ from X into a T-local space X_T is a localization at T if ϕ is a $\mathbb{Z}_T$-equivalence.

This prescribes a universal property. If $f: X \longrightarrow Z$ is any map from X to a T-local space Z, then there is a map $\tilde{f}$, unique up to homotopy, that makes the following diagram commute.

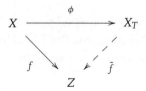

Therefore, localizations are unique up to homotopy if they exist. We shall prove in §19.3 that they do always exist, but we focus on nilpotent spaces for now.

REMARK 5.2.4. On the full subcategory of connected spaces in $\mathrm{Ho}\,\mathscr{T}$ that admit localizations at T, localization is automatically functorial (up to homotopy). For a map $f: X \longrightarrow Y$, there is a unique map $f_T: X_T \longrightarrow Y_T$ in Ho$\mathscr{T}$ such that $\phi \circ f = f_T \circ \phi$ in Ho$\mathscr{T}$, by the universal property.

When specialized to Eilenberg-Mac Lane spaces $K(A, 1)$, these definitions lead to alternative topological descriptions of T-local abelian groups and of the algebraic localizations of abelian groups at T. The proofs are exercises in the use of the representability of cohomology.

PROPOSITION 5.2.5. *An abelian group B is T-local if and only if the space $K(B, 1)$ is T-local.*

PROOF. If B is T-local and $\xi: X \longrightarrow Y$ is a $\mathbb{Z}_T$-equivalence, then

$$\xi^*: H^1(Y; B) \longrightarrow H^1(X; B)$$

is an isomorphism. Since this is the map

$$\xi^*: [Y, K(B, 1)] \longrightarrow [X, K(B, 1)],$$

$K(B, 1)$ is T-local. Conversely, if $K(B, 1)$ is T-local, then the identity map of $K(B, 1)$ is a $\mathbb{Z}_T$-equivalence to a T-local space and is thus a localization at T. However, the map $\phi: K(B, 1) \longrightarrow K(B_T, 1)$ that realizes $\phi: B \longrightarrow B_T$ on fundamental groups is also a $\mathbb{Z}_T$-equivalence, by Corollary 5.1.4 and (4.1.5). Therefore ϕ is also a localization at T. By the uniqueness of localizations, ϕ must be an equivalence and thus $\phi: B \longrightarrow B_T$ must be an isomorphism. □

COROLLARY 5.2.6. *An abelian group B is T-local if and only if the homomorphism $\xi^*: H^*(Y; B) \longrightarrow H^*(X; B)$ induced by any $\mathbb{Z}_T$-equivalence $\xi: X \longrightarrow Y$ is an isomorphism.*

PROOF. If B has the cited cohomological property, then $K(B, 1)$ is T-local by the representability of cohomology and B is T-local by the previous result. The converse holds by the definition of a $\mathbb{Z}_T$-equivalence. □

PROPOSITION 5.2.7. *A homomorphism $\phi: A \longrightarrow B$ of abelian groups is an algebraic localization at T if and only if the map, unique up to homotopy,*

$$\phi: K(A, 1) \longrightarrow K(B, 1)$$

that realizes ϕ on π_1 is a topological localization at T.

PROOF. By Corollary 5.1.4 and (4.1.5), if $\phi\colon A \longrightarrow B$ is an algebraic localization at T, then $\phi\colon K(A,1) \longrightarrow K(B,1)$ is a $\mathbb{Z}_T$-equivalence. Since $K(B,1)$ is T-local by the previous result, this proves that ϕ is a topological localization. Conversely, if $\phi\colon K(A,1) \longrightarrow K(B,1)$ is a localization at T and C is any T-local abelian group, then the isomorphism

$$\phi^*\colon H^1(K(B,1),C) \longrightarrow H^1(K(A,1),C)$$

translates by the representability of cohomology and passage to fundamental groups into the isomorphism

$$\phi^*\colon \mathrm{Hom}(B,C) \longrightarrow \mathrm{Hom}(A,C).$$

This adjunction expresses the universal property of algebraic localization. $\square$

By induction on n, we can now localize Eilenberg-Mac Lane spaces $K(A,n)$.

THEOREM 5.2.8. *If B is a T-local abelian group, then $K(B,n)$ is a T-local space and $\tilde{H}_*(K(B,n);\mathbb{Z})$ is T-local in each degree. For any abelian group A, the map $\phi\colon K(A,n) \longrightarrow K(A_T,n)$, unique up to homotopy, that realizes the localization $\phi\colon A \longrightarrow A_T$ on π_n is a localization at T.*

PROOF. If $\xi\colon X \longrightarrow Y$ is a $\mathbb{Z}_T$-equivalence, then

$$\xi^*\colon [Y, K(B,n)] \longrightarrow [X, K(B,n)]$$

is the isomorphism induced on the nth cohomology group. Thus $K(B,n)$ is T-local.

For $n \geq 2$, we may write $\Omega K(A,n) = K(A,n-1)$, and the map ϕ induces a map of path space fibrations

$$
\begin{array}{ccccc}
K(A,n-1) & \longrightarrow & PK(A,n) & \longrightarrow & K(A,n) \\
\Omega\phi \downarrow & & \downarrow & & \downarrow \phi \\
K(A_T,n-1) & \longrightarrow & PK(A_T,n) & \longrightarrow & K(A_T,n).
\end{array}
$$

This induces a map of Serre spectral sequences converging to the homologies of contractible spaces. The map $\Omega\phi$ on fibers realizes ϕ on π_{n-1}, and we assume inductively that it is a $\mathbb{Z}_T$-equivalence. By the comparison theorem for spectral sequences, Theorem 24.6.1 below, it follows that the map ϕ on base spaces is also a $\mathbb{Z}_T$-equivalence. This proves the last statement.

Taking $A = B$ to be T-local, we prove that $\tilde{H}_*(K(B, n), \mathbb{Z})$ is T-local by inductive comparison of the $\mathbb{Z}$ and $\mathbb{Z}_T$ homology Serre spectral sequences of the displayed path space fibrations. Starting with the case $n = 1$ in Theorem 5.1.3, we find that

$$\tilde{H}_*(K(B, n), \mathbb{Z}) \cong \tilde{H}_*(K(B, n), \mathbb{Z}_T). \qquad \square$$

5.3. Localizations of nilpotent spaces

Our construction is based on a special case of the dual Whitehead theorem. Take $\mathscr{A}$ in Theorem 3.3.9 to be the collection of $\mathbb{Z}_T$-modules. Then that result takes the following form, which generalizes the fact that $K(B, n)$ is a T-local space if B is a T-local abelian group.

THEOREM 5.3.1. *Every $\mathbb{Z}_T$-tower is a T-local space.*

We use this result to construct localizations of nilpotent spaces.

THEOREM 5.3.2. *Every nilpotent space X admits a localization $\phi : X \longrightarrow X_T$.*

PROOF. In view of Theorem 3.5.4, we may assume without loss of generality that X is a Postnikov tower $\lim X_i$ constructed from maps $k_i : X_i \longrightarrow K(A_i, n_i + 1)$, where A_i is an abelian group, $n_{i+1} \geq n_i \geq 1$, and only finitely many $n_i = n$ for any $n \geq 1$. Here $X_0 = *$, and we let $(X_0)_T = *$. Assume inductively that a localization $\phi_i : X_i \longrightarrow (X_i)_T$ has been constructed and consider the following diagram, in which we write $K(A_i, n_i) = \Omega K(A_i, n_i + 1)$.

$$
\begin{array}{ccccccc}
K(A_i, n_i) & \longrightarrow & X_{i+1} & \longrightarrow & X_i & \overset{k_i}{\longrightarrow} & K(A_i, n_i + 1) \\
\Omega\phi \downarrow & & \phi_{i+1} \downarrow & & \phi_i \downarrow & & \phi \downarrow \\
K((A_i)_T, n_i) & \longrightarrow & (X_{i+1})_T & \longrightarrow & (X_i)_T & \underset{(k_i)_T}{\longrightarrow} & K((A_i)_T, n_i + 1)
\end{array}
$$

By Theorem 5.3.1, since ϕ_i is a $\mathbb{Z}_T$-equivalence and $K((A_i)_T, n_i + 1)$ is a $\mathbb{Z}_T$-local space there is a map $(k_i)_T$, unique up to homotopy that makes the right square commute up to homotopy. The space X_{i+1} is the fiber Fk_i, and we define $(X_{i+1})_T$ to be the fiber $F(k_i)_T$.

By Lemma 1.2.3, there is a map ϕ_{i+1} that makes the middle square commute and the left square commute up to homotopy. By Theorem 5.3.1, $(X_{i+1})_T$ is

T-local since it is a $\mathbb{Z}_T$-tower. We claim that ϕ_{i+1} induces an isomorphism on homology with coefficients in $\mathbb{Z}_T$ and is thus a localization at T. Applying the Serre spectral sequence to the displayed fibrations, we obtain spectral sequences

$$E^2_{p,q} \cong H_p(X_i; H_q(K(A_i, n_i); \mathbb{Z}_T)) \implies H_{p+q}(X_{i+1}; \mathbb{Z}_T)$$

$$E^2_{p,q} \cong H_p((X_i)_T; H_q(K((A_i)_T, n_i); \mathbb{Z}_T)) \implies H_{p+q}((X_{i+1})_T; \mathbb{Z}_T)$$

and a map between them. By Theorem 5.2.8, the induced map on the homology of fibers is an isomorphism. Since ϕ_i is a localization at T and thus a $\mathbb{Z}_T$-equivalence, the map on E^2 terms is an isomorphism. It follows that ϕ_{i+1} is a $\mathbb{Z}_T$-equivalence, as claimed.

Let $X_T = \lim (X_i)_T$ and $\phi = \lim \phi_i \colon X \longrightarrow X_T$. Then ϕ is a $\mathbb{Z}_T$-equivalence by Proposition 2.5.9 and is thus a localization of X at T. $\qquad\square$

Our explicit "cocellular" construction of localizations of Postnikov towers allows us to be more precise about functoriality than in Remark 5.2.4.

THEOREM 5.3.3. *Let X and Y be Postnikov towers and let $\psi \colon X \longrightarrow Y$ be a cocellular map. Choose cocellular localizations at T of X and Y. Then there exists a cocellular map $\psi_T \colon X_T \longrightarrow Y_T$, unique up to cocellular homotopy, such that $\psi_T \circ \phi$ is homotopic to $\phi \circ \psi$.*

PROOF. We construct ψ_T and $h \colon \psi_T \circ \phi \simeq \phi \circ \psi$ by inductive application of coHELP, Theorem 3.3.7, to the following diagrams and passage to limits.

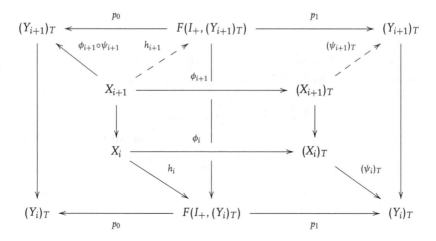

Now let ψ_T and η be maps $X_T \longrightarrow Y_T$ such that both $\psi_T \circ \phi$ and $\eta \circ \phi$ are homotopic to $\phi \circ \psi$. Let P_{i+1} be the pullback of $F(I_+, (Y_i)_T)$ and $F(\{0,1\}_+, (Y_{i+1})_T)$ over $F(\{0,1\}_+, (Y_i)_T)$ and observe that the fibration $(Y_{i+1})_T \longrightarrow (Y_i)_T$ and cofibration $\{0,1\} \longrightarrow I$ induce a canonical fibration

$$F(I_+, (Y_{i+1})_T) \longrightarrow P_{i+1}.$$

We construct the following diagram inductively and apply coHELP to it.

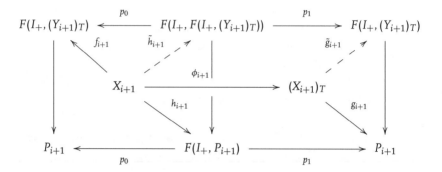

The map $f_{i+1} \colon X_{i+1} \longrightarrow F(I_+, (Y_{i+1})_T)$ is a composite of homotopies

$$\psi_T \circ \phi \simeq \phi \circ \psi \simeq \eta \circ \phi.$$

The map $g_{i+1} \colon (X_{i+1})_T \longrightarrow P_{i+1}$ is the map into P_{i+1} defined by the maps

$$(X_{i+1})_T \longrightarrow (X_i)_T \xrightarrow{\tilde{g}_i} F(I_+, (Y_i)_T)$$

and

$$\alpha \colon (X_{i+1})_T \longrightarrow F(\{0,1\}_+, (Y_{i+1})_T),$$

where $\alpha(x)(0) = \psi_T(x)$ and $\alpha(x)(1) = \eta(x)$. It is not difficult to check inductively that this pair of maps does define a map into the pullback. Similarly, the map $h_{i+1} \colon X_{i+1} \longrightarrow F(I_+, P_{i+1})$ is defined by the pair of maps

$$X_{i+1} \longrightarrow X_i \xrightarrow{\tilde{h}_i} F(I_+, F(I_+, (Y_i)_T))$$

and

$$X_{i+1} \longrightarrow (X_{i+1})_T \xrightarrow{\beta} F(I_+, F(\{0,1\}_+, (Y_{i+1})_T)),$$

where $\beta(x)(t)(0) = \psi_T(x)$ and $\beta(x)(t)(1) = \eta(x)$. On passage to limits, the maps $\tilde{g}_i$ give the desired homotopy from ψ_T to η. $\qquad\square$

We have analogous conclusions for quotient towers and pullbacks.

PROPOSITION 5.3.4. *Let W be a quotient tower of a Postnikov tower X with projection $\pi: X \longrightarrow W$. Then there are cocellular localizations X_T of X and W_T of W such that W_T is a quotient tower of X_T whose projection satisfies $\pi_T \circ \phi = \phi \circ \pi$. If $\pi^{-1}(*)$ is connected, the map $\phi: \pi^{-1}(*) \longrightarrow (\pi_T)^{-1}(*)$ obtained by restricting $\phi: X \to X_T$ to fibers is again a localization at T. If, further, Y is a Postnikov tower, $\theta: Y \longrightarrow W$ is a cocellular map, and $\theta_T: Y_T \longrightarrow W_T$ is chosen as in Theorem 5.3.3, then the pullback $X_T \times_{W_T} Y_T$ of π_T and θ_T is a cocellular localization of the pullback $X \times_W Y$ of π and θ.*

PROOF. The first statement is an easy induction based on the definition of a quotient tower. For the second statement, note that $\pi^{-1}(*)$ is a Postnikov tower with one cocell for each cocell of X that does not factor through W. For the last statement, note that $X \times_W Y$ can be viewed as a Postnikov tower with one cocell for each cocell of X that does not factor through W and each cocell of Y. □

Taking $W = *$, we have the following special case.

COROLLARY 5.3.5. *$X_T \times Y_T$ is a cocellular localization of $X \times Y$.*

Either applying Proposition 5.3.4 to the path space fibration or arguing directly, we obtain a similar result for loop spaces.

COROLLARY 5.3.6. *If X is a simply connected Postnikov tower, then $\Omega(X_T)$ is a cocellular localization of ΩX.*

5.4. Localizations of nilpotent groups

Let q be a prime. A group G is said to be uniquely q-divisible if the qth power function $G \longrightarrow G$ is a bijection.

REMARK 5.4.1. Of course, an abelian group B is uniquely q-divisible if and only if the multiplication homomorphism $q: B \longrightarrow B$ is an isomorphism. In turn, this holds if and only if $B \otimes \mathbb{F}_q = 0$ and $\text{Tor}(B, \mathbb{F}_q) = 0$.

REMARK 5.4.2. If G is uniquely q-divisible for all primes $q \notin T$ and $g \in G$ is an element of finite order prime to T, so that $g^r = 1$ for some product r of primes not in T, then $g = 1$.

DEFINITION 5.4.3. A T-local group is a nilpotent group that is uniquely q-divisible for all primes q not in T.

Thus, for us, a T-local group is necessarily nilpotent. The localization of a nilpotent group G at T is defined the same way as for abelian groups. It is a map $\phi \colon G \longrightarrow G_T$ to a T-local group G_T that is universal among such maps. However, it is no longer obvious how to construct such a map, and we will use topology rather than algebra to do so. We need some preliminaries that allow us to recast the definitions just given in terms of $\mathbb{Z}_T$-nilpotency. We give a complete proof of the following generalization of Lemma 5.1.1 to help familiarize the reader with the relevant kinds of algebraic arguments, but the result itself will be superceded by Corollary 5.4.11 below, which shows that the centrality assumption is unnecessary. Thus the result can be viewed as scaffolding leading toward that generalization.

LEMMA 5.4.4. *Let*

$$1 \longrightarrow G' \overset{\iota}{\longrightarrow} G \overset{\psi}{\longrightarrow} G'' \longrightarrow 1$$

be a central extension of groups. If any two of G', G, and G'' are uniquely q-divisible, then so is the third.

PROOF. First assume that G' and G'' are uniquely q-divisible. To show that the qth power map of G is surjective, let $x \in G$. Then $\psi(x) = z^q$ for some $z \in G''$ since G'' is uniquely q-divisible. Since ψ is surjective, $\psi(y) = z$ for some $y \in G$. Therefore $x = y^q \iota(y')$ for some $y' \in G'$. Since G' is uniquely q-divisible, $y' = x'^q$ for some $x' \in G'$ and thus $x = y^q \iota(x')^q = (y\iota(x'))^q$ since our extension is central. To show that the qth power map of G is injective, let $x^q = y^q$ for $x, y \in G$. Then $\psi(x)^q = \psi(y)^q$ and, since G'' is uniquely q-divisible, $\psi(x) = \psi(y)$. By exactness, $x = y\iota(x')$ for some $x' \in G'$ and so $x^q = y^q \iota(x'^q)$. Since ι is injective $x'^q = 1$. Since G' is uniquely q-divisible, $x' = 1$ and $x = y$.

Next, assume that G' and G are uniquely q-divisible. The qth power map of G'' is surjective since the qth power map of G is surjective. Suppose that $\psi(x)^q = \psi(y)^q$. Then $x^q = \iota(z)y^q$ for some $z \in \iota(G')$. Since G' is uniquely q-divisible, $z = w^q$ for some w. By centrality, $x^q = \iota(w)^q y^q = (\iota(w)y)^q$. Since G is uniquely q-divisible, $x = \iota(w)y$ and thus $\psi(x) = \psi(y)$.

Finally, assume that G and G'' are uniquely q-divisible. The qth power map of G' is injective since the qth power map of G is injective. Let $x \in G'$. In G, $\iota(x) = y^q$ for some y, hence $\psi(y)^q = 1$. This implies that $\psi(y) = 1$ and thus $y = \iota(z)$ for some $z \in G'$, and $z^q = x$ since $\iota(z^q) = \iota(x)$. $\qquad\square$

It is convenient to use this result in conjunction with the following observation. Let $Z(G)$ denote the center of a group G. Of course, $Z(G)$ is nontrivial if G is nilpotent.

LEMMA 5.4.5. *If G is a uniquely q-divisible nilpotent group, then $Z(G)$ is a uniquely q-divisible abelian group.*

PROOF. The qth power operation on $Z(G)$ is injective because that is true in G. If $z \in Z(G)$, then $z = y^q$ for a unique $y \in G$. For any $g \in G$,

$$z = y^q = g^{-1}y^q g = (g^{-1}yg)^q$$

and therefore $y = g^{-1}yg$, so that $y \in Z(G)$. Thus the qth power operation on $Z(G)$ is also surjective. □

Using these results, we can generalize our alternative descriptions of what it means for an abelian group to be T-local to the nilpotent case.

LEMMA 5.4.6. *Let G be a nilpotent group. Then G is T-local if and only if G is $\mathbb{Z}_T$-nilpotent.*

PROOF. We prove both implications by induction on the nilpotency class of G, starting with the abelian case.

Suppose first that G is T-local. By the previous lemma, $Z(G)$ is a nonzero T-local abelian group (and thus a $\mathbb{Z}_T$-module). Applying Lemma 5.4.4 to the central extension

$$1 \longrightarrow Z(G) \longrightarrow G \longrightarrow G/Z(G) \longrightarrow 1,$$

we conclude that $G/Z(G)$ is T-local. By the induction hypothesis, this implies that $G/Z(G)$ is $\mathbb{Z}_T$-nilpotent. Therefore, by Lemma 3.1.3, G is also $\mathbb{Z}_T$-nilpotent.

Conversely, suppose that G is $\mathbb{Z}_T$-nilpotent with a $\mathbb{Z}_T$-central series

$$1 = G_q \subset G_{q-1} \subset \cdots \subset G_1 \subset G_0 = G$$

of minimal length. Then G_{q-1} is central in G, and G_{q-1} and G/G_{q-1} are T-local by the induction hypothesis. Therefore, by Lemma 5.4.4, G is also T-local. □

This allows us to generalize Propositions 5.2.5 and 5.2.7 to nilpotent groups.

PROPOSITION 5.4.7. *A nilpotent group G is T-local if and only if the space $K(G, 1)$ is T-local.*

PROOF. If G is T-local and thus $\mathbb{Z}_T$-nilpotent, then Theorem 3.5.4 constructs $K(G, 1)$ as a Postnikov $\mathbb{Z}_T$-tower and thus as a T-local space. Conversely, if

$K(G, 1)$ is T-local, then it is equivalent to its cocellular localization, which is a $\mathbb{Z}_T$-tower and thus displays G as a $\mathbb{Z}_T$-nilpotent group on passage to π_1. □

PROPOSITION 5.4.8. *A homomorphism $\phi \colon G \longrightarrow H$ between nilpotent groups is an algebraic localization at T if and only if the map, unique up to homotopy,*

$$\phi \colon K(G, 1) \longrightarrow K(H, 1)$$

that realizes ϕ on π_1 is a topological localization at T.

PROOF. If $\phi \colon K(G, 1) \longrightarrow K(H, 1)$ is a localization, then specialization of its universal property to target spaces $Z = K(J, 1)$ and passage to fundamental groups shows that $\phi \colon G \longrightarrow H$ satisfies the universal property required of an algebraic localization at T. For the converse, we know by the proof of Theorem 5.3.2 that $K(G, 1)$ has a topological localization $K(G_T, 1)$. If $\phi \colon G \longrightarrow H$ is an algebraic localization at T, then H must be isomorphic to G_T, since both are algebraic localizations, and therefore $K(H, 1)$ must also be a topological localization of $K(G, 1)$. □

In the course of proving that localizations exist, we implicitly proved the following homological result, which generalizes Corollary 5.1.4.

PROPOSITION 5.4.9. *If $\phi \colon G \longrightarrow G_T$ is the localization of a nilpotent group G, then $\phi_* \colon H_*(G; \mathbb{Z}_T) \longrightarrow H_*(G_T; \mathbb{Z}_T)$ is an isomorphism and therefore*

$$\phi^* \colon H^*(G_T; B) \longrightarrow H^*(G; B)$$

is an isomorphism for all T-local abelian groups B.

The exactness of localization also generalizes from abelian to nilpotent groups.

PROPOSITION 5.4.10. *If $1 \longrightarrow G' \longrightarrow G \longrightarrow G'' \longrightarrow 1$ is an exact sequence of nilpotent groups, then*

$$1 \longrightarrow G'_T \longrightarrow G_T \longrightarrow G''_T \longrightarrow 1$$

is an exact sequence.

PROOF. The given exact sequence implies that the homotopy fiber of the evident map $K(G, 1) \longrightarrow K(G'', 1)$ is a $K(G', 1)$. Using a given central series for $K(G, 1)$ and the quotient central series for $K(G'', 1)$, the arguments of the previous section show that we can construct $K(G, 1) \longrightarrow K(G'', 1)$ as the projection of a tower onto a quotient tower. As in Proposition 5.3.4, we can then choose localizations such that $K(G_T, 1) \longrightarrow K(G''_T, 1)$ is also a projection onto a quotient tower and the induced map on fibers is a localization $K(G', 1) \longrightarrow K(G'_T, 1)$. This gives us a fibration

$$K(G'_T, 1) \longrightarrow K(G_T, 1) \longrightarrow K(G''_T, 1).$$

Its long exact sequence of homotopy groups reduces to the claimed short exact sequence. □

This implies the promised generalization of Lemma 5.4.4 to non-central extensions.

COROLLARY 5.4.11. *Let*

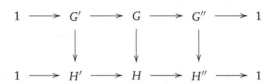

be a commutative diagram of nilpotent groups with exact rows. If any two of G', G, and G'' are T-nilpotent, then so is the third. If any two of the vertical arrows are localizations at T, then so is the third.

PROOF. First let the bottom row be the localization of the top row. If two of G', G, and G'' are T-nilpotent, then two of the vertical arrows are isomorphisms, hence the third vertical arrow is also an isomorphism by the five lemma. Therefore the third group is T-nilpotent. Returning to the general diagram and taking two of its vertical arrows to be localizations at T, we have just shown that the bottom row is an exact sequence of T-local groups. Therefore the map from the top exact sequence to the bottom exact sequence factors through the localization at T of the top sequence. Two of the resulting new vertical arrows are isomorphisms, hence so is the third, and therefore the third vertical arrow of the original diagram is a localization at T. □

5.5. Algebraic properties of localizations of nilpotent groups

We have constructed localizations of nilpotent groups topologically. While that gives the most efficient exposition, it obscures elementwise algebraic properties. We here describe some results that give a better algebraic understanding of localizations of nilpotent groups and of maps between them, focusing primarily on just what we shall need later and proving only those results that play an essential role. Considerably more information appears in both the topological and algebraic literature; see for example [60, 62, 141]. It is convenient to introduce the following notation.

DEFINITION 5.5.1. A T'-number is a product of primes not in T. An element g of a group G is T'-torsion if $g^r = 1$ for some T'-number r; G is a T'-torsion group if all of its elements are T'-torsion. A homomorphism $\alpha: G \longrightarrow H$ is

(i) a T-monomorphism if its kernel is a T'-torsion subgroup;
(ii) a T-epimorphism if for each element $h \in H$, there is an element $g \in G$ and a T'-number r such that $h = \alpha(g)^r$; this implies that the cokernel of α is a T'-torsion group; and
(iii) a T-isomorphism if it is a T-monomorphism and a T-epimorphism.

Observe that the proof of Theorem 5.3.2 shows that the nilpotency class of G_T is at most the nilpotency class of G.

PROPOSITION 5.5.2. *Let G be a nilpotent group and let $\phi: G \longrightarrow G_T$ be its localization at T.*

(i) *ϕ is a T-monomorphism; its kernel is the set of all T'-torsion elements of G, hence this set is a normal subgroup of G.*
(ii) *ϕ is a T-epimorphism; that is, for every element $h \in G_T$, there is an element $g \in G$ and a T'-number r such that $h^r = \phi(g)$.*
(iii) *Every element of G is T'-torsion if and only if $G_T = 1$.*

PROOF. The proof is by induction on the nilpotency class of G. We first show (i) and (ii) for abelian groups and then deduce them for nilpotent groups.

For an abelian group A, $\phi: A \longrightarrow A_T$ is part of the exact sequence

$$ 0 \longrightarrow \mathrm{Tor}\,(A, \mathbb{Z}_T/\mathbb{Z}) \longrightarrow A \xrightarrow{\ \phi\ } A_T \longrightarrow A \otimes (\mathbb{Z}_T/\mathbb{Z}) \longrightarrow 0 $$

that arises from the short exact sequence

$$ 0 \longrightarrow \mathbb{Z} \longrightarrow \mathbb{Z}_T \longrightarrow \mathbb{Z}_T/\mathbb{Z} \longrightarrow 0. $$

All elements of $\mathbb{Z}_T/\mathbb{Z}$ are T'-torsion, so the cokernel $A \otimes \mathbb{Z}_T/\mathbb{Z}$ and kernel $\mathrm{Tor}(A, \mathbb{Z}_T/\mathbb{Z})$ of ϕ are T'-torsion. The conclusion follows easily.

For a nilpotent group G, there is a central series

$$1 = G_q \subset G_{q-1} \subset \cdots \subset G_1 \subset G_0 = G.$$

Since localization is exact, we have a commutative diagram with exact rows

$$
\begin{array}{ccccccccc}
1 & \longrightarrow & G_1 & \longrightarrow & G & \longrightarrow & G/G_1 & \longrightarrow & 1 \\
 & & \phi \downarrow & & \phi \downarrow & & \phi \downarrow & & \\
1 & \longrightarrow & (G_1)_T & \longrightarrow & G_T & \longrightarrow & (G/G_1)_T & \longrightarrow & 1.
\end{array}
$$

By induction, we may assume that the conclusion holds for G_1 and G/G_1. From here, the proof is diagram chasing reminiscent of the proof of the five lemma and of Lemma 5.4.4 above. We leave the details as an exercise for the reader.

For (iii), if $G_T = 1$, then all elements of G are in the kernel of ϕ and G is a T'-torsion group by (i). For the converse, let $h \in G_T$. By (ii), there is an element $g \in G$ and a T'-number r such that $h^r = \phi(g)$. Since G is a T'-torsion group, there is also a T'-number s such that $g^s = 1$. Then $h^{rs} = \phi(g^s) = 1$, hence $h = 1$ by Remark 5.4.2. $\qquad\square$

COROLLARY 5.5.3. *Let $\alpha \colon G \longrightarrow H$ be a homomorphism from a nilpotent group G to a T-local group H. Then α is a localization at T if and only if it is a T-isomorphism.*

PROOF. The forward implication is given by parts (i) and (ii) of Proposition 5.5.2, and the converse follows from (iii), which shows that the localizations at T of the kernel and cokernel of α are trivial. $\qquad\square$

With a little more work, one can prove the following more general result.

PROPOSITION 5.5.4. *Let $\alpha \colon G \longrightarrow H$ be a homomorphism between nilpotent groups and let $\alpha_T \colon G_T \longrightarrow H_T$ be its localization.*

(i) α_T is a monomorphism if and only if α is a T-monomorphism.
(ii) α_T is an epimorphism if and only if α is a T-epimorphism.
(iii) α_T is an isomorphism if and only if α is a T-isomorphism.

SKETCH PROOF. One would like to say that since localization is exact, the sequence

$$1 \longrightarrow (\ker \alpha)_T \longrightarrow G_T \longrightarrow H_T \longrightarrow (\operatorname{coker} \alpha)_T \longrightarrow 1$$

is exact, so that the conclusions follow from Proposition 5.5.2(iii). However, in our non-abelian situation, we must take into account that the two implicit short exact sequences do not splice, since the image of α is not a normal subgroup of H. Part (i) works naively but part (ii) needs more a little more work to show how the notion of T-epimorphism circumvents this problem. Details are given in [60, Cor. 6.4]. The key element of the proof is the following group theoretic observation, which is [60, Thm. 6.1]. We shall have another use for it shortly. □

Recall that the lower central series of G is defined by $\Gamma^1(G) = G$ and, inductively, $\Gamma^{j+1}(G) = [G, \Gamma^j(G)]$.

LEMMA 5.5.5. *Let g and h be elements of a group G. If $h^q = 1$, then*

$$(gh)^{q^j} \equiv g^{q^i} \mod \Gamma^{j+1}(G).$$

Therefore, if G is nilpotent of nilpotency class c, then $(gh)^{q^c} = g^{q^c}$.

The following observations will play a role in proving the fracture theorems for localization. Recall that localization commutes with finite products. It does not commute with infinite products in general, but we have the following observation.

LEMMA 5.5.6. *If G_i is a T-local group for all elements of an indexing set I, then $\prod_{i \in I} G_i$ is T-local.*

PROOF. A group G is T-local if and only if the qth-power function $G \longrightarrow G$ is a bijection for all q not in T. A product of bijections is a bijection. □

LEMMA 5.5.7. *Localization at T commutes with pullbacks.*

PROOF. Let G and H be the pullbacks displayed in the following commutative diagram, where γ is obtained by the universal property of pullbacks.

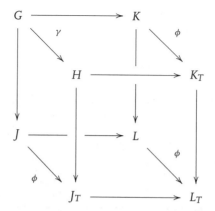

The claim is that γ is a localization of G at T. When J, K, and L are abelian, G is the kernel of the difference map $J \times K \longrightarrow H$ and the conclusion follows by the exactness of localization. A different proof applies in general. For a prime $q \notin T$, the qth power function on $H \subset J_T \times K_T$ is a bijection since $(x, y) \in H$ if and only if $(x^q, y^q) \in H$. Therefore H is T-local. We now use Corollary 5.5.3.

Let $(j, k) \in G$. If $\gamma(j, k) = 1$, then $\phi(j) = 1 \in J_T$ and $\phi(k) = 1 \in K_T$, hence $j^r = 1$ and $k^s = 1$ for T'-numbers r and s. Therefore $(j, k)^{rs} = 1$. This implies that the kernel of γ is the T'-torsion subgroup of G, so that γ is a T-mono-morphism. If $(x, y) \in H$, we must have $x^q = \phi(j') \in J_T$ and $y^r = \phi(k') \in K_T$ for some $j' \in J$, $k' \in K$ and T'-numbers q and r. Let $s = qr$, $j = (j')^r$, and $k = (k')^q$. Then $x^s = \phi(j)$ and $y^s = \phi(k)$. Let m and n denote the images of j and k in L. These may not be equal, but $\ell = m^{-1}n$ maps to 1 in L_T, hence there is a T'-number u such that $\ell^u = 1$. Of course, $n = m\ell$. Lemma 5.5.5 shows that $n^{u^c} = m^{u^c}$, where c is the nilpotency class of L. Therefore (j^{u^c}, k^{u^c}) is in G and γ maps it to $(x, y)^{su^c}$ in H. This proves that γ is a T-epimorphism. $\quad\square$

We round out our discussion with some discussion of the behavior of local-ization with respect to various central series of a nilpotent group G. We first record a direct consequence of our cocellular construction of localizations.

PROPOSITION 5.5.8. *For any central series*

$$\{1\} = G_q \subset G_{q-1} \subset \cdots \subset G_0 = G$$

of G, the localization $\phi\colon G \longrightarrow G_T$ passes to subquotients to give localizations

$$G_j/G_{j+1} \longrightarrow (G_j/G_{j+1})_T, \quad G_j \longrightarrow (G_j)_T \quad and \quad G/G_j \longrightarrow (G/G_j)_T$$

for $1 \leq j$.

This should make the following pair of results not too surprising. For complete proofs, see [60, Thms. 5.6, 5.9]. Note that the lower central series is functorial in G. Using that together with the two out of three result in Corollary 5.4.11, we obtain an inductive proof of the following result.

PROPOSITION 5.5.9. *The localization* $\phi : G \longrightarrow G_T$ *at T passes to subquotients to give localizations*

$$\Gamma^j(G) \longrightarrow \Gamma^j(G_T) \quad and \quad G/\Gamma^j(G) \longrightarrow G_T/\Gamma^j(G_T)$$

for $1 \leq j$.

Recall that the upper central series of G is defined inductively by letting $Z_0(G) = 1$, letting $Z_1(G)$ be the center of G, and letting $Z_{j+1}(G)/Z_j(G)$ be the center of $G/Z_j(G)$. Note that since the center is not functorial, neither is the upper central series. Proposition 4.5.9 gives the starting point for the following result, and we shall prove part of it in Lemma 5.6.6 below.

PROPOSITION 5.5.10. *If H is T-local, then so is $Z_j(H)$ for $j \geq 1$. The localization $\phi : G \longrightarrow G_T$ restricts to maps $Z_j(G)$ to $Z_j(G_T)$ and, if G is finitely generated, these restrictions are localizations for $1 \leq j$.*

5.6. Finitely generated T-local groups

It is often necessary to restrict attention to finitely generated modules over the principal ideal domain $\mathbb{Z}_T$, and we need the appropriate notion of finite generation for T-local groups. Although the ideas are clear enough, we have not found a treatment adequate for our purposes in either the algebraic or the topological literature. We prove what we need in this section, leaving some details to the algebraic literature. Our purpose is to say just enough to lay the groundwork for the later fracture theorems.

Recall that an $f\mathbb{Z}_T$-nilpotent group G is one that admits a $\mathbb{Z}_T$-central series whose subquotients are finitely generated $\mathbb{Z}_T$-modules. It is immediate from our cocellular construction of localizations that G_T is $f\mathbb{Z}_T$-nilpotent if G is f-nilpotent. We gave several equivalent conditions for a group to be f-nilpotent in Proposition 4.5.9, and we shall prove a T-local analogue of that result. We shall use it to characterize $f\mathbb{Z}_T$-nilpotent spaces in Theorem 6.1.4, in analogy with our characterization of f-nilpotent spaces in Theorem 4.5.2.

Remember that we require T-local groups to be nilpotent and not just uniquely q-divisible for primes not in T. We shall need the following definition.

DEFINITION 5.6.1. For a subgroup H of a group G, define H_T^G to be the set of elements $g \in G$ such that $g^r \in H$ for some T'-number r.[1]

The following result is due to Warfield [141, Thm. 3.25]; we shall not repeat its proof.

LEMMA 5.6.2. *For any subgroup H of a nilpotent group G, the set H_T^G is a subgroup of G.*

For a subgroup H of a T-local group G, H_T^G can be characterized as the smallest T-local subgroup of G that contains H. By Proposition 5.5.2, if $\phi \colon G \longrightarrow G_T$ is a localization of G, then $G_T = \phi(G)_T^G$. We adopt the following terminology.

DEFINITION 5.6.3. If H is the subgroup of a nilpotent group G generated by a set S, then we call H_T^G the subgroup of G T-generated by S. We say that G is T-generated by S if $G = H_T^G$. We say that G is finitely T-generated if it has a finite set of T-generators.

By an easy clearing of denominators argument, a finitely generated $\mathbb{Z}_T$-module is a finitely T-generated abelian group. The following analogue of Proposition 4.5.9 admits a similar proof. However, since the details are not obvious and the result is not in the literature, we shall give the proof. We will need the following notion.

DEFINITION 5.6.4. A group G is T-supersolvable if it has a finite normal series of subgroups that are uniquely q-divisible for $q \notin T$ and whose successive subquotients are cyclic $\mathbb{Z}_T$-modules.

PROPOSITION 5.6.5. *Let G be a T-local group. Then the following statements are equivalent.*

(i) *G is $f\mathbb{Z}_T$-nilpotent.*
(ii) *$G/([G, G]_T^G)$ is finitely T-generated.*
(iii) *G is finitely T-generated.*
(iv) *Every T-local subgroup of G is finitely T-generated.*

---

1. In [78, 141], H_T^G is called the T'-isolator of H, where T' is the set of primes not in T.

PROOF. Write $G' = [G, G]_T^G$. It is easily checked that G' is a normal subgroup of G, and we let $\pi: G \longrightarrow G/G'$ be the quotient homomorphism. Certainly (i) implies (ii). Assume (ii). To prove (iii), choose a finite set S of elements h_i of G such that the elements $\pi(h_i)$ T-generate G/G', let H be the subgroup of G generated by S and let $H_T^G \supset H$ be the subgroup of G T-generated by S. The restriction of π to H_T^G is an epimorphism. Indeed, if $k \in G/G'$, then there is a T'-number r such that k^r is in the subgroup generated by the $\pi(h_i)$, say $k^r = \pi(h)$ for $h \in H$. Since G is T-local, there is a unique $j \in G$ such that $j^r = h$, and then $\pi(j) = k$ since $\pi(j)^r = k^r$. It follows exactly as in the proof of Proposition 4.5.9 that $G = G'H_T^G$. By Lemma 5.6.7 below, this implies that $H_T^G = G$.

The proof that (iii) implies (iv) is a modification of the proof in [56, Thm. 10.2.4] of the corresponding implication of Proposition 4.5.9. It proceeds in two steps. First, by Lemma 5.6.8 below, if G is finitely T-generated, then G is T-supersolvable. Second, by Lemma 5.6.9 below, every T-local subgroup of a T-supersolvable group is T-supersolvable and therefore T-finitely generated. Finally, if (iv) holds, then the center $Z(G)$ is T-local by Lemma 5.4.5 and is therefore finitely T-generated. Since G is finitely T-generated, so is $G/Z(G)$. By induction on the nilpotency class, we may assume that $G/Z(G)$ is $f\mathbb{Z}_T$-nilpotent. Therefore G is $f\mathbb{Z}_T$-nilpotent. $\square$

We must prove the lemmas quoted in the proof just given. We use the upper central series, and we usually abbreviate $Z_j(G)$ to Z_j. It is standard group theory that $Z_q = G$ if and only if G is nilpotent of nilpotency class q. The following observation is false for the lower central series since $[G, G]$ need not be T-local when G is T-local.

LEMMA 5.6.6. *If G is nilpotent, then G is $\mathbb{Z}_T$-nilpotent if and only if each Z_j is $\mathbb{Z}_T$-nilpotent or, equivalently, each Z_j/Z_{j-1} is a T-local abelian group.*

PROOF. By induction on the nilpotency class of G, this is immediate from Lemmas 5.4.4 and 5.4.5. To see this, it helps to observe that

$$Z_{j-1}(G/Z_1(G)) = Z_j(G)/Z_1(G). \qquad \square$$

LEMMA 5.6.7. *Let G be T-local and suppose that $G = [G, G]_T^G J$, where $J \subset G$ is T-local. Then $J = G$.*

PROOF. Let $J_0 = J$ and $J_{i+1} = Z_{i+1}J_i$. Then J_i is a normal subgroup of J_{i+1}. Indeed, for $z \in Z_{i+1}$ and $g \in G$ $[z, g] = zgz^{-1}g^{-1}$ is in Z_i. When $g \in J_i$, this gives that $zgz^{-1} = [z, g]g$ is in $Z_iJ_i = J_i$, the last equality holding since $J_i \supset Z_i$. Suppose that J is a proper subgroup of G. Since $Z_m = G$ for some m, there is an i such that $J_{i+1} = G$ but $J_i \neq G$. The quotient G/J_i is a T-local abelian group, since it is isomorphic to $Z_{i+1}/J_i \cap Z_{i+1}$, and it follows that $J_i \supset [G, G]_T^G$. Therefore

$$G = [G, G]_T^G J \subset [G, G]_T^G J_i = J_i,$$

which is a contradiction. $\qquad\square$

LEMMA 5.6.8. *A finitely T-generated T-local group G is T-supersolvable.*

PROOF. Let H be the subgroup generated (in the usual sense) by a finite set of T-generators, so that $G = H_T^G$. Since G is nilpotent, so is H, by Lemma 3.1.3. By Proposition 4.5.9, $Z(H)$ is finitely generated. We claim that $Z(G) = Z(H)_T^G$ and is therefore finitely T-generated. To see this, let $z \in Z(H)$ and $g \in G$. There is a T'-number r such that $g^r \in H$, and then $g = zgz^{-1}$ since $g^r = zg^r z^{-1} = (zgz^{-1})^r$. This shows that $Z(H) \subset Z(G)$, and therefore $Z(H) = Z(G) \cap H$. It follows that

$$Z(H)_T^G = (Z(G) \cap H)_T^G = Z(G)_T^G \cap H_T^G = Z(G) \cap G = Z(G).$$

The second equality holds since it is easily checked that $(J \cap H)_T^G = J_T^G \cap H_T^G$ for any pair of subgroups of a T-local group, and the third equality holds by Lemma 5.4.5. Now $Z(G)$ is a finitely generated $\mathbb{Z}_T$-module and is thus a finite direct sum of cyclic $\mathbb{Z}_T$-modules, so $Z(G)$ is obviously T-supersolvable. We may assume by induction on the nilpotency class that $G/Z(G)$ is T-supersolvable, and it follows that G is T-supersolvable. $\qquad\square$

LEMMA 5.6.9. *Let G be a T-supersolvable group. Then G is finitely T-generated. If H is a subgroup that is uniquely q-divisible for $q \notin T$, then H is T-supersolvable. If H is also normal in G, then G/H is T-supersolvable.*

PROOF. Consider a finite normal series

$$\{1\} = G_m \subset G_{m-1} \subset \cdots \subset G_0 = G$$

such that the G_i are uniquely q-divisible for $q \notin T$ and the G_{i+1}/G_i are cyclic $\mathbb{Z}_T$-modules. Inductively, G_1 is T-generated by $m - 1$ elements, and these elements together with an element that projects to a $\mathbb{Z}_T$-generator of G/G_1 T-generate G. The intersections $H \cap G_i$ give a similar normal series for H and, if H is normal, the images of the G_i in G/H give a similar normal series for G/H. The essential point is just that submodules and quotient modules of cyclic $\mathbb{Z}_T$-modules are cyclic. $\qquad\square$

REMARK 5.6.10. The algebra of T-local and especially finitely T-generated T-local groups deserves more algebraic study than it has yet received. For just one example of a further result that seems not to be in the literature, one can prove by the methods of [78, §67] that the normalizer of a T-local subgroup of a T-local group is again T-local.

6

CHARACTERIZATIONS AND PROPERTIES OF LOCALIZATIONS

We give several characterizations of localizations in §6.1, and we use these to study the homotopical behavior of localization with respect to standard constructions on based spaces in §§6.2–6.4. Here §6.2 deals with limits and fibrations, §6.3 deals with function spaces, and §6.4 deals with colimits and cofiber sequences. The commutation relations in §6.4 lead to a dual cellular construction of localizations of simply connected spaces, as we explain in §6.5. We give still other constructions of localizations of H-spaces and co-H-spaces in §6.6, and we show that localizations preserve such structure. Finally, in §6.7, we discuss rationalization. In particular, we calculate the rationalizations of spheres and use the result to give a quick proof of Serre's theorem about the finiteness of the homotopy groups of spheres.

6.1. Characterizations of localizations of nilpotent spaces

In order to understand the behavior of various space-level constructions with respect to localization, we need to show that several alternative conditions on a map are equivalent to its being a localization. We state two omnibus theorems for ease of reference and then proceed to their proofs.

THEOREM 6.1.1. *The following properties of a nilpotent space Z are equivalent, and they hold if and only if Z is T-local.*

(i) *Z is a $\mathbb{Z}_T$-nilpotent space.*
(ii) *$\xi^*: [Y, Z] \longrightarrow [X, Z]$ is a bijection for every $\mathbb{Z}_T$-equivalence $\xi: X \longrightarrow Y$.*
(iii) *Each $\pi_n Z$ is a T-local group (nilpotent if $n = 1$, abelian if $n > 1$).*
(iv) *Each $\tilde{H}_n(Z; \mathbb{Z})$ is a T-local abelian group.*

THEOREM 6.1.2. *For a nilpotent space X, the following properties of a map $\phi: X \longrightarrow Y$ from X to a T-local space Y are equivalent. There exists one and, up to homotopy, only one such map, namely the localization $X \longrightarrow X_T$.*

111

(i) $\phi^*: [Y, Z] \longrightarrow [X, Z]$ is an isomorphism for all T-local spaces Z.

(ii) ϕ is a $\mathbb{Z}_T$-equivalence.

(iii) $\phi_*: \pi_n X \longrightarrow \pi_n Y$ is localization at T for $n \geq 1$.

(iv) $\phi_*: \tilde{H}_n(X; \mathbb{Z}) \longrightarrow \tilde{H}_n(Y; \mathbb{Z})$ is localization at T for $n \geq 1$.

In Theorem 6.1.1, (ii) is the definition of what it means to be T-local. In Theorem 6.1.2, (ii) is the definition of what it means for ϕ to be a localization at T, and we have already proven the existence and uniqueness of such a localization. Thus in both results, it suffices to prove the equivalence of (ii) with the remaining properties. It is noteworthy in all three results that the actions of fundamental groups on higher homotopy groups are not mentioned. In particular, the implication (iii) $\implies$ (i) in Theorem 6.1.1 shows that if the groups $\pi_n Z$ are T-local, then $\pi_1 Z$ must act $\mathbb{Z}_T$-nilpotently on them. We comment on the meaning of (iv), which will have no analogue in the case of completions, and interpolate an important variant of Theorem 6.1.1 before proceeding to the proofs.

REMARK 6.1.3. Since $\mathbb{Z}_T$ is $\mathbb{Z}$-flat, Tor $(A, \mathbb{Z}_T) = 0$ for all abelian groups A and the universal coefficient theorem gives an isomorphism

$$\alpha: \tilde{H}_n(X; \mathbb{Z}) \otimes \mathbb{Z}_T \longrightarrow \tilde{H}_n(X; \mathbb{Z}_T).$$

Moreover, under this isomorphism the canonical map

$$\beta: \tilde{H}_*(X; \mathbb{Z}) \longrightarrow \tilde{H}_*(X; \mathbb{Z}_T)$$

induced by $\mathbb{Z} \longrightarrow \mathbb{Z}_T$ coincides with the localization homomorphism. Therefore (iv) in Theorem 6.1.1 is equivalent to the assertion that β is an isomorphism when X is T-local. By the naturality of α and β applied to the map ϕ, this implies that (iv) in Theorem 6.1.2 is equivalent to the assertion that

$$\phi_*: \tilde{H}_*(X; \mathbb{Z}_T) \longrightarrow \tilde{H}_*(X_T; \mathbb{Z}_T)$$

is an isomorphism.

THEOREM 6.1.4. *The following properties of a nilpotent space Z are equivalent.*

(i) Z is an $f \mathbb{Z}_T$-nilpotent space.

(ii) Each $\pi_n(Z)$ is a finitely T-generated T-local group.

(iii) $H_i(Z; \mathbb{Z})$ is a finitely generated $\mathbb{Z}_T$-module for each $i \geq 1$.

REMARK 6.1.5. We shall develop a theory of "T-CW complexes" in §6.5, but only for simply connected T-local spaces. With that theory in place, we can prove as in Theorem 4.5.2 that a simply connected space is $f \mathbb{Z}_T$-nilpotent if and only if it has the weak homotopy type of a T-CW complex with finite skeleta. We say that a space weakly equivalent to such a space is simply connected of finite T-type.

PROOF OF THEOREM 6.1.1. We proceed step by step.

(i) $\Longrightarrow$ *(ii)*. Since a $\mathbb{Z}_T$-nilpotent space is weakly equivalent to a Postnikov $\mathbb{Z}_T$-tower, this is a special case of Theorem 5.3.1.

(ii) $\Longrightarrow$ *(i)*. Since Z is T-local, its localization $\phi: Z \longrightarrow Z_T$ must be a weak equivalence, by the uniqueness of localizations. By our cocellular construction, Z_T is a Postnikov $\mathbb{Z}_T$-tower and is therefore $\mathbb{Z}_T$-nilpotent.

(i) $\Longrightarrow$ *(iii)*. If Z is $\mathbb{Z}_T$-nilpotent, then $\pi_n Z$ is a $\mathbb{Z}_T$-nilpotent $\pi_1 Z$-group for $n \geq 1$ and is thus T-local by Lemma 5.4.6 if $n = 1$ and by Lemma 5.1.1 if $n > 1$.

(iii) $\Longrightarrow$ *(i)*. We can prove this algebraically or topologically. Algebraically, since each $\pi_n Z$ is T-local, $\pi_1 Z$ is a $\mathbb{Z}_T$-nilpotent $\pi_1 Z$-group by Lemma 5.4.6 and $\pi_n Z$ for $n > 1$ is a $\mathbb{Z}_T$-nilpotent $\pi_1 Z$-group by Lemmas 4.1.1 and 4.1.2. Topologically, since Z is nilpotent, it has a cocellular localization $\phi: Z \to Z_T$ in which Z_T is $\mathbb{Z}_T$-nilpotent. By (iii) of Theorem 6.1.2, proven below, $\phi_*: \pi_n Z \longrightarrow \pi_n Z_T$ is localization at T and is thus an isomorphism. Therefore ϕ is a weak equivalence and Z is $\mathbb{Z}_T$-nilpotent.

(i) $\Longrightarrow$ *(iv)*. When $Z = K(B, n)$ for a $\mathbb{Z}_T$-module B, (iv) holds by Theorem 5.2.8. For the general case, we may assume that Z is a Postnikov $\mathbb{Z}_T$-tower $Z = \lim Z_i$, where Z_{i+1} is the fiber of a map $k_i: Z_i \longrightarrow K(B_i, n_i + 1)$ and B_i is a $\mathbb{Z}_T$-module. Using the form of (iv) given in Remark 6.1.3, the map from the Serre spectral sequence

$$H_p(Z_i; H_q(K(B_i, n_i); \mathbb{Z})) \Longrightarrow H_{p+q}(Z_{i+1}; \mathbb{Z})$$

to the Serre spectral sequence

$$H_p(Z_i; H_q(K(B_i, n_i); \mathbb{Z}_T)) \Longrightarrow H_{p+q}(Z_{i+1}; \mathbb{Z}_T)$$

shows that (iv) for Z_i implies (iv) for Z_{i+1}. On passage to limits, we see by Proposition 2.5.9 that (iv) holds for Z.

(iv) $\Longrightarrow$ *(i)*. We have the localization $\phi: Z \longrightarrow Z_T$. We shall shortly prove the implication (ii) $\Longrightarrow$ (iv) in Theorem 6.1.2, and this gives that

$$\phi_*: \tilde{H}_*(Z; \mathbb{Z}) \longrightarrow \tilde{H}_*(Z_T; \mathbb{Z})$$

is localization at T. By our assumption (iv), the domain here is T-local, so that this localization is an isomorphism. Since Z and Z_T are nilpotent spaces, ϕ is a weak equivalence by Theorem 3.3.8 and therefore Z, like Z_T, is $\mathbb{Z}_T$-nilpotent. $\qquad\square$

PROOF OF THEOREM 6.1.4. Proposition 4.5.9 implies that (i) and (ii) are equivalent. One can see that (i) implies (iii) by inductive use of the Serre spectral sequences of the stages in a Postnikov tower of X. Changing the ground ring from $\mathbb{Z}$ to $\mathbb{Z}_T$, the proof that (iii) implies (ii) is the same as the proof that (iv) implies (iii) in Theorem 4.5.2. $\qquad\square$

PROOF OF THEOREM 6.1.2. Again, we proceed step by step.

(ii) $\Longrightarrow$ (i). Since T-local spaces are weakly equivalent to Postnikov T-towers, this is a special case of Theorem 5.3.1.

(i) $\Longrightarrow$ (ii). Since $K(B, n)$ is T-local for a $\mathbb{Z}_T$-module B, this implication is immediate from the representability of cohomology.

(ii) $\Longrightarrow$ (iii) and (iv). By the uniqueness of localizations, it is enough to prove that our cocellular localization $\phi\colon X \longrightarrow X_T$ satisfies (iii) and (iv). Thus we assume that X is a Postnikov tower $\lim X_i$ constructed from maps $k_i\colon X_i \longrightarrow K(A_i, n_i + 1)$, where A_i is an abelian group, $n_{i+1} \geq n_i \geq 1$, and only finitely many $n_i = n$ for each $n \geq 1$. Construct X_T by Theorem 5.3.2. On passage to homotopy groups, the map of fibrations constructed in the proof of Theorem 5.3.2 gives a map of short exact sequences

$$
\begin{array}{ccccccccc}
1 & \longrightarrow & A_i & \longrightarrow & \pi_{n_i} X_{i+1} & \longrightarrow & \pi_{n_i} X_i & \longrightarrow & 1 \\
& & \downarrow{\scriptstyle \phi} & & \downarrow{\scriptstyle \phi_{i+1*}} & & \downarrow{\scriptstyle \phi_{i*}} & & \\
1 & \longrightarrow & (A_i)_T & \longrightarrow & \pi_{n_i}[(X_{i+1})_T] & \longrightarrow & \pi_{n_i}[(X_i)_T] & \longrightarrow & 1.
\end{array}
$$

By construction, the groups in the lower sequence are T-local. Since the left and right vertical arrows are localizations at T, so is the middle arrow, by Corollary 5.4.11. Inductively, this proves that (iii) holds. To see that (iv) holds in the form given in Remark 6.1.3, we just repeat the proof of Theorem 5.3.2 using homology rather than cohomology.

(iv) $\Longrightarrow$ (ii). This is an application of the universal coefficient theorem. Let C be T-local and consider the map of exact sequences

$$0 \longrightarrow \text{Ext}(H_{n-1}(X_T; \mathbb{Z}_T), C) \longrightarrow H^n(X_T; C) \longrightarrow \text{Hom}(H_n(X_T; \mathbb{Z}_T), C) \longrightarrow 0$$

$$0 \longrightarrow \text{Ext}(H_{n-1}(X; \mathbb{Z}_T), C) \longrightarrow H^n(X; C) \longrightarrow \text{Hom}(H_n(X; \mathbb{Z}_T), C) \longrightarrow 0$$

induced by ϕ. By assumption and Remark 6.1.3, the left and right vertical maps are isomorphisms, hence the middle vertical map is also an isomorphism and ϕ is a $\mathbb{Z}_T$-equivalence.

(iii) $\Longrightarrow$ *(ii)*. Assuming that $\phi_* \colon \pi_n X \longrightarrow \pi_n Y$ is a localization at T for every $n \geq 1$, we must prove that ϕ is a $\mathbb{Z}_T$-equivalence. This is given by Proposition 5.4.8 when $X = K(G, 1)$ for a nilpotent group G and by Theorem 5.2.8 when $X = K(A, n)$ for an abelian group A.

We first deal with the case when X and Y are simple spaces and then use universal covers to deal with the general case. Thus suppose that X and Y are simple. We may assume that they are simple Postnikov towers and that ϕ is cocellular. Then $X = \lim X_i$ is defined by k-invariants $k^{i+2} \colon X_i \longrightarrow K(\pi_{i+1}X, i+2)$, and similarly for Y. The map ϕ induces maps of fibration sequences.

$$
\begin{array}{ccccc}
K(\pi_{i+1}X, i+1) & \longrightarrow & X_{i+1} & \longrightarrow & X_i \\
\downarrow & & \downarrow & & \downarrow \\
K(\pi_{i+1}Y, i+1) & \longrightarrow & Y_{i+1} & \longrightarrow & Y_i
\end{array}
$$

Let B be a $\mathbb{Z}_T$-module. We have Serre spectral sequences of the form

$$E_2^{p,q} = H^p(X_i; H^q(K(\pi_{i+1}X, i+1); B)) \Longrightarrow H^{p+q}(X_{i+1}; B)$$

and similarly for the Y_i. Since the base spaces are simple, the local systems are trivial. By induction and the case of Eilenberg-Mac Lane spaces, the induced map of E_2 terms is an isomorphism and therefore so is the map $H^*(Y_{i+1}; B) \longrightarrow H^*(X_{i+1}; B)$. Passing to limits, we conclude from Proposition 2.5.9 that ϕ is a $\mathbb{Z}_T$-equivalence. Now consider the general case. We have a map of fibrations

$$
\begin{array}{ccccc}
\tilde{X} & \longrightarrow & X & \longrightarrow & K(\pi_1 X, 1) \\
\downarrow & & \downarrow & & \downarrow \\
\tilde{Y} & \longrightarrow & Y & \longrightarrow & K(\pi_1 Y, 1).
\end{array}
$$

The Serre spectral sequence for X has the form

$$E_2^{p,q} = H^p(K(\pi_1 X, 1); H^q(\tilde{X}; B)) \implies H^{p+q}(X; B)$$

and similarly for Y. Here the action of the fundamental group of the base space on the cohomology of the fiber can be nontrivial. By the case of $K(G, 1)$'s,

$$\phi^*: H^*(K(\pi_1 Y, 1); B) \longrightarrow H^*(K(\pi_1 X, 1); B)$$

is an isomorphism for all $\mathbb{Z}_T$-modules B. By Proposition 4.2.1, $H^q(\tilde{Y}; B)$ is a $\mathbb{Z}_T$-nilpotent $\pi_1 Y$-module. By Lemma 4.1.6, this implies that

$$\phi^*: H^*(\pi_i Y, H^p(\tilde{Y}; B)) \longrightarrow H^*(\pi_1 X, H^p(\tilde{Y}; B))$$

is an isomorphism, where $\pi_1 X$ acts on $H^q(\tilde{Y}; B)$ through ϕ. By the previous step, $\phi^*: H^q(\tilde{Y}; B) \longrightarrow H^q(\tilde{X}; B)$ is an isomorphism, and the actions of $\pi_1 X$ are the same on the source and target. Therefore the map

$$\phi^*: H^*(\pi_1 Y, H^*(\tilde{Y}; B)) \longrightarrow H^*(\pi_1 X, H^*(\tilde{X}; B))$$

of E_2 terms is an isomorphism and ϕ is a $\mathbb{Z}_T$-equivalence. $\square$

6.2. Localizations of limits and fiber sequences

The characterizations of localizations imply numerous basic commutation relations between localization and familiar topological constructions. We gave some such results using the explicit cocellular construction of localizations in Proposition 5.3.4 and Corollaries 5.3.5 and 5.3.6. Their homotopical versions can be proven either directly from the characterizations or by approximating given nilpotent spaces and maps by Postnikov towers and cocellular maps. The latter approach leads to the following homotopical observation. Recall the notion of a $\mathbb{Z}_T$-map from Definition 4.3.2.

LEMMA 6.2.1. *If $f: X \longrightarrow Y$ is a map between nilpotent spaces, then its localization $f_T: X_T \longrightarrow Y_T$ is a $\mathbb{Z}_T$-map.*

PROOF. By Theorem 3.5.4 we may assume that X and Y are Postnikov towers and that f is a cocellular map. We may then construct $\phi_X: X \longrightarrow X_T$ and $\phi_Y: Y \longrightarrow Y_T$ by Theorem 5.3.2 and construct f_T by Theorem 5.3.3, so that it too is a cocellular map. Since a map of abelian groups between T-local abelian groups is a map of $\mathbb{Z}_T$-modules, by Lemma 4.1.2, the conclusion follows. $\square$

REMARK 6.2.2. Recall Notations 4.5.1 and 4.3.3. There are precise analogues for f-nilpotent and $f\mathbb{Z}_T$-nilpotent spaces for all results in this section and the next. That is, if we start with finitely generated input in any of our results, then we obtain finitely generated output. On the fundamental group level, this relies on Proposition 5.6.5.

We first record the homotopical versions of commutation results that we have already seen cocellularly and then give a stronger result about fibrations.

PROPOSITION 6.2.3. *If X and Y are nilpotent spaces, then $(X \times Y)_T$ is naturally equivalent to $X_T \times Y_T$.*

PROPOSITION 6.2.4. *If X is nilpotent and $\Omega_0(X)$ denotes the basepoint component of ΩX, then $(\Omega_0 X)_T$ is naturally equivalent to $\Omega_0(X_T)$.*

PROOF. This is immediate by inspection of homotopy groups. Alternatively, observe that $\Omega_0 X$ is equivalent to $\Omega \tilde{X}$, where $\tilde{X}$ is the universal cover of X, and apply the cocellular version to $\tilde{X}$. □

We state the following result in terms of homotopy pullbacks, as defined in Definition 2.2.1, rather than fibrations. The conclusion is that localization commutes with homotopy pullbacks. This will play a key role in the fracture theorems for localization.

PROPOSITION 6.2.5. *Let $f: X \longrightarrow A$ and $g: Y \longrightarrow A$ be maps between nilpotent spaces, let $N_0(f, g)$ be the basepoint component of the homotopy pullback $N(f, g)$, and let f_T and g_T be localizations of f and g at T.*

(i) $N_0(f, g)$ is nilpotent.
(ii) If $N(f, g)$ is connected, then $N(f_T, g_T)$ is connected.
(iii) $N_0(f, g)$ is T-local if X, Y, and A are T-local.
(iv) $N_0(f_T, g_T)$ is a localization at T of $N_0(f, g)$.

PROOF. Part (i) holds by the case $\mathscr{C} = \mathscr{A}b$ of Proposition 4.4.3. For part (ii), Corollary 2.2.3 shows how to determine connectivity by the tail end of an exact sequence, and the exactness of localization of nilpotent groups gives the conclusion. Since a space is $\mathbb{Z}_T$-nilpotent if and only if it is T-local and nilpotent, we have a choice of proofs for part (iii), depending on whether or not we want to use the notion of a $\mathbb{Z}_T$-map. While the homotopy groups of a homotopy pullback are not the pullbacks of the corresponding homotopy groups in general, the description given in Corollary 2.2.3, together with Lemma 5.5.7, implies that the homotopy groups of $N_0(f,g)$ are T-local. Alternatively, we can use the case $\mathscr{C} = \mathscr{A}_{\mathbb{Z}_T}$ of Proposition 4.4.3 to prove directly that $N(f,g)$ is $\mathbb{Z}_T$-nilpotent. Using Corollary 5.4.11, part (iv) follows by comparison of the long exact sequences of homotopy groups for $N_0(f,g)$ and $N_0(f_T,g_T)$ given in Corollary 2.2.3. □

THEOREM 6.2.6. *Let $f\colon X \longrightarrow Y$ be a map to a connected space Y such that Y and all components of X are nilpotent. Let $F = Ff$. Then each component of F is nilpotent and there is a homotopy commutative diagram*

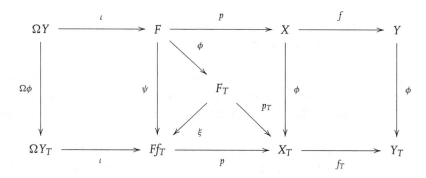

with the following properties.

(i) *The map $\phi\colon Y \longrightarrow Y_T$ is a localization at T.*

(ii) *The maps $\phi\colon X \longrightarrow X_T$ and $\phi\colon F \longrightarrow F_T$ are the disjoint unions of localizations at T of the components of X and F defined using any (compatible) choices of basepoints in these components.*

(iii) *The rows are canonical fiber sequences.*

(iv) *The restriction of ψ to a map from a component of F to the component of its image is a localization at T.*

(v) *The map $\xi\colon F_T \longrightarrow Ff_T$ is an equivalence to some of the components of Ff_T.*

(vi) *Fix $x \in X$, let $y = f(x) \in Y$, and assume that the images of*

$$f_*\colon \pi_1(X,x) \longrightarrow \pi_1(Y,y) \quad \text{and} \quad f_{T_*}\colon \pi_1(X_T,\phi(x)) \longrightarrow \pi_1(Y_T,\phi(y))$$

are normal subgroups. Then

(a) *the quotient group $\tilde{\pi}_0(F)$ is nilpotent;*

(b) *the quotient group $\tilde{\pi}_0(Ff_T)$ is $\mathbb{Z}_T$-nilpotent; and*

(c) *$\psi_*\colon \tilde{\pi}_0(F) \longrightarrow \tilde{\pi}_0(Ff_T)$ is a localization at T.*

PROOF. We may as well replace f by its mapping path fibration. We see that the fiber $F = Ff$ is nilpotent by Proposition 4.4.1 or 4.4.2. We start the construction of the diagram with the upper fibration sequence and chosen localizations ϕ of Y and the components of X and F. By the universal property, applied one component of X and F at a time, there are maps $f_T \colon X_T \longrightarrow Y_T$ and $p_T \colon F_T \longrightarrow X_T$, unique up to homotopy, such that $f_T \circ \phi \simeq \phi \circ f$ and $p_T \circ \phi \simeq \phi \circ p$. The map f_T then gives rise to the bottom fibration sequence. Again by Proposition 4.4.1 or 4.4.2, the components of Ff_T are nilpotent. Moreover, the groups in (vi) are nilpotent by Lemmas 3.1.3 and 4.3.4. We see that Ff_T and $\tilde{\pi}_0(Fp_T)$ are T-local by Corollary 5.4.11. Let ψ be any fill-in making the left and middle squares commute up to homotopy. Then a comparison of long exact sequences of homotopy groups shows that $\psi \colon F \longrightarrow Ff_T$ restricts to localizations of components and the last clause of (vi) holds. By the uniqueness of localization, there results a componentwise equivalence ξ such that $\xi \circ \phi \simeq \psi$, and $p \circ \xi \simeq p_T$ by the uniqueness of p_T. □

The previous result simplifies when Y is simply connected and therefore F is connected. In that case, we can ignore the interior of the central square and parts (v) and (vi), concluding simply that $\psi \colon F \longrightarrow Ff_T$ is a localization of F at T.

6.3. Localizations of function spaces

We first record an essentially obvious consequence of the general theory of localizations. All of our spaces have given basepoints, and we let X_* denote the component of the basepoint of X.

LEMMA 6.3.1. *Let X be nilpotent and Y be T-local and nilpotent. Then*

$$\phi^* \colon F(X_T, Y)_* \longrightarrow F(X, Y)_*$$

is a weak homotopy equivalence.

PROOF. We state this in terms of weak equivalence since $F(X, Y)$ need not have the homotopy type of a CW complex even under our standing assumption that all given spaces have the homotopy types of CW complexes. Since X is connected,

$$\pi_n(F(X, Y)_*) = [S^n, F(X, Y)_*] \cong [X, (\Omega^n Y)_*]$$

for $n > 0$, and similarly with X replaced by X_T. The space $(\Omega^n Y)_*$ is T-local since Y is T-local, and the conclusion follows from the universal property of the localization $\phi: X \longrightarrow X_T$. $\qquad\square$

Our main interest in this section is to use our study of fibrations to examine the interaction between localization at T and function spaces. The results here will play a key role in the proofs of the fracture theorems for localization. Changing notation, we take X to be the target space and K to be the source. We are interested in finite based CW complexes K, not necessarily nilpotent. For such K, $F(K, X)$ does have the homotopy type of CW complex, by [103]. Without changing its homotopy type, we can arrange that K has a single vertex and based attaching maps. By a based CW complex, we mean one of this sort. We form the function space $F(K, X)$ of based maps $K \longrightarrow X$. It has a canonical basepoint, namely the trivial map, which we denote by $*$, but we generally ignore this fact.

THEOREM 6.3.2. *Let X be a nilpotent space and K be a finite based connected CW complex. Let $f \in F(K, X)$, and let $F(K, X)_f$ denote the component of $F(K, X)$ that contains f. Let K^i denote the i-skeleton of K and define $[K, X]_f$ to be the set of all $g \in [K, X]$ such that $g|K^{n-1} = f|K^{n-1}$ in $[K^{n-1}, X]$, where n is the dimension of K. Let $\phi: X \longrightarrow X_T$ be a localization of X at T. Then the following statements hold.*

(i) *$F(K, X)_f$ is a nilpotent space, $F(K, X_T)_{\phi \circ f}$ is a $\mathbb{Z}_T$-nilpotent space, and $\phi_*: F(K, X)_f \longrightarrow F(K, X_T)_{\phi \circ f}$ is a localization of spaces at T.*

(ii) *$[K, X]_f$ is a nilpotent group, $[K, X_T]_{\phi \circ f}$ is a $\mathbb{Z}_T$-nilpotent group, and $\phi_*: [K, X]_f \longrightarrow [K, X_T]_{\phi \circ f}$ is a localization at T.*

PROOF. First consider the case when K is a finite wedge of i-spheres, where $i \geq 1$. Here $F(K, X)_f = F(\vee S^i, X)_f$ is a component of a finite product of copies of $\Omega^i X$ and is thus a simple space. Since $F(K, X)$ is a loop space and thus a group up to homotopy, its components are all homotopy equivalent. Therefore $\pi_n(F(K, X)_f) \cong \times \pi_n(\Omega^i X) \cong \times \pi_{n+i}(X)$ for $n \geq 1$, and similarly with X replaced by X_T. Since $\pi_{n+i}(X_T)$ is a localization of $\pi_{n+i}(X)$, we see that $F(K, X_T)_{\phi \circ f}$ is T-local and ϕ_* is a localization of $F(K, X)_f$ at T by use of the homotopical characterizations in Theorems 6.1.1 and 6.1.2. Similarly, (ii) holds since the skeleton K^{i-1} is a point, so that $[K, X]_f = [K, X]$ is a finite product of copies of $\pi_i(X)$ and $[K, X_T]_{\phi \circ f} = [K, X_T]$ is the corresponding product of copies of $\pi_i(X_T)$.

Now assume that (i) holds for K^{n-1}, where K has dimension n with $n \geq 2$. Let K be the cofiber of a wedge of attaching maps $j \colon J \longrightarrow K^{n-1}$ where J is a finite wedge of $(n-1)$-spheres. Since K can be identified with Cj, the fiber Fj^*, $j^* = F(j, \mathrm{id})$, is homeomorphic to $F(K, X)$ by Lemma 1.1.6. (Here the fiber Fj^* is defined with respect to the canonical basepoint of $F(J, X)$). Thus, for any map $f \colon K \longrightarrow X$ with restriction e to K^{n-1}, we have a restriction

$$F(K, X)_f \longrightarrow F(K^{n-1}, X)_e \longrightarrow F(J, X)_*$$

to components of a canonical fibration sequence. By the first case and the induction hypothesis, the components $F(K^{n-1}, X)_e$ and $F(J, X)_*$ are nilpotent with localizations $F(K^{n-1}, X_T)_{\phi \circ e}$ and $F(J, X_T)_*$. Now (i) follows directly from Theorem 6.2.6(i). Note that we can take $\psi = F(\mathrm{id}, \phi)$ in this specialization of that result, so that its diagram takes a more canonical form.

To prove (ii), it suffices to identify the set $[K, X]_f$ with the group $\tilde{\pi}_0(F(K, X), f)$, and similarly with X replaced by X_T. We have the required normality conditions since $F(J, X)_*$ is a loop space, so that its fundamental group is abelian and the image of the fundamental group of $F(K^{n-1}, X)_e$ is necessarily a normal subgroup. By Theorem 6.2.6(ii),

$$\psi_* \colon \tilde{\pi}_0(F(K, X), f) \longrightarrow \tilde{\pi}_0(F(K, X_T), \phi \circ f)$$

is a localization at T for each $f \in F(K, X)$, so that these identifications will imply (ii). Thus consider the exact sequence of pointed sets

$$\pi_1(F(K^{n-1}, X), e) \xrightarrow{j^*} \pi_1(F(J, X), f \circ j) \xrightarrow{\partial} \pi_0(F(K, X), f) \xrightarrow{i^*} \pi_0(F(K^{n-1}, X), e).$$

Here $[K, X]$ can be identified with $\pi_0(F(K, X), f)$, and $[K, X]_f$ can be identified with the kernel of i^*. This exact sequence restricts to the sequence

$$0 \longrightarrow \tilde{\pi}_0(F(K, X), f) \xrightarrow{\partial} \pi_0(F(K, X), f) \xrightarrow{i^*} \pi_0(F(K^{n-1}, X), e)$$

and this identifies $\tilde{\pi}_0(F(K, X), f)$ with the kernel $[K, X]_f$ of i^*. $\qquad\square$

6.4. Localizations of colimits and cofiber sequences

Using the homological characterization of localizations, we obtain analogues of the results in §6.2 for wedges, suspensions, cofiber sequences, and smash products. However, in the non-simply connected case, the required preservation of nilpotency is not automatic. Wedges behave badly, for example. The wedge $S^1 \vee S^1$ is not nilpotent since a free group on two generators is not nilpotent, and the wedge $S^1 \vee S^2$ is not nilpotent since π_1 does not act nilpotently on π_2. In the following series of results, appropriate natural maps are obtained from the universal properties or functoriality of the specified constructions.

PROPOSITION 6.4.1. *If X, Y, and $X \vee Y$ are nilpotent spaces, then $(X \vee Y)_T$ is naturally equivalent to $X_T \vee Y_T$.*

PROPOSITION 6.4.2. *If X is nilpotent, then $(\Sigma X)_T$ is naturally equivalent to $\Sigma(X_T)$.*

PROPOSITION 6.4.3. *Let $i : A \longrightarrow X$ be a cofibration and $f : A \longrightarrow Y$ be a map, where A, X, Y, X/A, and $X \cup_A Y$ are nilpotent. If we choose localizations such that $i_T : A_T \longrightarrow X_T$ is a cofibration, then $X_T \cup_{A_T} Y_T$ is a localization of $X \cup_A Y$.*

PROOF. We first note that $Ci_T \simeq X_T/A_T$ is a localization of X/A and then use the long exact sequences of $\mathbb{Z}_T$-local homology groups of the cofibration sequences $Y \longrightarrow X \cup_A Y \longrightarrow X/A$ and $Y_T \longrightarrow X_T \cup_{A_T} Y_T \longrightarrow X_T/A_T$. $\quad\square$

PROPOSITION 6.4.4. *Let $f : X \longrightarrow Y$ be a map such that X, Y, and Cf are nilpotent and let $\phi : X \longrightarrow X_T$ and $\phi : Y \longrightarrow Y_T$ be localizations. Then any fill-in ϕ in the map of canonical cofiber sequences*

$$
\begin{array}{ccccccc}
X & \xrightarrow{\ f\ } & Y & \xrightarrow{\ i\ } & Cf & \xrightarrow{\ \pi\ } & \Sigma X \\
\downarrow{\scriptstyle\phi} & & \downarrow{\scriptstyle\phi} & & \downarrow{\scriptstyle\phi} & & \downarrow{\scriptstyle\Sigma\phi} \\
X_T & \xrightarrow[\ f_T\ ]{} & Y_T & \xrightarrow[\ i\ ]{} & Cf_T & \xrightarrow[\ \pi\ ]{} & \Sigma Y_T
\end{array}
$$

is a localization of Cf at T.

PROPOSITION 6.4.5. *Let X be the colimit of a sequence of cofibrations $X_i \longrightarrow X_{i+1}$ between nilpotent spaces and choose localizations $(X_i)_T \longrightarrow (X_{i+1})_T$ that are cofibrations. Then $\operatorname{colim}(X_i)_T$ is naturally a localization of X.*

PROPOSITION 6.4.6. *If X, Y, and $X \wedge Y$ are nilpotent, then $(X \wedge Y)_T$ is naturally equivalent to $X_T \wedge Y_T$.*

PROOF. The Künneth theorem and the homological characterization of T-local spaces imply that $X_T \wedge Y_T$ is T-local. The Künneth theorem also implies that the smash product of localizations is a localization. The universal property of $(X \wedge Y)_T$ gives the conclusion. □

6.5. A cellular construction of localizations

We have constructed localizations of nilpotent spaces at T using a cocellular method, and we have proven a number of properties of localizations that do not depend on the particular construction. Other constructions can also be useful. In this section, we describe a dual cellular construction of the localizations of simply connected spaces. The construction does not generalize to nilpotent spaces, but it has redeeming features. For example, it gives rise to local cellular chain complexes for the computation of T-local homology.

Since we are interested in simply connected spaces, the logical first step in the construction is to pick a fixed localization for S^2. However, it is easier to start with S^1. We can construct a localization $\phi\colon S^1 \longrightarrow S_T^1$ by taking S_T^1 to be $K(\mathbb{Z}_T, 1)$ for some particular construction of this Eilenberg-Mac Lane space; the map $\phi\colon S^1 \longrightarrow S_T^1$ is then induced by the inclusion $\mathbb{Z} \longrightarrow \mathbb{Z}_T$. We gave an explicit construction in the proof of Theorem 5.1.3.

Now take $S^2 = \Sigma S^1$ and $S_T^2 = \Sigma(S_T^1)$. Inductively, define S^i to be ΣS^{i-1} and define S_T^i to be ΣS_T^{i-1}. Define $\phi\colon S^i \longrightarrow S_T^i$ to be $\Sigma\phi$. This gives a localization by Proposition 6.4.2. Similarly, we localize wedges of spheres, pushouts along attaching maps, and sequential colimits by the evident and natural use of Propositions 6.4.1, 6.4.3, and 6.4.5. Note that we must start with S^2 rather than S^1 since otherwise taking wedges of spheres loses nilpotency. This is where the restriction to simply connected spaces enters.

Now recall the definition of $\mathscr{J}$-complexes from Definition 3.3.3. It specializes to give $\mathscr{J}S^2$-complexes and $\mathscr{J}S_T^2$-cell complexes. In both cases, we can define CW complexes by requiring cells to be attached only to cells of lower dimension, and then we can arrange that the sequential filtration that describes when cells are attached coincides with the skeletal filtration; note that the 1-skeleton of such a CW complex is just the basepoint. We refer to cell complexes and CW complexes in $\mathscr{J}S_T^2$ as T-cell complexes and T-CW complexes. The following theorem hardly requires proof since it follows directly from the results of the previous section. However, we give some details to illustrate the duality with the cocellular construction.

THEOREM 6.5.1. *There is a cellular localization functor from the homotopy category of $\mathscr{J}S^2$-cell complexes to the homotopy category of $\mathscr{J}S_T^2$-cell complexes that takes CW complexes to T-CW complexes.*

PROOF. Describe $X \in \mathscr{J}S^2$ as in Definition 3.3.3. Thus X is the colimit of a sequence of cofibrations $X_n \longrightarrow X_{n+1}$ such that $X_0 = *$, X_{n+1} is the pushout of the cofibration $J_n \longrightarrow CJ_n$ and a map $j_n \colon J_n \longrightarrow X_n$, where J_n is a wedge of spheres of dimension at least two. When X is a CW complex, we may take J_n to be a wedge of n-spheres and then X_n is the n-skeleton of X. We define $(X_T)_n$ inductively, starting with $(X_T)_0 = *$. Given $(X_T)_n$, consider the following diagram.

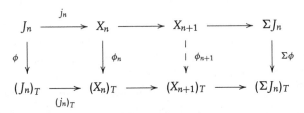

The upper row is a cofiber sequence, and $(J_n)_T$ is the wedge of the localizations of the wedge summands S^q of J_n. There is a map $(j_n)_T$, unique up to homotopy, such that the left square commutes up to homotopy by the universal property of localizations. We define $(X_{n+1})_T$ to be its cofiber and then the lower row is also a cofiber sequence. There is a fill-in map ϕ_{n+1} that makes the middle square commute and the right square commute up to homotopy, by Lemma 1.2.1. As noted in the previous section, the homological characterizations of T-local spaces and localizations at T imply that ϕ_{n+1} is a localization at T and that if X_T is the colimit of the $(X_n)_T$, then the map $\phi \colon X \longrightarrow X_T$ obtained by passage to colimits is also a localization at T. A quick summary of the construction is that we can construct $(X_T)_n$ as $(X_n)_T$. $\square$

In the previous result, we understand the relevant homotopy categories to be full subcategories of $\mathrm{Ho}\mathscr{T}$, and then the functoriality of the construction is immediate from the functoriality of localization in general. However, we can refine this by dualizing the cocellular functoriality of Theorem 5.3.3. Recall from [93, p. 74] that, for CW complexes X and Y, any map $X \longrightarrow Y$ is homotopic to a cellular map and any two homotopic cellular maps are cellularly homotopic. The proof works equally well to give precisely the same conclusion for CW complexes in $\mathscr{J}S^2$ or $\mathscr{J}S_T^2$. Therefore, restricting to CW complexes (which is no restriction on homotopy types), the full subcategories just mentioned are the same as the categories of CW complexes and cellular homotopy classes of cellular maps.

In the construction above, if we start with a CW complex X, we obtain a T-CW complex X_T. If $f \colon X \longrightarrow X'$ is a cellular map, we can construct

$f_T \colon X_T \longrightarrow X'_T$ to be a cellular map, and it will be unique up to cellular homotopy.

We can define the cellular chains of a T-CW complex Y in the same way as for ordinary CW complexes, letting

6.5.2 $$C_q(Y) = H_q(Y^q, Y^{q-1}; \mathbb{Z}_T)$$

and letting the differential come from the connecting homomorphism of the triple (Y^q, Y^{q-1}, Y^{q-2}), as in [93, p. 117]. In positive degrees, we get isomorphic chains if we replace $\mathbb{Z}$ by $\mathbb{Z}_T$. If $Y = X_T$ is the cellularly constructed localization of a CW complex X, then

6.5.3 $$C_q(X_T) \cong C_q(X; \mathbb{Z}_T).$$

The cellular chains of T-CW complexes are functorial on cellular maps, and the isomorphism (6.5.3) is natural with respect to cellular localizations of maps. We remark parenthetically that when Y is of dimension n, the corresponding cochains satisfy

6.5.4 $$\tilde{C}^q(Y; \pi) \cong \pi_{n-q} F(Y^q / Y^{q-1}, K(\pi, n))$$

for any $\mathbb{Z}_T$-module π, as in Exercise 3.3.4.

6.6. Localizations of H-spaces and co-H-spaces

Recall that an H-space, or Hopf space, X is a space together with a product $X \times X \longrightarrow X$ such that the basepoint $* \in X$ is a two-sided unit up to homotopy. We may assume that the basepoint is a strict unit, and we often denote it by e. Of course, topological monoids and loop spaces provide the canonical examples. Nonconnected examples are often of interest, but we continue to take X to be connected. Recall from Corollary 1.4.5 that X is a simple space.

The elementary construction of S^1_T in the proof of Theorem 5.1.3 used the product on S^1, and we can use the product on any H-space Y to obtain a precisely similar construction of Y_T. Again, we order the primes q_i not in T by size and define $r_1 = q_1$ and $r_i = q^i_1 \cdots q_i$, so that $\mathbb{Z}_T$ is the colimit over i of the maps $r_i \colon \mathbb{Z} \longrightarrow \mathbb{Z}$. Applied pointwise, the product on Y gives a product between based maps $S^n \longrightarrow Y$. This product is homotopic to the product induced by the pinch map $S^n \longrightarrow S^n \vee S^n$, as we leave to the reader to check. The latter product induces the addition on homotopy groups, hence so does the former. An H-space is homotopy associative if its product satisfies the associative law up to homotopy. We do not require this. However, associating the product in any fixed order, we can define iterated products. In particular, restricting such an r-fold product to the image of the diagonal $Y \longrightarrow Y^r$, we obtain an rth-power map $r\colon Y \longrightarrow Y$. It induces multiplication by r on each group $\pi_n(Y)$.

PROPOSITION 6.6.1. *For an H-space Y, the localization Y_T can be constructed as the telescope of the sequence of r_i^{th}-power maps $Y \longrightarrow Y$.*

PROOF. Writing Y_T for the telescope, we have $\pi_n(Y_T) = \operatorname{colim} \pi_n(Y)$, where the colimit is taken with respect to the homomorphisms given by multiplication by r_i, and this colimit is $\pi_n(Y)_T$. The inclusion of the base of the telescope is a map $Y \longrightarrow Y_T$ that induces localization on homotopy groups. $\qquad\square$

The following basic result is more important than this construction and does not depend on it.

PROPOSITION 6.6.2. *If Y is an H-space with product μ, then Y_T is an H-space with product μ_T such that the localization $\phi\colon Y \longrightarrow Y_T$ is a map of H-spaces.*

PROOF. Consider the diagram

$$
\begin{array}{ccc}
Y \times Y & \overset{\mu}{\longrightarrow} & Y \\
{\scriptstyle \phi \times \phi}\big\downarrow & & \big\downarrow{\scriptstyle \phi} \\
Y_T \times Y_T & \underset{\mu_T}{\longrightarrow} & Y_T.
\end{array}
$$

The map $\phi \times \phi$ is a localization of $Y \times Y$ at T, by Theorem 6.1.1. By the universal property, there is a map μ_{T_i} making the diagram commute up to homotopy. To see that the basepoint of Y_T is a homotopy unit for μ_{T_i}, consider the diagram

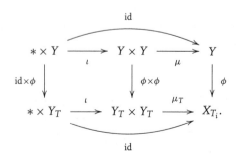

The identity map of Y_T and the map $\mu_T \iota$ are both localizations of id $= \mu\iota$ on Y, hence they are homotopic, and similarly for the right unit property. $\qquad\square$

The converse does not hold. There are many interesting spaces that are not *H*-spaces but have localizations that are *H*-spaces. In fact, in 1960, well before localizations were constructed in general, Adams [2] observed that S_T^n is an *H*-space for all odd n and all sets T of odd primes. In fact, there is a quite simple construction of a suitable product.

EXERCISE 6.6.3. [130, p. 14]. Let n be odd. Give $S^n \subset D^{n+1}$ the basepoint $* = (1, 0, \ldots, 0)$. Define $f\colon S^n \times S^n \longrightarrow S^n$ by $f(x, y) = y - (2\Sigma x_i y_i)x$. Observe that f does have image in S^n and show that the degrees of the restrictions $S^n \longrightarrow S^n$ of f to $x = *$ and $y = *$ are 2 and -1, respectively.

That is, f has bidegree $(2, -1)$. Since $\pi_n(S_T^n) = [S_T^n, S_T^n] = \mathbb{Z}_T$ and $1/2 \in \mathbb{Z}_T$, we have a map of degree $1/2$ on S_T^n. We also have a map of degree -1. The required product on S_T^n is the composite

$$S_T^n \times S_T^n \xrightarrow{\ (1/2, -1)\ } S_T^n \times S_T^n \xrightarrow{\quad f_T\quad} S_T^n.$$

This map has bidegree $(1, 1)$, which means that it gives S_T^n an *H*-space structure. Adams observed further that S_T^n is homotopy commutative and is homotopy associative if $3 \notin T$. These can be checked from the construction.

In contrast, S^n itself is an *H*-space only if $n = 0$, 1, 3, or 7, by Adams' solution to the Hopf invariant one problem [1]. Here the Lie group S^3 is not homotopy commutative and the *H*-space S^7 given by the unit Cayley numbers is not homotopy associative. There is a large literature on finite T-local *H*-spaces, especially when $T = \{p\}$. In the simply connected case, finite here is best understood as meaning homotopy equivalent to a T-CW complex that has finitely many cells.

The study breaks into two main variants. In one of them, one allows general finite T-local *H*-spaces, not necessarily homotopy associative or commutative, and asks what possible underlying homotopy types they might have. In the other, one studies finite T-local loop spaces, namely spaces X that are of the homotopy type of finite T-CW complexes and are also homotopy equivalent to ΩBX for some T-local space BX. Such X arise as localizations of finite loop spaces. One asks, typically, how closely such spaces resemble compact Lie groups and what limitations the *H*-space structure forces on the homology and homotopy groups. The structure theorems for Hopf algebras that we give later provide a key starting point for answering such questions.

Although of less interest, it should be observed that Propositions 6.6.1 and 6.6.2 have analogues for co-H-spaces. A co-H-space Z is a space with a coproduct $\delta: Z \longrightarrow Z \vee Z$ such that $\nabla \circ \delta$ is homotopic to the identity, where $\nabla: Z \vee Z \longrightarrow Z$ is the folding map. Suspensions ΣX with their pinch maps provide the canonical examples. We can define r^{th} copower maps $Z \longrightarrow Z$ by iterating the coproduct in some fixed order and then applying the r-fold iteration of ∇. Assuming for simplicity that Z is simply connected, to avoid dealing with the group $\pi_1(Z)$, we can check that this map too induces multiplication by r on homotopy groups. From here the proof of the following result is the same as the proof of Proposition 6.6.1.

PROPOSITION 6.6.4. *For a simply connected co-H-space Z, the localization Z_T can be constructed as the telescope of the sequence of r_i^{th} copower maps $Z \longrightarrow Z$.*

PROPOSITION 6.6.5. *If Z is a simply connected co-H-space with coproduct δ, then Z_T is a co-H-space with coproduct δ_T such that the localization $\phi: Z \longrightarrow Z_T$ is a map of co-H-spaces.*

PROOF. Since Z is simply connected, the wedge $Z_T \vee Z_T$ is a localization of $Z \vee Z$, and the conclusion follows by use of the universal property. □

6.7. Rationalization and the finiteness of homotopy groups

Localization at the empty set of primes is called rationalization. Logically it should be denoted $X_\emptyset$, but it is usually denoted X_0. It will play a special role in the fracture theorems since rationalization $X \longrightarrow X_0$ factors up to equivalence as the composite $X \longrightarrow X_T \longrightarrow X_0$ of localization at T and rationalization for every set of primes T. It is also of considerable interest in its own right. We give a few examples here and return to the rationalization of H-spaces in Chapter 9.

Many results in algebraic topology that preceded the theory of localization are conveniently proven using the newer theory. We illustrate this with a proof of a basic theorem of Serre on the finiteness of the homotopy groups of spheres. Serre proved the result using (Serre) classes of abelian groups [125]. The proof using rationalization is simpler and more illuminating.

THEOREM 6.7.1 (SERRE). *For $n \geq 1$, the homotopy groups $\pi_q(S^n)$ are finite with the exceptions of $\pi_n(S^n) = \mathbb{Z}$ for all n and $\pi_{2n-1}(S^n) = \mathbb{Z} \oplus F_n$ for n even, where F_n is finite.*

The fundamental class $\iota_n \in H^n(K(\mathbb{Q}, n); \mathbb{Q})$ is represented by the identity map of $K(\mathbb{Q}, n)$, and we have the following easy calculation.

PROPOSITION 6.7.2. *The cohomology algebra $H^*(K(\mathbb{Q}, n); \mathbb{Q})$ is the exterior algebra on ι_n if n is odd and the polynomial algebra on ι_n if n is even.*

PROOF. For $n = 1$ and $n = 2$, this is clear from Theorem 5.2.8 and the fact that $S^1 = K(\mathbb{Z}, 1)$ and $\mathbb{C}P^\infty = K(\mathbb{Z}, 2)$. We proceed by induction on n, using the Serre spectral sequence of the path space fibration

$$K(\mathbb{Q}, n) \longrightarrow PK(\mathbb{Q}, n+1) \longrightarrow K(\mathbb{Q}, n+1).$$

Here ι_n transgresses (via d_{n+1}) to ι_{n+1}. For n even, the Leibnitz rule implies that $d_{n+1}(\iota_n^q) = q\iota_{n+1}\iota_n^{q-1}$, and the spectral sequence is concentrated on the 0th and $(n+1)^{st}$ columns. For n odd, the Leibnitz rule implies that $d_{n+1}(\iota_{n+1}^q \iota_n) = \iota_{n+1}^{q+1}$, and the spectral sequence is concentrated on the 0th and n^{th} rows. $\square$

PROOF OF THEOREM 6.7.1. Starting with a representative $k \colon S^n \longrightarrow K(\mathbb{Z}, n)$ for the fundamental class of S^n and arguing as in the cocellular construction of localizations, we obtain a homotopy commutative diagram

$$
\begin{array}{ccccccc}
K(\mathbb{Z}, n-1) & \longrightarrow & S^n\langle n\rangle & \longrightarrow & S^n & \overset{k}{\longrightarrow} & K(\mathbb{Z}, n) \\
\phi\downarrow & & \phi\downarrow & & \phi\downarrow & & \phi\downarrow \\
K(\mathbb{Q}, n-1) & \longrightarrow & S^n\langle n\rangle_0 & \longrightarrow & S^n_0 & \underset{k_0}{\longrightarrow} & K(\mathbb{Q}, n)
\end{array}
$$

in which the rows are canonical fiber sequences and the maps ϕ are rationalizations.

If n is odd, k_0 induces an isomorphism in rational cohomology and is therefore an equivalence. This implies that all homotopy groups $\pi_q(S^n)$, $q > n$, are in the kernel of rationalization. That is, they are torsion groups. Since they are finitely generated by Theorem 4.5.4, they are finite.

If n is even, the Serre spectral sequence of k_0 implies that $H^*(S^n\langle n\rangle_0; \mathbb{Q})$ is an exterior algebra on a class ι_{2n-1} of degree $2n - 1$ that transgresses to ι_n^2. Here $d_{2n}(\iota_n^q \iota_{2n-1}) = \iota_n^{q+2}$, the spectral sequence is concentrated on the 0th and $(2n - 1)^{st}$ rows, and ι_n survives to the fundamental class of S^n_0. Therefore the Hurewicz dimension of $S^n\langle n\rangle_0$ is $2n - 1$, and a map $S^{2n-1}_0 \longrightarrow S^n\langle n\rangle$ that represents ι_{2n-1} must be an equivalence. Thus the rationalization of the homotopy group $\pi_q(S^n)$, $q > 0$, is 0 if $q \neq 2n - 1$ and is $\mathbb{Q}$ if $q = 2n - 1$. Since

$\pi_{2n-1}(S^n)$ is a finitely generated abelian group with rationalization $\mathbb{Q}$, it must be the direct sum of $\mathbb{Z}$ and a finite group. $\qquad\square$

We point out an implication of the proof just given.

COROLLARY 6.7.3. *Consider the rationalization* $k_0 \colon S_0^n \longrightarrow K(\mathbb{Q}, n)$ *of the canonical map* $k \colon S^n \longrightarrow K(\mathbb{Z}, n)$. *If* n *is odd,* k_0 *is an equivalence. If* n *is even, the fiber of* k_0 *is* $K(\mathbb{Q}, 2n - 1)$.

6.8. The vanishing of rational phantom maps

In this brief section, we give an observation that shows, in effect, that phantom maps are usually invisible to the eyes of rational homotopy theory.

LEMMA 6.8.1. *Let* X *be a connected CW complex with finite skeleta. If* Z *is* $f\mathbb{Q}$-*nilpotent, then*

$$\lim{}^1 [\Sigma X^n, Z] = 0$$

and

$$[X, Z] \to \lim [X^n, Z]$$

is a bijection.

PROOF. We claim that each $[\Sigma X^n, Z]$ is an $f\mathbb{Q}$-nilpotent group. The proof is by induction on n. Note first that if J is a finite wedge of i-spheres, $i \geq 1$, then $[\Sigma J, Z]$ is a finite dimensional $\mathbb{Q}$-vector space and $[J, Z]$ is an $f\mathbb{Q}$-nilpotent group.

We may assume that $X^0 = *$. Let J_n, $n \geq 1$, be a wedge of n-spheres such that X^{n+1} is the cofiber of a map $\mu_n \colon J_n \to X^n$. Then there is an exact sequence

$$\cdots \ \xrightarrow{(\Sigma^2 \mu_n)^*} \ [\Sigma^2 J_n, Z] \ \xrightarrow{\delta} \ [\Sigma X^{n+1}, Z] \ \xrightarrow{\iota^*} \ [\Sigma X^n, Z] \ \xrightarrow{(\Sigma \mu_n)^*} \ [\Sigma J_n, Z].$$

By Lemma 1.4.6(v), the image of δ is central in $[\Sigma X^{n+1}, Z]$.

Assume inductively that $[\Sigma X^n, Z]$ is an $f\mathbb{Q}$-nilpotent group. By Lemma 5.1.2, $(\Sigma^2 \mu_n)^*$ is a map of $\mathbb{Q}$-vector spaces and $(\Sigma \mu_n)^*$ is a $\mathbb{Q}$-map. By Lemma 4.3.4, $\operatorname{im} \delta \cong \operatorname{coker}(\Sigma^2 \mu_n)^*$ is a finite dimensional $\mathbb{Q}$-vector space and $\operatorname{coker} \delta \cong \ker(\Sigma \mu_n)^*$ is an $f\mathbb{Q}$-nilpotent group. Then Lemma 5.1.2 and the exact sequence

$$0 \to \operatorname{im} \delta \to [\Sigma X^{n+1}, Z] \to \operatorname{coker} \delta \to 0$$

imply that $[\Sigma X^{n+1}, Z]$ is $\mathbb{Q}$-nilpotent. By Proposition 5.6.5, im δ and coker δ both have finite sets of $\emptyset$-generators. The images of the $\emptyset$-generators for im δ and a choice of inverse images of the $\emptyset$-generators for coker δ give a finite set of $\emptyset$-generators for $[\Sigma X^{n+1}, Z]$. Proposition 5.6.5 then implies that $[\Sigma X^{n+1}, Z]$ is $f\mathbb{Q}$-nilpotent.

Now the following easy observation, which is trivial in the abelian case, implies that the sequence $[\Sigma X^n, Z]$ satisfies the Mittag-Leffler condition described in Section 2.3, so that the result follows from Theorem 2.3.3(i). □

LEMMA 6.8.2. *Any descending chain of $f\mathbb{Q}$-nilpotent groups has finite length.*

7

FRACTURE THEOREMS FOR
LOCALIZATION: GROUPS

In Chapter 5, we described how to construct localizations of nilpotent spaces. In the next chapter, we go in the opposite direction and describe how to start with local spaces and construct a "global space" and how to reconstruct a given global space from its localizations. Results such as these are referred to as fracture theorems. In contrast to Chapter 5, where we constructed localizations of nilpotent groups from localizations of nilpotent spaces, we first prove fracture theorems for abelian and nilpotent groups in this chapter and then use the results of Chapter 6 to extend the algebraic fracture theorems to nilpotent spaces in the next.

Although much of this material can be found in [21, 62, 133], there seem to us to be significant gaps and oversights in the literature, including some quite misleading incorrect statements, and there is no single place to find a full account. There are also some ways to proceed that are correct but give less complete answers; we shall say little about them here. The new concept of a "formal localization" plays a central role in our exposition, and that concept leads us (in §7.5) to a new perspective on the "genus" of a nilpotent group, namely the set of isomorphism classes of nilpotent groups whose localizations at each prime are isomorphic to those of the given group. In the next chapter, we will find an analogous perspective on the genus of a nilpotent space.

Throughout this chapter, let I be an indexing set and let T_i be a set of primes, one for each $i \in I$. Let $S = \bigcap_{i \in I} T_i$ and $T = \bigcup_{i \in I} T_i$; it is sensible to insist that $T_i \cap T_j = S$ for $i \neq j$ and that $T_i \neq S$ for all i, and we assume that this holds. Thus the $T_i - S$ give a partition of the primes in $T - S$.

We are mainly interested in the case when S is empty and localization at S is rationalization. We are then starting with a partition of the set of primes in T, and we are most often interested in the case when T is the set of all primes. For example, I might be the positive integers and T_i might be the set consisting of just the ith prime number p_i. A common situation is when $I = \{1, 2\}$,

$T_1 = \{p\}$, and T_2 is the set of all other primes. For example, spaces often look very different when localized at 2 and when localized away from 2.

Until §7.6, all given groups in this chapter are to be nilpotent. We have two kinds of results. In one, we start with a T-local group G and ask how to reconstruct it from its localizations at the T_i. We call these global to local results and treat them in §7.2, after developing perspectives and preliminaries in §7.1. We give a conceptual proof that works simultaneously for nilpotent and abelian groups, but in §7.6 we give a general group theoretical result that allows an alternative proof by induction on the nilpotency class and that will be needed later to prove the analogous global to local result for completions.

In the other, we are given T_i-local groups and ask how to construct a T-local group from them. We call these local to global results and treat them in §7.4, after developing the notion of a formal localization in §7.3. In both, we are concerned with certain basic pullback diagrams, and it turns out that there are simplifying features when the indexing set I is finite or the given groups are finitely generated.

As a matter of philosophy or psychology, the global to local and local to global perspectives should be thought of as two ways of thinking about essentially the same phenomenon. We either start with a global object and try to reconstruct it up to equivalence from its local pieces or we start with local pieces and try to construct a global object with equivalent local pieces. These processes should be inverse to each other. In all cases the global to local results are actually implied by the local to global results, but for purposes of exposition we prefer to think of first localizing and then globalizing, rather than the other way around.

7.1. Global to local pullback diagrams

For nilpotent groups G, and in particular for T-local groups G, we have the localizations

$$\phi \colon G \longrightarrow G_S, \quad \phi_i \colon G \longrightarrow G_{T_i}, \quad \text{and} \quad \psi_i \colon G_{T_i} \longrightarrow G_S.$$

Since G_S is T_i-local for each i, we can and do choose ψ_i to be the unique homomorphism such that $\psi_i \phi_i = \phi$ for each i. We also let $\phi_S \colon \prod_i G_{T_i} \longrightarrow (\prod_i G_{T_i})_S$ denote a localization at S. We fix these notations throughout this section. We are headed toward a description of a T-local group G in terms of its localizations. We have the following two commutative diagrams, in which ϕ_S is a localization at S, P and Q are pullbacks, and α and β are given by the universal property of pullbacks.

7.1.1

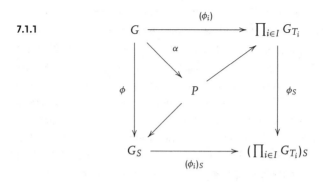

7.1.2

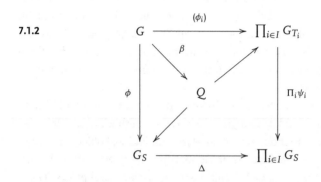

Here (ϕ_i) denotes the map with coordinates ϕ_i and $(\phi_i)_S$ denotes its localization at S. By Lemma 5.5.6, $\prod_i G_S$ is S-local and we may identify it with its localization at S. Applying Lemma 5.5.7 to commute pullbacks with localizaton at T and at T_k, we obtain the following conclusion.

PROPOSITION 7.1.3. *The groups P and Q are T-local, and for each $k \in I$ localization at T_k gives pullback diagrams*

$$
\begin{array}{ccc}
P_{T_k} & \longrightarrow & (\prod_{i\in I} G_{T_i})_{T_k} \\
\downarrow & & \downarrow {\scriptstyle (\phi_S)_{T_k}} \\
G_S & \xrightarrow{(\phi_i)_S} & (\prod_{i\in I} G_{T_i})_S
\end{array}
\quad and \quad
\begin{array}{ccc}
Q_{T_k} & \longrightarrow & (\prod_{i\in I} G_{T_i})_{T_k} \\
\downarrow & & \downarrow {\scriptstyle (\Pi_i\psi_i)_{T_k}} \\
G_S & \xrightarrow{\Delta} & \prod_{i\in I} G_S.
\end{array}
$$

We have a comparison diagram that relates the T-local groups P and Q. If $\pi_i \colon \prod G_{T_i} \longrightarrow G_{T_i}$ is the projection, there is a unique map $\tilde{\pi}_i \colon (\prod G_{T_i})_S \longrightarrow G_S$ such that $\tilde{\pi}_i \circ \phi_S = \psi_i \circ \pi_i$. The map $(\tilde{\pi}_i) \colon (\prod G_{T_i})_S \longrightarrow \prod_i G_S$ with coordinates $\tilde{\pi}_i$ is the localization $(\psi_i)_S$ of the map $(\psi_i) \colon \prod G_{T_i} \longrightarrow \prod_i G_S$.

7.1.4

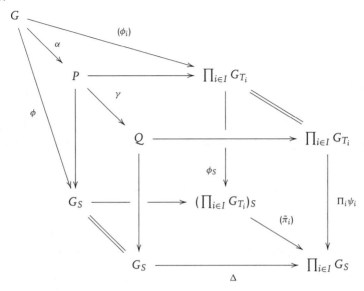

Here γ is given by the universal property of the pullback Q. It is immediate from the diagram that $\gamma \circ \alpha = \beta$. Since localization commutes with finite products, the map $(\tilde{\pi}_i)$ is an isomorphism and the distinction between diagrams (7.1.1) and (7.1.2) disappears when I is finite.

LEMMA 7.1.5. *If I is finite, the pullback diagrams (7.1.1) and (7.1.2) may be identified and the map $\gamma \colon P \longrightarrow Q$ is an isomorphism.*

We are interested in determining when α and β are isomorphisms. Since $\gamma\alpha = \beta$, they can both be isomorphisms only if γ is an isomorphism. This can easily happen even when $(\tilde{\pi}_i)$ is not an isomorphism and the pullback diagrams cannot be identified. Indeed, we have the following observation.

LEMMA 7.1.6. *Suppose that $(\tilde{\pi}_i)\colon (\prod_{i \in I} G_{T_i})_S \longrightarrow \prod_{i \in I} G_S$ is a monomorphism. Then γ is an isomorphism, hence α is an isomorphism if and only if β is an isomorphism.*

PROOF. Let $g \in G_S$ and $h \in \prod_i G_{T_i}$. If $(g, h) \in P$ and $\gamma(g, h) = (1, 1)$, then of course the coordinates must be $g = 1$ and $h = 1$. That is, γ is always a monomorphism. Now suppose that $(g, h) \in Q$. Then $g = \psi_i(h) = \tilde{\pi}_i \phi_S(h)$ for all i. If $\bar{g}$ denotes the image of g in $(\prod_i G_{T_i})_S$, then we also have $g = \tilde{\pi}_i(\bar{g})$ for all i. Since $(\tilde{\pi}_i)$ is a monomorphism, $\bar{g} = \phi_S(h)$ and $(g, h) \in P$. Thus γ is an epimorphism. $\qquad\square$

Before giving our general results, we record an important elementary example where the two pullback diagrams do not coincide, but their pullback groups do.

PROPOSITION 7.1.7. *Let T_i be the ith prime p_i. If A is a finitely generated abelian group, then both of the following diagrams are pullbacks.*

$$
\begin{array}{ccc}
A & \xrightarrow{(\phi_{p_i})} & \prod_i A_{(p_i)} \\
{\scriptstyle\phi_0}\downarrow & & \downarrow{\scriptstyle\phi_0} \\
A_0 & \xrightarrow[(\phi_{p_i})_S]{} & (\prod_i A_{(p_i)})_0
\end{array}
\qquad and \qquad
\begin{array}{ccc}
A & \xrightarrow{(\phi_{p_i})} & \prod_{i\in I} A_{(p_i)} \\
{\scriptstyle\phi_0}\downarrow & & \downarrow{\scriptstyle\Pi_i\psi_i} \\
A_0 & \xrightarrow[\Delta]{} & \prod_i A_0
\end{array}
$$

However, the difference map

$$(\phi_{p_i})_S - \phi_0\colon A_0 \times \prod_i A_{(p_i)} \longrightarrow (\prod_i A_{(p_i)})_0,$$

whose kernel in the left diagram is A is an epimorphism, but the difference map

$$\Delta - \Pi_i\psi_i\colon A_0 \times \prod_i A_{(p_i)} \longrightarrow \prod_i A_0,$$

whose kernel in the right diagram is also A is only an epimorphism in the trivial case when A is finite.

PROOF. Since A is a finite direct sum of cyclic groups, it suffices to prove this when $A = \mathbb{Z}/p^n$ for some prime p and when $A = \mathbb{Z}$. The first case is trivial. We leave the second as an illuminating exercise for the reader. $\qquad\square$

In some of the early literature, the focus is on the pullback Q and the resulting map $\beta\colon G \longrightarrow Q$, but then there are counterexamples that show that β is not always an isomorphism and thus the pullback Q does not always recover the original group. In fact, that is usually the case when the indexing set I is infinite. Of course, such counterexamples carry over to topology. Moreover, even when $\gamma\colon P \longrightarrow Q$ is an isomorphism, its topological analogue will not induce an equivalence of homotopy pullbacks in general. The reader who looks back at Corollary 2.2.3 will see the relevance of the observation about epimorphisms in the previous result.

We shall prove in the next section that α is always an isomorphism. The proof will use the following observation about the detection of isomorphisms.

LEMMA 7.1.8. *A homomorphism* $\alpha\colon G \longrightarrow H$ *between T-local groups is an isomorphism if and only if* $\alpha_{T_k}\colon G_{T_k} \longrightarrow H_{T_k}$ *is an isomorphism for all* $k \in I$.

PROOF. This is proven by two applications of Proposition 5.5.4. Since α_{T_k} is an isomorphism, α is a T_k-isomorphism. Since T is the union of the T_k, it follows that α is a T-isomorphism. In turn, this implies that α_T is an isomorphism. Since G and H are T-local, α itself is an isomorphism. $\quad\square$

However, perhaps the most interesting aspect of the proof will be the central role played by the following categorical observation about pullbacks. It applies to any category that has categorical products. Such categories are said to be cartesian monoidal. Examples include the categories of abelian groups, groups, spaces, and sets. Less obviously, the homotopy category $\text{Ho}\mathcal{T}$ is another example, even though pullbacks do not generally exist in $\text{Ho}\mathcal{T}$.

LEMMA 7.1.9. *In any cartesian monoidal category, a commutative diagram of the following form is a pullback.*

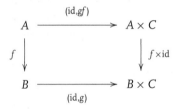

PROOF. A proof using elements, if we have them, just observes that if $a \in A$, $b \in B$, and $c \in C$ satisfy $(b, g(b)) = (f(a), c)$, then $b = f(a)$ and $c = g(f(a))$. However, it is an easy categorical exercise to verify the result directly from the universal properties that define products and pullbacks (e.g., [93, p. 16]). $\quad\square$

7.2. Global to local: abelian and nilpotent groups

Here, finally, is our main algebraic global to local result.

THEOREM 7.2.1. *Let G be a T-local group.*

(i) $(\phi_i)\colon G \longrightarrow \prod_{i \in I} G_{T_i}$ *is a monomorphism.*

(ii) *The following diagram is a pullback.*

$$\begin{array}{ccc}
G & \xrightarrow{(\phi_i)} & \prod_{i\in I} G_{T_i} \\
\phi \downarrow & & \downarrow \phi_S \\
G_S & \xrightarrow[(\phi_i)_S]{} & (\prod_{i\in I} G_{T_i})_S
\end{array}$$

Moreover, every element $z \in (\prod_{i\in I} G_{T_i})_S$ is a product $z = \phi_S(x)(\phi_i)_S(y)$ for some $x \in \prod_{i\in I} G_{T_i}$ and $y \in G_S$.

(iii) If G is finitely T-generated, the following diagram is also a pullback.

$$\begin{array}{ccc}
G & \xrightarrow{(\phi_i)} & \prod_{i\in I} G_{T_i} \\
\phi \downarrow & & \downarrow \Pi_i \psi_i \\
G_S & \xrightarrow[\Delta]{} & \prod_{i\in I} G_S
\end{array}$$

PROOF.

(i). This holds by two applications of Proposition 5.5.2. The group $\ker (\phi_i)$ is the intersection of the kernels of the localizations $\phi_i\colon G \longrightarrow G_{T_i}$ and all elements of $\ker \phi_i$ are T_i'-torsion. Since T is the union of the T_i, all elements of the intersection are T'-torsion. Since G is T-local, $\ker (\phi_i)$ is trivial.

(ii). Let $\alpha\colon G \longrightarrow P$ be the map given in (7.1.1). We must prove that α is an isomorphism. Since P is T-local, by Proposition 7.1.3, it suffices to show that $\alpha_{T_k}\colon G_{T_k} \longrightarrow P_{T_k}$ is an isomorphism for all $k \in I$. Again by Proposition 7.1.3, P_{T_k} is the pullback of the localizations at T_k of the maps $(\phi_i)_S$ and ϕ_S. It suffices to show that G_{T_k} is also the pullback of these localizations at T_k. To get around the fact that localization does not commute with infinite products, we think of $\prod_{i\in I} G_{T_i}$ as the product of the two groups G_{T_k} and $\prod_{j\neq k} G_{T_j}$. Localization at T_k commutes with finite products and is the identity on T_k-local groups, such as G_S. Thus, after localization at T_k, our maps are

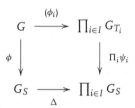

$$\begin{array}{ccc}
 & & G_{T_k} \times (\prod_{j\neq k} G_{T_j})_{T_k} \\
 & & \downarrow (\psi_k, \psi_k) \\
G_S & \xrightarrow[(\mathrm{id},(\phi_j)_S)]{} & G_S \times (\prod_{j\neq k} G_{T_j})_S.
\end{array}$$

We claim that the group $H = (\prod_{j \neq k} G_{T_j})_{T_k}$ is S-local, so that the second component ψ_k of the right-hand map is the identity. By Lemma 7.1.9, this will imply that G_{T_k} is the pullback of these two maps and will thus complete the proof. To prove the claim, let q be a prime that is not in S. If q is not in T_k, then the qth power function $H \longrightarrow H$ is a bijection since H is T_k-local. If q is in T_k, then q is not in any T_j with $j \neq k$ since q is not in $S = T_j \cap T_k$. The qth power function is then an isomorphism on each G_{T_j} and therefore also on H. This proves the claim.

An alternative argument is possible. One can first prove this in the abelian case by the argument just given and then argue by induction on the degree of nilpotency of G. The argument is not uninteresting but is longer. We give a general group theoretical result that specializes to the required inductive step in Lemma 7.6.2.

We must still prove the last statement in (ii). We show this when $G = A$ is abelian here. Using the exactness of localization at T_i and S and the fact that a product of exact sequences is exact, the proof when G is nilpotent is completed by induction up a central series, using Lemma 7.6.1 below. In the abelian case, let C be the cokernel of the difference map

$$(\phi_i)_S - \phi_S \colon A_S \times \prod_i A_{T_i} \longrightarrow (\prod_i A_{T_i})_S,$$

whose kernel is A. It suffices to prove that $C = 0$. Since C is a $\mathbb{Z}_T$-module, this holds if the localization of C at each prime $p \in T$ is zero (e.g., [6, Prop. 3.8]). In turn, since p is in some T_k, this holds if the localization of C at T_k is zero. But we have just shown that after localization at T_k we obtain a pullback diagram of the simple form displayed in Lemma 7.1.9, and it is obvious that the difference map for such a pullback of modules over any ring is an epimorphism.

(iii). In this case we do not have a direct argument that works for general nilpotent groups. We first use the structure theorem for finitely generated modules over the PID $\mathbb{Z}_T$ to prove the result in the abelian case. Here the result reduces to the case of cyclic T-modules, where the proof is no more difficult than that of Proposition 7.1.7. A few more details will be given in the local to global analogue, Proposition 7.4.4 below. The general case follows by induction on the degree of nilpotency of G, using that T-local subgroups of finitely T-generated T-local groups are T-local, by Proposition 5.6.5, and that the center of a T-local group is T-local, by Lemma 5.4.5.

The inductive step can be viewed as a specialization of Lemma 7.6.2 below, but special features of the case at hand allow a simpler argument. Thus let

$$1 \longrightarrow H \stackrel{\iota}{\longrightarrow} G \stackrel{\rho}{\longrightarrow} K \longrightarrow 1$$

be an exact sequence of finitely T-generated T-local groups and assume that the conclusion holds for H and K. Let $g_i \in G_{T_i}$ and $g_0 \in G_S$ be given such that $\psi_i(g_i) = g_0$ for all $i \in I$. We must show that there is a unique element $g \in G$ such that $\phi_i(g) = g_i$ for all i and therefore $\phi(g) = \psi_i \phi_i(g) = g_0$. There is a unique element $k \in K$ such that $\phi_i(k) = \rho_i(g_i)$ for all i, where ρ_i denotes the localization of ρ at T_i. Choose $g' \in G$ such that $\rho(g') = k$. Then $\rho_i(g_i) = \phi_i \rho(g') = \rho_i \phi_i(g')$, hence $g_i = \phi_i(g') \iota_i(h_i)$ where $h_i \in H_{T_i}$ and, again, ι_i denotes the localization of ι at T_i. Since $\psi_i(g_i) = g_0$ and $\psi_i \phi_i = \phi$, $\psi_i(h_i)$ is independent of i. Therefore there is a unique element $h \in H$ such that $\phi_i(h) = h_i$ for all i. Let $g = g' \iota(h)$. Then

$$\phi_i(g) = \phi_i(g') \phi_i \iota(h) = \phi_i(g') \iota_i(h_i) = g_i$$

for all i. To prove that g is unique, it suffices to show that 1 is the unique element $g \in G$ such that $\phi_i(g) = 1 \in G_{T_i}$ for all i. For such a g, $\rho(g) = 1$ by the uniqueness in K, so that we can write $g = \iota(h)$. Then we must have $\phi_i(h) = 1 \in H_{T_i}$ for all i, hence $h = 1$ by the uniqueness in H. $\quad\square$

REMARK 7.2.2. Theorem 7.2.1(ii) was proven under a finite generation hypothesis in Hilton, Mislin, and Roitberg [62], and Hilton and Mislin later noticed that the hypothesis can be removed [61]. That fact is not as well-known as it should be. We learned both it and most of the elegant proof presented here from Bousfield.

7.3. Local to global pullback diagrams

In §7.2, we started with a T-local group and showed that it was isomorphic to the pullback of some of its localizations. In the next section, the results go in the opposite direction. We start with local groups with suitable compatibility, and we use these to construct a "global" group. Again, all given groups are to be nilpotent. However, we need some preliminaries since, a priori, we do not have an analogue of the pullback diagram (7.1.1).

Quite generally, we do have an analogue of the pullback diagram (7.1.2). Let I be an indexing set and suppose that we are given groups H and G_i for $i \in I$

together with homomorphisms $\psi_i \colon G_i \longrightarrow H$. We understand the pullback of the ψ_i's to be the pullback Q displayed in the diagram

7.3.1
$$
\begin{array}{ccc}
Q & \xrightarrow{\ (\delta_i)\ } & \prod_{i \in I} G_i \\
{\scriptstyle \varepsilon} \downarrow & & \downarrow {\scriptstyle \Pi_i \psi_i} \\
H & \xrightarrow[\ \Delta\ ]{} & \prod_{i \in I} H.
\end{array}
$$

Now return to the standing assumptions in this chapter, so that we have an indexing set I together with sets of primes T_i for $i \in I$ such that $T = \bigcup T_i$, $T_i \cap T_j = S$ for $i \neq j$, and $T_i \neq S$ for $i \in I$. We consider the pullback diagram (7.3.1) when G_i is T_i-local, H is S-local, and $\psi_i \colon G_i \longrightarrow H$ is localization at S for each i. Since it is a pullback of T-local groups, the group Q is then T-local. In general, we cannot expect the coordinates δ_k to all be localizations at k. By comparison with §7.2, we expect to encounter difficulties when the indexing set I is infinite, and that is indeed the case. However, in view of Proposition 7.1.7, we also expect these difficulties to diminish under finite generation hypotheses. That, however, is less true than we might expect.

The diagram (7.1.2) displays the special case of (7.3.1) that we obtain when we start with a T-local group G and consider its localizations at the T_i and S. As in §7.2, we would like to have a companion pullback diagram

7.3.2
$$
\begin{array}{ccc}
P & \xrightarrow{\ (\mu_i)\ } & \prod_{i \in I} G_i \\
{\scriptstyle \nu} \downarrow & & \downarrow {\scriptstyle \phi_S} \\
H & \xrightarrow[\ \omega\ ]{} & \left(\prod_{i \in I} G_i \right)_S,
\end{array}
$$

where ϕ_S is a localization at S. However, since we do not start with a global group G and its localizations ϕ_i, as in (7.1.1), we do not, a priori, have a map ω. This suggests the following definition. Observe that if $\pi_i \colon \prod G_i \to G_i$ is the projection, then there is a unique map $\tilde{\pi}_i \colon (\prod G_i)_S \to H$ such that $\tilde{\pi}_i \circ \phi_S = \psi_i \circ \pi_i$, and $(\tilde{\pi}_i)$ is the localization $(\psi_i)_S$ of the map $(\psi_i) \colon \prod G_i \longrightarrow \prod_i H$.

DEFINITION 7.3.3. Let G_i be a T_i-local group and H be an S-local group, and let $\psi_i \colon G_i \longrightarrow H$ be a localization at S for each i. Let $\phi_S \colon \prod_i G_i \longrightarrow (\prod_i G_i)_S$

be a localization at S and let $\tilde{\pi}_i\colon (\prod_{i\in I} G_i)_S \longrightarrow H$ be the unique map such that $\tilde{\pi}_i \circ \phi_S = \psi_i \circ \pi_i$ for all i. A formal localization associated to the ψ_i is a map $\omega\colon H \longrightarrow (\prod_{i\in I} G_i)_S$ such that the composite $\tilde{\pi}_i \circ \omega$ is the identity map of H for each $i \in I$.

The name comes from the fact that when ω exists, it turns out that the map $\mu_i\colon P \longrightarrow G_i$ in (7.3.2) is a localization at T_i, hence the map ν is a localization at S and ω is the localization of (μ_i) at S. There must be a formal localization ω whenever the maps $\delta_k\colon Q \longrightarrow G_k$ in (7.3.1) are localizations at T_k for all k. Indeed, the composites $\psi_k \circ \delta_k = \varepsilon\colon Q \longrightarrow H$ are then localizations at S for all k. By Theorem 7.2.1(ii), if we define P to be the pullback of ϕ_S and the localization $\omega = (\delta_i)_S$ such that $\omega \circ \varepsilon = \phi_S \circ (\delta_i)$, then the resulting canonical map $\alpha\colon Q \longrightarrow P$ that we obtain must be an isomorphism. We conclude that a general local to global construction that recovers the local groups that we start with must incorporate the existence of a map ω as in (7.3.2).

When we are given a formal localization ω, we obtain a comparison diagram and a comparison map $\gamma\colon P \longrightarrow Q$ analogous to (7.1.4), but with G_{T_i} and G_S there replaced by G_i and H.

7.3.4

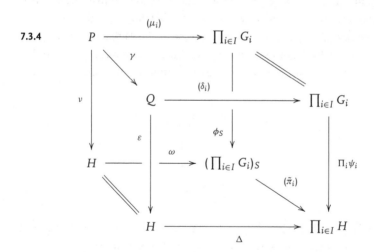

In view of the following observation, the notion of a formal localization is only needed when the indexing set I is infinite.

LEMMA 7.3.5. *Let I be finite. Then $(\tilde{\pi}_i)$ is an isomorphism and the composite $\omega = (\tilde{\pi}_i)^{-1} \circ \Delta$ is the unique formal localization associated to the $\psi_i\colon G_i \longrightarrow H$, hence the diagrams (7.3.2) and (7.3.1) are canonically isomorphic.*

7.4. Local to global: abelian and nilpotent groups

With the notations of (7.3.2), here is our main algebraic local to global theorem.

THEOREM 7.4.1. *Let G_i be a T_i-local group and H be an S-local group, and let $\psi_i \colon G_i \longrightarrow H$ be a localization at S for each $i \in I$. If I is finite or, more generally, if I is infinite and the ψ_i have a formal localization ω, then P is T-local and the induced map $\tilde{\mu}_k \colon P_{T_k} \longrightarrow G_k$ is an isomorphism for each $k \in I$.*

PROOF. Despite the difference in context, the proof is exactly the same as the proof of part (ii) of Theorem 7.2.1, with G_{T_i} and G_S there replaced by G_i and H and with $(\phi_i)_S$ there replaced by ω. Thus P is T-local since it is a pullback of T-local groups, and the maps $\tilde{\mu}_k$ are isomorphisms since Lemma 7.1.9 applies to show that G_k agrees with the pullback obtained by localizing the pullback diagram that defines P at T_k. $\qquad\square$

REMARK 7.4.2. In fact, the global to local result Theorem 7.2.1(ii) can be viewed as a direct corollary of the local to global result Theorem 7.4.1. To see that, we apply Theorem 7.4.1 to G_{T_i} and G_0 for a given T-local group G and then use Lemma 7.1.8 to conclude that $\alpha \colon G \longrightarrow P$ is an isomorphism.

In many of the applications, the indexing set I is finite and of course it is then most natural to work directly with Q rather than introduce formal localizations. In that case, the following addendum is important. Recall the characterization of $f\mathbb{Z}_T$-nilpotent groups from Proposition 5.6.5.

PROPOSITION 7.4.3. *Assume in Theorem 7.4.1 that I is finite and G_i is $f\mathbb{Z}_{T_i}$-nilpotent. Then Q is $f\mathbb{Z}_T$-nilpotent.*

PROOF. Assume first that $G_i = A_i$ is abelian. Then A_i is a finitely generated $\mathbb{Z}_{T_i}$-module. Multiplying any given generators by scalars to clear denominators, we can assume that the generators of A_i are in the image of the $\mathbb{Z}_T$-module Q. Let $Q' \subset Q$ be the $\mathbb{Z}_T$-submodule generated by pre-images of the generators of the A_i for all $i \in I$. The localization of Q/Q' at each prime $p \in T$ is zero, hence $Q/Q' = 0$ and Q is a finitely generated $\mathbb{Z}_T$-module.

The nilpotent case is proven by induction on the least common bound q of the lower central series

$$\{1\} = G_{i,q} \subset G_{i,q-1} \subset \cdots \subset G_{i,1} \subset G_{i,0} = G_i$$

of the G_i. We prove inductively that G is $f\mathbb{Z}_T$-nilpotent. It is therefore finitely T-generated by Proposition 5.6.5. We use the lower central series since it is functorial, so that the lower central series of each G_i maps to the lower central series

$$\{1\} = H_q \subset H_{q-1} \subset \cdots \subset H_1 \subset H_0 = H$$

of H. We have induced maps of central extensions

$$
\begin{array}{ccccccccc}
1 & \longrightarrow & G_{i,j}/G_{i,j+1} & \longrightarrow & G_i/G_{i,j+1} & \longrightarrow & G_i/G_{i,j} & \longrightarrow & 1 \\
& & \downarrow & & \downarrow & & \downarrow & & \\
1 & \longrightarrow & H_j/H_{j+1} & \longrightarrow & H/H_{j+1} & \longrightarrow & H/H_j & \longrightarrow & 1.
\end{array}
$$

For each j, we obtain three pullbacks as in (7.3.1), and they assemble to a central extension of pullbacks by Lemma 7.6.2(ii), whose key epimorphism hypothesis is satisfied by the abelian case of the last statement of Theorem 7.2.1(ii). □

There is sometimes an analogue of Theorem 7.4.1 for the pullback Q, rather than the pullback P, even when I is infinite. When this holds, $\gamma : P \longrightarrow Q$ is an isomorphism. However, the analogous statement for homotopy pullbacks in topology will fail in general.

PROPOSITION 7.4.4. *Let A_i be a finitely generated $\mathbb{Z}_{T_i}$-module and B be a finitely generated $\mathbb{Z}_S$-module, and let $\psi_i : A_i \longrightarrow B$ be a localization at S for each $i \in I$. Assume that A_i has no $(T_i - S)$-torsion for all but finitely many i. Then the induced map $Q_{T_k} \longrightarrow A_k$ is an isomorphism for all $k \in I$.*

PROOF. We use the structure theory for modules over a PID. Since the A_i all localize to B at S, we find that

$$A_i \cong F_{T_i} \oplus C_i \oplus D \quad \text{and} \quad B \cong F_T \oplus D$$

for some finitely generated free abelian group F, some finite $(T_i - S)$-torsion abelian groups C_i, and some finite S-torsion abelian group D. Here all but finitely many of the C_i are zero. Under these isomorphisms, ψ_i is the sum of a localization $F_{T_i} \longrightarrow F_S$ at S, the zero homomorphism on C_i, and an isomorphism $D \longrightarrow D$. From this, we *cannot* conclude that

$$Q \cong F_T \oplus \prod_{i \in I} C_i \oplus D.$$

However, observing as in the proof of Theorem 7.2.1(iii) that $(\prod_{j\neq k} A_j)_{T_k}$ is S-local, we can conclude that for each k

$$Q_{T_k} \cong F_{T_k} \oplus C_k \oplus D. \qquad \square$$

REMARK 7.4.5. In contrast with Proposition 7.4.3, we shall see in the next section that Q need not be finitely T-generated even when F is free on one generator and all C_i and D are zero.

REMARK 7.4.6. One might conjecture that a generalization of Proposition 7.4.4 would apply when given $f\mathbb{Z}_{T_i}$-nilpotent groups G_i with a common bound on their nilpotency class. The outline of proof would follow the proof of Proposition 7.4.3. However, application of Lemma 7.6.2(ii) fails since its key epimorphism hypothesis usually fails, as noted in Proposition 7.1.7.

7.5. The genus of abelian and nilpotent groups

It is natural to ask how many groups can have isomorphic localizations at each set of primes T_i. In general, this relates to the question of how unique formal localizations are. This in turn raises the question of how unique the localizations $\psi_i\colon G_i \longrightarrow H$ are in the local to global context. For definiteness and familiarity, we assume that T is the set of all primes, S is the empty set, and T_i is the set consisting of just the ith prime number p_i.

A first thoughtless answer is that the ψ_i are unique since the rationalization $\psi_i\colon G_i \longrightarrow H$ of a p_i-local group G_i is unique. But of course it is only unique up to a universal property, and if one composes ψ_i with an isomorphism $\xi_i\colon H \longrightarrow H$, then the resulting composite $\psi_i' = \xi_i\psi_i$ again satisfies the universal property. If we have formal localizations associated to the ψ_i and the ψ_i' and form the associated pullbacks P and P', then P and P' need not be isomorphic, but they do have isomorphic localizations at each prime p_i. This leads to the following definition.

DEFINITION 7.5.1. Let G be a nilpotent group. The extended genus of G is the collection of isomorphism types of nilpotent groups G' such that the localizations G_p and G_p' are isomorphic for all primes p. If G is finitely generated, then the genus of G is the set of isomorphism classes of finitely generated nilpotent groups in the extended genus of G.

There is an extensive literature on these algebraic notions, and we shall not go into detail. Rather, we shall explain how to calculate the extended genus under a simplifying hypothesis that holds when G is finitely generated and serves to eliminate the role of formal localizations. The actual genus can then be sought inside the extended genus, one method being to exploit partitions of the set of primes into finite sets and make use of Proposition 7.4.3. This gives a partial blueprint for the analogous topological theory.

Let $\mathrm{Aut}\,(G)$ denote the group of automorphisms of G. We show that elements of the extended genus are usually in bijective correspondence with double cosets

$$\mathrm{Aut}\,(G_0)\backslash \prod_i \mathrm{Aut}\,(G_0)/ \prod_i \mathrm{Aut}\,(G_i).$$

Any nilpotent group G is isomorphic to a pullback as displayed in (7.1.1). Here we start from localizations $\phi_i \colon G \longrightarrow G_i$ of G at p_i and rationalizations $\psi_i \colon G_i \to G_0$ and use the resulting rationalization $\omega = (\phi_i)_0 \colon G_0 \to (\prod_i G_i)_0$, which is a formal rationalization of G_0. We can reconstruct a representative group in each element of the extended genus of G starting from these fixed groups G_i and G_0, using pullbacks as displayed in (7.3.2). However, the rationalizations $\psi_i \colon G_i \longrightarrow G_0$ used in the specification of the relevant formal completion $\omega \colon G_0 \longrightarrow (\prod_i G_i)_0$ can vary, the variation being given by an automorphism ξ_i of the rational nilpotent group G_0. Similarly, the rationalization of $\prod_i G_i$ can vary by an automorphism of the rational nilpotent group $(\prod_i G_i)_0$.

To be more precise about this, observe that, up to isomorphism, any two groups G and G' in the same extended genus can be represented as pullbacks P and P' as displayed in the top triangles of commutative diagrams

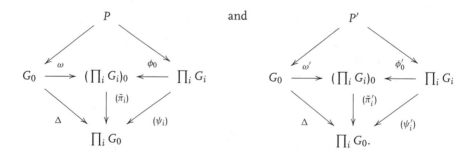

Here $\tilde{\pi}_i$ is the unique map $(\prod_i G_i)_0 \longrightarrow G_0$ such that $\tilde{\pi}_i \circ \phi_0 = \psi_i \circ \pi_i$, and similarly for $\tilde{\pi}_i'$. Usually the pullbacks P and P' of (ω, ϕ_0) and (ω', ϕ_0') are not isomorphic. We fix a reference pullback $P \cong G$ and have the following result.

PROPOSITION 7.5.2. *Assume that $(\tilde{\pi}_i)$ is a monomorphism. Then the extended genus of G is in bijective correspondence with*

$$\operatorname{Aut}(G_0)\backslash \prod_i \operatorname{Aut}(G_0)/ \prod_i \operatorname{Aut}(G_i).$$

PROOF. Since $(\tilde{\pi}_i)$ is a monomorphism, ω is uniquely determined and, by Lemma 7.1.6, the canonical map $\gamma: P \longrightarrow Q$ from the pullback P of (ω, ϕ_0) to the pullback Q of $(\Delta, (\psi_i))$ is an isomorphism. Since localizations are unique up to automorphisms of their targets, we see that $(\tilde{\pi}_i')$ is also a monomorphism and that the analogous canonical map $\gamma': P' \longrightarrow Q'$ is also an isomorphism. Thus the monomorphism hypothesis allows us to ignore formal localizations and concentrate on pullbacks of diagrams of the form $(\Delta, (\psi_i'))$.

The double cosets are defined with respect to $\Delta: G_0 \longrightarrow \prod_i \operatorname{Aut}(G_0)$ and the homomorphisms $\operatorname{Aut}(G_i) \longrightarrow \operatorname{Aut}(G_0)$ that send an automorphism ζ_i of G_i to the unique automorphism $\tilde{\zeta}_i$ of G_0 such that $\tilde{\zeta}_i \circ \psi_i = \psi_i \circ \zeta_i$. We emphasize that this definition refers to the ψ_i of the fixed reference pullback $P \cong Q$.

For any rationalizations $\psi_i': G_i \longrightarrow G_0$, there are automorphisms ξ_i' of G_0 such that $\xi_i' \circ \psi_i = \psi_i'$. Sending the automorphism $\xi' = (\xi_i')$ to the isomorphism class of the pullback Q' of

$$G_0 \xrightarrow{\Delta} \prod_i G_0 \xleftarrow{(\psi_i')} \prod_i G_i$$

gives a surjection from the set $\operatorname{Aut}(\prod_i G_0)$ to the extended genus of G. Suppose we have an isomorphism $\zeta: Q' \longrightarrow Q''$ between two such pullbacks and consider the following diagram, in which the front and back squares are pullbacks.

7.5.3

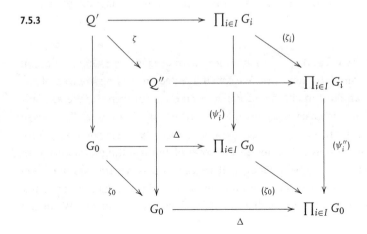

The unlabeled left vertical arrows are rationalizations and the coordinates of the unlabeled top horizontal arrows are localizations at p_i. The universal property of localizations gives automorphisms ζ_i of G_i and ζ_0 of G_0 making the top and left squares commute. The bottom square obviously commutes, and then the right square must also commute, so that $\zeta_0 \circ \psi_i' = \psi_i'' \circ \zeta_i$. As above, we have a unique automorphism $\tilde{\zeta}_i$ of G_0 such that $\tilde{\zeta}_i \circ \psi_i = \psi_i \circ \zeta_i$. Writing $\psi_i' = \xi_i' \circ \psi_i$ and $\psi_i'' = \xi_i'' \circ \psi_i$, these equalities imply that $\xi_i'' = \zeta_0 \circ \xi_i' \circ \tilde{\zeta}_i^{-1}$, so that the automorphisms $\xi' = (\xi_i')$ and $\xi'' = (\xi_i'')$ are in the same double coset. Conversely, if ξ' and ξ'' are in the same double coset, then we obtain an isomorphism ζ as displayed in (7.5.3). $\qquad\square$

REMARK 7.5.4. If G is abelian, then $(\tilde{\pi}_i)$ is a monomorphism if and only if all but finitely many of the p_i-local abelian groups G_i are torsion free. It follows that $(\tilde{\pi}_i)$ is a monomorphism when G is f-nilpotent or, equivalently by Proposition 4.5.9, finitely generated.

We give an illuminating elementary example. Above, we started with a given group G together with fixed localizations $G_i = G_{p_i}$ and a fixed rationalization G_0. If we start out with an abelian group $A = A \otimes \mathbb{Z}$, we have the canonical localizations $A_i = A \otimes \mathbb{Z}_{(p_i)}$ and the canonical rationalization $A_0 = A \otimes \mathbb{Q}$. The inclusions $\mathbb{Z} \subset \mathbb{Z}_{(p_i)}$, $\mathbb{Z} \subset \mathbb{Q}$, and $\mathbb{Z}_{(p_i)} \subset \mathbb{Q}$ induce canonical localizations and rationalizations

$$\phi_i: A \longrightarrow A_i, \quad \phi: A \longrightarrow A_0, \quad \text{and} \quad \psi_i: A_i \longrightarrow A_0$$

such that $\psi_i \phi_i = \phi$. This gives a canonical pullback diagram with pullback A.

EXAMPLE 7.5.5. Let $A = \mathbb{Z}$. An automorphism of $\mathbb{Q}$ is just a choice of a unit in $\mathbb{Q}$, and similarly for $\mathbb{Z}_{(p_i)}$. We are interested in varying choices of ψ_i', and these amount to choices of nonzero rational numbers ξ_i. We are only interested in the coset of ξ_i modulo the action of the units of $\mathbb{Z}_{(p_i)}$ given by multiplication, hence we may as well take $\xi_i = p^{-r_i}$ for some $r_i \geq 0$ for each i. We are only interested in the coset of the resulting automorphism $(p_i^{-r_i})$ modulo action by units of $\mathbb{Q}$. If we can clear denominators, the coset is that of (1_i), so that to obtain noncanonical pullback diagrams up to isomorphism, we must assume that $r_i > 0$ for all but finitely many i. We obtain a diagram

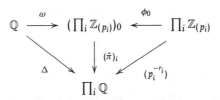

by letting ϕ_0 be the canonical rationalization, $\phi_0(1_i) = (1_i)$, so that $\tilde{\pi}_i(1_i) = p_i^{-r_i}$. Then ω can and must be defined by setting $\omega(1) = (p_i^{r_i})$. The pullback that we obtain is the subgroup of $\mathbb{Q}$ consisting of fractions (in reduced form) whose denominators are not divisible by $p_i^{r_i+1}$ for any i. These are infinitely generated groups when $r_i > 0$ for all but finitely many i. Multiplication by a fixed nonzero rational number gives an isomorphism from one of these groups to another. The resulting isomorphism classes give the extended genus of $\mathbb{Z}$, which is uncountable.

In contrast, the structure theory for finitely generated abelian groups has the following immediate consequence.

PROPOSITION 7.5.6. *The isomorphism class of A is the only element of the genus of a finitely generated abelian group A.*

EXAMPLE 7.5.7. In contrast, non-isomorphic finitely generated nilpotent groups can be in the same genus, and this can already happen when the nilpotency class is two. A class of examples due to Milnor is described in [62, p. 32]. If r and s are relatively prime integers, let $G_{r/s}$ be the group with four generators g_1, g_2, h_1, h_2 and with relations specified by letting

$$[g_1, g_2]^s = 1 \quad \text{and} \quad [g_1, g_2]^r = [h_1, h_2]$$

and letting all triple commutators be 1. Then $G_{r/s}$ and $G_{r'/s'}$ are non-isomorphic groups that are in the same genus if and only if either $r \equiv r' \bmod s$ or $rr' \equiv \pm 1 \bmod s$.

REMARK 7.5.8. The word "genus" is due to Mislin [105], following an analogy due to Sullivan [133], and has nothing to do with the use of the word elsewhere in mathematics. Rather, the analogy is with genetics or, perhaps better, taxonomy. Think of a group as an animal, isomorphic groups as animals in the same species, and groups in the same genus as animals in the same genus.

Since groups in the same extended genus can be quite unlike each other, they might be thought of as animals in the same family.

7.6. Exact sequences of groups and pullbacks

We prove two results about pullback diagrams and exact sequences here. Exceptionally, we do not require our groups to be nilpotent. Our first result was used in the proof of Theorem 7.2.1(ii) and will be used again in Chapter 12.

LEMMA 7.6.1. *Suppose that the rows in the following commutative diagram are exact and that the image of ρ_1 is a central subgroup of G_1.*

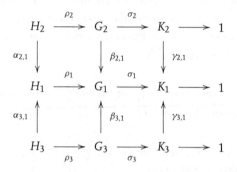

Assume that for each element $h_1 \in H_1$ there are elements $h_2 \in H_2$ and $h_3 \in H_3$ such that $h_1 = \alpha_{2,1}(h_2)\alpha_{3,1}(h_3)$ and similarly for the groups K_i. Then for each $g \in G_1$, there are elements $g_2 \in G_2$ and $g_3 \in G_3$ such that $g_1 = \beta_{2,1}(g_2)\beta_{3,1}(g_3)$.

PROOF. Let $g_1 \in G_1$. There are elements $k_2 \in K_2$ and $k_3 \in K_3$ such that $\sigma_1(g_1) = \gamma_{2,1}(k_2)\gamma_{3,1}(k_3)$. There are also elements $g_2' \in G_2$ and $g_3' \in G_3$ such that $\sigma_2(g_2') = k_2$ and $\sigma_3(g_3') = k_3$. Then $\sigma_1(g_1^{-1}\beta_{2,1}(g_2')\beta_{3,1}(g_3')) = 1$, so there is an element $h_1 \in H_1$ such that $\rho_1(h_1^{-1}) = g_1^{-1}\beta_{2,1}(g_2')\beta_{3,1}(g_3')$ and therefore, by centrality, $g_1 = \beta_{2,1}(g_2')\rho_1(h_1)\beta_{3,1}(g_3')$. Moreover, there are elements $h_2 \in H_2$ and $h_3 \in H_3$ such that $h_1 = \alpha_{2,1}(h_2)\alpha_{3,1}(h_3)$ and thus $\rho_1(h_1) = \beta_{2,1}\rho_2(h_2)\beta_{3,1}\rho_3(h_3)$. Let $g_2 = g_2'\rho_2(h_2)$ and $g_3 = \rho_3(h_3)g_3'$. Then $g_1 = \beta_{2,1}(g_2)\beta_{3,1}(g_3)$. Note that we do not assume and do not need any of the ρ_i to be monomorphisms in this proof. $\square$

The first part of the following result is relevant to global to local results. It allows alternative inductive proofs of some of our results in this chapter and is used in Chapter 12. The second part is relevant to local to global results. It was used to prove Proposition 7.4.3.

LEMMA 7.6.2. *Consider a commutative diagram of groups*

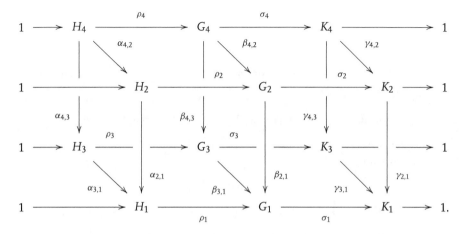

Assume that the three rows

$$1 \longrightarrow H_i \xrightarrow{\rho_i} G_i \xrightarrow{\sigma_i} K_i \longrightarrow 1,$$

$1 \le i \le 3$ *are exact. Consider the fourth row and the three squares*

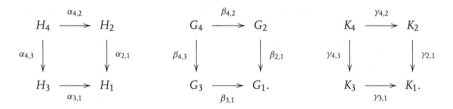

(i) *If the fourth row is a central extension and the left and right squares are pullbacks, then the middle square is a pullback.*

(ii) *If all three squares are pullbacks and every element $h_1 \in H_1$ is a product $\alpha_{2,1}(h_2)\alpha_{3,1}(h_3)$ for some $h_2 \in H_2$ and $h_3 \in H_3$, then the fourth row is exact.*

PROOF.

(i) We verify the universal property required for the middle square to be a pullback. Let $\tau_2 \colon G \longrightarrow G_2$ and $\tau_3 \colon G \longrightarrow G_3$ be homomorphisms such that $\beta_{2,1}\tau_2 = \beta_{3,1}\tau_3$. We must show there is a unique homomorphism $\tau_4 \colon G \longrightarrow G_4$ such that $\beta_{4,2}\tau_4 = \tau_2$ and $\beta_{4,3}\tau_4 = \tau_3$.

Note that $\gamma_{2,1}\sigma_2\tau_2 = \sigma_1\beta_{2,1}\tau_2 = \sigma_1\beta_{3,1}\tau_3 = \gamma_{3,1}\sigma_3\tau_3$. Since the square formed by the K_i's is a pullback there is an induced homomorphism

$$\omega \colon G \longrightarrow K_4$$

such that $\gamma_{4,2}\omega = \sigma_2\tau_2$ and $\gamma_{4,3}\omega = \sigma_3\tau_3$.

Since σ_4 is surjective, for each $g \in G$ there is a $g_4 \in G_4$ such that $\sigma_4(g_4) = \omega(g)$. Then

$$\sigma_2(\beta_{4,2}(g_4) \cdot \tau_2(g^{-1})) = \gamma_{4,2}\sigma_4(g_4) \cdot \sigma_2\tau_2(g^{-1}) = \gamma_{4,2}\omega(g) \cdot \sigma_2\tau_2(g^{-1}) = 1,$$

hence there is a unique $h_2 \in H_2$ such that $\rho_2(h_2) = \beta_{4,2}(g_4) \cdot \tau_2(g^{-1})$. Similarly,

$$\sigma_3(\beta_{4,3}(g_4) \cdot \tau_3(g^{-1})) = 1,$$

hence there is a unique $h_3 \in H_3$ such that $\rho_3(h_3) = \beta_{4,3}(g_4) \cdot \tau_3(g^{-1})$. Note that

$$\rho_1(\alpha_{2,1}(h_2) \cdot \alpha_{3,1}(h_3^{-1})) = \beta_{2,1}\rho_2(h_2) \cdot \beta_{3,1}\rho_3(h_3^{-1})$$

$$= \beta_{2,1}\beta_{4,2}(g_4) \cdot \beta_{2,1}\tau_2(g^{-1}) \cdot \beta_{3,1}\tau_3(g) \cdot \beta_{3,1}\beta_{4,3}(g_4^{-1})$$

$$= 1.$$

Since ρ_1 is injective, $\alpha_{2,1}(h_2) = \alpha_{3,1}(h_3)$. Since the square formed by the H_i's is a pullback, there is a unique element $h_4 \in H_4$ such that $\alpha_{4,2}(h_4) = h_2$ and $\alpha_{4,3}(h_4) = h_3$. Then, using that $\rho_4(H_4)$ is central in G_4 and $\beta_{4,2}\rho_4 = \rho_2\alpha_{4,2}$,

$$\beta_{4,2}(g_4 \cdot \rho_4(h_4^{-1})) = \beta_{4,2}(\rho_4(h_4^{-1}) \cdot g_4)$$

$$= \rho_2(h_2^{-1}) \cdot \beta_{4,2}(g_4)$$

$$= \tau_2(g) \cdot \beta_{4,2}(g_4^{-1}) \cdot \beta_{4,2}(g_4) = \tau_2(g).$$

Similarly,

$$\beta_{4,3}(g_4 \cdot \rho_4(h_4^{-1})) = \tau_3(g).$$

To each element $g \in G$ this associates an element $\tau_4(g) = g_4 \cdot \rho_4(h_4^{-1}) \in G_4$ such that $\beta_{4,2}(\tau_4(g)) = \tau_2(g)$ and $\beta_{4,3}(\tau_4(g)) = \tau_3(g)$. To show that τ_4 is a well-defined function $G \longrightarrow G_4$, we must show that it is independent of the choice of g_4. Again using that ρ_4 factors through the center of G_4, it will follow that τ_4 is a homomorphism.

Suppose that g_4' is another element of G_4 such that $\sigma_4(g_4') = \omega(g)$. As above, we obtain elements h_2', h_3', and h_4' such that

$$\rho_2(h_2') = \beta_{4,2}(g_4') \cdot \tau_2(g^{-1}), \quad \rho_3(h_3') = \beta_{4,3}(g_4') \cdot \tau_3(g^{-1}),$$

$$\alpha_{4,2}(h_4') = h_2', \quad \text{and} \quad \alpha_{4,3}(h_4') = h_3'.$$

Again as above, this implies that

$$\beta_{4,2}(g_4' \cdot \rho_4((h_4')^{-1})) = \tau_2(g) \quad \text{and} \quad \beta_{4,3}(g_4' \cdot \rho_4((h_4')^{-1})) = \tau_3(g).$$

We must show that $g_4 \cdot \rho_4(h_4^{-1}) = g_4' \cdot \rho_4((h_4')^{-1})$. Since $\sigma_4 \rho_4$ is trivial,

$$\sigma_4(g_4 \cdot \rho_4(h_4^{-1}) \cdot \rho_4(h_4') \cdot (g_4')^{-1}) = \sigma_4(g_4) \cdot \sigma_4((g_4')^{-1}) = 1.$$

Therefore, there is an element $x \in H_4$ such that $\rho_4(x) = g_4 \cdot \rho_4(h_4^{-1}) \cdot \rho_4(h_4') \cdot (g_4')^{-1}$. Then, using that $\rho_2 \alpha_{4,2} = \beta_{4,2} \rho_4$,

$$\rho_2 \alpha_{4,2}(x) = \beta_{4,2}(g_4) \cdot \beta_{4,2} \rho_4(h_4^{-1}) \cdot \beta_{4,2} \rho_4(h_4') \cdot \beta_{4,2}((g_4')^{-1})$$

$$= \tau_2(g) \cdot \tau_2(g^{-1}) = 1.$$

Since ρ_2 is injective $\alpha_{4,2}(x) = 1$. Similarly, $\alpha_{4,3}(x) = 1$. Therefore $x = 1$. This implies that the map $\tau_4 : G \longrightarrow G_4$ is a well-defined homomorphism.

A similar argument shows that τ_4 is unique. Suppose we have another homomorphism $\tau_4' : G \longrightarrow G_4$ such that $\beta_{4,2} \tau_4' = \tau_2$ and $\beta_{4,3} \tau_4' = \tau_3$. Little diagram chases show that

$$\gamma_{4,2} \sigma_4 \tau_4 = \gamma_{4,2} \sigma_4 \tau_4' \quad \text{and} \quad \gamma_{4,3} \sigma_4 \tau_4 = \gamma_{4,3} \sigma_4 \tau_4'.$$

Since the square formed by the K_i's is a pullback, this implies that $\sigma_4 \tau_4 = \sigma_4 \tau_4'$. Therefore $\tau_4(g)(\tau_4')^{-1}(g) = \rho_4(x)$ for some $x \in H_4$. Further little diagram chases and the fact that ρ_2 and ρ_3 are injective imply that $\alpha_{4,2}(x) = 1$ and $\alpha_{4,3}(x) = 1$. Since the square formed by the H_i's is a pullback, this implies that $x = 1$ and therefore that $\tau_4 = \tau_4'$.

(ii) It is clear that ρ_4 is a monomorphism, $\sigma_4 \rho_4$ is trivial, and $\ker(\sigma_4) = \mathrm{im}(\rho_4)$. We must show that the map σ_4 is an epimorphism. Let $(k_2, k_3) \in K_4$, so that $\gamma_{2,1}(k_2) = \gamma_{3,1}(k_3)$, and choose $g_2' \in G_2$ and $g_3' \in G_3$ such that $\sigma_2(g_2') = k_2$ and $\sigma_3(g_3') = k_3$. Then $\sigma_1(\beta_{2,1}(g_2')\beta_{3,1}(g_3')^{-1}) = 1$, hence $\beta_{2,1}(g_2')(\beta_{3,1}(g_3'))^{-1} = \rho_1(h_1)$ and $h_1 = \alpha_{2,1}(h_2^{-1})\alpha_{3,1}(h_3)$ for some $h_i \in H_i$. Let $g_2 = \rho_2(h_2)g_2'$ and $g_3 = \rho_3(h_3)g_3'$. Then $(g_2, g_3) \in G_4$ and $\sigma_4(g_2, g_3) = (k_2, k_3)$. $\qquad \square$

8

FRACTURE THEOREMS
FOR LOCALIZATION: SPACES

In this chapter we extend the results of the previous chapter to nilpotent spaces. We first consider fracture theorems for maps from finite CW complexes into nilpotent spaces and then consider fracture theorems for nilpotent spaces themselves. For the former, we induct up the skeleta of the domain. For the latter, we could use the results of the previous chapter to conclude corresponding results for Eilenberg-MacLane spaces and then induct up Postnikov towers to extend the conclusions to nilpotent spaces. However, we shall see that more elegant proofs that directly mimic those of the previous chapter are available. As usual, all given spaces are taken to be based, connected, and of the homotopy types of CW complexes.

As in the previous chapter, let T_i, $i \in I$, be sets of primes, and let $T = \bigcup_{i \in I} T_i$, and $S = \bigcap_{i \in I} T_i$. We assume that $T_i \cap T_j = S$ if $i \neq j$, and that $T_i \neq S$ for $i \in I$.

8.1. Statements of the main fracture theorems

For ease of reference we record the main results of this chapter, which are analogues of the main results of the previous chapter. For our global to local results, we let

$$\phi \colon X \longrightarrow X_S, \quad \phi_i \colon X \longrightarrow X_{T_i}, \quad \text{and} \quad \psi_i \colon X_{T_i} \longrightarrow X_S$$

be localizations of a nilpotent space X such that $\psi_i \phi_i \simeq \phi$ for each $i \in I$. We also let $\phi_S \colon \prod_i X_{T_i} \longrightarrow (\prod_i X_{T_i})_S$ denote a localization at S.

THEOREM 8.1.1. *Let X be a T-local space and let K be a finite CW complex. Then the function*

$$(\phi_{i*}) \colon [K, X] \longrightarrow \prod_{i \in I} [K, X_{T_i}]$$

is an injection and the following diagram is a pullback of sets.

8.1.2

$$\begin{array}{ccc} [K, X] & \xrightarrow{(\phi_i)_*} & [K, \prod_{i \in I} X_{T_i}] \\ \phi_* \downarrow & & \downarrow \phi_{S_*} \\ [K, X_S] & \xrightarrow{((\phi_i)_S)_*} & [K, (\prod_{i \in I} X_{T_i})_S] \end{array}$$

THEOREM 8.1.3. *Let X be a T-local space. Then the following diagram is a homotopy pullback of spaces.*

$$\begin{array}{ccc} X & \xrightarrow{(\phi_i)} & \prod_{i \in I} X_{T_i} \\ \phi \downarrow & & \downarrow \phi_S \\ X_S & \xrightarrow{(\phi_i)_S} & (\prod_{i \in I} X_{T_i})_S \end{array}$$

ADDENDUM 8.1.4. If the indexing set I is finite, then both theorems remain valid if we replace $(\prod_{i \in I} X_{T_i})_S$ by the equivalent space $\prod_{i \in I} X_S$ in the lower right corner of the pullback diagrams. Now suppose that I is infinite and that X is $f\mathbb{Z}_T$-nilpotent.

(i) In Theorem 8.1.1, the conclusion remains true if we replace $(\prod_{i \in I} X_{T_i})_S$ by $\prod_{i \in I} X_S$ in the lower right corner.

(ii) In Theorem 8.1.3, the conclusion generally fails if we replace $(\prod_{i \in I} X_{T_i})_S$ by $\prod_{i \in I} X_S$ in the lower right corner.

For our local to global results, let X_i be a T_i-local nilpotent space, let Y be an S-local space, and let $\psi_i \colon X_i \longrightarrow Y$ and $\phi_S \colon \prod_{i \in I} X_i \longrightarrow (\prod_{i \in I} X_i)_S$ be localizations at S. Let $\tilde{\pi}_i \colon (\prod_{i \in I} X_i)_S \longrightarrow Y$ be the map, unique up to homotopy, such that $\tilde{\pi}_i \circ \phi_S \simeq \psi_i \circ \pi_i$. Then $(\tilde{\pi}_i) \colon (\prod_{i \in I} X_i)_S \longrightarrow \prod_i Y$ is a localization of (ψ_i) at S.

DEFINITION 8.1.5. A formal localization associated to the maps $\psi_i \colon X_i \longrightarrow Y$ is a homotopy class of maps $\omega \colon Y \longrightarrow (\prod_{i \in I} X_i)_S$ that satisfies the following two properties.

(i) The composite of ω and $\tilde{\pi}_i$ is homotopic to the identity map for each $i \in I$.

(ii) Each element $z \in \pi_1((\prod_{i \in I} X_i)_S)$ is the product of an element $\phi_{S*}(x)$ and an element $\omega_*(y)$, where $x \in \pi_1(\prod_{i \in I} X_i)$ and $y \in \pi_1(Y)$.

As will emerge shortly, ω exists and is unique if I is finite. Given ω, let P be the homotopy pullback in the diagram

8.1.6

$$
\begin{array}{ccc}
P & \xrightarrow{\mu} & \prod_{i \in I} X_i \\
\downarrow{\scriptstyle \nu} & & \downarrow{\scriptstyle \phi_S} \\
Y & \xrightarrow{\omega} & (\prod_{i \in I} X_i)_S.
\end{array}
$$

By Corollary 2.2.3, property (ii) is equivalent to requiring P to be connected.

THEOREM 8.1.7. *Let X_i be T_i-local and Y be S-local, and let $\psi_i: X_i \longrightarrow Y$ be a localization at S for each $i \in I$. Assume that either I is finite or I is infinite and the ψ_i have an associated formal localization ω. Then P is T-local and the induced map $P_{T_k} \longrightarrow X_k$ is an equivalence for all $k \in I$.*

COROLLARY 8.1.8. *The coordinate $P \longrightarrow X_k$ of μ in (8.1.6) is a localization at T_k, hence ν in (8.1.6) is a localization at S and ω is the localization of μ at S.*

We have phrased our results in terms of P, but much of the literature focuses instead on the homotopy pullback Q in the diagram

8.1.9

$$
\begin{array}{ccc}
Q & \longrightarrow & \prod_{i \in I} X_i \\
\downarrow & & \downarrow{\scriptstyle \Pi_i \psi_i} \\
Y & \xrightarrow{\Delta} & \prod_{i \in I} Y.
\end{array}
$$

Here there is no formal localization, and the following observation, whose proof is the same as that of Lemma 7.3.5, shows that the notion of a formal localization is only needed when the indexing set I is infinite.

LEMMA 8.1.10. *Let I be finite. Then $(\tilde{\pi}_i)$ is an equivalence and the composite $\omega = (\tilde{\pi}_i)^{-1} \circ \Delta$ is the unique formal localization associated to the $\psi_i: X_i \longrightarrow Y$, hence the diagrams (8.1.6) and (8.1.9) are canonically equivalent.*

By Proposition 7.4.3 and Theorem 4.5.2, Theorem 8.1.7 has the following important refinement when the indexing set I is finite.

COROLLARY 8.1.11. *If I is finite and X_i is $f\mathbb{Z}_{T_i}$-nilpotent for each $i \in I$, then $Q \simeq P$ is $f\mathbb{Z}_T$-nilpotent. In particular, if T is the set of all primes, then Q is f-nilpotent.*

WARNING 8.1.12. The space Q is always T-local. However, even if each X_i is $f\mathbb{Z}_{T_i}$-nilpotent and simply connected, the induced map $Q_{T_k} \longrightarrow X_k$ is generally not an equivalence for all $k \in I$ (since otherwise $P \longrightarrow Q$ would be an equivalence) and the homotopy groups of Q are generally not finitely generated $\mathbb{Z}_T$-modules (as will become clear in our discussion of the extended genus in §8.5). This contradicts claims made in several important early papers on the subject.

We prove Theorem 8.1.1 in §8.2, Theorem 8.1.3 in §8.3, and Theorem 8.1.7 in §8.4. The latter two proofs are direct and conceptual, but we explain alternative proofs by induction up Postnikov towers in §8.6. This depends on the general observation that homotopy pullbacks of homotopy pullbacks are homotopy pullbacks, which is a topological analogue of Lemma 7.6.2. We note that it would be possible to instead first prove Theorem 8.1.7 and then deduce Theorem 8.1.3 from it, following Remark 7.4.2. The starting points for all of our proofs are the characterizations of T-local spaces and of localizations at T in terms of homotopy groups that are given in Theorems 6.1.1 and 6.1.2. In analogy with §7.5, in §8.5 we use the notion of a formal localization to describe the genus of a nilpotent space, namely the equivalence classes of spaces whose localizations at all primes p are equivalent to those of the given space.

8.2. Fracture theorems for maps into nilpotent spaces

We prove Theorem 8.1.1 in this section. Thus let K be a finite CW complex and X be a T-local nilpotent space. We emphasize that even if K and X are simply connected, the proofs here require the use of nilpotent groups.

THEOREM 8.2.1. *The function*
$$(\phi_{i*}) \colon [K, X] \longrightarrow \prod_{i \in I} [K, X_{T_i}]$$
is an injection.

PROOF. When $K = S^n$, the claim is that $(\phi_{i*})\colon [S^n, X] \longrightarrow \prod_{i \in I}[S^n, X_{T_i}]$ is a monomorphism. This means that

$$(\phi_{i*})\colon \pi_n X \longrightarrow \prod_{i \in I} (\pi_n X)_{T_i}$$

is a monomorphism. The claim follows from Theorem 7.2.1(i) since the groups $\pi_n X$ are T-local.

The case $K = \vee_J S^n$ reduces to showing that

$$(\phi_{i*})\colon \prod_J [S^n, X] \longrightarrow \prod_{i \in I} \left(\prod_J [S^n, X_{T_i}] \right)$$

is a monomorphism, and this follows from the case $K = S^n$.

We now argue by induction. Thus assume that the result holds for K^{n-1}, where the dimension of K is n, and recall the notation $[K, X]_f$ from Theorem 6.3.2. Let $f, g \in [K, X]$ and assume that $\phi_i \circ f \simeq \phi_i \circ g$ for all i. By induction, $f|K^{n-1} \simeq g|K^{n-1}$ and therefore $g \in [K, X]_f$.

By Theorem 6.3.2 and the assumption that X is T-local, $[K, X]_f$ is a $\mathbb{Z}_T$-nilpotent group, $[K, X_{T_i}]_{\phi_i \circ f}$ is a $\mathbb{Z}_{T_i}$-nilpotent group, and $[K, X]_f \longrightarrow [K, X_{T_i}]_{\phi_i \circ f}$ is localization at T_i. By Theorem 7.2.1(i), the map

$$[K, X]_f \longrightarrow \prod_{i \in I}[K, X_{T_i}]_{\phi_i \circ f}$$

is a monomorphism. Since $\phi_i \circ g = \phi_i \circ f = \mathrm{id}$ in $[K, X_{T_i}]_{\phi_i \circ f}$ for each i, $f \simeq g$. □

REMARK 8.2.2. Except that we change the group theoretic starting point, making use of Theorem 7.2.1(iii), the proof of (i) in Addendum 8.1.4 is exactly the same.

EXAMPLE 8.2.3. The assumption that K is finite is essential. An easy counterexample otherwise goes as follows. Let $\phi\colon S^n \longrightarrow S^n_T$ be localization at T, where $n \geq 2$, and let T_i be the ith prime in T. Let K be the cofiber of ϕ, so that we have a cofiber sequence

$$S^n \xrightarrow{\phi} S^n_T \xrightarrow{i} K \xrightarrow{\pi} S^{n+1} \xrightarrow{\Sigma\phi} S^{n+1}_T.$$

For each prime $p \in T$, $\phi_p\colon S^n_p \longrightarrow (S^n_T)_p$ is an equivalence, hence so is $\Sigma\phi_p$. Since the localization at p of the cofiber sequence is again a cofiber sequence, π_p must be null homotopic. However, π itself is not null homotopic since, if it were, the map $\Sigma\phi$ would have a left homotopy inverse and $\mathbb{Z}$ would be a direct summand of $\mathbb{Z}_T$.

When K has localizations, their universal property has the following consequence, whose algebraic precursor is recorded in Lemma 7.1.8. It says that to check whether or not two maps are homotopic, it suffices to check whether or not they become homotopic after localization at each prime p.

COROLLARY 8.2.4. *Let $f, g\colon K \longrightarrow X$ be maps, where K is a nilpotent finite CW complex and X is a nilpotent space. Then $f \simeq g$ if and only if $f_p \simeq g_p$ for all primes p.*

PROOF. By the theorem, $(\phi_{p_*})\colon [K, X] \longrightarrow \prod_p[K, X_p]$ is injective. Since K is nilpotent, we have $\phi_p^*\colon [K, X_p] \cong [K_p, X_p]$. Therefore the product of localizations

$$[K, X] \longrightarrow \prod_p[K_p, X_p]$$

is injective.   □

Retaining the notations of Theorem 8.2.1, observe that since $\psi_i \circ \phi_i \simeq \psi$ for each i the image of $(\phi_i)_*$ in $[K, \prod_{i \in I} X_{T_i}]$ factors through the pullback of sets constructed from the lower and right legs of the diagram (8.1.2). For the purposes of this proof, we give this pullback the abbreviated notation $P[K, X]$.

THEOREM 8.2.5. *The function*

$$(\phi_i)_*\colon [K, X] \longrightarrow P[K, X]$$

is a bijection of sets.

PROOF. By Theorem 8.2.1, $(\phi_i)_*$ is injective. We must show that it is surjective. First consider the case $K = S^n$. We must show that $(\phi_i)_*$ maps $\pi_n(X)$ isomorphically onto the pullback of the diagram

$$\prod_{i \in I} \pi_n(X_{T_i})$$

$$\downarrow$$

$$\pi_n(X_S) \xrightarrow{\ (\phi_{i*})_S\ } (\prod_{i \in I} \pi_n(X_{T_i}))_S.$$

This holds by Theorem 7.2.1(ii) since the groups $\pi_n X$ are T-local. The result for a wedge of spheres again reduces to the case of a single sphere.

Assume that the result holds for K^{n-1}, where K has dimension n. Consider an element $(g_i) \in P[K, X]$. By the induction hypothesis there is a map

$e \in [K^{n-1}, X]$ such that $\phi_i \circ e = g_i|K^{n-1}$ for all i. Let J be a wedge of $(n-1)$-spheres, chosen so that K is the cofiber of a map $j: J \longrightarrow K^{n-1}$. Since J is a wedge of $(n-1)$-spheres,

$$(\phi_i)_* : [J, X] \longrightarrow P[J, X]$$

is an isomorphism. Consider the following diagram.

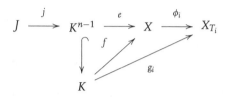

Since $\phi_i \circ e$ extends to g_i, $\phi_i \circ e \circ j$ is trivial by the long exact sequence for a cofibration. For maps from J, (ϕ_{i*}) is a bijection by the previous case. Therefore, $e \circ j$ is trivial and e extends to a map $f \in [K, X]$. Note that $\phi_i f|_{K^{n-1}}$ is homotopic to $g_i|_{K^{n-1}}$, but $\phi_i f$ is not necessarily homotopic to g_i on all of K. However, since $\phi_i f|_{K^{n-1}} \cong g_i|_{K^{n-1}}$ for all i, the maps (g_i) define an element of the nilpotent group given by the pullback of the diagram

$$\prod_{i \in I}[K, X_{T_i}]_{\phi_i \circ f}$$
$$\downarrow \scriptstyle{\prod_i \psi_{i*}}$$
$$[K, X_S] \xrightarrow{\ (\phi_{i*})_S\ } (\prod_{i \in I}[K, X_{T_i}]_{\phi \circ f})_S.$$

By Theorems 6.3.2 and 7.2.1, there is an element $g \in [K, X]_f$ such that $\phi_i \circ g = g_i$ for all i. □

EXAMPLE 8.2.6. The assumption that K is finite is again essential. Let $T = T_1 \cup T_2$ be a partition of the set of all primes. If $K = \mathbb{C}P^\infty$ and $X = S^3$, the square

$$\begin{array}{ccc}
[\mathbb{C}P^\infty, S^3] & \longrightarrow & [\mathbb{C}P^\infty, S^3_{T_1}] \\
\downarrow & & \downarrow \\
[\mathbb{C}P^\infty, S^3_{T_2}] & \longrightarrow & [\mathbb{C}P^\infty, S^3_0]
\end{array}$$

is not a pullback because the canonical map of $\lim^1$ terms

$$\lim^1 [\Sigma\mathbb{C}P^n, S^3] \longrightarrow \lim^1 [\Sigma\mathbb{C}P^n, S^3_{T_1}] \oplus \lim^1 [\Sigma\mathbb{C}P^n, S^3_{T_2}]$$

is not a monomorphism. The details are similar to those of §2.4; see [21, V.7.7].

While less important than the previous results, we record the following observation since it does not appear in the literature.

PROPOSITION 8.2.7. *If I is finite, then the formal sum*

$$\prod_{i\in I}[\Sigma K, X_{T_i}] \longrightarrow [\Sigma K, X_S]$$

is a surjection.

PROOF. The claim is that every element of the target is a product of elements $\psi_{i*}(x_i)$ from the source. Since $[\Sigma K, Y] \cong \pi_1(F(K, Y), *)$ for any Y, this follows from Theorem 6.3.2 by inducting up the skeleta of K and using Theorem 7.2.1. □

REMARK 8.2.8. It is possible to relax the hypothesis on the space K in Theorems 8.2.1 and 8.2.5 and in Proposition 8.2.7. If we take K to be a space with finitely generated integral homology, then there is a finite CW complex $\bar{K}$ and a map $f : \bar{K} \longrightarrow K$ such that

$$f^* : [K, Y] \cong [\bar{K}, Y];$$

see, for example, [62, II.4.2, II.4.3]. Therefore, we may apply the cited results to any space K with finitely generated integral homology.

8.3. Global to local fracture theorems: spaces

In this section, we follow §7.2 and prove the analogous global to local result, namely Theorem 8.1.3. Thus let X be a T-local nilpotent space. We must prove that the diagram

8.3.1

$$
\begin{array}{ccc}
X & \xrightarrow{(\phi_i)} & \prod_{i\in I} X_{T_i} \\
\phi \downarrow & & \downarrow \phi_S \\
X_S & \xrightarrow[(\phi_i)_S]{} & (\prod_{i\in I} X_{T_i})_S
\end{array}
$$

is always a homotopy pullback. Recall that part (ii) of Addendum 8.1.4 states that the analogous diagram

8.3.2

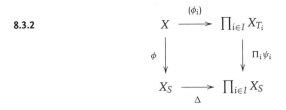

is generally not a homotopy pullback, even when X is $f\mathbb{Z}_T$-nilpotent. Remember that these diagrams are equivalent when the indexing set I is finite.

Recall too that Proposition 2.2.2 and Corollary 2.2.3 tell us how to compute the homotopy groups of homotopy pullbacks. In particular, let $\alpha: G_1 \longrightarrow G_0$ and $\beta: G_2 \longrightarrow G_0$ be homomorphisms of groups with pullback G. Consider the corresponding homotopy pullback X of Eilenberg-Mac Lane spaces $K(-, n)$, taking the given groups to be abelian when $n > 1$. Let $G_1 \times G_2$ act on G_0 by $g_0(g_1, g_2) = \alpha(g_1)^{-1} g_0 \beta(g_2)$ with orbit set J. Then

$$\pi_{n-1}(X) = J \quad \text{and} \quad \pi_n(X) = G,$$

with $\pi_q(X) = 0$ otherwise. Of course, for a product of groups G_i, we have $K(\prod G_i, n) \simeq \prod_i K(G_i, n)$. Now Theorem 7.2.1(ii) implies the following result since it shows that J is trivial in this case.

LEMMA 8.3.3. *The diagram (8.3.1) is a homotopy pullback when X is a T-local Eilenberg-Mac Lane space.*

EXAMPLE 8.3.4. The diagram (8.3.2) is not a homotopy pullback in the case when $X = K(\mathbb{Z}, n)$ and T_i is the ith prime number since, with additive notation, Proposition 7.1.7 shows that J is nonzero in this case.

From here, we can prove that the diagram (8.3.1) is a homotopy pullback by inducting up the Postnikov tower of X, as we will explain in §8.6. However, we have a simpler alternative proof that directly mimics the proof in algebra. It starts with the analogue of Lemma 7.1.8.

LEMMA 8.3.5. *Let X and Y be T-local nilpotent spaces. A map $f: X \longrightarrow Y$ is an equivalence if and only if $f_{T_k}: X_{T_k} \longrightarrow Y_{T_k}$ is an equivalence for each $k \in I$.*

PROOF. It suffices to show that $f_*: \pi_*(X) \longrightarrow \pi_*(Y)$ is an isomorphism if and only if $(f_{T_k})_*: \pi_*(X_{T_k}) \longrightarrow \pi_*(Y_{T_k})$ is an isomorphism for all k. This means that

$$(f_*)_{T_k}: \pi_*(X)_{T_k} \longrightarrow \pi_*(Y)_{T_k}$$

is an isomorphism for all k. Lemma 7.1.8 gives the conclusion. $\square$

Recall from Proposition 6.2.5 that localization preserves homotopy pull-backs.

THEOREM 8.3.6. *The diagram (8.3.1) is a homotopy pullback for any T-local nilpotent space X.*

PROOF. Let P be the homotopy pullback displayed in the diagram

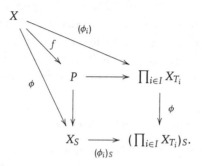

The outer square commutes up to homotopy, hence there is a map f that makes the diagram commute up to homotopy. We must show that f is an equivalence. By Lemma 8.3.5 it suffices to show that f_{T_k} is an equivalence for all $k \in I$. By Proposition 6.2.5, we obtain another homotopy pullback diagram after localizing at T_k.

$$
\begin{array}{ccc}
P_{T_k} & \longrightarrow & (\prod_{i\in I} X_{T_i})_{T_k} \\
\downarrow & & \downarrow \\
(X_S)_{T_k} & \longrightarrow & ((\prod_{i\in I} X_{T_i})_S)_{T_k}
\end{array}
$$

Arguments exactly like those in the proof of Theorem 7.2.1(ii) imply that this diagram is equivalent to the diagram

$$
\begin{array}{ccc}
P_{T_k} & \longrightarrow & X_{T_k} \times (\prod_{i\neq k} X_{T_i})_S \\
\downarrow & & \downarrow{\psi_k\times\mathrm{id}} \\
X_S & \xrightarrow{(\mathrm{id},(\phi_i)_S)} & X_S \times (\prod_{i\neq k} X_{T_i})_S.
\end{array}
$$

To interpret this homotopy pullback, we assume or arrange that ψ_k is a fibra-tion and then take the actual pullback. Here we are using actual identity maps where indicated, and we can apply Lemma 7.1.9 in the cartesian monoidal category of spaces to conclude that the homotopy pullback is equivalent to X_{T_k}. Parenthetically, we can also apply Lemma 7.1.9 in the homotopy category of spaces, and we conclude that in this case the homotopy pullback is in fact a pullback in the homotopy category. In any case, it follows that f_{T_k} is an equivalence. $\qquad\square$

REMARK 8.3.7. We repeat Remark 7.2.2, since it applies verbatim here. Theorem 8.1.3 was proven under a finite generation hypothesis in Hilton, Mislin, and Roitberg [62], and Hilton and Mislin later noticed that the hypothesis can be removed [61]. That fact is not as well-known as it should be. We learned both it and most of the elegant proof presented here from Bousfield.

8.4. Local to global fracture theorems: spaces

We here prove Theorem 8.1.3. Thus let X_i be a T_i-local nilpotent space and Y be an S-local nilpotent space. Let $\psi_i \colon X_i \longrightarrow Y$ be a localization at S for each i. When the indexing set I is infinite, we assume that we have a formal localization $\omega \colon Y \longrightarrow (\prod_i X_i)_S$. We then have the homotopy pullbacks P and Q of (7.1.1) and (7.1.2). These are both homotopy pullbacks of diagrams of T-local spaces and are therefore T-local if they are connected, by Proposition 6.2.5. Again, remember that these diagrams are equivalent when the indexing set I is finite. In that case, it is clear from algebra that P and Q are connected, and our definition of a formal localization ensures that P is connected in general. However, the algebraic description of the homotopy groups of a homotopy pullback in Proposition 2.2.2 shows that the homotopy groups of Q are quite badly behaved when I is infinite: quotients of the huge $\mathbb{Z}_S$-module $\prod_{i \in I} \pi_n(Y)$ appear with a shift of degree. In fact, Q is rarely a connected space.

We must prove that the induced map $P_{T_k} \longrightarrow X_k$ is an equivalence for all $k \in I$. The proof follows the outline of the proofs in §7.4 and §8.3. As in the latter section, we could start by using the algebraic result, Theorem 7.4.1, to prove the following topological analogue.

LEMMA 8.4.1. The maps $P_{T_k} \longrightarrow X_k$ are equivalences when the X_i are T_i-local Eilenberg-Mac Lane spaces.

However, an argument precisely like the proofs of Theorem 7.4.1(ii) and Theorem 8.1.3 gives the general conclusion directly.

THEOREM 8.4.2. *The maps $P_{T_k} \longrightarrow X_k$ are equivalences.*

PROOF. Let P be the homotopy pullback displayed in (8.1.6). We have assumed that P is connected, and by Proposition 6.2.5 we obtain another homotopy pullback diagram after localizing at T_k.

$$
\begin{array}{ccc}
P_{T_k} & \longrightarrow & (\prod_{i\in I} X_i)_{T_k} \\
\downarrow & & \downarrow \\
Y_{T_k} & \xrightarrow{\ \omega\ } & ((\prod_{i\in I} X_i)_S)_{T_k}
\end{array}
$$

Arguments exactly like those in the proof of Theorem 7.2.1(ii) imply that this diagram is equivalent to the diagram

$$
\begin{array}{ccc}
P_{T_k} & \longrightarrow & X_k \times (\prod_{i\neq k} X_i)_S \\
\downarrow & & \downarrow{\scriptstyle \psi_k \times \mathrm{id}} \\
Y & \xrightarrow{(\mathrm{id},\pi_2\omega)} & Y \times (\prod_{i\neq k} X_i)_S,
\end{array}
$$

where π_2 denotes the evident projection. The description of the right vertical arrow depends only on the ψ_i and not ω, but the assumed compatibility of ω with Δ ensures that its localization at T_k takes the required form $(\mathrm{id},\pi_2\omega)$. Lemma 7.1.9 applies to show that the homotopy pullback P_{T_k} is equivalent to X_k (and is a pullback in the homotopy category). $\qquad\square$

When I is infinite, we do not have an alternative inductive proof since we must start with a formal localization and it is not clear how those behave with respect to Postnikov towers. When I is finite, the formal localization is equivalent to $\Delta : Y \longrightarrow \prod_i Y$ and we do have such a proof, as we explain in §8.6.

8.5. The genus of nilpotent spaces

Much early work in the theory of localization focused on the concept of genus, which was introduced by Mislin [105] in the context of H-spaces. The literature is quite extensive and we refer the reader to the survey [100] of McGibbon for further information and many references. We first introduce the idea, state some key results, and give some examples, without proofs. We then describe how the notion of a formal localization applies to the analysis of the extended genus. As in §7.5, we assume that T is the set of all primes, S is the empty set, and T_i is the set consisting of just the ith prime number p_i.

DEFINITION 8.5.1. Let X be a nilpotent space. The extended genus of X, denoted $\mathbb{G}(X)$, is the collection of homotopy types of nilpotent spaces X' whose localizations at all primes p are equivalent to those of X. When X is f-nilpotent, the genus of X, denoted $G(X)$, is the collection of f-nilpotent homotopy types in $\mathbb{G}(X)$. A property of f-nilpotent spaces is said to be generic if it holds for all (or none) of the spaces in a given genus.

As we shall explain in §13.6, use of completions rather than localizations leads to two interesting variant notions of the genus of a space. A little thought about the structure of finitely generated abelian groups and the universal coefficient theorem gives the following consequence of the fact that the genus of any finitely generated abelian group has a single element. For the last statement, see [100, p. 82].

PROPOSITION 8.5.2. *Homology groups and homotopy groups, except for the fundamental group if it is non-abelian, are generic. While the integral cohomology groups are generic, the integral cohomology ring is not.*

EXAMPLE 8.5.3. Example 7.5.7 implies that the fundamental group is not generic.

Perhaps for this reason, but also because of the difficulty of computations, the study of the genus is generally restricted to simply connected spaces of finite type. The extended genus is generally very large, probably too large to be of interest in its own right. For example, Example 7.5.5 and Proposition 7.5.6 have the following consequence.

EXAMPLE 8.5.4. The extended genus of $K(\mathbb{Z}, n)$ is uncountable for $n \geq 1$. The genus of $K(A, n)$ has a single element for any finitely generated abelian group A.

McGibbon generalized this example to the following result [101, Thm. 2].

THEOREM 8.5.5. *Let X be simply connected of finite type and assume that either $H_n(X; \mathbb{Z}) = 0$ or $\pi_n(X) = 0$ for all but finitely many n. Then the extended genus of*

X is finite if and only if X_0 is contractible. If X_0 is not contractible, the extended genus of X is uncountable.

In contrast, Wilkerson [144] proved the following opposite conclusion for the genus; see also [100, Thm. 1].

THEOREM 8.5.6. *Let X be simply connected of finite type and assume that either $H_n(X; \mathbb{Z}) = 0$ or $\pi_n(X) = 0$ for all but finitely many n. Then the genus of X is finite.*

Having the homotopy type of a finite CW complex is not a generic property, by a counterexample of Mislin [106]. Being of the homotopy type of a space with finitely generated integral homology is generic, and that gives interest to Remark 8.2.8. Results of Zabrodsky [146, 2.9] and Mislin [107] give the following conclusion, and Zabrodsky's work in [146] gives a complete recipe for the computation of the genus in this case.

THEOREM 8.5.7. *The property of being a finite H-space is generic.*

A beautiful worked out example was given by Rector [117].

THEOREM 8.5.8. *The genus of $\mathbb{H}P^\infty$ is uncountably infinite.*

For comparison, McGibbon [99] computed the genus of the finite nilpotent projective spaces $\mathbb{R}P^{2n+1}$, $\mathbb{C}P^n$ and $\mathbb{H}P^n$ for $1 \le n < \infty$. For these spaces X, he uses pullbacks over X to give the set $G(X)$ a group structure and proves the following result.

THEOREM 8.5.9. *Let n be a positive integer.*

(i) $G(\mathbb{R}P^{2n+1}) = 1$,

(ii) $G(\mathbb{C}P^n) = 1$,

(iii) $G(\mathbb{H}P^n) \cong \mathbb{Z}/2 \oplus \cdots \oplus \mathbb{Z}/2$, where the number of factors equals the number of primes p such that $2 \le p \le 2n - 1$.

As in the case of nilpotent groups, asking how unique formal localizations are gives a starting point of the analysis of the extended genus, and one can then seek the actual genus inside that. Again, a general idea is to exploit finite partitions of the primes and Corollary 8.1.11, and that leads to many of the most interesting examples. We shall say just a little more about that in §9.4, where we consider fracture theorems for finite H-spaces. Although we shall not correlate this approach to the analysis of the genus with the existing literature in this book, the flavor is much the same. Many known calculations rely on an understanding of double cosets of homotopy automorphism groups, and we indicate their relevance, following §7.5. Roughly speaking, the conclusion is that, under suitable hypotheses, elements of the extended genus of an f-nilpotent space are in bijective correspondence with the double cosets

$$\text{hAut}(X_0)\backslash \prod_i \text{hAut}(X_0)/ \prod_i \text{hAut}(X_i),$$

where hAut denotes the group of self-homotopy equivalences of a space X.

Any nilpotent space X is equivalent to a homotopy pullback P constructed from rationalizations ψ_i of its localizations X_i at p_i and a formal localization $\omega\colon X_0 \longrightarrow (\prod_i X_i)_0$ of its rationalization X_0. Since we are only trying to classify homotopy types, we can use the same spaces X_i and X_0 to construct a representative for each homotopy type in the same genus as X. The ψ_i can vary, the variation being given by a self-homotopy equivalence ξ_i of the rational nilpotent space X_0. Similarly, the rationalization of $\prod_i X_i$ can vary by a self-equivalence of the rational nilpotent space $(\prod_i X_i)_0$. Notice that Theorem 7.2.1(ii) and Corollary 2.2.3 imply that the homotopy groups $\pi_n(P)$ are isomorphic to the pullbacks $\pi_n(X_0) \times_{\pi_n((\prod_i X_i)_0)} \pi_n(\prod_i X_i)$ for any space P in the same extended genus as X; any variation is in the maps that define these pullbacks. As already noted, the homotopy groups are generic when X is simply connected of finite type.

To be more precise, up to equivalence any two spaces X and X' in the same genus can be represented as homotopy pullbacks P and P' as displayed in the top triangles of commutative diagrams

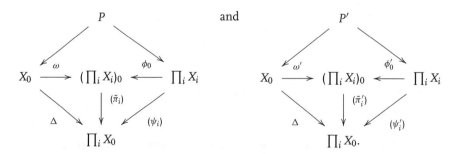

Here $\tilde{\pi}_i$ is the map $(\prod_i X_i)_0 \longrightarrow X_0$, unique up to homotopy, such that $\tilde{\pi}_i \circ \phi_0 \simeq \psi_i \circ \pi_i$, and similarly for $\tilde{\pi}'_i$. Usually the homotopy pullbacks of (ω, ϕ_0) and (ω', ϕ'_0) are not equivalent. We fix a reference pullback $P \simeq X$ and have the following result. It is less satisfactory than its algebraic analogue Proposition 7.5.2 but serves to give the idea.

PROPOSITION 8.5.10. *Assume that*

$$(\tilde{\pi}_i)_* : [X_0, (\prod_i X_i)_0] \longrightarrow [X_0, \prod_i X_0]$$

is a monomorphism. Then the extended genus of X is in bijective correspondence with a subset of

$$hAut(X_0) \backslash \prod_i hAut(X_0) / \prod_i hAut(X_i).$$

PROOF. Since $(\tilde{\pi}_i)_*$ is a monomorphism, the homotopy class of ω is uniquely determined by the requirement that $\tilde{\pi}_i \circ \omega \simeq \mathrm{id}$ for all i. Similarly, the homotopy class of the localization ϕ_0 is uniquely determined by the ψ_i. Since localizations are unique up to equivalences of their targets, we see that $(\tilde{\pi}_i')_*$ is also a monomorphism and ω' and ϕ_0' are also uniquely determined.

The double cosets of the statement are defined with respect to the diagonal $\Delta: h\,\mathrm{Aut}\,(X_0) \longrightarrow \prod_i h\,\mathrm{Aut}\,(X_0)$ and the homomorphisms $h\,\mathrm{Aut}\,(X_i) \longrightarrow h\,\mathrm{Aut}\,(X_0)$ that send a self-equivalence ζ_i of X_i to the self-equivalence $\tilde{\zeta}_i$ of X_0, unique up to homotopy, such that $\tilde{\zeta}_i \circ \psi_i = \psi_i \circ \zeta_i$. This definition refers to the ψ_i of the fixed-reference pullback P. We must restrict to the double cosets of K (which is not claimed to be a subgroup).

At this point the analysis diverges from the algebraic setting, since the homotopy pullbacks P and Q of (ω, ϕ_0) and $(\Delta, (\psi_i))$ are not only not equivalent but, in contrast to Proposition 8.5.2, they can have very different homotopy groups. This means that we cannot even take for granted the existence of formal localizations ω' for all choices of the ψ_i'. However, since we are trying to determine the extended genus, we are starting with given global spaces that determine the required formal localizations.

For each P' and each i, there is a self-equivalence ξ' of X_0 such that $\xi_i' \circ \psi_i \simeq \psi_i'$. Let K be the subset of $\prod_i hAut(X_0)$ consisting of those $\xi = (\xi')$ such that the ψ_i' admit a formal localization ω', necessarily unique. Sending ξ to the homotopy class of the pullback P', we obtain a surjection from K to the extended genus of X.

The double cosets of the statement are defined with respect to the diagonal $\Delta: h\,\mathrm{Aut}\,(X_0) \longrightarrow \prod_i h\,\mathrm{Aut}\,(X_0)$ and the homomorphisms $h\,\mathrm{Aut}\,(X_i) \longrightarrow h\,\mathrm{Aut}\,(X_0)$ that send a self-equivalence ζ_i of X_i to the self-equivalence $\tilde{\zeta}_i$ of X_0, unique up to homotopy, such that $\tilde{\zeta}_i \circ \psi_i = \psi_i \circ \zeta_i$. This definition refers to the ψ_i of the fixed reference pullback P. We must restrict to the double cosets of K[1].

---

1. We do not claim that K is a subgroup of $\prod_i h\,\mathrm{Aut}\,(X_0)$ in general, but we expect the two to be equal in reasonable examples.

Suppose we have a homotopy equivalence $\zeta : P' \longrightarrow P''$ between two such pullbacks and consider the following diagram, in which the front and back squares are pullbacks.

8.5.11

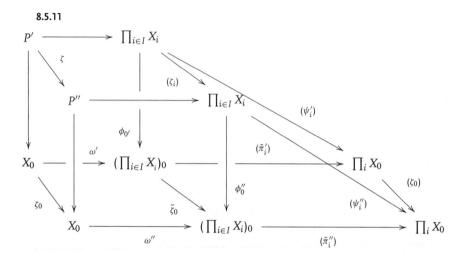

The unlabeled left vertical arrows are rationalizations and the coordinates of the unlabeled top horizontal arrows are localizations at p_i. The universal properties of localizations give self-equivalences ζ_i of X_i and ζ_0 of X_0 making the top and left squares homotopy commutative. There is a map $\tilde{\zeta}_0$, unique up to homotopy, that makes the bottom square commute up to homotopy, and then the right square must also commute up to homotopy. We have completed the diagram to the right to aid in identifying the double coset interpretation. As above, we have a self-equivalence $\tilde{\zeta}_i$ of X_0 such that $\tilde{\zeta}_i \circ \psi_i \simeq \psi_i \circ \zeta_i$. We also have self-equivalences ξ' and ξ'' of X_0 such that $\xi'_i \circ \psi_i \simeq \psi'_i$ and $\xi''_i \circ \psi_i \simeq \psi''_i$. Choosing homotopy inverses $\tilde{\zeta}_i^{-1}$, we see that $\xi''_i \simeq \zeta_0 \circ \zeta'_i \circ \tilde{\zeta}_i^{-1}$. Running the argument in reverse, we see that if ξ' and ξ'' are in the same double coset, then we obtain an equivalence ζ as in (8.5.11). $\qquad\square$

8.6. Alternative proofs of the fracture theorems

We explain parenthetically how to prove Theorems 8.1.3 and 8.1.7 by inducting up Postnikov towers. This uses an easy but useful general observation about homotopy pullbacks. We shall make essential use of it later.

PROPOSITION 8.6.1. *Consider the following homotopy commutative diagram.*

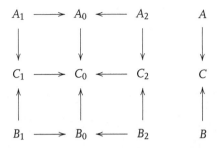

$$P_1 \longrightarrow P_0 \longleftarrow P_2$$

Let the column and row displayed to the right and at the bottom be obtained by passage to homotopy pullbacks from the corresponding rows and columns of the diagram. Let D be the homotopy pullback of the column at the right and P be the homotopy pullback of the row at the bottom. Then D and P are equivalent.

PROOF. If the diagram commutes and D and P are the actual pullbacks, then the conclusion holds by the well-known and easy categorical analogue that pullbacks of pullbacks are pullbacks [80, IX, §8]. To calculate the six given homotopy pullbacks, we must use the mapping path fibration to replace at least one of the two maps with a common target in each row and column of the diagram by a fibration and then take actual pullbacks, and a diagram chase shows that we can simultaneously make the diagram commute rather than just commute up to homotopy. We claim that judicious choices make one of the two maps in the induced column and row a fibration. That reduces the homotopy pullbacks to actual pullbacks to which the categorical analogue applies. We choose to replace the maps

$$A_1 \longrightarrow A_0, \ B_2 \longrightarrow B_0, \ A_2 \longrightarrow C_2, \ B_1 \longrightarrow C_1, \ C_1 \longrightarrow C_0, \text{ and } B_0 \longrightarrow C_0$$

by fibrations (the last two choices being arbitrary). With these choices, the induced maps of pullbacks $A \longrightarrow C$ and $P_2 \longrightarrow P_0$ are also fibrations. By symmetry, we need only prove the first of these. Thus suppose given a lifting problem

We use that $A_2 \longrightarrow C_2$ is a fibration to obtain a lift λ_2 in the following diagram, and we then use that $A_1 \longrightarrow A_0$ is a fibration to obtain a lift λ_1.

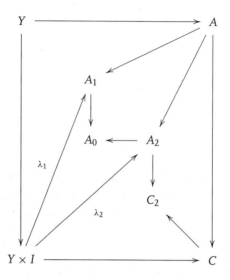

The lifts λ_1 and λ_2 induce the required lift λ into the pullback A. $\square$

EXAMPLE 8.6.2. As a perhaps amusing example, observe that ΩX is the homotopy pullback of the diagram $* \longrightarrow X \longleftarrow *$. If P is the homotopy pullback of maps $f: X \longrightarrow A$ and $g: Y \longrightarrow A$, then the proposition shows that ΩP is the homotopy pullback of the maps Ωf and Ωg. As a more serious example, when g is $* \longrightarrow A$, P is the fiber of f. We shall shortly specialize Proposition 8.6.1 to the case when one of the columns is trivial and we are considering pullbacks of fibers.

Proposition 8.6.1 specializes to give the inductive step of the promised alternative proof that the diagram (8.3.1) is a homotopy pullback. To see that, suppose that the conclusion holds for a T-local space X and let Y be the fiber of a map $k: X \longrightarrow K$, where, for definiteness, K is a simply connected T-local Eilenberg-Mac Lane space. Note that Y is the homotopy pullback of k and the trivial map $* \longrightarrow K$, or the actual pullback of k and the path space fibration $PK \longrightarrow K$. We apply Proposition 8.6.1 to the following homotopy commutative diagram. The homotopy pullbacks of its rows are as indicated in the column at the right since localizations preserve homotopy pullbacks by Proposition 6.2.5. The homotopy pullbacks of its columns are as indicated in the row at the bottom by Lemma 8.3.3 and asssumption.

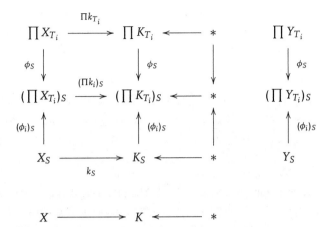

By Proposition 8.6.1, Y is equivalent to the homotopy pullback of the right column.

Finally, we show how to prove inductively that the maps $P_{T_k} \longrightarrow X_k$ of §8.4 are equivalences when I is finite. We can apply the following result to each stage of compatible Postnikov towers for the X_i and Y and then pass to limits.

LEMMA 8.6.3. *Let K_i be a T_i-local, simply connected Eilenberg-Mac Lane space, L be an S-local, simply connected Eilenberg-Mac Lane space, and $\psi_i : K_i \longrightarrow L$ be a localization at S. Let X_i be a T_i-local space, Y be an S-local space, and $\psi_i : X_i \longrightarrow Y$ be a localization at S. Let X_i' and Y' be the fibers of maps $k_i : X_i \longrightarrow K_i$ and a map $\ell : Y \longrightarrow L$ such that $\ell \psi_i \simeq \psi k_i$ for all i and consider the following diagram.*

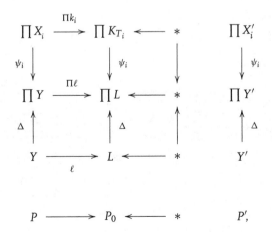

where P, P_0, and P' are the homotopy pullbacks of the columns above them. Assume that $P \longrightarrow X_k$ is a localization at T_k for each k. Then $P' \longrightarrow X_k'$ is a localization at T_k for each k.

PROOF. There is a map ℓ as stated because localization commutes with finite products. More precisely, this is a consequence of Lemma 8.1.10. By Proposition 8.6.1, P' is the fiber of the map $P \longrightarrow P_0$, and for each k there results a map of fiber sequences

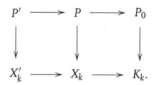

The space X'_k is T_k-local by Theorem 5.3.1. The middle and right vertical arrows are localizations at T_k by assumption and Lemma 8.4.1. Therefore, by the five lemma applied to homotopy groups, the left vertical arrow is a localization at k. □

9

RATIONAL H-SPACES AND
FRACTURE THEOREMS

We here describe the category of rational H-spaces and apply this to give more algebraically calculational fracture theorems for H-spaces. Rather tautologically, a major theme of algebraic topology is the algebraization of homotopy theory. In some cases, the algebraization is complete, and this is true for rational homotopy theory, as proven by Quillen [115] and Sullivan [134]; see also Bousfield and Gugenheim [20] and Félix, Halperin, and Thomas [48]. However, the algebraization of the rational homotopy theory of H-spaces is elementary, depending only on the structure theory for Hopf algebras that we shall develop in Chapter 22. The reader may want to look at that chapter before reading this one.

After describing the cited algebraization, we show how to give it more topological content via the Samelson product on homotopy groups. This gives a Lie algebra structure on $\pi_*(X)$ for a connected H-group X (as defined in Definition 9.2.1) such that the Lie algebra $\pi_*(X) \otimes \mathbb{Q}$ is determined by the Hopf algebra $H_*(X; \mathbb{Q})$. This leads to an all too brief discussion of Whitehead products, which are the starting point for serious work in unstable homotopy theory. We then return to fracture theorems and describe how such results can be algebraicized when restricted to H-spaces.

In this chapter, we agree to say that a rational space Y is of finite type if its integral, or equivalently rational, homology groups are finite dimensional vector spaces over $\mathbb{Q}$.

9.1. The structure of rational H-spaces

We shall prove the following basic result, which describes the homotopy types of rational H-spaces.

THEOREM 9.1.1. *If Y is a connected rational H-space of finite type, then Y is equivalent to a product $\times_n K(\pi_n(Y), n)$ of rational Eilenberg-Mac Lane spaces. If Y is finite (as a rational CW complex), then $\pi_n(Y) = 0$ for all even n.*

The finite type hypothesis is needed due to the topology rather than the algebra. The problem is the lack of a Künneth theorem for infinite products.

Since the *k*-invariants of X_0 are the rationalizations of the *k*-invariants of *X*, the theorem has the following implication.

COROLLARY 9.1.2. *The k-invariants of an H-space X that is rationally of finite type are torsion classes. If X is finite (as a CW complex), then X has the rational homotopy type of a finite product of odd dimensional spheres.*

When *X* is finite, the number of spherical factors is the *rank* of *X* and the list of their dimensions is the *type* of *X*. The classification problem for finite *H*-spaces considers those *X* with the same rank and type. When *X* is a compact Lie group, this notion of rank coincides with the classical one: the rank of *X* is the dimension of a maximal torus.

The multiplication on an *H*-space *X* determines a comultiplication on the rational cohomology of *X*. This is compatible with the multiplication induced on $H^*(X; \mathbb{Q})$ by the diagonal map and so $H^*(X; \mathbb{Q})$ is a commutative and associative quasi-Hopf algebra. Here "quasi" refers to the fact that the coproduct need not be coassociative since the product on *X* need not be homotopy associative. The proof of Theorem 9.1.1 is based on the following structure theorem for rational quasi-Hopf algebras, which is Theorem 22.4.1.

THEOREM 9.1.3. *If A is a commutative, associative, and connected quasi-Hopf algebra over $\mathbb{Q}$, then A is isomorphic as an algebra to the tensor product of an exterior algebra on odd degree generators and a polynomial algebra on even degree generators.*

We calculated the rational cohomology of $K(\mathbb{Q}; n)$ in Proposition 6.7.2. It is the polynomial algebra $P[\iota_n]$ if *n* is even and the exterior algebra $E[\iota_n]$ if *n* is odd.

PROOF OF THEOREM 9.1.1. By Theorem 9.1.3, the rational cohomology of *X* is a tensor product of exterior and polynomial algebras. We can choose representative maps $X \longrightarrow K(\mathbb{Q}, n)$ for the generators and take them as the coordinates of a map *f* from *X* to a product of $K(\mathbb{Q}, n)$'s. The map *f* induces an isomorphism on rational cohomology and therefore, by our finite type hypothesis, on rational homology. Since the source and target of f are simple and rational, the $\mathbb{Q}$-equivalence f must be an equivalence. $\square$

While our interest is in *H*-spaces, the proof makes clear that the conclusion applies to any simple rational space of finite type whose rational cohomology is a free commutative $\mathbb{Q}$-algebra.

We can elaborate the argument to obtain an equivalence of categories.

THEOREM 9.1.4. *Rational cohomology defines a contravariant equivalence from the homotopy category of rational connected H-spaces of finite type to the category of commutative, associative, and connected quasi-Hopf algebras of finite type.*

PROOF. The required contravariant functor $\mathbb{S}$ from *H*-spaces to Hopf algebras is given by $\mathbb{S}(Y) = H^*(Y; \mathbb{Q})$. The inverse functor $\mathbb{T}$ from Hopf algebras to *H*-spaces assigns to a Hopf algebra A the product of Eilenberg-Mac Lane spaces $K(\mathbb{Q}, n)$, one for each algebra generator of degree n. This fixes the space and therefore the cohomology algebra, but it does not fix the *H*-space structure. The coproduct $\psi: A \longrightarrow A \otimes A$ sends each generator x of degree n to an element of $A \otimes A$ of degree n. By the Künneth theorem, we may identify $A \otimes A$ with $H^*(Y \times Y; \mathbb{Q})$. By the representability of cohomology, the element $\psi(x)$ is represented by a map $Y \times Y$ to $K(\mathbb{Q}, n)$. These maps are the coordinates of the product $Y \times Y \longrightarrow Y$ that makes Y into an *H*-space. A moment's reflection will convince the reader that $\mathbb{S}\mathbb{T}(A)$ is isomorphic to A as a quasi-Hopf algebra, and it follows from Theorem 9.1.1 that $\mathbb{T}\mathbb{S}(Y)$ is equivalent to Y as an *H*-space. $\square$

It is clear from the proof that this equivalence of categories provides a dictionary for translating topological properties into algebraic properties. For example, we have the following elaborations.

PROPOSITION 9.1.5. *A rational H-space Y of finite type is homotopy associative (or homotopy commutative) if and only if $H^*(Y; \mathbb{Q})$ is coassociative (or cocommutative). If Y is of finite homological dimension, so that $H^*(Y; \mathbb{Q})$ is an exterior algebra, then Y is homotopy associative if and only if Y is equivalent as an H-space to a finite product of spaces $K(\mathbb{Q}, 2n - 1)$. In particular, it is then homotopy commutative.*

PROOF. The first statement is clear. For the second, if Y is of finite homological dimension, then, as we shall prove in Corollary 22.4.3, $H^*(Y;\mathbb{Q})$ is an exterior Hopf algebra. This means that its generators x are primitive, $\psi(x) = x \otimes 1 + 1 \otimes x$. Since the fundamental classes of Eilenberg-Mac Lane spaces are clearly primitive, the conclusion follows. For the last statement, recall from [93, p. 127] that Eilenberg-Mac Lane spaces can be constructed as commutative topological groups. □

9.2. The Samelson product and $H_*(X; \mathbb{Q})$

In this digressive section, which elaborates on [104, App.], we show that there is a conceptual topological way of interpreting the homology $H_*(X;\mathbb{Q})$ when X is a connected homotopy associative H-space, not necessarily of finite homological dimension. We begin with some preliminaries on the structure of H-spaces.

DEFINITION 9.2.1. An H-monoid is a homotopy associative H-space. An H-group is an H-monoid with a map χ providing inverses up to homotopy, so that

$$\mu \circ (\mathrm{id} \times \chi) \circ \Delta \simeq * \simeq \mu \circ (\chi \times \mathrm{id}) \circ \Delta,$$

where $*$ denotes the trivial map at the unit element e and μ is the product. If we abuse notation by writing $\chi(x) = x^{-1}$ and writing $\mu(x, y) = xy$, then the condition becomes "$xx^{-1} = e = x^{-1}x$" up to homotopy. An H-monoid is grouplike if $\pi_0(X)$ is a group under the product induced by the product on X.

More elegantly, an H-space X is an H-monoid if the functor $[-, X]$ is monoid–valued and is an H-group if the functor $[-, X]$ is group-valued. Using the uniqueness of inverses in a group, a formal argument shows that if X is an H-monoid, then it is an H-group if we have either of the homotopies displayed in the definition; the other will follow. Note in particular that the set $[S^0, X]$ of components is then a group, so that an H-group is necessarily grouplike. Less obviously, the converse often holds.

LEMMA 9.2.2. *A connected H-monoid is an H-group.*

PROOF. Define the "shearing map" $\xi: X \times X \longrightarrow X \times X$ by $\xi(x, y) = (x, xy)$. On $\pi_n(X) \times \pi_n(X)$, including $n = 1$, it induces the homomorphism of abelian groups that sends (a, b) to $(a, a + b)$. This is an isomorphism; its inverse sends (a, b) to $(a, b - a)$. Therefore ξ induces an isomorphism on homotopy groups and is an equivalence. Choose a homotopy inverse ξ^{-1} and define $\chi = \pi_2 \circ \xi^{-1} \circ \iota_1$, where $\iota_1(x) = (x, e)$ and $\pi_2(x, y) = y$. Then since ξ^{-1} is homotopy inverse to ξ, $\pi_1 \circ \xi^{-1}$ is homotopic to π_1, and $\mu \circ \xi^{-1}$ is homotopic to π_2. Using the first of these, we see that $(\mathrm{id} \times \chi) \circ \Delta = (\mathrm{id} \times \pi_2) \circ (\mathrm{id} \times \xi^{-1}) \circ (\mathrm{id} \times \iota_1) \circ \Delta$ is homotopic to $\xi^{-1} \circ \iota_1$. Using the second, we see that $\mu \circ \xi^{-1} \circ \iota_1$ is homotopic to the constant map. $\square$

Digressing from our usual assumption that X is connected, we have the following weaker analogue. We shall make use of it later, when proving Bott periodicity.

LEMMA 9.2.3. *If X is a grouplike H-monoid, then H is homotopy equivalent to $X_e \times \pi_0(X)$, where X_e denotes the component of the unit element e. If, further, X is homotopy commutative, then X is equivalent to $X_e \times \pi_0(X)$ as an H-space and therefore X is an H-group.*

PROOF. Choose a basepoint x_i in each component $[x_i]$, taking e in the component $[e] = X_e$. Write y_i for the basepoint in the component inverse to $[x_i]$ in $\pi_0(X)$. Define $\mu: X \longrightarrow X_e \times \pi_0(X)$ by $\mu(x) = (x y_i, [x_i])$ for $x \in [x_i]$ and define $\nu: X_e \times \pi_0(X) \longrightarrow X$ by $\nu(x, [x_i]) = x x_i$. Then μ and ν are inverse equivalences, and they are maps of H-spaces if X is homotopy commutative. $\square$

Now consider the rational homology of a connected H-group X. Since the Hopf algebra $H_*(X; \mathbb{Q})$ is cocommutative, it is primitively generated and is therefore isomorphic to the universal enveloping algebra $U(P)$, where P denotes the Lie algebra (under the commutator) of primitive elements in $H_*(X; \mathbb{Q})$. This statement is explained and proven in Theorem 22.3.1 and Corollary 22.3.3. The vector space of primitive elements depends only on the coproduct of $H_*(X; \mathbb{Q})$ and therefore depends only on the diagonal map of X, not on its product. Therefore, as a vector space, P can be identified with the primitive elements in the homology of the product of Eilenberg-Mac Lane spaces that is equivalent to the rationalization of X. However, a moment's thought makes clear that the Hurewicz homomorphism for a rational Eilenberg-Mac Lane space identifies its homotopy groups with the primitive elements in its rational homology. Therefore that is also true for the rationalization X_0. This proves the following result.

PROPOSITION 9.2.4. *The Hurewicz homomorphism*

$$h\colon \pi_*(X_0) \longrightarrow H_*(X_0; \mathbb{Z}) \cong H_*(X; \mathbb{Q})$$

is a monomorphism whose image is the vector subspace of primitive elements.

This raises the question of whether there is a homotopical construction of a Lie bracket on $\pi_*(X)$ for an H-group X that is compatible under the Hurewicz homomorphism with the commutator in $H_*(X; \mathbb{Z})$. The answer is that indeed there is. The relevant product on homotopy groups is called the Samelson product, and we shall define it shortly. The discussion just given will then have the following immediate implication. Here and below, when X is connected, we understand $\pi_*(X)$ to mean the graded abelian group consisting of the homotopy groups of X in positive degrees.

THEOREM 9.2.5. *Let X be a connected H-group of finite type. Then $H_*(X; \mathbb{Q})$ is isomorphic as a Hopf algebra to $U(\pi_*(X) \otimes \mathbb{Q})$, where $\pi_*(X)$ is regarded as a Lie algebra under the Samelson product.*

DEFINITION 9.2.6. Let X be an H-group. Write $\chi(x) = x^{-1}$ and define a map $\phi\colon X \times X \longrightarrow X$ by

$$\phi(x, y) = (xy)(x^{-1}y^{-1}).$$

As noted above Proposition 1.4.3, we may assume that e is a strict two-sided unit element, and a similar use of the nondegeneracy of the basepoint e shows that ϕ is homotopic to a map ϕ' that restricts to the trivial map on $X \vee X$ and thus factors through $X \wedge X$. For based spaces J and K, define the generalized Samelson product

$$\langle -, - \rangle \colon [J, X] \otimes [K, X] \longrightarrow [J \wedge K, X]$$

by $\langle f, g \rangle = [\phi' \circ (f \wedge g)]$. Specializing to $J = S^p$ and $K = S^q$, this gives the Samelson product

$$\langle -, - \rangle \colon \pi_p(X) \otimes \pi_q(X) \longrightarrow \pi_{p+q}(X).$$

PROPOSITION 9.2.7. *Let X be an H-group. If $x \in H_p(X; \mathbb{Z})$ and $y \in H_q(X; \mathbb{Z})$ are primitive, then*

$$\phi_*(x \otimes y) = xy - (-1)^{pq} yx \equiv [x, y].$$

Therefore the Hurewicz homomorphism $h\colon \pi_*(X) \longrightarrow \tilde{H}_*(X; \mathbb{Z})$ satisfies

$$h(\langle f, g \rangle) = [h(x), h(y)].$$

PROOF. The map ϕ is defined by the commutative diagram

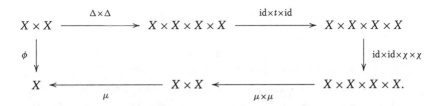

For a primitive element x, $\Delta_*(x) = x \otimes 1 + 1 \otimes x$ and $\chi(x) = -x$, by (21.3.4). From here, a simple chase of the displayed diagram gives the first statement. Since $h(f \wedge g) = h(f) \otimes h(g)$, the second statement follows. □

For connected rational H-groups Y, $h \colon \pi_*(Y) \longrightarrow \tilde{H}_*(Y; \mathbb{Q})$ is a monomorphism, and it follows that $\pi_*(Y)$ is a Lie algebra under the Samelson product. Since $\pi_*(X_0)$ is $\pi_*(X) \otimes \mathbb{Q}$, the rationalization of $\pi_*(X)$ is a Lie algebra for any connected H-group. It requires more work to show that $\pi_*(X)$ is itself a Lie algebra. We only sketch the proof. Complete details may be found in [143, X§5].

PROPOSITION 9.2.8. *For a connected H-group X, $\pi_*(X)$ is a Lie algebra under the Samelson product.*

SKETCH PROOF. We need three preliminaries, the first of which is group theoretical. Recall that the lower central series $\{\Gamma_i\}$ of a group G is given by $\Gamma_1 = G$ and, inductively, $\Gamma_{i+1} = [\Gamma_i, G]$. If $\{G_i\}$ is any central series of G, starting as usual with $G_0 = G$, then $\Gamma_i \subset G_{i-1}$. If G is nilpotent of class m, then $\Gamma_{m+1} = 0$, and it follows that all iterated commutators of length $m + 1$ are zero. By [148, pp. 82–84], for $x, y, z \in G$,

9.2.9 $$[x, yz] \equiv [x, y][x, z] \mod \Gamma_3$$

and

9.2.10 $$[x, [y, z]][y, [z, x]][z, [x, y]] \equiv 1 \mod \Gamma_4.$$

Second, we need the notion of the category of a space. For finite based connected CW complexes X_i, $1 \le i \le k$, filter the product $Y = X_1 \times \cdots \times X_k$ by letting $(x_1, \ldots, x_k)$ be in $F_j Y$ if at least $k - j$ of the coordinates x_i are basepoints. Thus $F_0 Y$ is a point, $F_1 Y$ is the wedge of the X_i, and so on, with $F_n Y = Y$. Taking all $X_i = X$, say that the category of X is less than k if the diagonal map $\Delta \colon (X, *) \longrightarrow (X^k, F_{k-1} X^k)$ is homotopic to a map into $F_{k-1} X^k$. It follows that X has category less than $k + 1$. The category of X, denoted cat(X), is defined to be the minimal k such that X has category less than $k+1$. When $k=0$, this means that X is contractible.

Third, we need the calculation of cat(Y) when Y is a product of spheres. A stratification of X of height k is a filtration by subcomplexes $F_j X$, $1 \le j \le k$, such that $F_0 X = *$, the boundary of each cell in $F_j X$ is contained in $F_{j-1} X$, and $F_k X = X$. If X has such a stratification, then cat(X) $\le k$; in particular, cat(X) $\le$ dim(X). For a lower bound on cat(X), one checks that if X has category less than k, then the product of any k elements in $\tilde{H}^*(X; \mathbb{Z})$ is zero. Now let $X_i = S^{p_i}$, $p_i \ge 1$, and $Y = X_1 \times \cdots \times X_k$. The filtration of Y above is a stratification of height k, hence cat(Y) $\le k$, and equality follows since the product of the k fundamental classes is nonzero.

Now return to our H-group X. Let K be a finite CW complex. If cat(X) $= k$, one proves that the group $G = [K, X]$ is nilpotent of class at most k [143, X.3.6]. For a more explicit result, suppose that K has a stratification $F_j K$, $1 \le j \le k$. Define $G_j \subset G$ to be the set of maps in $f \in [K, X]$ such that $f|_{F_j K}$ is null homotopic. Then $\{G_i\}$ is a central series for G [143, X.3.10].

We apply all of this with K taken to be a product of either 2 or 3 spheres, so that cat(K) $= 2$ or cat(K) $= 3$. We conclude that all double commutators $[x, [y, z]]$ vanish in the group $[S^p \times S^q, X]$ and all triple commutators vanish in the group $[S^p \times S^q \times S^r, X]$. Noticing that the maps $[K \wedge L, X] \longrightarrow [K \times L, X]$ are monomorphisms, and similarly for $K \wedge L \wedge M$, we see that we can check the required algebraic identities by working either in the group $[S^p \times S^q, X]$, in which $\Gamma_3 \subset G_2 = 0$, or in the group $[S^p \times S^q \times S^r, X]$, in which $\Gamma_4 \subset G_3 = 0$.

For example, to check that $\langle f, g+h \rangle = \langle f, g \rangle + \langle f, h \rangle$, for $f \in \pi_p(X)$ and $g, h \in \pi_q(X)$, we map these homotopy groups to the group $[S^p \times S^q, X]$, where the equation holds modulo $\Gamma_3 = 0$ by (9.2.9). The proof that $\langle f, g \rangle + (-1)^{pq} \langle g, f \rangle$ works similarly, using that $t: S^{p+q} = S^p \wedge S^q \longrightarrow S^q \wedge S^p \cong S^{q+p}$ has degree $(-1)^{pq}$. Similarly, for $f \in \pi_p(X), g \in \pi_q(X)$, and $h \in \pi_r(X)$, the Jacobi identity

$$(-1)^{pr} \langle f, \langle g, h \rangle \rangle + (-1)^{pq} \langle g, \langle h, f \rangle \rangle + (-1)^{rq} \langle h, \langle f, g \rangle \rangle = 0$$

can be deduced from the equation (9.2.10) in the group $[S^p \times S^q \times S^r, X]$. The signs enter from transpositions needed to arrange that all elements lie in this group. □

9.3. The Whitehead product

In this even more digressive section, we briefly describe the Whitehead product, which is fundamental in the deeper parts of unstable homotopy theory and is conspicuous by its absence from [93]. It is most easily defined as a special case of the Samelson product. That is the approach that we shall take, although it risks obscuring the importance of the definition.

For any based spaces J and X, we may identify $[\Sigma J, X]$ with $[J, \Omega X]$. It is relevant to signs that we write suspensions on the right, $\Sigma J = J \wedge S^1$. Since ΩX is an H-group under concatenation of paths, we have the generalized Samelson product

$$\langle -, - \rangle \colon [J, \Omega X] \otimes [K, \Omega X] \longrightarrow [J \wedge K, \Omega X].$$

We rewrite this as

$$[-, -] \colon [\Sigma J, X] \otimes [\Sigma K, X] \longrightarrow [\Sigma(J \wedge K), X]$$

and call it the generalized Whitehead product. Taking Taking $J = S^{p-1}$ and $K = S^{q-1}$, this specializes to the Whitehead product

$$[-, -] \colon \pi_p(X) \otimes \pi_q(X) \longrightarrow \pi_{p+q-1}(X).$$

Clearly this is natural in X. It is determined in general by knowledge of the $[i_p, i_q]$, where $i_p \in \pi_p(S^p)$ is the fundamental class.

More formally, in analogy with cohomology operations we define an r-ary homotopy operation of degree n to be a natural transformation

$$\Psi \colon \pi_{p_1}(X) \times \cdots \times \pi_{p_r}(X) \longrightarrow \pi_{p+n}(X), \quad p = p_1 + \cdots + p_r.$$

Such homotopy operations are in canonical bijective correspondence with elements $\psi \in \pi_{p+n}(S^{p_1} \vee \cdots \vee S^{p_r})$. The element ψ corresponding to Ψ is $\Psi(i_{p_1}, \cdots, i_{p_r})$, and the operation Ψ corresponding to ψ is given by

$$\Psi(f_1, \cdots, f_r) = ((\nabla \circ (f_1 \vee \cdots \vee f_r))_*(\psi).$$

Here $f_i \colon S^{p_i} \longrightarrow X$ is an element of $\pi_{p_i}(X)$, and ∇ is the fold map, which is the identity on each wedge summand of $X \vee \cdots \vee X$. The Whitehead products are the most important examples. From this point of view, the Whitehead product $[i_p, i_q]$ is thought of as a map $S^{p+q-1} \longrightarrow S^p \vee S^q$, and it is the attaching map for the construction of $S^p \times S^q$ from $S^p \vee S^q$.

The Whitehead products appear in the EHP-sequence, which is the most important tool for the study of unstable homotopy groups. Expositions may be found in Whitehead's 1978 book [143], which nowhere mentions localization, and Cohen's 1985 lecture notes [30], which assume familiarity with it. The latter is especially recommended as a follow-up to this book for the reader who is interested in learning more about classical homotopy theory.

9.4. Fracture theorems for H-spaces

Now return to the fracture theorem context of the previous chapter. As there, we let T_i, $i \in I$, be sets of primes. We now assume that $T_i \cap T_j = \varnothing$ for $i \neq j$, so that the set S of the fracture theorems is empty, and we assume that $T = \bigcup_i T_i$ is the set of all primes. Thus we are relating global spaces X to their rationalizations X_0 through intermediate localizations X_{T_i}.

THEOREM 9.4.1. *Let X be a nilpotent finite CW complex or, more generally, a nilpotent space with finitely generated integral homology.*

(i) *X is an H-space if and only if each X_{T_i} is an H-space and the coproducts on $H^*(X; \mathbb{Q})$ induced by the rationalization maps $\psi_i \colon X_{T_i} \longrightarrow X_0$ all coincide.*

(ii) *X is a homotopy associative (or commutative) H-space if and only if each X_{T_i} is a homotopy associative (or commutative) H-space.*

PROOF. First, let X be an H-space. By Proposition 6.6.2, each X_{T_i} has a unique H-space structure such that the localization $\phi_i \colon X \longrightarrow X_{T_i}$ is an H-map. Similarly, the resulting product on X_{T_i} induces a unique H-space structure on X_0 such that the localization $\psi_i \colon X_{T_i} \longrightarrow X_0$ is an H-map. Since the composite $\psi_i \phi_i$ is rationalization, each of these H-space structures on X_0 must coincide with the one induced by that of X. Thus they induce the same coproduct on $H^*(X; \mathbb{Q}) \cong H^*(X_0; \mathbb{Q})$. This implication does not depend on X having finitely generated integral homology.

Conversely, suppose that each X_{T_i} is an H-space and the induced coproducts on $H^*(X; \mathbb{Q})$ coincide. Using Remark 8.2.8, we can apply Theorem 8.2.5 and Addendum 8.1.4(i) with K taken to be the n-fold product X^n for any n, since these spaces are again nilpotent and have finitely generated integral homology groups. Using the notations for maps from the cited references, for purposes of the present proof we let $Q[X^n, X]$ denote the pullback displayed in the diagram

9.4.2
$$
\begin{array}{ccc}
Q[X^n, X] & \xrightarrow{\;(\phi_i)_*\;} & [X^n, \prod_{i \in I} X_{T_i}] \\[2mm]
{\scriptstyle \phi_*} \downarrow & & \downarrow {\scriptstyle (\prod_i \psi_i)_*} \\[2mm]
[X^n, X_0] & \xrightarrow[\;\Delta_*\;]{} & [X^n, \prod_{i \in I} X_0].
\end{array}
$$

We conclude from Theorem 8.2.5 and Addendum 8.1.4(i) that the canonical map

9.4.3 $$[X^n, X] \longrightarrow Q[X^n, X]$$

is a bijection. By the universal properties of $(X^n)_{T_i}$ and $(X^n)_0$ and the commutation of localization with finite products, the induced maps

$$[(X_{T_i})^n, X_{T_i}] \cong [(X^n)_{T_i}, X_{T_i}] \longrightarrow [X^n, X_{T_i}]$$

and

$$[(X_0)^n, X_0] \cong [(X^n)_0, X_0] \longrightarrow [(X^n)_{T_i}, X_0] \longrightarrow [X^n, X_0]$$

are bijections for all i. Putting these bijections together, we can rewrite (9.4.2) as

9.4.4

$$
\begin{array}{ccc}
Q[X^n, X] & \xrightarrow{(\phi_{i*})} & \prod_{i \in I}[(X_{T_i})^n, X_{T_i}] \\
\phi_* \downarrow & & \downarrow \phi_* \\
[(X_0)^n, X_0] & \xrightarrow{\Delta} & \prod_{i \in I}[(X_0)^n, X_0].
\end{array}
$$

By assumption, the product maps on X_{T_i} induce the same map on $H^*(X; \mathbb{Q})$ and, by Theorem 9.1.4, this implies that the induced products on X_0 are homotopic. Thus the assumptions give us a well-defined element of $Q[X^2, X]$. The corresponding element of $[X^2, X]$ is an H-space structure on X, the unit condition being obtained by two applications of (9.4.3) with $n = 1$.

If X is homotopy associative or homotopy commutative, then so is X_{T_i} since the multiplication on X_{T_i} is induced from that on X. Conversely, if the X_{T_i} are homotopy associative, we can apply (9.4.3) with $n = 3$ to see that

$$\mu \circ (\mu \times 1) \simeq \mu \circ (1 \times \mu) : X \times X \times X \longrightarrow X.$$

Similarly, if the X_{T_i} are homotopy commutative and $t: X \times X \longrightarrow X \times X$ is the interchange map, we can apply (9.4.3) with $n = 2$ to conclude that

$$\mu \circ t \simeq \mu : X \times X \longrightarrow X. \qquad \square$$

THEOREM 9.4.5. *Let I be finite. Let Y_i be a T_i-local H-space such that $H_*(Y_i; \mathbb{Z})$ is finitely generated over $\mathbb{Z}_{T_i}$. Let A be a quasi-Hopf algebra over $\mathbb{Q}$, and let*

$$\psi_i^* : A \longrightarrow H^*(Y_i; \mathbb{Q})$$

be an isomorphism of quasi-Hopf algebras. Then there exists one and, up to equivalence of H-spaces, only one H-space X such that X_{T_k} is equivalent as an H-space to Y_k for each $k \in I$. Moreover, X has finitely generated integral homology.

PROOF. By the equivalence of categories given in Theorem 9.1.4, there is a rational H-space Y corresponding to the Hopf algebra A and for each map ψ_i^* there is a map of H-spaces

$$\psi_i \colon Y_i \longrightarrow Y$$

that realizes the map ψ_i^* on rational cohomology. Let X be the homotopy pullback of the ψ_i. By Theorem 8.1.7, the canonical map $X_{T_i} \longrightarrow Y_i$ is an equivalence for each $i \in I$. Since the integral homology of Y_i is finitely generated over $\mathbb{Z}_{T_i}$, X has finitely generated integral homology by Proposition 7.4.3. As in the previous proof, we now have the bijection (9.4.3). The product on X is the element of $[X^2, X]$ that corresponds to the products on the X_{T_i} in $Q[X^2, X]$. Again, the unit condition is obtained by two applications of (9.4.3) with $n = 1$. The uniqueness follows from the uniqueness of Y and Theorem 8.1.1. □

The results of this section give the starting point for the subject of finite H-space theory. For example, one can build exotic finite H-spaces, ones not equivalent to compact Lie groups, S^7, and products thereof, by patching together localizations at different sets of primes of different global H-spaces that happen to be rationally equivalent. Returning to the taxonomic analogy of Remark 7.5.8, this is an application of a standard approach to the construction of interesting examples of global spaces with well-understood localizations that are in the same genus. Thinking of spaces, homotopy types, and spaces in the same genus as analogous to animals, animals in the same species, and animals in the same genus, algebraic topologists are expert at genetic modification to produce different species in the same genus. We usually modify spaces using finite sets I, especially partitions of the primes into two disjoint sets T_1 and T_2. In that case, the local to global fraction results go under the name of Zabrodsky mixing, following [145, 146, 147].

The idea is to take two spaces X_1 and X_2 that are equivalent rationally but have very different localizations at T_1 and T_2 and construct a hybrid beast by our local to global construction. We refer the interested reader to [2, p. 79] for an amusing discussion of the resulting bestiary. The historically first example was due to Hilton and Roitberg [63]. They constructed an H-space X that is in the same genus as the Lie group $Sp(2)$ but is not equivalent to it. Both X and $Sp(2)$ are equivalent to $S^3 \times S^7$ away from the primes 2 and 3. As is explained in [62, pp. 122–127], the three H-spaces in sight, X, $Sp(2)$, and $S^3 \times S^7$, are total spaces of bundles over S^7 with fiber S^3, and every simply connected finite H-space with rational cohomology $E[x_3, x_7]$ is equivalent to the total space of such a bundle. There is a large literature devoted to examples such as this.

For a beautiful concrete general result, we quote the following remarkable theorem of Hubbuck [69].

THEOREM 9.4.6. *Let X be a connected homotopy commutative finite H-space. Then X is homotopy equivalent to a torus $T = (S^1)^n$ for some n.*

Since T admits a unique H-space structure, it follows that the equivalence is necessarily an equivalence of H-spaces. Observe that it is not even assumed that X is homotopy associative, but the result implies that it is. The following corollary is essentially equivalent to the theorem.

COROLLARY 9.4.7. *A simply connected homotopy commutative finite H-space is contractible.*

PART 3

Completions of spaces
at sets of primes

10

COMPLETIONS OF NILPOTENT
GROUPS AND SPACES

We develop completion at T for abelian groups, nilpotent groups, and nilpotent spaces. We say right away that there is a choice here. It is usual to focus on a single prime p, and there is no loss of information in doing so since completion at T is the product over $p \in T$ of the completions at p, and similarly for all relevant algebraic invariants. We have chosen to work with sets of primes, but the reader may prefer to concentrate on a single fixed prime.

In contrast to localization, completions of abelian groups can sensibly be defined in different ways, and the most relevant definitions are not standard fare in basic graduate algebra courses. Here again we construct and study completions of nilpotent groups topologically rather than algebraically. We discuss various ways of completing abelian groups in §1. We define completions of spaces and connect the definition to the algebraic theory of completions in §2. We then construct completions of nilpotent spaces by induction up their Postnikov towers in §3. We specialize to obtain completions of nilpotent groups in §4.

Recall our notational conventions from the Introduction. In particular, T is a fixed and nonempty set of primes throughout this chapter. Maps ϕ will always denote completions.

10.1. Completions of abelian groups

10.1.1. p-*adic completion*
It is usual to define the completion of an abelian group A at a given prime p to be the p-adic completion

$$\hat{A}_p = \lim (A/p^r A),$$

where the limit is defined with respect to the evident quotient homomorphisms. For later reference, we recall that the limit can be displayed in the short exact sequence

10.1.1 $$0 \longrightarrow \hat{A}_p \longrightarrow \times_r A/p^r A \stackrel{\alpha}{\longrightarrow} \times_r A/p^r A \longrightarrow 0,$$

where α is the difference of the identity map and the map whose r^{th} coordinate is the composite of the projection to $A/p^{r+1}A$ and the quotient homomorphism $q \colon A/p^{r+1}A \longrightarrow A/p^r A$; since the maps q are epimorphisms, α is an epimorphism [93, p. 147]. This definition will not fully serve our purposes since p-adic completion is neither left nor right exact in general, and exactness properties are essential to connect up with the topology. The Artin-Rees lemma implies the following analogue of Lemma 5.1.2.

LEMMA 10.1.2. *When restricted to finitely generated abelian groups, the p-adic completion functor is exact.*

When $A = \mathbb{Z}$, we write $\mathbb{Z}_p$ instead of $\hat{\mathbb{Z}}_p$ for the ring of p-adic integers, and we abbreviate $\mathbb{Z}/n\mathbb{Z}$ to $\mathbb{Z}/n$. Observe that the p-adic completion functor takes values in the category of $\mathbb{Z}_p$-modules. The action is given by the evident natural maps

$$\lim \mathbb{Z}/p^r \otimes \lim A/p^r A \longrightarrow \lim (\mathbb{Z}/p^r \otimes A/p^r A) \cong \lim A/p^r A.$$

When A is finitely generated, p-adic completion is given by the map $\psi \colon A \longrightarrow A \otimes \mathbb{Z}_p$ specified by $\psi(a) = a \otimes 1$, this again being a consequence of the Artin-Rees lemma. In this case, the alternative notion of completion at p that we shall give shortly agrees with p-adic completion. Since $\mathbb{Z}_p$ is torsion free, it is a flat $\mathbb{Z}$-module, which gives us another way of seeing Lemma 10.1.2.

Even if we restrict to finitely generated abelian groups, we notice one key point of difference between localization and completion. While a homomorphism of abelian groups between p-local groups is necessarily a map of $\mathbb{Z}_{(p)}$-modules, a homomorphism of abelian groups between p-adically complete abelian groups need not be a map of $\mathbb{Z}_p$-modules.

10.1.2. Derived functors of p-adic completion

To overcome the lack of exactness of p-adic completion in general, we consider the left derived functors of the p-adic completion functor. For the knowledgable reader, we recall that left derived functors are usually defined only for right exact functors, in which case the 0th left derived functor agrees with the given functor. However, the definition still makes sense for functors that are not right exact. We shall not go into the general theory of derived functors since, for our present purposes, the abstract theory is less useful than a concrete description of the specific example at hand. The left derived functors of p-adic completion are given on an abelian group A by first taking a free resolution

$$0 \longrightarrow F' \longrightarrow F \longrightarrow A \longrightarrow 0$$

of A, then applying p-adic completion, and finally taking the homology of the resulting length two chain complex $\hat{F}'_p \longrightarrow \hat{F}_p$. Thus the left derived functors of p-adic completion are defined by

$$L_0(A) = \operatorname{coker}(\hat{F}'_p \longrightarrow \hat{F}_p) \quad \text{and} \quad L_1(A) = \ker(\hat{F}'_p \longrightarrow \hat{F}_p).$$

These groups are independent of the choice of resolution, as one checks by comparing resolutions, and they are functorial in A. The higher left derived functors are zero. We have a map of exact sequences

10.1.3

$$
\begin{array}{ccccccccc}
0 & \longrightarrow & F' & \longrightarrow & F & \longrightarrow & A & \longrightarrow & 0 \\
& & \downarrow & & \downarrow & & \downarrow & & \\
0 & \longrightarrow & L_1A & \longrightarrow & \hat{F}'_p & \longrightarrow & \hat{F}_p & \longrightarrow & L_0A & \longrightarrow & 0.
\end{array}
$$

It induces a natural map

$$\phi \colon A \longrightarrow L_0A.$$

Since kernels and cokernels of maps of $\mathbb{Z}_p$-modules are $\mathbb{Z}_p$-modules, since a free abelian group is its own free resolution, and since p-adic completion is exact when restricted to finitely generated abelian groups, we have the following observations.

LEMMA 10.1.4. *The functors L_0 and L_1 take values in $\mathbb{Z}_p$-modules. If A is either a finitely generated abelian group or a free abelian group, then $L_0A = \hat{A}_p$, $L_1A = 0$, and $\phi \colon A \longrightarrow L_0A$ coincides with p-adic completion.*

We usually work at a fixed prime, but we write L_0^p and L_1^p when we need to record the dependence of the functors L_i on the chosen prime p.

DEFINITION 10.1.5. Fix a prime p. We say that the completion of A at p is defined if $L_1A = 0$, and we then define the completion of A at p to be the homomorphism $\phi \colon A \longrightarrow L_0A$. We say that A is p-complete if $\phi \colon A \longrightarrow L_0A$ is an isomorphism. As we shall see in Proposition 10.1.18, if A is p-complete, then $L_1A = 0$.

EXAMPLE 10.1.6. We have seen that finitely generated and free abelian groups are completable and their completions at p coincide with their p-adic completions.

EXAMPLE 10.1.7. $\mathbb{Z}_p \otimes \mathbb{Z}_p$ and $\mathbb{Z}/p^\infty$ (see below) are $\mathbb{Z}_p$-modules that are not p-complete.

The essential exactness property of our derived functors, which is proven in the same way as the long exact sequences for Tor and Ext, reads as follows.

LEMMA 10.1.8. *For a short exact sequence of abelian groups*

$$0 \longrightarrow A' \longrightarrow A \longrightarrow A'' \longrightarrow 0,$$

there is a six term exact sequence of $\mathbb{Z}_p$-modules

$$0 \longrightarrow L_1 A' \longrightarrow L_1 A \longrightarrow L_1 A'' \longrightarrow L_0 A' \longrightarrow L_0 A \longrightarrow L_0 A'' \longrightarrow 0.$$

This sequence is natural with respect to maps of short exact sequences.

10.1.3. Reinterpretation in terms of Hom and Ext

These derived functors give a reasonable replacement for p-adic completion, but they may seem unfamiliar and difficult to compute. However, they can be replaced by isomorphic functors that are more familiar and sometimes more easily computed. Define $\mathbb{Z}/p^\infty$ to be the colimit of the groups $\mathbb{Z}/p^r$ with respect to the homomorphisms $p: \mathbb{Z}/p^r \longrightarrow \mathbb{Z}/p^{r+1}$ given by multiplication by p.

EXERCISE 10.1.9. Verify that $\mathbb{Z}/p^\infty \cong \mathbb{Z}[p^{-1}]/\mathbb{Z}$.

NOTATION 10.1.10. For a prime p and an abelian group A, define $\mathbb{E}_p A$ to be $\mathrm{Ext}\,(\mathbb{Z}/p^\infty, A)$ and define $\mathbb{H}_p A$ to be $\mathrm{Hom}\,(\mathbb{Z}/p^\infty, A)$.

Of course, $\mathbb{E}_p A = 0$ if A is a divisible and hence injective abelian group. Write $\mathrm{Hom}\,(\mathbb{Z}/p^r, A) = A_r$ for brevity. We may identify A_r with the subgroup of elements of A that are annihilated by p^r.

PROPOSITION 10.1.11. *There is a natural isomorphism*

$$\mathbb{H}_p A \cong \lim A_r,$$

where the limit is taken with respect to the maps $p: A_{r+1} \longrightarrow A_r$, and there is a natural short exact sequence

$$0 \longrightarrow \lim{}^1 A_r \longrightarrow \mathbb{E}_p A \xrightarrow{\xi} \hat{A}_p \longrightarrow 0.$$

PROOF. The exact sequence

$$0 \longrightarrow \mathbb{Z} \xrightarrow{p^r} \mathbb{Z} \longrightarrow \mathbb{Z}/p^r \longrightarrow 0$$

displays a free resolution of $\mathbb{Z}/p^r$, and the sum of these is a resolution of $\oplus_r \mathbb{Z}/p^r$. Since $\mathrm{Hom}\,(\mathbb{Z}, A) \cong A$, we may identify $\mathrm{Ext}\,(\mathbb{Z}/p^r, A)$ with $A/p^r A$. Moreover, we have maps of free resolutions

$$
\begin{array}{ccccccccc}
0 & \longrightarrow & \mathbb{Z} & \xrightarrow{\ p^{r+1}\ } & \mathbb{Z} & \longrightarrow & \mathbb{Z}/p^{r+1} & \longrightarrow & 0 \\
& & \ \downarrow{\scriptstyle p} & & \ \| & & \ \downarrow{\scriptstyle q} & & \\
0 & \longrightarrow & \mathbb{Z} & \xrightarrow[\ p^r\ ]{} & \mathbb{Z} & \longrightarrow & \mathbb{Z}/p^r & \longrightarrow & 0.
\end{array}
$$

The colimit $\mathbb{Z}/p^\infty$ fits into a short exact sequence

$$0 \longrightarrow \oplus_r \mathbb{Z}/p^r \xrightarrow{\iota} \oplus_r \mathbb{Z}/p^r \longrightarrow \mathbb{Z}/p^\infty \longrightarrow 0.$$

Writing 1_r for the image of 1 in $\mathbb{Z}/p^r$, $\iota(1_r) = 1_r - p1_{r+1}$. The resulting six term exact sequence of groups $\mathrm{Ext}\,(-, A)$ takes the form

10.1.12

$$0 \longrightarrow \mathbb{H}_p A \longrightarrow \times_r A_r \xrightarrow{\iota^*} \times_r A_r \longrightarrow \mathbb{E}_p A \longrightarrow \times_r A/p^r A \xrightarrow{\iota^*} \times_r A/p^r A \longrightarrow 0.$$

The first map ι^* is the difference of the identity map and the map whose r^{th} coordinate is $p \colon A_{r+1} \longrightarrow A_r$. Its kernel and cokernel are $\lim A_r$ and $\lim^1 A_r$, respectively. The second map ι^* is the map α of (10.1.1) whose kernel and cokernel are $\hat{A}_p$ and 0, respectively. $\qquad\square$

EXAMPLE 10.1.13. Any torsion abelian group A with all torsion prime to p satisfies $\mathbb{H}_p A = 0$ and $\mathbb{E}_p A = 0$.

EXAMPLE 10.1.14. $\mathbb{H}_p(\mathbb{Z}/p^\infty)$ is a ring under composition, and it is isomorphic to the ring $\mathbb{Z}_p$ by inspection of the limit system in the previous result; $\mathbb{E}_p(\mathbb{Z}/p^\infty) = 0$ since it is a quotient of $\mathrm{Ext}\,(\mathbb{Z}/p^\infty, \mathbb{Z}[p^{-1}]) = 0$.

The following immediate consequence of Proposition 10.1.11 shows that $\mathbb{E}_p A$ is isomorphic to $\hat{A}_p$ in the situations most often encountered in algebraic topology.

COROLLARY 10.1.15. *If the p-torsion of A is of bounded order, then* $\mathbb{H}_p A = 0$ *and* $\xi \colon \mathbb{E}_p A \longrightarrow \hat{A}_p$ *is an isomorphism.*

EXAMPLE 10.1.16. If $A = \oplus_{n \geq 1} \mathbb{Z}/p^n$, then $\mathbb{E}_p A$ is not a torsion group, the map $\xi \colon \mathbb{E}_p A \longrightarrow \hat{A}_p$ is not an isomorphism, and A is not p-complete.

PROPOSITION 10.1.17. *There are natural isomorphisms*

$$L_0(A) \cong \mathbb{E}_p A \quad \text{and} \quad L_1(A) \cong \mathbb{H}_p A.$$

Moreover $\phi \colon A \longrightarrow L_0 A$ *coincides with the connecting homomorphism*

$$\delta \colon A \cong \mathrm{Hom}\,(Z, A) \longrightarrow \mathrm{Ext}\,(\mathbb{Z}/p^\infty, A) = \mathbb{E}_p A$$

associated with the short exact sequence

$$0 \longrightarrow \mathbb{Z} \longrightarrow \mathbb{Z}[p^{-1}] \longrightarrow \mathbb{Z}/p^\infty \longrightarrow 0.$$

PROOF. Let

$$0 \longrightarrow F' \longrightarrow F \longrightarrow A \longrightarrow 0$$

be a free resolution of A. From this sequence we obtain the exact sequence

$$0 \longrightarrow L_1 A \longrightarrow \hat{F}'_p \longrightarrow \hat{F}_p \longrightarrow L_0 A \longrightarrow 0$$

of (10.1.3) and also, since $\mathbb{H}_p F = 0$, the exact sequence of Ext groups

$$0 \longrightarrow \mathbb{H}_p A \longrightarrow \mathbb{E}_p F' \longrightarrow \mathbb{E}_p F \longrightarrow \mathbb{E}_p A \longrightarrow 0.$$

Since $\mathbb{E}_p F \cong \hat{F}_p$ for free abelian groups F, we may identify these two exact sequences. The last statement follows since the diagram (10.1.3) has an easily checked analogue for $\mathbb{H}_p$ and $\mathbb{E}_p$. □

These isomorphisms may seem a little unnatural at first sight since Ext is a derived functor of Hom. It was first noted by Harrison [58] that these Ext groups give a homologically appropriate variant of the classical p-adic completion functor.

PROPOSITION 10.1.18. *Let A be an abelian group and let B be any of $\hat{A}_p$, $\mathbb{H}_p A$, and $\mathbb{E}_p A$. Then $\mathbb{H}_p B = 0$ and $\delta \colon B \longrightarrow \mathbb{E}_p B$ is an isomorphism. Equivalently, $L_1 B = 0$ and $\phi \colon B \longrightarrow L_0 B$ is an isomorphism. Therefore, if $\phi \colon A \longrightarrow L_0 A$ is an isomorphism, then $L_1 A \cong L_1 L_0 A = 0$.*

PROOF. Using the six term sequence of groups $\mathrm{Ext}\,(-, B)$ associated to the short exact sequence $0 \longrightarrow \mathbb{Z} \longrightarrow \mathbb{Z}[p^{-1}] \longrightarrow \mathbb{Z}/p^\infty \longrightarrow 0$, we see that $\mathbb{H}_p B = 0$ and $\delta \colon B \longrightarrow \mathbb{E}_p B$ is an isomorphism if and only if

10.1.19 $\mathrm{Hom}\,(\mathbb{Z}[p^{-1}], B) = 0$ and $\mathrm{Ext}\,(\mathbb{Z}[p^{-1}], B) = 0.$

This condition certainly holds if $p^r B = 0$ for any r, so it holds for all A_r and $A/p^r A$. If (10.1.19) holds for groups B_i, then it holds for their product $\times_i B_i$. Suppose given a short exact sequence $0 \longrightarrow B' \longrightarrow B \longrightarrow B'' \longrightarrow 0$. If (10.1.19) holds for B, then

$$\mathrm{Hom}\,(\mathbb{Z}[p^{-1}], B') = 0, \quad \mathrm{Hom}\,(\mathbb{Z}[p^{-1}], B'') \cong \mathrm{Ext}\,(\mathbb{Z}[p^{-1}], B'),$$

$$\text{and}\ \ \mathrm{Ext}\,(\mathbb{Z}[p^{-1}], B'') = 0.$$

If (10.1.19) holds for B'', then it holds for B' if and only if it holds for B. Now the short exact sequence (10.1.1) implies that (10.1.19) holds for $B = \hat{A}_p$, and the four short exact sequences into which the six term exact sequence (10.1.12) breaks up by use of kernels and cokernels implies that (10.1.19) holds for $B = \mathbb{E}_p A$ and $B = \mathbb{H}_p A$. $\square$

Further interesting group theoretical results on "cotorsion abelian groups", of which p-complete abelian groups are examples, were obtained by Harrison [58] long before their relevance to topology was noticed. His results were later summarized by Bousfield and Kan [21, pp. 181–182]. Since we will not have need of them, we will not recall them here.

10.1.4. The generalization to sets of primes

DEFINITION 10.1.20. Fix a nonempty set of primes T and recall that $\mathbb{Z}[T^{-1}]$ is obtained by inverting the primes in T, whereas $\mathbb{Z}_T$ is obtained by inverting the primes not in T. Define

$$\mathbb{H}_T A = \mathrm{Hom}\,(\mathbb{Z}[T^{-1}]/\mathbb{Z}, A) \quad \text{and} \quad \mathbb{E}_T A = \mathrm{Ext}\,(\mathbb{Z}[T^{-1}]/\mathbb{Z}, A).$$

We say that the completion of A at T is defined if $\mathbb{H}_T A = 0$, and we then define the completion of A at T to be the connecting homomorphism

$$\phi \colon A \cong \mathrm{Hom}\,(\mathbb{Z}, A) \longrightarrow \mathbb{E}_T A$$

that arises from the short exact sequence

$$0 \longrightarrow \mathbb{Z} \longrightarrow \mathbb{Z}[T^{-1}] \longrightarrow \mathbb{Z}[T^{-1}]/\mathbb{Z} \longrightarrow 0.$$

We say that B is T-complete if ϕ is an isomorphism. We let $\mathscr{A}_T$ denote the collection of all abelian groups that are completable at T and we let $\mathscr{B}_T \subset \mathscr{A}_T$ denote the collection of all T-complete abelian groups.

REMARK 10.1.21. The short exact sequence above gives rise to an exact sequence

$$0 \to \mathbb{H}_T A \to \text{Hom}\,(\mathbb{Z}[T^{-1}], A) \to A \to \mathbb{E}_T A \to \text{Ext}\,(\mathbb{Z}[T^{-1}], A) \to 0;$$

A is completable at T if $\text{Hom}\,(\mathbb{Z}[T^{-1}], A) = 0$ and B is T-complete if and only if

10.1.22 $\qquad \text{Hom}\,(\mathbb{Z}[T^{-1}], B) = 0 \quad \text{and} \quad \text{Ext}\,(\mathbb{Z}[T^{-1}], B) = 0.$

The inclusion $\mathbb{Z}[T^{-1}] \to \mathbb{Q}$ induces an isomorphism $\mathbb{Z}[T^{-1}]/\mathbb{Z} \to \mathbb{Q}/\mathbb{Z}_T$, and $\mathbb{Q}/\mathbb{Z}_T$ is isomorphic to the T-torsion subgroup of $\mathbb{Q}/\mathbb{Z}$. In turn, $\mathbb{Q}/\mathbb{Z}$ is isomorphic to the direct sum over all primes p of the groups $\mathbb{Z}/p^\infty$. These statements are well-known in the theory of infinite abelian groups, and we invite the reader to check them for herself. It follows that the definitions above generalize those given when T is a single prime. Indeed, we have the chains of isomorphisms

$$\begin{aligned}
\mathbb{H}_T A & = \text{Hom}\,(\mathbb{Z}[T^{-1}]/\mathbb{Z}, A) \\
& \cong \text{Hom}\,(\oplus_{p \in T}\, \mathbb{Z}[p^{-1}]/\mathbb{Z}, A) \\
& \cong \times_{p \in T} \text{Hom}\,(\mathbb{Z}[p^{-1}]/\mathbb{Z}, A) \\
& = \times_{p \in T}\, \mathbb{H}_p A
\end{aligned}$$

and

$$\begin{aligned}
\mathbb{E}_T A & = \text{Ext}^1\,(\mathbb{Z}[T^{-1}]/\mathbb{Z}, A) \\
& \cong \text{Ext}^1\,(\oplus_{p \in T}\, \mathbb{Z}[p^{-1}]/\mathbb{Z}, A) \\
& \cong \times_{p \in T} \text{Ext}^1\,(\mathbb{Z}[p^{-1}]/\mathbb{Z}, A) \\
& = \times_{p \in T}\, \mathbb{E}_p A.
\end{aligned}$$

Analogously, we define

$$\hat{A}_T = \times_{p \in T} \hat{A}_p.$$

By Lemma 10.1.4, all of these are modules over the ring $\hat{\mathbb{Z}}_T = \times_{p \in T} \mathbb{Z}_p$.

The results we have proven for a single prime p carry over to sets of primes. For example, Corollary 10.1.15 and Proposition 10.1.18 imply the following results.

PROPOSITION 10.1.23. *If A is a torsion-free or finitely generated $\mathbb{Z}_S$-module for any set of primes $S \supset T$, then $\mathbb{H}_T A = 0$ and the canonical map $\mathbb{E}_T A \longrightarrow \hat{A}_T$ is an isomorphism; its inverse can be identified with the map $A \otimes \hat{\mathbb{Z}}_T \longrightarrow \mathbb{E}_T A$ induced by the action of $\hat{\mathbb{Z}}_T$ on $\mathbb{E}_T A$. In particular, $\mathbb{E}_T$ restricts to an exact functor from finitely generated $\mathbb{Z}_T$-modules to $\hat{\mathbb{Z}}_T$-modules.*

PROPOSITION 10.1.24. *For any abelian group A, the groups $\hat{A}_T$, $\mathbb{H}_T A$, and $\mathbb{E}_T A$ are T-complete.*

10.2. The definition of completions of spaces at T

Recall that we take all spaces to be path connected. Recall too that we let $\mathbb{F}_T = \times_{p \in T} \mathbb{F}_p$. We have the following three basic definitions, which are written in precise parallel to the definitions in the case of localization. The equivalence in the following definition follows directly from Definition 3.3.10 and Proposition 3.3.11.

DEFINITION 10.2.1. A map $\xi \colon X \longrightarrow Y$ is said to be an $\mathbb{F}_T$-equivalence if $\xi_* \colon H_*(X; \mathbb{F}_p) \longrightarrow H_*(X; \mathbb{F}_p)$ is an isomorphism for all primes $p \in T$ or, equivalently, if $\xi^* \colon H^*(Y; B) \to H^*(X; B)$ is an isomorphism for all $\mathbb{F}_T$-modules B.

DEFINITION 10.2.2. A space Z is T-complete if $\xi^* \colon [Y, Z] \longrightarrow [X, Z]$ is a bijection for all $\mathbb{F}_T$-equivalences $\xi \colon X \longrightarrow Y$.

Diagrammatically, this says that for any map $f \colon X \longrightarrow Z$, there is a map $\tilde{f}$, unique up to homotopy, that makes the following diagram commute up to homotopy.

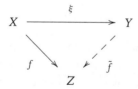

DEFINITION 10.2.3. A map $\phi \colon X \longrightarrow \hat{X}_T$ from X into a T-complete space $\hat{X}_T$ is a completion at T if ϕ is an $\mathbb{F}_T$-equivalence.

This prescribes a universal property. If $f \colon X \longrightarrow Z$ is any map from X to a T-complete space Z, then there is a map $\tilde{f}$, unique up to homotopy, that makes the following diagram commute.

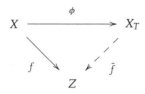

Therefore completions are unique up to homotopy if they exist. We shall prove in Chapter 19 that they do always exist, but we focus on nilpotent spaces for now.

REMARK 10.2.4. On the full subcategory of connected spaces in Ho$\mathscr{T}$ that admit completions at T, completion is automatically functorial (up to homotopy). For a map $f\colon X \longrightarrow Y$, there is a unique map $\hat{f}_T\colon \hat{X}_T \longrightarrow \hat{Y}_T$ in Ho$\mathscr{T}$ such that $\phi \circ f = \hat{f}_T \circ \phi$ in Ho$\mathscr{T}$, by the universal property.

The definitions just given do not mention any of the algebraic notions that we discussed in the previous section. However, they lead directly to consideration of the collection $\mathscr{B}_T$ of T-complete abelian groups, as the following analogue of Corollary 5.2.6 shows.

THEOREM 10.2.5. *The $\mathbb{F}_T$-equivalences coincide with the maps that induce isomorphisms on cohomology with coefficients in all groups in $\mathscr{B}_T$, and $\mathscr{B}_T$ is the largest collection of abelian groups for which this is true.*

PROOF. Let $\mathscr{C}_T$ denote the collection of all abelian groups C such that

$$\xi^*\colon H^*(Y; C) \longrightarrow H^*(X; C)$$

is an isomorphism for all $\mathbb{F}_T$-equivalences $\xi\colon X \longrightarrow Y$. Our claim is that $\mathscr{B}_T = \mathscr{C}_T$. The collection $\mathscr{C}_T$ has the following closure properties.

(i) If two terms of a short exact sequence of abelian groups are in $\mathscr{C}_T$, then so is the third term since a short exact sequence gives rise to a natural long exact sequence of cohomology groups.

(ii) If $p \in T$ and $p^r C = 0$, then $C \in \mathscr{C}_T$, as we see by (i) and induction on r; the case $r = 1$ holds by the definition of an $\mathbb{F}_T$-equivalence.

(iii) Any product of groups in $\mathscr{C}_T$ is also in $\mathscr{C}_T$ since $H^*(X; \times_i C_i)$ is naturally isomorphic to $\times_i H^*(X; C_i)$.

(iv) By (i), the limit of a sequence of epimorphisms $f_i\colon C_{i+1} \longrightarrow C_i$ between groups in $\mathscr{C}_T$ is a group in $\mathscr{C}_T$ since we have a natural short exact sequence

$$0 \longrightarrow \lim C_i \longrightarrow \times_i C_i \longrightarrow \times_i C_i \longrightarrow 0;$$

the $\lim^1$ error term is 0 on the right because the f_i are epimorphisms; see §2.3.

(v) All groups $\hat{A}_T$ are in $\mathscr{C}_T$, as we see by (ii), (iii), and (iv).

(vi) $\mathbb{E}_T A$ is in $\mathscr{C}_T$ if A is completable at T, as we see by (i), the exact sequence (10.1.3) and Proposition 10.1.17.

(vii) A is in $\mathscr{C}_T$ if A is T-complete since A is then completable and isomorphic to $\mathbb{E}_T A$.

This proves that $\mathscr{B}_T \subset \mathscr{C}_T$. For the opposite inclusion, we observe first that the unique map from $K(\mathbb{Z}[T^{-1}], 1)$ to a point is an $\mathbb{F}_T$-equivalence. Indeed, we have seen that $K(\mathbb{Z}[T^{-1}], 1)$ is a localization of $S^1 = K(\mathbb{Z}, 1)$ away from T. Its only nonzero reduced homology group is $\tilde{H}_1(K(\mathbb{Z}[T^{-1}], 1); \mathbb{Z}) \cong \mathbb{Z}[T^{-1}]$. Since multiplication by $p \in T$ is an isomorphism on this group, the universal coefficient theorem implies that $\tilde{H}_*(\mathbb{Z}[T^{-1}], \mathbb{F}_p) = 0$ for $p \in T$. For $C \in \mathscr{C}_T$, we conclude that $\tilde{H}^*(\mathbb{Z}[T^{-1}], C) = 0$. By the universal coefficient theorem again,

$$\mathrm{Hom}\,(\mathbb{Z}[T^{-1}], C) \cong H^1(\mathbb{Z}[T^{-1}], C) = 0$$

and

$$\mathrm{Ext}\,(\mathbb{Z}[T^{-1}], C) \cong H^2(\mathbb{Z}[T^{-1}], C) = 0.$$

By Remark 10.1.21, this means that C is T-complete. $\qquad\square$

Recall the discussion of profinite groups from §2.5 and say that a profinite group $B = \lim B_d$ is T-profinite if the B_d are T-torsion groups. It is clear from Theorem 10.2.5 and its proof that many T-profinite abelian groups are in $\mathscr{B}_T$ and are thus T-complete in our sense, but it is not clear that all of them are. However, Theorem 10.2.5 and the restriction of Theorem 2.6.1 to T-profinite abelian groups imply the following result.

COROLLARY 10.2.6. *All T-profinite abelian groups are in $\mathscr{B}_T$.*

Returning to topology, we can now relate $\mathscr{B}_T$ to Eilenberg-Mac Lane spaces.

COROLLARY 10.2.7. *If B is T-complete, then $K(B, n)$ is T-complete for all $n \geq 1$.*

PROOF. If $\xi \colon X \longrightarrow Y$ is an $\mathbb{F}_T$-equivalence, then

$$\xi^* \colon H^*(Y; B) \longrightarrow H^*(X; B)$$

is an isomorphism by Theorem 10.2.5 and thus

$$\xi^* : [Y, K(B, n)] \longrightarrow [X, K(B, n)]$$

is an isomorphism by the representability of cohomology. $\qquad \square$

By analogy with Proposition 5.2.5, we have an alternative topological description of the collection $\mathscr{B}_T$ of T-complete abelian groups. As in that result, we cannot prove this without first doing a little homological calculation, but we defer that to the next section.

PROPOSITION 10.2.8. *An abelian group B is T-complete if and only if the space $K(B, 1)$ is T-complete.*

PROOF. If B is T-complete, then $K(B, 1)$ is T-complete by the previous result. Suppose that $K(B, 1)$ is T-complete. Then the identity map of $K(B, 1)$ is a completion at T. Moreover,

$$\xi^* : 0 = [*, K(B, 1)] \longrightarrow [K(\mathbb{Z}[T^{-1}], 1), K(B, 1)]$$

is an isomorphism since $K(\mathbb{Z}[T^{-1}], 1) \longrightarrow *$ is an $\mathbb{F}_T$-cohomology isomorphism, as we observed in the proof of Theorem 10.2.5. Using the representability of cohomology, this gives that

$$\mathrm{Hom}\,(\mathbb{Z}[T^{-1}], B) \cong H^1(\mathbb{Z}[T^{-1}], B) = 0.$$

This implies that $\mathbb{H}_T B = 0$, so that B is completable at T. In Theorem 10.3.2 below, we shall show among other things that the map $\phi : K(B, 1) \longrightarrow K(\mathbb{E}_T B, 1)$ that realizes $\phi : B \longrightarrow \mathbb{E}_T B$ on fundamental groups is an $\mathbb{F}_T$-equivalence, and its target is T-complete since $\mathbb{E}_T B$ is T-complete by Proposition 10.1.24. Thus ϕ is also a completion of $K(B, 1)$ at T. By the uniqueness of completion, ϕ must be an equivalence and thus $\phi : B \longrightarrow \mathbb{E}_T B$ must be an isomorphism. $\qquad \square$

10.3. Completions of nilpotent spaces

We construct completions here, beginning with completions of Eilenberg-Mac Lane spaces. That was the easy step in the case of localizations, but it is the key step in the case of completions. We first record the relevant special case of the dual Whitehead theorem. Take $\mathscr{A}$ in Theorem 3.3.9 to be $\mathscr{B}_T$. Then that result takes the following form, which generalizes the fact that $K(B, n)$ is T-complete if B is T-complete.

THEOREM 10.3.1. *Every $\mathscr{B}_T$-tower is a T-complete space.*

We use this result to construct completions of nilpotent spaces, dealing separately with Eilenberg-Mac Lane spaces before proceeding to the general case.

THEOREM 10.3.2. *For each abelian group A and each $n \geq 1$, there is a completion $\phi\colon K(A, n) \longrightarrow K(A, n)_T^\wedge$. The space $K(A, n)_T^\wedge$ is $\mathscr{B}_T$-nilpotent, its only nonzero homotopy groups are*

$$\pi_n(K(A, n)_T^\wedge) = \mathbb{E}_T A$$

and

$$\pi_{n+1}(K(A, n)_T^\wedge) = \mathbb{H}_T A,$$

and $\phi_*\colon \pi_n(K(A, n)) \longrightarrow \pi_n(K(A, n)_T^\wedge)$ coincides with $\phi\colon A \longrightarrow \mathbb{E}_T A$.

PROOF. First consider a free abelian group F. Here we have $\mathbb{H}_T F = 0$ and $\mathbb{E}_T F \cong \hat{F}_T$. We claim that the map

$$\phi\colon K(F, n) \longrightarrow K(\hat{F}_T, n)$$

that realizes $\phi\colon F \longrightarrow \hat{F}_T$ is a completion at T. Since $\hat{F}_T$ is T-complete by Proposition 10.1.24, $K(\hat{F}_T, n)$ is T-complete by Corollary 10.2.7. Thus we only need to prove that ϕ_* is an isomorphism on mod p homology for $p \in T$. We proceed by induction on n, and we first consider the case $n = 1$.

The projection $\hat{F}_T \longrightarrow \hat{F}_p$ induces an isomorphism on mod p homology since its kernel is local away from p. We now use the LHS spectral sequence, Proposition 24.5.3, of the quotient group $\hat{F}_p/F$. The spectral sequence has the form

$$E_{p,q}^2 = H_p(\hat{F}_p/F; H_q(F; \mathbb{F}_p)) \Longrightarrow H_{p+q}(\hat{F}_p; \mathbb{F}_p).$$

The group $\hat{F}_p/F$ is uniquely p-divisible. One can see this, for example, by noting that the canonical map $F \longrightarrow \hat{F}_p$ is a monomorphism of torsion-free abelian groups that induces an isomorphism upon reduction mod p. Alternatively, writing elements of F in terms of a basis for F and writing integer coefficients in p-adic form, we see that elements of $F/p^r F$ can be written in the form $f + pg$, where the coefficients appearing in f satisfy $0 \leq a < p$. If we have an element $(f_r + pg_r)$ of $\lim F/p^r F \subset \times_r F/p^r F$ with components written in this form, then compatibility forces (f_r) to come from an element $f \in F$, and it follows that our given element is congruent to $p(g_r)$ mod F. It follows that the terms with $p > 0$ are zero and the spectral sequence collapses to the edge isomorphism

$$\phi_* : H_*(F; \mathbb{F}_p) \cong E^2_{0,*} = E^\infty_{0,*} \cong H_*(\hat{F}_p; \mathbb{F}_p).$$

For $n > 1$, take $K(F, n-1) = \Omega K(F, n)$ and consider the map of path space fibrations

$$
\begin{array}{ccccc}
K(F, n-1) & \longrightarrow & PK(F, n) & \longrightarrow & K(F, n) \\
\downarrow & & \downarrow & & \downarrow \\
K(\hat{F}_T, n-1) & \longrightarrow & PK(\hat{F}_T, n) & \longrightarrow & K(\hat{F}_T, n).
\end{array}
$$

By the Serre spectral sequence, the induction hypothesis, and the comparison theorem, Theorem 24.6.1, the map

$$\phi_* : H_q(K(F, n), \mathbb{F}_p) \longrightarrow H_q(K(\hat{F}_T, n), \mathbb{F}_p)$$

is an isomorphism and therefore ϕ is a completion at T.

Now consider a general abelian group A. Write A as a quotient F/F' of free abelian groups and let $i \colon F' \longrightarrow F$ be the inclusion. We construct a map of fibration sequences

$$
\begin{array}{ccccccc}
K(F, n) & \longrightarrow & K(A, n) & \longrightarrow & K(F', n+1) & \overset{i}{\longrightarrow} & K(F, n+1) \\
\Omega\phi \downarrow & & \phi \downarrow & & \phi \downarrow & & \phi \downarrow \\
K(\hat{F}_T, n) & \longrightarrow & K(A, n)^\wedge_T & \longrightarrow & K(\hat{F}'_T, n+1) & \overset{i^\wedge_T}{\longrightarrow} & K(\hat{F}_T, n+1).
\end{array}
$$

Here the map i realizes the algebraic map i on passage to π_{n+1} and can be viewed as the map from the fiber to the total space of a fibration with base space $K(A, n+1)$. We take $K(A, n)$ to be the fiber Fi and take $K(F, n) = \Omega K(F, n+1)$. The two completion maps on the right have been constructed, and that on the left is the loops of that on the right. The map $i^\wedge_T$ is the map, unique up to homotopy, that makes the right square commute up to homotopy, and it realizes the algebraic map $i^\wedge_T$ on passage to π_{n+1}. We define $K(A, n)^\wedge_T$ to be its fiber. By Lemma 1.2.3, there is a dotted arrow map ϕ that makes the middle square commute and the left square commute up to homotopy. This map induces an isomorphism on mod p homology for $p \in T$ by the map of Serre spectral sequences induced by the map of fibrations given by the left two squares. To show that ϕ is a completion of $K(A, n)$ at T it remains to show that $K(A, n)^\wedge_T$ is complete. Since $K(A, n)^\wedge_T$ is visibly a $\mathscr{B}_T$-tower, this holds by Theorem 10.3.1.

The bottom fibration sequence above gives a long exact sequence

$$\cdots \longrightarrow \pi_{n+1}(K(\hat{F}_T, n)) \longrightarrow \pi_{n+1}(K(A, n)_T^\wedge) \longrightarrow \pi_{n+1}(K(\hat{F}'_T, n+1))$$

$$\longrightarrow \pi_n(K(\hat{F}_T, n)) \longrightarrow \pi_n(K(A, n)_T^\wedge) \longrightarrow \pi_n(K(\hat{F}'_T, n+1)) \longrightarrow \cdots.$$

By the case of free abelian groups this simplifies to

$$0 \longrightarrow \pi_{n+1}(K(A, n)_T^\wedge) \longrightarrow \hat{F}'_T \xrightarrow{\ i_T^\wedge\ } \hat{F}_T \longrightarrow \pi_n(K(A, n)_T^\wedge) \longrightarrow 0.$$

The map $i_T^\wedge$ is the product over $p \in T$ of the maps $i_p^\wedge$, and our algebraic definitions and results give exact sequences

$$0 \longrightarrow \mathbb{H}_p A \longrightarrow \hat{F}'_p \longrightarrow \hat{F}_p \longrightarrow \mathbb{E}_p A \longrightarrow 0.$$

The product over $p \in T$ of these exact sequences is isomorphic to the previous exact sequence, and this gives the claimed identification of homotopy groups. Comparing the map on homotopy groups given by our map of fibration sequences to the diagram (10.1.3), we see that the map on n^{th} homotopy groups induced by ϕ is the algebraic map ϕ. $\qquad\square$

In view of Example 10.1.14, we have an interesting explicit example where homotopy groups shift dimension.

EXAMPLE 10.3.3. For a prime p, $K(\mathbb{Z}/p^\infty, n)_p^\wedge$ is an Eilenberg-Mac Lane space $K(\mathbb{Z}_p, n+1)$.

This is not an exotic example. Analogous dimension-shifting examples play a central role in comparing the algebraic K-theory of an algebraically closed field, which is concentrated in odd degrees, to topological K-theory, which is concentrated in even degrees [89, 113, 116].

The generalization from Eilenberg-Mac Lane spaces to nilpotent spaces works in precisely the same way as the construction of localizations. We need only replace the localizations $K(A_T, n)$ by the completions $K(A, n)_T^\wedge$. The fact that the latter are not Eilenberg-Mac Lane spaces does not change the details of the construction.

THEOREM 10.3.4. *Every nilpotent space X admits a completion $\phi\colon X \longrightarrow \hat{X}_T$.*

PROOF. Exactly as in the proof of Theorem 5.3.2, we may assume that X is a Postnikov tower $\lim X_i$ constructed from maps $k_i\colon X_i \longrightarrow K(A_i, n_i + 1)$, where A_i is an abelian group, $n_{i+1} \geq n_i \geq 1$, and only finitely many $n_i = n$

for any $n \geq 1$. Here $X_0 = *$, and we let $(X_0)^\wedge_T = *$. Assume that a completion $\phi_i \colon X_i \longrightarrow (X_i)^\wedge_T$ has been constructed and consider the following diagram, in which we write $K(A_i, n_i) = \Omega K(A_i, n_i + 1)$.

$$
\begin{array}{ccccccc}
K(A_i, n_i) & \longrightarrow & X_{i+1} & \longrightarrow & X_i & \xrightarrow{\;k_i\;} & K(A_i, n_i + 1) \\
\Omega\phi \downarrow & & \phi_{i+1} \downarrow & & \phi_i \downarrow & & \phi \downarrow \\
K(A_i, n_i)^\wedge_T & \longrightarrow & (X_{i+1})^\wedge_T & \longrightarrow & (X_i)^\wedge_T & \xrightarrow[(k_i)^\wedge_T]{} & K(A_i, n_i + 1)^\wedge_T
\end{array}
$$

By Theorem 10.3.1, since ϕ_i is an $\mathbb{F}_T$-equivalence and $K(A_i, n_i + 1)^\wedge_T$ is a T-complete space there is a map $(k_i)^\wedge_T$, unique up to homotopy, that makes the right square commute up to homotopy. The space X_{i+1} is the fiber Fk_i, and we define $(X_{i+1})^\wedge_T$ to be the fiber $F(k_i)^\wedge_T$.

By Lemma 1.2.3, there is a map ϕ_{i+1} that makes the middle square commute and the left square commute up to homotopy. By Theorem 10.3.1, $(X_{i+1})^\wedge_T$ is T-complete since it is a $\mathscr{B}_T$-tower. To see that ϕ_{i+1} is a completion at T it remains to show that it induces an isomorphism on homology with coefficients in $\mathbb{F}_p$ for $p \in T$. The proof is a comparison of Serre spectral sequences exactly like that in the proof of Theorem 5.3.2. We define $X^\wedge_T = \lim (X_i)^\wedge_T$ and $\phi = \lim \phi_i \colon X \to X^\wedge_T$. Then ϕ is an $\mathbb{F}_T$-equivalence by Proposition 2.5.9 and is thus a completion of X at T. $\qquad\square$

Similarly, the proofs of the following analogues of Theorem 5.3.3, Proposition 5.3.4, and Corollaries 5.3.5 and 5.3.6 concerning the functoriality of our cocellular constructions are virtually identical to the proofs of those results.

THEOREM 10.3.5. *Let X and Y be Postnikov towers and let $\psi \colon X \longrightarrow Y$ be a cocellular map. Choose cocellular completions at T of X and Y. Then there exists a cocellular map $\psi^\wedge_T \colon X^\wedge_T \longrightarrow Y^\wedge_T$, unique up to cocellular homotopy, such that $\psi^\wedge_T \circ \phi$ is homotopic to $\phi \circ \psi$.*

PROPOSITION 10.3.6. *Let W be a quotient tower of a Postnikov tower X with projection $\pi \colon X \longrightarrow W$. Then there are cocellular completions $X^\wedge_T$ of X and $W^\wedge_T$ of W such that $W^\wedge_T$ is a quotient tower of $X^\wedge_T$ whose projection satisfies $\pi^\wedge_T \circ \phi = \phi \circ \pi$. If $\pi^{-1}(*)$ is connected, the map $\phi \colon \pi^{-1}(*) \longrightarrow (\pi^\wedge_T)^{-1}(*)$ obtained by restricting $\phi \colon X \longrightarrow X^\wedge_T$ to fibers is again a completion at T. If, further,*

Y is a Postnikov tower, $\theta: Y \longrightarrow W$ is a cocellular map, and $\theta_T^\wedge: Y_T^\wedge \longrightarrow W_T^\wedge$ is chosen as in Theorem 10.3.5, then the pullback $X_T^\wedge \times_{W_T^\wedge} Y_T^\wedge$ of $\pi_T^\wedge$ and $\theta_T^\wedge$ is a cocellular completion of the pullback $X \times_W Y$ of π and θ.

COROLLARY 10.3.7. *$X_T^\wedge \times Y_T^\wedge$ is a cocellular completion of $X \times Y$.*

COROLLARY 10.3.8. *If X is a simply connected Postnikov tower, then $\Omega(X_T^\wedge)$ is a cocellular completion of ΩX.*

10.4. Completions of nilpotent groups

As in the case of localization we can extend the definition of completion from abelian groups to nilpotent groups by using the completion at T of nilpotent spaces. By Theorem 10.3.4, for any nilpotent group G there is a completion $K(G, 1)_T^\wedge$ of $K(G, 1)$. By construction, $\pi_n(K(G, 1)_T^\wedge) = 0$ for $n \geq 3$. We define

$$\mathbb{E}_T G = \pi_1(K(G, 1)_T^\wedge)$$

and

$$\mathbb{H}_T G = \pi_2(K(G, 1)_T^\wedge),$$

and we let $\phi: G \longrightarrow \mathbb{E}_T G$ be the homomorphism induced on π_1 by the completion $\phi: K(G, 1) \longrightarrow K(G, 1)_T^\wedge$. Of course, as a second homotopy group, $\mathbb{H}_T G$ is abelian. By the functoriality of topological completion, $\mathbb{H}_T$ and $\mathbb{E}_T$ are functors and ϕ is a natural transformation. We say that G is completable at T if $\mathbb{H}_T G = 0$, and we then call ϕ the completion of G at T. We say that G is T-complete if ϕ is an isomorphism; as in the abelian case, this implies that $\mathbb{H}_T G = 0$. The universal property of topological completions specializes to show that completion at T is universal among homomorphisms $G \longrightarrow H$ of nilpotent groups such that G is completable at T and H is T-complete.

The following three results are the analogues for completion of Propositions 5.4.7, 5.4.8, and 5.4.9. The proofs of the second and third of them are identical to the proofs in the case of localization, but the proof of the first must take account of the fact that not every nilpotent group is completable.

LEMMA 10.4.1. *A nilpotent group G is T-complete if and only if G is $\mathscr{B}_T$-nilpotent.*

PROOF. If G is T-complete, then both its identity homomorphism and the homomorphism obtained by passage to π_1 from the inductive construction of completion at T in Theorem 10.3.4 are completions of G at T. By unique-ness, they agree up to isomorphism under G, and the latter description dis-plays G as a $\mathscr{B}_T$-nilpotent group. Conversely, suppose that G is a $\mathscr{B}_T$-nilpotent group. Then $K(G, 1)$ is a $\mathscr{B}_T$-nilpotent space and is equivalent to a Post-nikov $\mathscr{B}_T$-tower. Since we only have a fundamental group to construct, each $n_i = 1$ and the groups A_i are all T-complete in the inductive diagram that appears in the proof of Theorem 10.3.4. The maps ϕ_i defined inductively there are all equivalences between Eilenberg-Mac Lane spaces $K(-, 1)$, hence $\phi\colon K(G, 1) \longrightarrow K(G, 1)_T^\wedge$ is an equivalence. Therefore, $K(G, 1)_T^\wedge$ must be a space $K(\mathbb{E}_T G, 1)$ and $\phi\colon G \longrightarrow \mathbb{E}_T G$ must be an isomorphism. □

PROPOSITION 10.4.2. *A homomorphism $\phi\colon G \longrightarrow H$ between completable nil-potent groups is an algebraic completion at T if and only if the map, unique up to homotopy,*

$$\phi\colon K(G, 1) \longrightarrow K(H, 1)$$

that realizes ϕ on π_1 is a topological completion at T.

PROPOSITION 10.4.3. *If $\phi\colon G \longrightarrow \hat{G}_T$ is the completion of a completable nilpo-tent group G, then*

$$\phi_*\colon H_*(G; \mathbb{F}_p) \longrightarrow H_*(\hat{G}_T; \mathbb{F}_p)$$

is an isomorphism for all primes $p \in T$.

Proposition 5.4.10 and Corollary 5.4.11 also have analogues for comple-tions.

PROPOSITION 10.4.4. *Let $1 \longrightarrow G' \longrightarrow G \longrightarrow G'' \longrightarrow 1$ be an exact sequence of nilpotent groups. Then the induced maps give a fibration sequence*

$$K(G', 1)_T^\wedge \longrightarrow K(G, 1)_T^\wedge \longrightarrow K(G'', 1)_T^\wedge,$$

and the resulting long exact sequence of homotopy groups has the form

$$1 \longrightarrow \mathbb{H}_T G' \longrightarrow \mathbb{H}_T G \longrightarrow \mathbb{H}_T G'' \longrightarrow \mathbb{E}_T G' \longrightarrow \mathbb{E}_T G \longrightarrow \mathbb{E}_T G'' \longrightarrow 1.$$

PROOF. We can choose a central series for G that begins with a central series for G' and ends with the inverse image of a central series for G''. We can construct corresponding Postnikov towers, so $K(G'', 1)$ is a quotient tower of $K(G, 1)$ with fiber $K(G', 1)$. Then, by Theorem 10.3.5, we can arrange our completions so that the map $K(G, 1)_T^\wedge \longrightarrow K(G'', 1)_T^\wedge$ is the projection onto a quotient tower and the map on the fiber is completion at T. □

COROLLARY 10.4.5. *Suppose given a map of short exact sequences of completable nilpotent groups*

$$
\begin{array}{ccccccccc}
1 & \longrightarrow & G' & \longrightarrow & G & \longrightarrow & G'' & \longrightarrow & 1 \\
& & \downarrow & & \downarrow & & \downarrow & & \\
1 & \longrightarrow & H' & \longrightarrow & H & \longrightarrow & H'' & \longrightarrow & 1
\end{array}
$$

in which the groups on the bottom row are T-complete. If any two of the three vertical arrows are completions at T, then so is the third.

We give some algebraic properties of completion at T in the rest of the section. Some of these properties are direct generalizations from the abelian case, but others give further information even in that case.

LEMMA 10.4.6. *For any nilpotent group G, $\mathbb{H}_T G$ and $\mathbb{E}_T G$ are T-complete. Moreover, the map $\phi\colon G \longrightarrow \mathbb{E}_T G$ gives rise to isomorphisms*

$$
\mathbb{H}_T(\ker \phi) \cong \mathbb{H}_T G \quad and \quad \mathbb{E}_T G \cong \mathbb{E}_T(\operatorname{im} \phi),
$$

and $\mathbb{E}_T(\ker \phi) = 0$, $\mathbb{H}_T(\operatorname{im} \phi) = 0$, $\mathbb{E}_T(\operatorname{coker} \phi) = 0$, and $\mathbb{H}_T(\operatorname{coker} \phi) = 0$.

PROOF. Since $K(G,1)^{\wedge}_T$ is nilpotent, it can be constructed as a Postnikov tower with quotient tower $K(\mathbb{E}_T G, 1)$. The fiber of the quotient map is a space $K(\mathbb{H}_T G, 2)$. Labeling the quotient map π and the fiber inclusion map ι, Proposition 10.3.6 gives a map of fibration sequences

$$
\begin{array}{ccccc}
K(\mathbb{H}_T G, 2) & \xrightarrow{\iota} & K(G,1)^{\wedge}_T & \xrightarrow{\pi} & K(\mathbb{E}_T G, 1) \\
\phi \downarrow & & \downarrow \phi & & \downarrow \phi \\
K(\mathbb{H}_T G, 2)^{\wedge}_T & \xrightarrow{\iota^{\wedge}_T} & (K(G,1)^{\wedge}_T)^{\wedge}_T & \xrightarrow{\pi^{\wedge}_T} & K(\mathbb{E}_T G, 1)^{\wedge}_T.
\end{array}
$$

The vertical arrows ϕ are completions, and the middle arrow is an equivalence since the identity map of $K(G,1)^{\wedge}_T$ is also a completion. By construction, $K(\mathbb{H}_T G, 2)^{\wedge}_T$ is simply connected and $\pi_n(K(\mathbb{E}_T G, 1)^{\wedge}_T) = 0$ for $n > 2$. Comparing the long exact sequences of homotopy groups, we see that

(i) $\mathbb{H}_T \mathbb{H}_T G \cong \pi_3(K(\mathbb{H}_T G, 2)^{\wedge}_T) = 0$;

(ii) $\pi_2(\pi^{\wedge}_T) = 0$ and hence $\mathbb{H}_T \mathbb{E}_T G = \pi_2(K(\mathbb{E}_T G, 1)^{\wedge}_T) = 0$; and

(iii) $\pi_2(\iota^{\wedge}_T)$ and $\pi_1(\pi^{\wedge}_T)$ are isomorphisms.

Therefore the left and right vertical arrows ϕ induce isomorphisms of homotopy groups. This proves the first statement. The second statement is left as an exercise for the reader. It is shown by applying Proposition 10.4.4 to the two short exact sequences obtained by factoring ϕ through its image. The proof entails diagram chasing of the two resulting six term exact sequences, using the fact that the composite $\mathbb{E}_T G \longrightarrow \mathbb{E}_T \mathrm{im}\phi \longrightarrow \mathbb{E}_T \mathbb{E}_T G$ is the isomorphism $\mathbb{E}_T \phi$. □

We will not make direct use of the following result, but we will make a little use of some of its consequences. To prove it, we drop our attempt to be algebraically as well as topologically self-contained and use some group theory that can be found in standard sources, such as Kurosh [78].

PROPOSITION 10.4.7. *Let G be a nilpotent group.*

(i) $\mathbb{H}_T G = \mathbb{H}_T(_T G)$, *where $_T G$ is the T-torsion subgroup of G.*
(ii) $\mathbb{H}_T G = 0$ *if the p-torsion elements of G are of bounded order for $p \in T$.*
(iii) $\mathbb{E}_T G = 0$ *if and only if G is p-divisible for $p \in T$.*

PROOF. The quotient group $G/_T G$ is T-torsion free, hence so are the abelian subquotients of its upper central series [78, II, pp. 245, 247]. This implies that $\mathbb{H}_T(G/_T G) = 0$, and (i) now follows from Proposition 10.4.4. Since the abelian subquotients of any central series of $_T G$ will inherit the boundedness property in (ii), (i) and the abelian group case of (ii) imply that (ii) holds in general.

For (iii), assume first that $\mathbb{E}_T G = 0$. Consider the fiber $F\phi$ of the map

$$\phi\colon K(G, 1) \longrightarrow K(G, 1)^\wedge_T.$$

The space $K(G, 1)^\wedge_T$ is nilpotent, and its reduced mod p homology is zero for $p \in T$. Therefore the integral homology of $F\phi$ is local away from p and so $F\phi$ is local away from T, hence the nilpotent group $\pi_1(F\phi)$ is local away from p. Since $\mathbb{E}_T G = 0$, G is a quotient of $\pi_1(F\phi)$ and is therefore p-divisible for $p \in T$.

Conversely, assume that G is p-divisible for $p \in T$. We have not yet proven the abelian case of the claim, so we consider that first. If G is abelian and T-torsion free, then it is a $\mathbb{Z}[T^{-1}]$-module. If G is abelian and a T-torsion group, then it is a direct sum of copies of $\mathbb{Z}/p^\infty$ for $p \in T$ [78, I, p. 165]. In either case, $\mathrm{Hom}\,(\mathbb{Z}[T^{-1}], G) \longrightarrow \mathrm{Hom}\,(\mathbb{Z}, G)$ is an epimorphism and

$\phi\colon G \longrightarrow \mathbb{E}_T G$ is the zero homomorphism. Therefore the isomorphism $\mathbb{E}_T \phi\colon \mathbb{E}_T G \longrightarrow \mathbb{E}_T \mathbb{E}_T G$ is zero and thus $\mathbb{E}_T G = 0$. For general nilpotent p-divisible groups G, an argument in [78, II, p. 237] shows that G admits a central series all of whose abelian subquotients are of one of the two types just considered, so that $\mathbb{E}_T G = 0$. $\square$

REMARK 10.4.8. By the previous two results $\mathbb{H}_T G = \mathbb{H}_T K$, where K is the T-torsion subgroup of $\ker \phi$. Using Proposition 10.4.4 and Lemma 10.4.6, we find that $\mathbb{E}_T K = 0$. By the previous result, K is thus a divisible torsion nilpotent group. It is therefore abelian [78, II, p. 235], in accordance with our definition of $\mathbb{H}_T G$ as an abelian group. When $T = \{p\}$,

$$\mathbb{H}_p G = \mathbb{H}_p K = \operatorname{Hom}(\mathbb{Z}/p^\infty, K) = \operatorname{Hom}(\mathbb{Z}/p^\infty, G).$$

The last Hom refers to the category of groups, and the last equality is the observation that any group homomorphism $\mathbb{Z}/p^\infty \longrightarrow G$ factors uniquely through K.

COROLLARY 10.4.9. For any nilpotent group G, $\mathbb{H}_T G$ is a torsion-free abelian group, and $\operatorname{Ext}(\mathbb{H}_T G, \mathbb{E}_T A) = 0$ for all abelian groups A.

PROOF. By inspection, $\mathbb{H}_T G$ is torsion free when G is abelian, and the general case follows from the previous remark. The second statement follows since $\mathbb{E}_T A$ is an Ext group, $\operatorname{Tor}(B, C) = 0$ if B is torsion free [25, VII.4.2], and

$$\operatorname{Ext}(B, \operatorname{Ext}(C, D)) \cong \operatorname{Ext}(\operatorname{Tor}(B, C), D)$$

for all abelian groups B, C, and D [25, VI.3.5a]. $\square$

REMARK 10.4.10. We shall use this together with the fact that, for abelian groups A and B, $\operatorname{Ext}(B, A)$ classifies extensions $0 \longrightarrow A \longrightarrow C \longrightarrow B \longrightarrow 0$ of abelian groups [79, p. 68]. Thus $\operatorname{Ext}(B, A) = 0$ implies that every such extension splits in the form $C \cong A \oplus B$.

The results above are less complete than in the abelian case in that we have not yet considered T-adic completion of nilpotent groups. We will never make later use of such a notion, but we sketch how the theory goes, without striving for rigor. There are several equivalent ways to define T-adic completion, and we give the one that best fits our way of thinking about completion.

DEFINITION 10.4.11. For a nilpotent group G, define the T-adic completion of G to be the inverse limit of the $\mathcal{B}_T$-nilpotent groups under G. That is, the inverse system runs over the commutative diagrams of homomorphisms

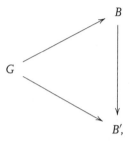

where B and B' are T-nilpotent. This gives a functor since for $f: H \longrightarrow G$ we can map an element of $\hat{H}_T$ to the element of $\hat{G}_T$ given by the coordinates indexed on composites $H \longrightarrow G \longrightarrow B$. By the definition, we have a natural map $\phi: G \longrightarrow \hat{G}_T$. Applying the functor $\mathbb{E}_T$ and using that the groups B in the limit system are T-complete, we obtain a natural map $\psi: \mathbb{E}_T G \longrightarrow \hat{G}_T$ such that $\psi \circ \phi = \phi$. That is, the T-adic completion ϕ factors through the T-completion ϕ.

The definition is not quite rigorous because we have not shown that the inverse system can be restricted to a cofinal set of homomorphisms $G \longrightarrow B$. The standard way around this is to restrict attention to epimorphisms, taking the limit of T-nilpotent images of G, as we did implicitly for p-adic completion of abelian groups. However $\hat{G}_T$ is defined, we will have $\hat{G}_T = \times_p \hat{G}_p$, and we could start with that as part of the definition. We did that in the abelian case and, as there, we could redefine $\hat{G}_p$ to be $\lim G/G^{p^r}$, where G^{p^r} denotes the subgroup, necessarily normal, generated by all elements g^{p^r} for $g \in G$. This gives a well-defined functor, and it is the definition of choice in the algebraic literature (see, for example, [141, p. 52]), but now we must check that each G/G^{p^r} is p-complete. Alternatively, we can redefine $\hat{G}_p$ using epimorphisms to $\mathbb{F}_p$-nilpotent groups. When G is finitely generated, we can replace $\mathcal{B}_p$-nilpotent groups by finite p-groups in the limit system, and then $\hat{G}_p$ is isomorphic to the classical profinite completion of G at p [126, p. I-5].

PROPOSITION 10.4.12. *If the p-torsion elements of G are of bounded order for $p \in T$, then $\psi: \mathbb{E}_T G \longrightarrow \hat{G}_T$ is an isomorphism.*

SKETCH PROOF. The hypothesis on the p-torsion is inherited by all subquotients of G, so $\mathbb{H}_T$ vanishes on all groups in sight. Profinite completion at p is an exact functor on finitely generated nilpotent groups, and one can generalize to show that this remains true without finite generation under our hypothesis on the p-torsion. In the case of abelian groups, this is a p-adic generalization of the Artin-Rees lemma that we proved implicitly in the first section of this chapter. Given the exactness, $\hat{G}_T$ is p-complete and ψ is an isomorphism by induction on the nilpotency class of G, using Proposition 10.4.4 and Corollary 10.4.5. For closely related results, see [141, Thms. 7.4, 7.6]. $\qquad\square$

11

CHARACTERIZATIONS AND PROPERTIES
OF COMPLETIONS

We give several characterizations of completions in §1. Starting with §2, we restrict attention to completions at a single prime p, and we study the homotopical behavior of completion at p with respect to standard constructions on based spaces. The treatment runs parallel to that in Chapter 6, and we focus on points where completion behaves differently from localization.

11.1. Characterizations of completions of nilpotent spaces

We show that several alternative conditions on a map are equivalent to its being a completion at T. We have the following pair of omnibus theorems.

THEOREM 11.1.1. *The following properties of a nilpotent space Z are equivalent, and they hold if and only if Z is T-complete.*

(i) *Z is a $\mathscr{B}_T$-nilpotent space.*
(ii) *$\xi^*: [Y, Z] \longrightarrow [X, Z]$ is a bijection for every $\mathbb{F}_T$-equivalence $\xi: X \longrightarrow Y$.*
(iii) *Each $\pi_n Z$ is a T-complete group (nilpotent if $n = 1$, abelian if $n > 1$).*

THEOREM 11.1.2. *For a nilpotent space X, the following properties of a map $\phi: X \longrightarrow Y$ from X to a T-complete space Y are equivalent. There exists one and, up to homotopy, only one such map, namely the the completion $X \longrightarrow \hat{X}_T$.*

(i) *$\phi^*: [Y, Z] \longrightarrow [X, Z]$ is an isomorphism for all T-complete spaces Z.*
(ii) *ϕ is an $\mathbb{F}_T$-equivalence.*
    Moreover, for each $n \geq 1$, there is a natural and splittable exact sequence

$$0 \longrightarrow \mathbb{E}_T \pi_n X \longrightarrow \pi_n Y \longrightarrow \mathbb{H}_T \pi_{n-1} X \longrightarrow 0,$$

and the composite $\pi_n X \xrightarrow{\phi} \mathbb{E}_T \pi_n X \longrightarrow \pi_n Y$ is ϕ_*. If X is $\mathscr{A}_T$-nilpotent, so that each $\mathbb{H}_T \pi_n X = 0$, the following condition is equivalent to (i) and (ii).

(iii) *For $n \geq 1$, $\phi_* : \pi_n X \longrightarrow \pi_n Y$ is completion at T.*

In Theorem 11.1.1, (ii) is the definition of what it means to be T-complete. In Theorem 11.1.2, (ii) is the definition of what it means for ϕ to be a completion at T, and we have already proven the existence and uniqueness of such a completion. Therefore, in both results, it suffices to prove the equivalence of (ii) with the remaining properties. Note that the hypothesis for criterion (iii) of Theorem 11.1.2 is essential. For example, $\pi_n(K(\mathbb{Z}/p^\infty, n)\hat{_p}) = 0$.

PROOF OF THEOREM 11.1.1. This is very nearly the same as the proof of Theorem 6.1.1. We use our characterization of $\mathscr{B}_T$ in Theorem 10.2.5 and the dual Whitehead theorem for the collection $\mathscr{B}_T$ to see that (i) $\Longrightarrow$ (ii). The implications (ii) $\Longrightarrow$ (i) and (i) $\Longrightarrow$ (iii) are proven as in Theorem 6.1.1. The topological proof of (iii) $\Longrightarrow$ (i) in Theorem 6.1.1 works just as well here, although the algebraic proof does not. $\qquad\square$

PROOF OF THEOREM 11.1.2. Again, much of this is nearly the same as the proof of Theorem 6.1.2. The equivalence of (i) and (ii) follows from Theorem 10.3.1 and the representability of cohomology. The proof that (iii) $\Rightarrow$ (ii) when X is $\mathscr{A}_T$-nilpotent is the same as the corresponding implication, (iii) $\Longrightarrow$ (ii), of Theorem 6.1.1. Conversely, to see that (ii) implies (iii) when X is $\mathscr{A}_T$-nilpotent, it suffices to check the general statement about homotopy groups. For that purpose, we may use our cocellular completion ϕ. The conclusion holds when X is an Eilenberg-Mac Lane space by Theorem 10.3.2. The claimed exact sequence of homotopy groups and the description of ϕ_* follow inductively by chasing the maps of exact sequences of homotopy groups associated to the maps of fibration sequences in the inductive construction of $\hat{X}_T$ in Theorem 10.3.4. The chase uses the exact sequences displayed in Lemma 10.1.8 and, more explicitly, Proposition 10.4.4. With the notations of the proof of Theorem 3.5.4, these give exact sequences of the form

$$0 \longrightarrow \mathbb{H}_T A_i \longrightarrow \mathbb{H}_T(G/G_{n,j+1}) \longrightarrow \mathbb{H}_T(G/G_{n,j})$$

$$\longrightarrow \mathbb{E}_T A_i \longrightarrow \mathbb{E}_T(G/G_{n,j+1}) \longrightarrow \mathbb{E}_T(G/G_{n,j}) \longrightarrow 1$$

determined by the central series used to build up $G = \pi_n X$. At each stage of the inductive construction of $\hat{X}_T$, we are building one of these exact sequences in the $(n+1)^{st}$ and n^{th} homotopy groups of $\hat{X}_T$, and by the time we have finished building up the nth homotopy group we have also built the summand $\mathbb{H}_T\pi_n X$ of the $(n+1)^{st}$ homotopy group. At each stage of the construction, we have splittings in view of Corollary 10.4.9 and Remark 10.4.10. □

Recall that $\mathbb{H}_T A \cong \times_{p\in T}\mathbb{H}_p A$ and $\mathbb{E}_T A \cong \times_{p\in T}\mathbb{E}_p A$. Together with Theorem 11.1.2, these observations have the following consequence, which was promised at the very beginning of the previous chapter.

COROLLARY 11.1.3. *For a nilpotent space X, the canonical natural map*

$$(\pi_p)\colon \hat{X}_T \longrightarrow \times_{p\in T}\hat{X}_p$$

is an isomorphism in Ho$\mathcal{T}$.

PROOF. Observe that $\hat{X}_p$ is a T-complete space since $\mathscr{B}_p$ is contained in $\mathscr{B}_T$, so that any $\mathscr{B}_p$-nilpotent space is $\mathscr{B}_T$-nilpotent. By the universal property of $\hat{X}_T$ the completions $X \to \hat{X}_p$ factor through canonical natural maps $\pi_p\colon \hat{X}_T \to \hat{X}_p$ for $p \in T$:

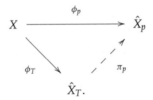

The map $(\pi_p)\colon \hat{X}_T \longrightarrow \times_{p\in T}\hat{X}_p$ induces an isomorphism on homotopy groups and is thus a weak equivalence or, equivalently, an isomorphism in Ho$\mathcal{T}$. □

The following result makes clear that completion at T can be thought of as a refinement of localization at T.

PROPOSITION 11.1.4. *The completion at T of a nilpotent space X is the composite of its localization at T and the completion at T of the localization X_T.*

PROOF. Since $\mathbb{F}_p$ is a p-local abelian group it is also T-local. Therefore any $\mathbb{F}_T$-equivalence is a $\mathbb{Z}_T$-equivalence. By the definitions of T-local and T-complete spaces, this implies that any T-complete space is T-local. By the universal property of localization at T, we obtain a map $\tilde{\phi}$ making the following diagram commute.

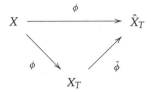

Since $\tilde{\phi}$ is clearly an $\mathbb{F}_T$-equivalence, it is a completion of X_T at T. $\qquad\square$

REMARK 11.1.5. Observe that we have no analogues of the homological criteria in parts (iv) of Theorems 6.1.1 and 6.1.2. In fact, the integral homology of completions is so poorly behaved that it is almost never used in practice. The groups $\tilde{H}_n(\hat{X}_T; \mathbb{Z})$ are always T-local, so they are uniquely q-divisible for $q \notin T$, but little more can be said about them in general. Observe that, by Remark 5.4.1 and the universal coefficient theorem, if $\tilde{H}_n(\hat{X}_T; \mathbb{F}_p) = 0$ and $\tilde{H}_{n+1}(\hat{X}_T; \mathbb{F}_p) = 0$ for $p \in T$, then $\tilde{H}_n(\hat{X}_T; \mathbb{Z})$ is a rational vector space.

One might naively hope that, at least if X is f-nilpotent, $\tilde{H}_*(\hat{X}_T; \mathbb{Z})$ might be isomorphic to $\tilde{H}_*(X; \mathbb{Z}) \otimes \hat{\mathbb{Z}}_T$, in analogy with what is true for localization. However, as observed in [21, VI.5.7], that is already false when $X = S^n$. For $q > n$, the groups $H_q(\hat{S}_T^n; \mathbb{Z})$ are rational vector spaces. Let n be odd. Then Corollary 6.7.3 implies that the rationalization of $\hat{S}_T^n$ is a space $K(\mathbb{Q} \otimes \mathbb{Z}_T, n)$ so that, for $q > n$,

$$H_q(\hat{S}_T^n; \mathbb{Z}) \cong H_q((S_T^n)_0; \mathbb{Q}) \cong H_q(K(\mathbb{Q} \otimes \mathbb{Z}_T, n); \mathbb{Q}).$$

For a $\mathbb{Q}$-vector space V, $H_*(K(V, n); \mathbb{Q})$ behaves homologically as if it were a graded exterior algebra generated by $H_n(K(V, n); \mathbb{Q}) = V$. In particular, $H_{qn}(S_T^n; \mathbb{Z})$ is an uncountable $\mathbb{Q}$-vector space for $q \geq 2$.

11.2. Completions of limits and fiber sequences

Since completions see one prime at a time, by Corollary 11.1.3, we now fix a prime p and only consider completion at p henceforward. This allows us to work with the Noetherian ring $\mathbb{Z}_p$ rather than the ring $\hat{\mathbb{Z}}_T$, which is not Noetherian if the set T is infinite.

This section is analogous to the corresponding section, §6.2, for localization. The main difference is that it is necessary to be more careful here since exactness properties are more subtle and since the characterizations of completions are weaker. To begin with, we do not have a general analogue for completions of the result that a p-local nilpotent group is $\mathbb{Z}_{(p)}$-nilpotent. This is related to the fact that a homomorphism between p-complete abelian groups need not have p-complete kernel and cokernel and need not be a homomorphism of $\mathbb{Z}_p$-modules. Recall the notion of a $\mathbb{Z}_p$-map from Definition 4.3.2.

LEMMA 11.2.1. *If* $f: X \longrightarrow Y$ *is a map between nilpotent spaces, then its completion* $\hat{f}_p: \hat{X}_p \longrightarrow \hat{Y}_p$ *is a* $\mathbb{Z}_p$*-map.*

PROOF. By Theorem 3.5.4 we may assume that X and Y are Postnikov towers and that f is a cocellular map. We may then construct $\phi_X: X \longrightarrow \hat{X}_p$ and $\phi_Y: Y \longrightarrow \hat{Y}_p$ by Theorem 10.3.4 and construct $\hat{f}_p$ by Theorem 10.3.5, so that it too is a cocellular map. Since the functors $\mathbb{H}_p$ and $\mathbb{E}_p$ that describe homotopy groups take values in the category of $\hat{\mathbb{Z}}_p$-modules, the conclusion follows. $\square$

While this result works in full generality, it is only useful to us when we obtain $f\mathbb{Z}_p$-maps. The problem is that, while the kernel and cokernel of a map of $\mathbb{Z}_p$-modules between p-complete abelian groups are $\mathbb{Z}_p$-modules, they still need not be p-complete. However, they are so when the given modules are finitely $\mathbb{Z}_p$-generated. Recall Notations 4.5.1 and 4.3.3. The proof above works to give the following refinement.

LEMMA 11.2.2. *If* $f: X \longrightarrow Y$ *is a map between* $f\mathbb{Z}_T$*-nilpotent spaces for any set of primes* T *such that* $p \in T$, *then its completion* $\hat{f}_p: \hat{X}_p \longrightarrow \hat{Y}_p$ *is an* $f\mathbb{Z}_p$*-map.*

The following two results work without f-nilpotency hypotheses on our spaces.

PROPOSITION 11.2.3. *If* X *and* Y *are nilpotent spaces, then* $(X \times Y)_{\hat{p}}$ *is naturally equivalent to* $\hat{X}_p \times \hat{Y}_p$.

PROPOSITION 11.2.4. *If* X *is nilpotent and* $\Omega_0(X)$ *denotes the basepoint component of* ΩX, *then* $(\Omega X)_{\hat{p}}$ *is naturally equivalent to* $\Omega_0(\hat{X}_p)$.

PROOF. As noted in the proof of Proposition 6.2.4, $\Omega_0 X$ is equivalent to $\Omega \tilde{X}$. The cocellular version of the statement applies to $\tilde{X}$. $\square$

Our methods do not give the most general possible forms of the next two results, but their f-nilpotency hypotheses are satisfied in the applications and would shortly become necessary in any case.

PROPOSITION 11.2.5. *Let $f\colon X \longrightarrow A$ and $g\colon Y \longrightarrow A$ be maps between f-nilpotent spaces, let $N_0(f,g)$ be the basepoint component of the homotopy pullback $N(f,g)$, and let $\hat{f}_p$ and $\hat{g}_p$ be p-completions of f and g.*

(i) If $N(f,g)$ is connected, then $N(\hat{f}_p, \hat{g}_p)$ is connected.
(ii) $N_0(f,g)$ is p-complete if X, Y, and A are p-complete.
(iii) $N_0(\hat{f}_p, \hat{g}_p)$ is a p-completion of $N_0(f,g)$.

PROOF. Recall from Proposition 6.2.5 that $N_0(f,g)$ is nilpotent. We mimic that result for the rest. The f-nilpotent hypothesis is not needed for (i) since Corollary 2.2.3 shows how to determine connectivity by the tail end of an exact sequence, and the right exactness of the functor $\mathbb{E}_p$ gives the conclusion. For (ii), Proposition 4.4.3, applied to the abelian category $f\mathscr{A}_{\mathbb{Z}_p}$ of finitely generated $\mathbb{Z}_p$-modules, shows that $N_0(\hat{f}_p, \hat{g}_p)$ is $f\mathbb{Z}_p$-nilpotent and is therefore p-complete. Using Corollary 10.4.5, part (iii) follows by comparison of the long exact sequences of homotopy groups for $N_0(f,g)$ and $N_0(f_T, g_T)$ given in Corollary 2.2.3. $\qquad\square$

Similarly, the proof of the following theorem uses the results just cited, and also Lemmas 3.1.3 and 4.3.4, exactly as in the proof of its analogue, Theorem 6.2.6, for localizations.

THEOREM 11.2.6. *Let $g\colon X \longrightarrow Y$ be a map to a connected space Y such that Y and all components of X are f-nilpotent. Let $F = Fg$. Then each component of F is f-nilpotent and there is a homotopy commutative diagram*

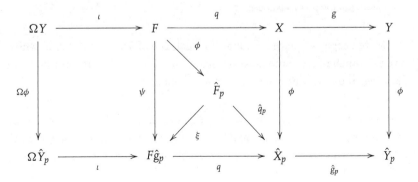

with the following properties.

(i) *The map* $\phi: Y \longrightarrow \hat{Y}_p$ *is a completion at p.*

(ii) *The maps* $\phi: X \longrightarrow \hat{X}_p$ *and* $\phi: F \longrightarrow \hat{F}_p$ *are the disjoint unions over the components of X and F of completions at T, defined using any (compatible) choices of base points in these components.*

(iii) *The rows are canonical fiber sequences.*

(iv) *The restriction of ψ to a map from a component of F to the component of its image is a completion at p.*

(v) *The map* $\xi: \hat{F}_p \longrightarrow F\hat{g}_p$ *is an equivalence to some of the components of* $\hat{F}_p$.

(vi) *Fix* $x \in X$, *let* $y = g(x) \in Y$, *and assume that the images of* $g_*: \pi_1(X, x) \to \pi_1(Y, y)$ *and* $\hat{g}_{p*}: \pi_1(\hat{X}_p, \phi(x)) \to \pi_1(\hat{Y}_p, \phi(y))$ *are normal subgroups. Then the quotient group* $\tilde{\pi}_0(F)$ *is f-nilpotent, the quotient group* $\tilde{\pi}_0(F\hat{g}_p)$ *is $f \mathbb{Z}_p$-nilpotent, and* $\psi_*: \tilde{\pi}_0(F) \longrightarrow \tilde{\pi}_0(F_p)$ *is a completion at p.*

As in the local case, the result simplifies when Y is simply connected and therefore F is connected. We can then ignore the interior of the central square and part (ii), concluding simply that the fill-in $\psi: F \longrightarrow F\hat{g}_p$ is a completion of F at p.

11.3. Completions of function spaces

We first record an essentially obvious consequence of the general theory of completions. We wrote the proof of Lemma 6.3.1 in such a way that it applies with minor changes of notation to prove the following analogue.

LEMMA 11.3.1. *Let X be nilpotent and Y be p-complete. Then*

$$\phi^*: F(\hat{X}_p, Y)_* \longrightarrow F(X, Y)_*$$

is a weak homotopy equivalence.

More deeply, we have the following analogue of Theorem 6.3.2, in which we use much of the same notation that we used there. This result will play a key role in the fracture theorems for completion.

THEOREM 11.3.2. *Let X be an f-nilpotent space and K be a finite based connected CW complex. Let $g \in F(K, X)$, and let $F(K, X)_g$ denote the component of $F(K, X)$ that contains g. Let K^i denote the i-skeleton of K and define $[K, X]_g$ to be the set of all $h \in [K, X]$ such that $h|K^{n-1} = g|K^{n-1} \in [K^{n-1}, X]$, where n is the dimension* of K. Let $\phi\colon X \longrightarrow \hat{X}_p$ be a completion of X at p. Then the following statements hold.

(i) $F(K,X)_g$ is an f-nilpotent space, $F(K,\hat{X}_p)_{\phi \circ g}$ is an $f\mathbb{Z}_p$-nilpotent space, and
$\phi_*\colon F(K,X)_g \longrightarrow F(K,\hat{X}_p)_{\phi \circ g}$ is a completion of spaces at p.

(ii) $[K,X]_g$ is an f-nilpotent group, $[K,\hat{X}_p]_g$ is an $f\mathbb{Z}_p$-nilpotent group, and
$\phi_*\colon [K,X]_g \longrightarrow [K,\hat{X}_p]_{\phi \circ g}$ is a completion at p.

The proof of Theorem 6.3.2 applies verbatim. Note that even if we could manage to eliminate f-nilpotency hypotheses in our results on fibrations in the previous section, we would still have to restrict to f-nilpotent spaces in this result, as the example of $K = S^n$ and $X = K(\mathbb{Z}/p^\infty, n)$ makes clear.

11.4. Completions of colimits and cofiber sequences

The analogues for completion of the results of §6.4 are intrinsically less satisfactory since the relevant constructions fail to preserve p-complete spaces. Notably, it is not true that ΣX is p-complete when X is p-complete, as we saw in Remark 11.1.5. As there, the problem is that we have no homological characterizations like those in Theorems 6.1.1 and 6.1.2 to rely on. However, using the characterization of completion in terms of mod p homology, we can obtain correct statements simply by completing the constructions that fail to be complete. We obtain the following conclusions.

PROPOSITION 11.4.1. *If X, Y, and $X \vee Y$ are nilpotent spaces, then $(X \vee Y)_{\hat{p}}$ is naturally equivalent to $(\hat{X}_p \vee \hat{Y}_p)_{\hat{p}}$.*

PROOF. $\phi \vee \phi\colon X \vee Y \longrightarrow \hat{X}_p \vee \hat{Y}_p$ is an $\mathbb{F}_p$-equivalence, but its target need not be p-complete. The composite of the displayed map with a completion of its target is a completion of its source. □

The proofs of the next few results are of exactly the same form.

PROPOSITION 11.4.2. *If X is nilpotent, then $(\Sigma X)_{\hat{p}}$ is naturally equivalent to $(\Sigma \hat{X}_p)_{\hat{p}}$.*

PROPOSITION 11.4.3. *Let $i\colon A \longrightarrow X$ be a cofibration and $f\colon A \longrightarrow Y$ be a map, where A, X, Y, X/A, and $X \cup_A Y$ are nilpotent. If we choose completions such that $\hat{i}_p\colon \hat{A}_p \longrightarrow \hat{X}_p$ is a cofibration, then $(\hat{X}_p \cup_{\hat{A}_p} \hat{Y}_p)_{\hat{p}}$ is a completion of $X \cup_A Y$.*

PROPOSITION 11.4.4. *Let $f: X \longrightarrow Y$ be a map such that X, Y, and Cf are nilpotent and let ψ be a fill-in in the map of canonical cofiber sequences*

$$
\begin{array}{ccccccc}
X & \xrightarrow{\ f\ } & Y & \xrightarrow{\ i\ } & Cf & \xrightarrow{\ \pi\ } & \Sigma X \\
{\scriptstyle \phi}\downarrow & & {\scriptstyle \phi}\downarrow & & \downarrow{\scriptstyle \psi} & & \downarrow{\scriptstyle \Sigma\phi} \\
\hat{X}_p & \xrightarrow[\ \hat{f}_p\ ]{} & \hat{Y}_p & \xrightarrow[\ i\ ]{} & \hat{Cf}_p & \xrightarrow[\ \pi\ ]{} & \Sigma\hat{X}_p
\end{array}
$$

in which the given maps ϕ are completions at p. Then the composite of ψ with a completion of its target is a completion of Cf.

PROPOSITION 11.4.5. *If X is the colimit of a sequence of cofibrations $X_i \longrightarrow X_{i+1}$ between nilpotent spaces, and if completions are so chosen that the completions $(X_i)\hat{_p} \longrightarrow (X_{i+1})\hat{_p}$ are cofibrations, then $(\operatorname{colim}(X_i)\hat{_p})\hat{_p}$ is a completion of X.*

PROPOSITION 11.4.6. *If X, Y, and $X \wedge Y$ are nilpotent, then $(X \wedge Y)\hat{_p}$ is naturally equivalent to $(\hat{X}_p \wedge \hat{Y}_p)\hat{_p}$.*

Clearly, in view of these results, it is unreasonable to expect to have a cellular construction of completions analogous to the cellular construction of localizations given in §6.5.

11.5. Completions of *H*-spaces

It is also unreasonable to expect to have naive constructions of completions of *H*-spaces and co-*H*-spaces analogous to the constructions for localizations given in §6.6, and we cannot expect completions of co-*H*-spaces to be co-*H*-spaces. However, completions of *H*-spaces behave well.

PROPOSITION 11.5.1. *If Y is an H-space with product μ, then $\hat{Y}_p$ is an H-space with product $\hat{\mu}_p$ such that $\phi: Y \longrightarrow \hat{Y}_p$ is a map of H-spaces.*

PROOF. The map $\phi \times \phi: Y \times Y \longrightarrow \hat{Y}_p \times \hat{Y}_p$ is a completion at p, so there is a map $\hat{\mu}_p$, unique up to homotopy, such that $\hat{\mu}_p \circ (\phi \times \phi)$ is homotopic to $\phi \circ \mu$. Left and right multiplication by the basepoint of $\hat{Y}_p$ are each homotopic to the identity by another application of the universal property. $\square$

There is a large body of interesting work on p-complete H-spaces. Here again, some of the interest is in seeing how much like compact Lie groups they are. For that comparison, one wants them to satisfy some reasonable finiteness condition, but in the absence of a cell structure it is not entirely obvious how to specify this. One also wants them to be equivalent to loop spaces. Of course, this holds for any topological group G, since G is equivalent to the loops on its classifying space BG. This leads to the following notion.

DEFINITION 11.5.2. A p-compact group is a triple (X, BX, ε), where BX is a p-complete space, $\varepsilon\colon X \longrightarrow \Omega BX$ is a homotopy equivalence, and the mod p cohomology of X is finite dimensional. It is often assumed that BX is simply connected, so that X is connected.

This notion was introduced and studied by Dwyer and Wilkerson [44], who showed how remarkably similar to compact Lie groups these X are. The completion of a compact Lie group is an example, but there are many others. Like compact Lie groups, p-compact groups have versions of maximal tori, normalizers of maximal tori, and Weyl groups. A complete classification, analogous to the classification of compact Lie groups, has recently been obtained [4, 5].

11.6. The vanishing of p-adic phantom maps

In parallel with §6.8, we give an observation that shows, in effect, that phantom maps are usually invisible to the eyes of p-adic homotopy theory. The proof relies on results from the literature about the vanishing of higher-derived functors of lim. Their proofs are not hard, but they would take us too far afield to give full details here.

LEMMA 11.6.1. Let X be a connected CW complex of finite type. If Z is $f\hat{\mathbb{Z}}_T$-nilpotent then
$$\lim{}^1[\Sigma X_i, Z] = 0$$
and
$$[X, Z] \to \lim[X_i, Z]$$
is a bijection.

The conclusion is similar to that of Lemma 6.8.1, but that result was proven using a cellular decomposition of X, whereas this result is proven using a cocellular decomposition of Z. Recall the definition of a $\mathscr{K}$-tower from Definition 3.3.1.

PROOF OF LEMMA 11.6.1. By Corollary 11.1.3 we may assume without loss of generality that $T = \{p\}$. Let $f\mathscr{A}_p$ denote the collection of finite abelian p-groups; we could equally well replace $f\mathscr{A}_p$ by the single group $\mathbb{Z}/p$ in the argument to follow.

We can construct the spaces $K(\mathbb{Z}/p^q, n)$ and $K(\mathbb{Z}_p, n)$ as $f\mathscr{A}_p$-towers. To be precise about this, we use that the group $H^{n+1}(K(\mathbb{Z}/p^q, n), \mathbb{F}_p)$ is a copy of $\mathbb{F}_p$ generated by $\beta_q(\iota_n)$, where $\iota_n \in H^n(K(\mathbb{Z}/p^q, n); \mathbb{F}_p)$ is the fundamental class and

$$\beta_q \colon H^n(-; \mathbb{Z}/p) \longrightarrow H^{n+1}(-; \mathbb{Z}/p)$$

is the qth Bockstein operation. That operation is obtained as the connecting homomorphism associated as in [93, p. 181, #3] to the short exact sequence

$$0 \longrightarrow \mathbb{Z}/p^q \longrightarrow \mathbb{Z}/p^{q+1} \longrightarrow \mathbb{Z}/p \longrightarrow 0,$$

followed by reduction mod p. Viewing $\beta_q(\iota_n)$ as a map

$$K(\mathbb{Z}/p^q, n) \longrightarrow K(\mathbb{Z}/p, n+1),$$

its fiber is a space $K(\mathbb{Z}/p^{q+1}, n)$. The limit of the resulting fibrations

$$K(\mathbb{Z}/p^{q+1}, n) \longrightarrow K(\mathbb{Z}/p^q, n)$$

is a space $K(\mathbb{Z}_p, n)$.

In view of Remark 3.3.2, it follows that for any finitely generated $\mathbb{Z}_p$-module B, the space $K(B, n)$ can be constructed as a $\mathscr{K}f\mathscr{A}_p$-tower with countably may cocells. By Theorem 3.5.4, the $f\mathbb{Z}_p$-nilpotent space Z can be taken to be a Postnikov $f\mathbb{Z}_p$-tower. Using Remark 3.3.2 again, it follows that all terms of the tower and Z itself are $\mathscr{K}f\mathscr{A}_p$-towers with countably many cocells. The commutations with sequential limits used in the construction give the more precise information that Z is the limit of countably many quotient towers W_j, each of which has finite homotopy groups. For a finite complex K, $[K, W_j]$ is finite and, by Theorem 2.3.3, $\lim^1$ vanishes on inverse sequences of finite groups. Thus, for each fixed i,

$$\lim_j^1 [\Sigma^1 X_i, W_j] = 0.$$

If we assume that the groups $[\Sigma X_i, W_j]$ are abelian, then a result of Roos [119, Thm. 3] (see also [120]) gives a spectral sequence that converges from

$$E_2^{p,q} = \lim_i^p \lim_j^q [\Sigma X_i, W_j]$$

to the derived functors $\lim^n$ of the bi-indexed system $\{[\Sigma X_i, W_j]\}$ of finite groups. Since our limit systems are sequential or have cofinal sequential subsystems, the E_2 terms with $p > 1$ or $q > 1$ are zero, as are the terms with $q = 1$, and the $\lim^n$ groups to which the spectral sequence converges are zero for $n \geq 1$. This forces $E_2 = E_\infty$ and $E_2^{1,0} = 0$, that is $\lim_i^1 [\Sigma X_i, Z] = 0$. A direct adaptation of Roos's arguments starting from the explicit definition of $\lim^1$ given in Definition 2.1.8 allows us to draw the same conclusion even when the groups $[\Sigma X_i, W_j]$ are not abelian. $\qquad\square$

12

FRACTURE THEOREMS FOR
COMPLETION: GROUPS

In this chapter we describe how to construct a global nilpotent group from a complete nilpotent group, a rational nilpotent group, and a compatibility condition. The compatibility condition involves a notion of formal completion, and we also give a brief discussion of what we call the adèlic genus of a nilpotent group. We describe the corresponding constructions for nilpotent spaces in the next chapter. The results in these chapters parallel those in Chapters 7 and 8, but the proofs are quite different. For example, at one key spot in this chapter we use a space-level result from the next chapter to prove a general algebraic result about completions of nilpotent groups. In turn, we use that algebraic result to prove a general topological result about completions of spaces in the next chapter, carefully avoiding circularity. This contrasts with Chapters 7 and 8, where we consistently used results for groups to prove the corresponding results for spaces.

More fundamentally, the algebra is here much less predictive of the topology. Many of the algebraic results require completability restrictions on the given groups, whereas the analogous topological results apply without any such restriction. Conceptually, the point is that the group theory knows only about Eilenberg-Mac Lane spaces $K(G, 1)$, but the topology knows how to use two-stage Postnikov towers to construct completions of Eilenberg-Mac Lane spaces $K(G, 1)$ for nilpotent groups G that are not completable algebraically. Since we are interested primarily in the topology, we shall not be overly thorough in our treatment of the algebra. However, we shall be quite carefully pedantic about those results that are not well documented in other sources, since it is quite hard to determine from the literature precisely what is and is not true.

We let T be any nonempty set of primes. The focus is on the case when T is the set of all primes and the case when T consists of a single prime p. Since completions at T split as the products of the completions at $p \in T$, the reader may wonder why we don't just work one prime at a time. The answer is that since we are relating global to local phenomena, these splittings do not imply that our results for the primes $p \in T$ imply our results for T itself. We shall say a bit more about this at the end of §12.2.

All given groups are to be nilpotent in this chapter, even when we neglect to say so. Recall from Lemma 5.4.6 that T-local groups are the same as $\mathbb{Z}_T$-nilpotent groups, as defined in Notation 4.5.1. Similarly, recall from Proposition 5.6.5 that finitely T-generated T-local groups are the same as $f\mathbb{Z}_T$-nilpotent groups.

12.1. Preliminaries on pullbacks and isomorphisms

We begin by showing that completion preserves certain pullbacks. This is in contrast to the case of localization, where all pullbacks are preserved. Of course, the difference is a consequence of the failure of exactness for completion. We then give a criterion for a map between T-local groups to be an isomorphism and show by example that its restrictive hypotheses cannot be eliminated.

LEMMA 12.1.1. *Let*

$$
\begin{array}{ccc}
A & \xrightarrow{f} & C \\
{\scriptstyle g}\downarrow & & \downarrow{\scriptstyle k} \\
B & \xrightarrow{h} & D
\end{array}
$$

be a pullback diagram of abelian groups, where D is rational. Then $\mathbb{E}_T A$ is isomorphic to $\mathbb{E}_T B \oplus \mathbb{E}_T C$ and, if B and C are completable at T, then so is A.

PROOF. Since D is rational, $\mathbb{E}_T D = 0$ by Proposition 10.1.11 and the first claim can be viewed as saying that the functor $\mathbb{E}_T$ preserves the displayed pullback. We are given the exact sequence

$$
0 \longrightarrow A \xrightarrow{(f,g)} B \oplus C \xrightarrow{h-k} D,
$$

and we let $I \subset D$ be the image of $h - k$. We then have a short exact sequence

$$0 \longrightarrow A \xrightarrow{(f,g)} B \oplus C \xrightarrow{h-k} I \longrightarrow 0.$$

By Proposition 10.4.4 (taking products over $p \in T$), it gives rise to an exact sequence

$$0 \to \mathbb{H}_T(A) \to \mathbb{H}_T(B \oplus C) \to \mathbb{H}_T(I) \to \mathbb{E}_T(A) \to \mathbb{E}_T(B \oplus C) \to \mathbb{E}_T(I) \to 0.$$

Since D is rational, so is I, hence $\mathbb{H}_T(I) = 0$ and $\mathbb{E}_T(I) = 0$, by Proposition 10.1.11. Therefore the sequence says that $\mathbb{E}_T(A) \cong \mathbb{E}_T(B) \oplus \mathbb{E}_T(C)$ and $\mathbb{H}_T(A) = 0$ if $\mathbb{H}_T(B \oplus C) = 0$. $\qquad\square$

LEMMA 12.1.2. *Let $\psi \colon H \longrightarrow G$ be a homomorphism between T-local groups, where G is completable at T. Then ψ is an isomorphism if and only if*

$$\psi_0 \colon H_0 \to G_0 \quad and \quad \mathbb{E}_T \psi \colon \mathbb{E}_T H \to \mathbb{E}_T G$$

are isomorphisms.

PROOF. The forward implication is obvious. Assume that ψ_0 and $\mathbb{E}_T \psi$ are isomorphisms. Since ψ_0 is an isomorphism, Proposition 5.5.4 implies that the kernel and cokernel of ψ are T-torsion groups.

Let $\overline{\mathrm{im}}(\psi)$ be the kernel of the cokernel of ψ, that is, the normal subgroup of G generated by $\mathrm{im}\,(\psi)$. The exact sequence

$$1 \longrightarrow \overline{\mathrm{im}}(\psi) \longrightarrow G \longrightarrow \mathrm{coker}\,(\psi) \longrightarrow 1$$

gives an exact sequence

$$1 \to \mathbb{H}_T \overline{\mathrm{im}}(\psi) \to \mathbb{H}_T(G) \to \mathbb{H}_T \mathrm{coker}\,(\psi) \to \mathbb{E}_T \overline{\mathrm{im}}(\psi) \to \mathbb{E}_T(G) \to \mathbb{E}_T \mathrm{coker}\,(\psi) \to 1.$$

Since the isomorphism $\mathbb{E}_T \psi$ factors through the map $\mathbb{E}_T(\overline{\mathrm{im}}(\psi)) \to \mathbb{E}_T(G)$, this map must be an epimorphism, hence the epimorphism $\mathbb{E}_T G \to \mathbb{E}_T \mathrm{coker}\,(\psi)$ must be trivial. This implies that $\mathbb{E}_T \mathrm{coker}\,(\psi) = 1$. By Proposition 10.4.7, $\mathrm{coker}\,(\psi)$ is T-divisible. Since it is also a T-torsion group, it is trivial. Thus ψ is an epimorphism. Now the exact sequence

$$1 \longrightarrow \ker(\psi) \longrightarrow H \longrightarrow G \longrightarrow 1$$

gives an exact sequence

$$1 \longrightarrow \mathbb{H}_T \ker(\psi) \longrightarrow \mathbb{H}_T H \longrightarrow \mathbb{H}_T G \longrightarrow \mathbb{E}_T \ker(\psi) \longrightarrow \mathbb{E}_T H \xrightarrow{\cong} \mathbb{E}_T G \longrightarrow 1.$$

Since $\mathbb{H}_T(G) = 0$ by hypothesis, this exact sequence shows that $\mathbb{E}_T \ker(\psi) = 1$. By Proposition 10.4.7 again, $\ker(\psi)$ is T-divisible. Since it is also a T-torsion group, it is trivial and ψ is an isomorphism. □

EXAMPLE 12.1.3. The completability hypothesis is essential. The abelian group $B = \oplus_{p \in T} \mathbb{Z}/p^{\infty}$ satisfies $B_0 = 0$ and $\mathbb{E}_T B = 0$. Therefore, for any abelian group A, for example $A = 0$, the inclusion and projection $A \longrightarrow A \oplus B \longrightarrow A$ are examples of maps that induce isomorphisms after rationalization and application of $\mathbb{E}_T$ but are not themselves isomorphisms.

12.2. Global to local: abelian and nilpotent groups

We agree to write ϕ_0 for rationalization and $\hat{\phi}$ for completion at T in this section. Example 12.1.3 shows that we cannot expect to recover a global T-local group G from G_0 and $\mathbb{E}_T G$ unless $\mathbb{H}_T G = 0$. Because the technical hypotheses differ, we consider monomorphism, isomorphism, and epimorphism conditions separately.

LEMMA 12.2.1. *If an abelian group A is completable at T, then the kernel of the completion $\hat{\phi}: A \to \mathbb{E}_T A$ is Hom $(\mathbb{Z}[T^{-1}], A)$. If G is an $f\mathbb{Z}_T$-nilpotent group, then the completion $\hat{\phi}: G \to \mathbb{E}_T G$ is a monomorphism.*

PROOF. The first statement holds by Remark 10.1.21. While Hom $(\mathbb{Z}[T^{-1}], A)$ is often nonzero, for example when $A = \mathbb{Q}$, it is easily seen to be zero when A is a finitely generated $\mathbb{Z}_T$-module. The second statement follows by the five lemma since all of the subquotients of any central series of an $f\mathbb{Z}_T$-nilpotent group are finitely T-generated, by Proposition 5.6.5. □

Finite generation hypotheses will shortly enter for another reason. Recall that quotients of T-completable groups need not be T-completable in general, as the example $B \cong \mathbb{Z}_T/\mathbb{Z}$ illustrates. However, we have the following observation.

LEMMA 12.2.2. *For any set of primes $T' \supset T$, all subquotients of all $f\mathbb{Z}_{T'}$-nilpotent groups are completable at T. In particular, all subquotients of $f\mathbb{Z}$-nilpotent groups are completable at T.*

PROOF. Since completion at T is the composite of localization at T and completion at T, this is implied by Proposition 5.6.5. □

The following ad hoc definition encodes greater generality.

DEFINITION 12.2.3. A nilpotent group G is nilpotently completable at T if it has a central series all of whose subquotients (including G itself) are completable at T.

In the abelian case, the following result is best possible. In the nilpotent case, the completability hypothesis is not actually essential, as we shall see from an alternative proof in the next section, but it is a natural condition to assume and is needed for our first proof.

THEOREM 12.2.4. *If G is a T-local group that is nilpotently completable at T, then the diagram*

$$
\begin{array}{ccc}
G & \xrightarrow{\;\hat{\phi}\;} & \mathbb{E}_T G \\
{\scriptstyle \phi_0}\downarrow & & \downarrow{\scriptstyle \phi_0} \\
G_0 & \xrightarrow[\hat{\phi}_0]{} & (\mathbb{E}_T G)_0
\end{array}
$$

is a pullback.

PROOF. Changing notation to $G = A$, we first prove this when G is abelian. Let B be the pullback displayed in the diagram

$$
\begin{array}{ccc}
B & \longrightarrow & \mathbb{E}_T A \\
\downarrow & & \downarrow{\scriptstyle \phi_0} \\
A_0 & \xrightarrow[(\hat{\phi})_0]{} & (\mathbb{E}_T A)_0.
\end{array}
$$

The universal property gives a map $\psi : A \to B$ that factors both the completion $A \longrightarrow \mathbb{E}_T A$ and the rationalization $A \longrightarrow A_0$. Since rationalization is exact,

$$
\begin{array}{ccc}
B_0 & \longrightarrow & (\mathbb{E}_T A)_0 \\
\downarrow & & \downarrow \\
A_0 & \longrightarrow & (\mathbb{E}_T A)_0
\end{array}
$$

is a pullback whose right vertical arrow is an isomorphism. Therefore, $B_0 \to A_0$ is an isomorphism and so is $\psi_0 \colon A_0 \to B_0$. Similarly, since $\mathbb{E}_T(A_0) = 0$, Lemma 12.1.1 shows that B is completable at T and $\mathbb{E}_T \psi \colon \mathbb{E}_T A \to \mathbb{E}_T B$ is an isomorphism. By Lemma 12.1.2, ψ is an isomorphism.

Reverting to our original notation, the general case is proven by induction on the nilpotency class of G, using Lemma 7.6.2. To see this, choose a central series

$$\{1\} = G_q \subset G_{q-1} \subset \ldots \subset G_0 = G$$

for G where G_{q-1} and G/G_{q-1} are completable at T. The base case and the inductive hypothesis imply that the diagrams

$$
\begin{array}{ccc}
(G/G_{q-1}) & \longrightarrow & \mathbb{E}_T(G/G_{q-1}) \\
\downarrow & & \downarrow \\
(G/G_{q-1})_0 & \longrightarrow & (\mathbb{E}_T(G/G_{q-1}))_0
\end{array}
\qquad
\begin{array}{ccc}
(G_{q-1}) & \longrightarrow & \mathbb{E}_T(G_{q-1}) \\
\downarrow & & \downarrow \\
(G_{q-1})_0 & \longrightarrow & (\mathbb{E}_T(G_{q-1}))_0
\end{array}
$$

are pullbacks. Since rationalization is exact,

$$1 \longrightarrow (G_{q-1})_0 \longrightarrow G_0 \longrightarrow (G/G_{q-1})_0 \longrightarrow 1$$

is exact. Since G/G_{q-1} is completable at T,

$$1 \longrightarrow \mathbb{E}_T(G_{q-1}) \longrightarrow \mathbb{E}_T(G) \longrightarrow \mathbb{E}_T(G/G_{q-1}) \longrightarrow 1$$

is exact. This exact sequence and the exactness of rationalization imply that

$$1 \longrightarrow (\mathbb{E}_T(G_{q-1}))_0 \longrightarrow (\mathbb{E}_T(G))_0 \longrightarrow (\mathbb{E}_T(G/G_{q-1}))_0 \longrightarrow 1$$

is exact. The claimed pullback follows from Lemma 7.6.2(i). $\qquad\square$

The next result is the one proven topologically.[1] In turn, this algebraic result will later be used in the proof of our topological global to local fracture theorem.

---

1. The result is stated in [39, Prop. 3.5], but without details of proof.

PROPOSITION 12.2.5. *For any T-local group G, every element $z \in (\mathbb{E}_T G)_0$ is a product $z = \phi_0(x)\hat{\phi}_0(y)$ for some $x \in \mathbb{E}_T G$ and $y \in G_0$.*

PROOF. In contrast to the analogue for localization, we prove this using topology. Again, we first prove this in the abelian case, writing $G = A$. We have $K(A, 2)$ available to us, and we let P be the homotopy pullback displayed in the diagram

$$
\begin{array}{ccc}
P & \xrightarrow{\ \mu\ } & K(A, 2)_T^{\wedge} \\
{\scriptstyle \nu}\downarrow & & \downarrow {\scriptstyle \phi_0} \\
K(A, 2)_0 & \xrightarrow[(\hat{\phi})_0]{} & (K(A, 2)_T^{\wedge})_0.
\end{array}
$$

Since the other three spaces in the diagram are simply connected, the description of the homotopy groups of P in Corollary 2.2.3 ensures that P is connected. By Theorem 13.1.5 (or Lemma 13.2.3) below, ν is a rationalization and μ is a completion at T, hence the induced map $\alpha\colon K(A, 2) \to P$ becomes an equivalence upon rationalization and completion at T. Therefore, by Corollary 13.2.2 below, α is an equivalence. In particular, $\pi_1(P) = 0$. The conclusion follows from Corollary 2.2.3 and the description of the homotopy groups of completions in Theorem 11.1.2.

The general case is again proven topologically, by specializing Lemma 13.2.4 below. Briefly, let Y be the fiber of a map $X \longrightarrow K(A, 2)$, where X is one stage in the inductive construction of $K(G, 1)$ and Y is the next stage. Let P be the homotopy pullback displayed in the diagram

$$
\begin{array}{ccc}
P & \longrightarrow & \hat{Y}_T \\
\downarrow & & \downarrow {\scriptstyle \phi_0} \\
Y_0 & \xrightarrow[\hat{\phi}_0]{} & (\hat{Y}_T)_0.
\end{array}
$$

By Lemma 13.2.4, the canonnical map $Y \longrightarrow P$ is an equivalence. Since Y is connected, so is P, and the conclusion again follows from Corollary 2.2.3 and Theorem 11.1.2. The reader may feel, as the authors do, that this is a rather mysterious way to prove something as concrete and algebraic as the result we are after. She might prefer an algebraic proof that just applies the elementary Lemma 7.6.1. Choosing a central series for G as in the proof of

Theorem 12.2.4 above and using the right exactness of the functor $\mathbb{E}_T$, we do obtain a commutative diagram with exact rows

$$
\begin{array}{ccccccc}
\mathbb{E}_T G_i/G_{i+1} & \longrightarrow & \mathbb{E}_T G/G_{i+1} & \longrightarrow & \mathbb{E}_T G/G_i & \longrightarrow & 1 \\
\downarrow & & \downarrow & & \downarrow & & \\
(\mathbb{E}_T G_i/G_{i+1})_0 & \longrightarrow & \mathbb{E}_T(G/G_{i+1})_0 & \longrightarrow & (\mathbb{E}_T G/G_i)_0 & \longrightarrow & 1 \\
\uparrow & & \uparrow & & \uparrow & & \\
(G_i/G_{i+1})_0 & \longrightarrow & (G/G_{i+1})_0 & \longrightarrow & (G/G_i)_0 & \longrightarrow & 1.
\end{array}
$$

However, to be able to quote Lemma 7.6.1, we need an intuitively obvious but technically elusive detail. We leave it as an exercise for the dissatisfied reader. □

EXERCISE 12.2.6. Let $0 \to A \to G \longrightarrow H \to 1$ be a central extension of nilpotent (or T-local) groups. Then the image of the induced map $\mathbb{E}_T A \longrightarrow \mathbb{E}_T G$ is a central subgroup, hence so is the image of the induced map $(\mathbb{E}_T A)_0 \longrightarrow (\mathbb{E}_T G)_0$.

It is natural to think of completion at p as the composite of localization at p and completion at p, and it is also natural to ask how Theorem 12.2.4 correlates with its analogue for localization, part (ii) of Theorem 7.2.1. Again assuming that G is T-local and using notations and constructions cognate with those in Theorem 7.2.1(ii), we have the following commutative diagram. It should be compared with the key diagram (7.1.4) of §7.1, in which G is compared with the pullbacks P and Q that are implicit in the top left square and its composite with the triangle in the diagram.

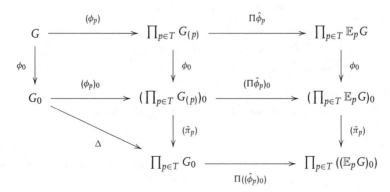

The vertical maps ϕ_0 are rationalizations. The map $\tilde{\pi}_p$ at the bottom right is obtained by rationalizing the projection $\Pi_{p \in T} \mathbb{E}_p G \longrightarrow \mathbb{E}_p G$, and the map $\tilde{\pi}_p$ at the bottom center was defined similarly above (7.1.4). The two vertical composites $\tilde{\pi}_p \circ \phi_0$ are rationalizations $G_{(p)} \longrightarrow G_0$ and $\mathbb{E}_p G \longrightarrow (\mathbb{E}_p G)_0$.

The upper left square is a pullback, by Theorem 7.2.1(ii), and so sometimes is the left part of the diagram with its middle horizontal arrow $(\phi_p)_0$ erased, by Proposition 7.1.7 (see also Proposition 7.4.4 and Remark 7.4.6). The right part of the diagram with its middle horizontal arrow $(\hat{\phi}_p)_0$ erased is a pullback when each $G_{(p)}$ satisfies the hypothesis of Theorem 12.2.4 (for the case $T = \{p\}$). This certainly holds when G is finitely T-generated, which is the main case of interest. We conclude that Theorem 12.2.4 for T is often but not always implied by Theorem 12.2.4 applied to the singleton sets $T = \{p\}$. As we have seen in Chapter 8, the divergence between the two pullbacks, P and Q, induced by the left part of the diagram becomes greater in the topological analogues.

12.3. Local to global: abelian and nilpotent groups

Our algebraic local to global result reads as follows. Its hypotheses seem to be minimal, but it is instructive to compare it with Theorem 13.3.1, where the topology allows us to generalize to groups that are not completable at T.

THEOREM 12.3.1. *Let*

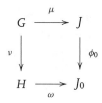

be a pullback square of nilpotent groups such that

  (i) J is T-complete;
  (ii) $\phi_0 : J \longrightarrow J_0$ is a rationalization of J; and
  (iii) H is rational.

Then G is T-local and completable at T, $\mu : G \to J$ is a completion at T, and $\nu : G \to H$ is a rationalization. Therefore ω is the rationalization of μ.

PROOF. Since J, J_0, and H are T-local, G is T-local by Lemma 5.5.7. We first prove the rest in the abelian case, and we change notations in accord with Lemma 12.1.1, letting $G = A$, $H = B$, and $J = C$. Since B is rational, $\mathbb{H}_T B = 0$

and $\mathbb{E}_T B = 0$. Since C is T-complete, $\mathbb{H}_T C = 0$ and $\mathbb{E}_T C \cong C$. Therefore, by Lemma 12.1.1, A is completable at T and $\mathbb{E}_T A \longrightarrow \mathbb{E}_T C \cong C$ is an isomorphism. Similarly, since rationalization preserves pullbacks, rationalization of the given pullback square shows that $A_0 \longrightarrow B_0 \cong B$ is an isomorphism. This proves the result in the abelian case. We can now change our point of view and think of the pullback A as a given T-local and T-completable abelian group that gives rise to our original pullback diagram via rationalization and completion at T. Therefore the abelian case of Proposition 12.2.5 applies. We will use that in our proof of the generalization to the nilpotent case.

Returning to the original notations, we prove the nilpotent case by induction on the larger of the nilpotency classes of J and H. The argument is similar to the proof of Proposition 7.4.3. To exploit naturality, we start with the lower central series of J and H. Rationalization gives a map from the lower central series of J to its rationalization, which is a lower central series of J_0 and so contains termwise the lower central series of J_0. The lower central series of H is already rational, and ω maps it into the lower central series of J_0 and thus into the rationalization of the lower central series of J. Since rationalization commutes with quotients, for each j we obtain a commutative diagram

$$
\begin{array}{ccccccccc}
1 & \longrightarrow & H_j/H_{j+1} & \longrightarrow & H/H_{j+1} & \longrightarrow & H/H_j & \longrightarrow & 1 \\
& & \downarrow & & \downarrow & & \downarrow & & \\
1 & \longrightarrow & (J_j/J_{j+1})_0 & \longrightarrow & (J/J_{j+1})_0 & \longrightarrow & (J/J_j)_0 & \longrightarrow & 1 \\
& & \uparrow & & \uparrow & & \uparrow & & \\
1 & \longrightarrow & J_j/J_{j+1} & \longrightarrow & J/J_{j+1} & \longrightarrow & J/J_j & \longrightarrow & 1
\end{array}
$$

of central extensions. We denote the resulting sequence of pullbacks by

$$
1 \longrightarrow A[j] \longrightarrow G[j+1] \longrightarrow G[j] \longrightarrow 1.
$$

By Lemma 7.6.2(ii), whose key epimorphism hypothesis we verified in our discussion of the abelian case, this is an exact sequence and in fact a central extension. We know the result for $A[j]$ and assume it inductively for $G[j]$. This clearly implies that $\mathbb{H}_T G[j+1] = 0$, and five lemma arguments give that the associated maps ν and μ for $G[j+1]$ are a rationalization and a completion at T. $\qquad\square$

As promised, this allows us to reprove and improve Theorem 12.2.4.

THEOREM 12.3.2. *If G is a T-local group that is completable at T, then the following diagram is a pullback.*

$$
\begin{array}{ccc}
G & \xrightarrow{\hat{\phi}} & \mathbb{E}_T G \\
\phi_0 \downarrow & & \downarrow \phi_0 \\
G_0 & \xrightarrow{\hat{\phi}_0} & (\mathbb{E}_T G)_0
\end{array}
$$

PROOF. Take $H = G_0$ and $J = \mathbb{E}_T G$, with ϕ_0 a rationalization of J and $\omega = \hat{\phi}_0$. Let P be the pullback of ϕ_0 and ω. The universal property of P gives a map $\alpha\colon G \longrightarrow P$ that induces an isomorphism upon rationalization and completion at T. Since P is completable at T, α is an isomorphism by Lemma 12.1.2. $\qquad\square$

12.4. Formal completions and the adèlic genus

We have emphasized the fully general nature of the local to global fracture theorems. That is both a virtue and a defect. For example, ω in Theorem 12.3.1 can have a large kernel, or can even be zero. In the latter case, letting $K = \ker(\phi_0\colon J \longrightarrow J_0)$, we have $G = H \times K$, $G_0 \cong H$, and $\mathbb{E}_T G \cong \mathbb{E}_T K$. Nothing like that can happen in the cases of greatest interest, where all groups in sight satisfy finite generation conditions over the appropriate ground ring and H and J are related by nontrivial rational coherence data.

We first develop conditions on the input that ensure that our local to global fracture theorem delivers $f\hat{\mathbb{Z}}_T$-nilpotent groups as output. This is subtle since finite generation conditions on the input are not always sufficient. There are finitely generated T-complete groups that cannot be realized as the completions of finitely generated T-local groups.

Specializing to the set of all primes, we then define and say just a little about the calculation of "adèlic" and "complete" variants of the (local) genus that we defined in §7.5. Here, assuming that we are given an $f\hat{\mathbb{Z}}$-nilpotent group that can be realized as the completion of an f-nilpotent group, we ask whether such a realization is unique and how to classify all such realizations.

To give a naive framework for dealing with these questions, we introduce an analogue of the notion of a formal localization of a rational nilpotent group that is naturally dictated by our cocellular constructions of localizations and completions. That motivation takes a global to local point of view, but the essential point is that the definition formalizes what is needed to go from finitely generated local data to finitely generated global data. Let $\hat{\mathbb{Q}}_T$ denote the ring $\hat{\mathbb{Z}}_T \otimes \mathbb{Q}$. Written that way, we think of first completing and then rationalizing. But we have $\hat{\mathbb{Q}}_T \cong \mathbb{Q} \otimes \hat{\mathbb{Z}}_T$. Written that way, we think of first rationalizing and then rationalizing completion maps. That makes good sense even though first rationalizing and then completing groups gives the trivial group.

Recall the notions of an R-map of R-nilpotent π-groups and of an fR-map of fR-nilpotent π-groups from Definition 4.3.1 and Notation 4.3.3. By the cocellular functoriality of our constructions of localizations and completions, localizations and completions of maps are R-maps for the appropriate ground ring R, as observed in Lemmas 6.2.1, 11.2.1, and 11.2.2. A formal completion is a particular kind of $\mathbb{Q}$-map. Before giving the full definition, we note the following analogue of the cited lemmas, which deals with a subsidiary part of the definition.

LEMMA 12.4.1. *If J is $\hat{\mathbb{Z}}_T$-nilpotent, then its rationalization $\phi_0 \colon J \longrightarrow J_0$ is a $\hat{\mathbb{Z}}_T$-map. Moreover, if J is $f\hat{\mathbb{Z}}_T$-nilpotent and its torsion subgroup is finite, then J and J_0 admit central series $\{J_i\}$ and $\{K_i\}$ such that $\phi_0(J_i) \subset K_i$ and the induced map of $\hat{\mathbb{Z}}_T$-modules $J_i/J_{i+1} \longrightarrow K_i/K_{i+1}$ is isomorphic as a $\hat{\mathbb{Q}}_T$-module under the $\hat{\mathbb{Z}}_T$-module J_i/J_{i+1} to the canonical map $\eta \colon J_i/J_{i+1} \longrightarrow J_i/J_{i+1} \otimes \mathbb{Q}$. That is, there is a commutative diagram*

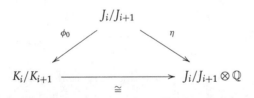

of $\hat{\mathbb{Z}}_T$-modules in which the isomorphism is a map of $\hat{\mathbb{Q}}_T$-modules.

PROOF. The first statement is easily proven by induction, using our cocellular constructions. For the second statement, the kernel of ϕ_0 is the torsion subgroup of J, which is finite by assumption and therefore an $f\mathbb{Z}_T$-nilpotent group. We may start our central series for J and J_0 with a central series for $\ker \phi_0$ and the constant central series $K_i = K$ of the same length. Thus we may as well replace J by $J/\ker \phi_0$ and so assume that ϕ_0 is a monomorphism.

Then the subquotients J_i/J_{i+1} are finitely generated free $\hat{\mathbb{Z}}_T$-modules and the conclusion follows. □

Of course, even if J is $f\hat{\mathbb{Z}}_T$-nilpotent, ϕ_0 is not an $f\hat{\mathbb{Z}}_T$-map since its target is not $f\hat{\mathbb{Z}}_T$-nilpotent. It is $f\hat{\mathbb{Q}}_T$-nilpotent, and we must use the ring $\hat{\mathbb{Q}}_T$ for algebraic understanding. The group π in the following definition plays no role in this section, but the generality will be relevant to the space-level analogue. We are only interested in the cases when either π is trivial or all given groups are abelian.

DEFINITION 12.4.2. Let π be a group, H be an $f\mathbb{Q}$-nilpotent π-group, and J be an $f\hat{\mathbb{Z}}_T$-nilpotent π-group with rationalization $\phi_0: J \longrightarrow J_0$. A formal completion of H at T associated to ϕ_0 is a homomorphism $\omega: H \longrightarrow J_0$ of π-groups with the following property. There exists an $f\mathbb{Q}$-central π-series $\{H_i\}$ for H, an $f\hat{\mathbb{Z}}_T$-central π-series $\{J_i\}$ for J, and an $f\hat{\mathbb{Q}}_T$-central π-series $\{K_i\}$ for J_0 such that the π-equivariant conclusion of Lemma 12.4.1 holds for ϕ_0 and each induced homomorphism $H_i/H_{i+1} \longrightarrow K_i/K_{i+1}$ of $\mathbb{Q}[\pi]$-modules is isomorphic as a $\hat{\mathbb{Q}}_T[\pi]$-module under the $\mathbb{Q}[\pi]$-module H_i/H_{i+1} to the natural homomorphism

$$\eta: H_i/H_{i+1} \cong H_i/H_{i+1} \otimes \mathbb{Z}_T \longrightarrow H_i/H_{i+1} \otimes \hat{\mathbb{Z}}_T.$$

Thus $\omega(H_i) \subset K_i$ and we have a commutative diagram

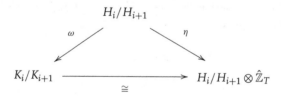

of maps of $\mathbb{Q}[\pi]$-modules such that the isomorphism is a map of $\hat{\mathbb{Q}}_T[\pi]$-modules. Observe that a formal completion ω is necessarily a monomorphism. Observe too that when the given groups are abelian, the only requirement on the map ω is that there must be a commutative diagram of $\mathbb{Q}[\pi]$-modules

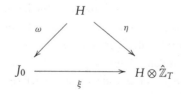

in which ξ is an isomorphism of $\hat{\mathbb{Q}}_T[\pi]$-modules. No compatibility with the map ϕ_0 is required.

From a global to local point of view, we have the following observation. As in Lemma 12.2.2, we use that completion at T factors through localization at T.

PROPOSITION 12.4.3. *For any $T' \supset T$, the rationalization $(\hat{\phi})_0$ of the completion $\hat{\phi}\colon G \longrightarrow \mathbb{E}_T G$ at T of an $f\mathbb{Z}_{T'}$-nilpotent group G is a formal completion $G_0 \longrightarrow (\mathbb{E}_T G)_0$.*

From a local to global point of view, we have the following addendum to Theorem 12.3.1.

THEOREM 12.4.4. *Let*

$$
\begin{array}{ccc}
G & \xrightarrow{\ \mu\ } & J \\
{\scriptstyle \nu}\big\downarrow & & \big\downarrow{\scriptstyle \phi_0} \\
H & \xrightarrow[\ \omega\ ]{} & J_0
\end{array}
$$

be a pullback square of nilpotent groups such that

   (i) J is $f\hat{\mathbb{Z}}_T$-complete and its torsion subgroup is finite;
   (ii) $\phi_0\colon J \longrightarrow J_0$ is a rationalization of J;
   (iii) H is rational; and
   (iv) ω is a formal completion associated to ϕ_0.

Then G is $f\mathbb{Z}_T$-nilpotent.

PROOF. The kernel of ϕ_0 is the torsion subgroup of J, which is finite by assumption and therefore an $f\mathbb{Z}_T$-nilpotent group. It coincides with the kernel of ν, and, arguing as in Proposition 12.4.3 we may as well replace G and J by their quotients by $\ker \phi_0$. That is, there is no loss of generality if we assume that G and J are torsion free.

We look first at the abelian case. Then H is just a finite dimensional rational vector space with an inclusion $\omega\colon H \longrightarrow J_0$ of abelian groups and thus of rational vector spaces that is equivalent under H to the canonical inclusion $H \longrightarrow H \otimes \hat{\mathbb{Z}}_T$. Then the pullback G is a torsion-free $\mathbb{Z}_T$-module whose rationalization is H and whose completion at T is J. We may choose basis vectors for

H that are in G and are not divisible by any $p \in T$. Let F be the free $\mathbb{Z}_T$-module on these generators and $\iota \colon F \longrightarrow G$ the canonical map of $\mathbb{Z}_T$-modules. Since ι rationalizes to the identity map of H, G is torsion free, and G/F is torsion free, ι is an isomorphism. By a pedantically careful diagram chase that we leave to the reader, our pullback diagram in this case is isomorphic to the canonical pullback diagram

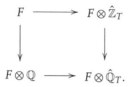

For the inductive step, we choose central series $\{H_i\}$, $\{J_i\}$, and $\{K_i\}$ for H, J, and J_0, respectively, as in the definition of a formal completion. As usual, these assemble into a sequence of pullback diagrams of central extensions. The sequence of pullback groups induced by each of these pullback diagrams is short exact by Lemma 7.6.2(ii); the epimorphism hypothesis in that result is satisfied by the abelian case of Proposition 12.2.5, which is quite easy to reprove algebraically under the finite generation hypotheses we have here. These extensions show that the pullbacks G_i of the diagrams $H_i \longrightarrow K_i \longleftarrow J_i$ display a central $\mathbb{Z}_T$-series for G with finitely generated subquotients, and the last statement follows by inspection. $\qquad \square$

REMARK 12.4.5. The notion of formal completion used in Theorem 12.4.4 may seem fussy. However, some such hypothesis is needed since Belfi and Wilkerson [10] have given an example of a finitely generated $\hat{\mathbb{Z}}_p$-nilpotent group J with mod p homology of finite type over $\mathbb{F}_p$ and with nilpotency class two such that there is no finitely generated $\mathbb{Z}_{(p)}$-nilpotent group G whose completion $\hat{G}_p$ is isomorphic to J, hence there is no finitely generated nilpotent group G whose completion is isomorphic to J.

We now specialize T to be the set of all primes. We have two alternative notions of genus.

DEFINITION 12.4.6. The adèlic genus of a finitely generated nilpotent group G is the set of isomorphism classes of finitely generated nilpotent groups G' such that G_0 is isomorphic to G'_0 and $\hat{G}_p$ is isomorphic to $\hat{G}'_p$ for all primes p. The complete genus of G is defined by dropping the requirement that G_0 be isomorphic to G'_0.

The name "adèlic" is suggested by Sullivan's analogy [133, 135] with the theory of adèles in number theory. There do not seem to be standard names for these notions in the literature, although the adèlic genus has been studied, in particular by Pickel [112]. The previous proof gives the analogue of Proposition 7.5.6.

PROPOSITION 12.4.7. *The isomorphism class of A is the only element of the adèlic genus of a finitely generated abelian group A.*

REMARK 12.4.8. In analogy with Example 7.5.7, Belfi and Wilkerson [10, 4.2] have given examples of non-isomorphic finitely generated nilpotent groups G and G' that are in the same adèlic genus but whose localizations G_p and G'_p are not isomorphic for some prime p, so that G and G' are not in the same genus.

We sketch briefly a naive approach to the study of the adèlic genus of a fixed f-nilpotent group G. We ignore the question of change of chosen central series in the definition of a formal completion. We may as well fix $J = \hat{G}$ and $H = G_0$. Then G is the pullback of a certain formal completion

$$H \xrightarrow{\omega} J_0 \xleftarrow{\phi_0} J.$$

The results of this section imply that we can construct any other group G' in the same adèlic genus as a pullback of a formal completion

$$H \xrightarrow{\omega'} J_0 \xleftarrow{\phi'_0} J.$$

There is a unique automorphism $\xi \colon J_0 \longrightarrow J_0$ of $\hat{\mathbb{Q}}$-modules such that $\xi \circ \phi'_0 = \phi_0$, and the pullback of

$$H \xrightarrow{\xi \circ \omega'} J_0 \xleftarrow{\phi_0} J$$

is isomorphic to G'. Thus we may as well fix ϕ_0 and consider all possible choices of ω. Two choices differ by a $\hat{\mathbb{Q}}$-automorphism of J_0. Let $\mathrm{Aut}(J_0)$ denote the group of such automorphisms. We send an automorphism ξ to the isomorphism class $[\xi]$ of the pullback of $\xi \circ \omega$ and ϕ_0. This is a well-defined surjective function from $\mathrm{Aut}(J_0)$ to the adèlic genus of G. A $\hat{\mathbb{Z}}$-automorphism ζ of J induces a $\hat{\mathbb{Q}}$-automorphism $\tilde{\zeta}$ of J_0 such that $\tilde{\zeta} \circ \phi_0 = \phi_0 \circ \zeta$. It follows that $[\tilde{\zeta} \circ \xi] = [\xi]$. We would like to say that a $\hat{\mathbb{Q}}$-automorphism ε of H induces a $\hat{\mathbb{Q}}$-automorphism $\tilde{\varepsilon}$ of J_0 such that $\tilde{\varepsilon} \circ \omega = \omega \circ \varepsilon$. It would follow that $[\xi \circ \tilde{\varepsilon}] = [\xi]$. The conclusion would be that the adèlic genus of G is in bijective correspondence with the double cosets

$$\operatorname{Aut} H \backslash \operatorname{Aut}(J_0) / \operatorname{Aut} J.$$

REMARK 12.4.9. This sketch is incomplete since we have not shown that ω is functorial on automorphisms, but the conclusion is correct by results of Pickel [112]. Moreover, combining with results of Auslander and Baumslag [7, 8] and Borel [13], one can prove the remarkable result that the adèlic genus and complete genus of G are both finite sets. A summary of how the argument goes is given in [144, §1].

13

FRACTURE THEOREMS FOR COMPLETION: SPACES

In this chapter we prove analogues of the results of the previous chapter for nilpotent spaces. As in Chapter 8, we begin with a fracture theorem for maps from finite CW complexes into f-nilpotent spaces. Aside from its restriction to f-nilpotent spaces, some such restriction being necessary, its proof is so precisely similar to the analogous arguments of §8.2 that we feel comfortable in leaving the details to the reader.

We then consider fracture theorems for nilpotent spaces, stating them in §13.1 and proving them in §13.2 and §13.3. The exposition follows the order given in the analogue for localization, starting with global to local results. However, these results are deduced from local to global results that are proven topologically, making minimal use of the corresponding results for groups. More straightforward inductive proofs based on the results for groups work under completability assumptions.

The last three sections are of a different character. In §13.4, we give an informal notion of the tensor product of a space and a ring. We have seen two examples. All localizations fit into this framework, and completions of f-nilpotent spaces do too. This gives a context in which to discuss Sullivan's formal completions in §13.5. These are extensions of tensor products with the rings $\hat{\mathbb{Z}}_T$ from simple spaces of finite type to more general simple spaces. Using these preliminaries, we return to the notion of genus in §13.6, where we describe two variants of the notion of genus that we discussed in §8.5.

Throughout this chapter, T denotes a fixed set of primes. The set of all primes and the set consisting of just one prime are the most interesting cases, and the reader may prefer to focus on those. We let $\hat{\phi}$ denote completion at T and ϕ_0 denote rationalization. All given spaces are to be nilpotent, even when we neglect to say so, and we understand T-local and T-complete spaces to be nilpotent. We may identify $\hat{X}_T$ with $\prod_{p \in T} \hat{X}_p$.

13.1. Statements of the main fracture theorems

The following result is a consequence of Theorems 11.1.2, 11.3.2, and 12.3.2 via arguments exactly like those in §8.2.

THEOREM 13.1.1. *Let X be an $f\mathbb{Z}_T$-nilpotent space and K be a finite CW complex. Then the function*

$$\hat{\phi}_* : [K, X] \longrightarrow [K, \hat{X}_T]$$

is an injection and the function

$$(\hat{\phi}, \phi_0) : [K, X] \longrightarrow [K, \hat{X}_T] \times_{[K,(\hat{X}_T)_0]} [K, X_0]$$

is a bijection. Moreover, the formal sum

$$[\Sigma K, \hat{X}_T] \times [\Sigma K, X_0] \longrightarrow [\Sigma K, (\hat{X}_T)_0]$$

is a surjection.

The examples $K = S^n$ and $X = K(\mathbb{Z}/p^\infty, n)$ or $X = K(\mathbb{Q}, n)$ show that the injectivity statement no longer holds when X is $\mathbb{Z}_T$-nilpotent (= T-local), rather than $f\mathbb{Z}_T$-nilpotent. The surjectivity statement is analogous to Proposition 8.2.7. The following consequence of the injectivity statement is analogous to Corollary 8.2.4.

COROLLARY 13.1.2. *Let $f, g : K \longrightarrow X$ be maps, where K is a nilpotent finite CW complex and X is an f-nilpotent space. Then $f \simeq g$ if and only if $\hat{f}_p \simeq \hat{g}_p$ for all primes p.*

REMARK 13.1.3. As in Remark 8.2.8, the previous results apply more generally, with K taken to be any space with finitely generated integral homology.

Surprisingly, the fracture theorems for spaces, as opposed to maps, require no $f\mathbb{Z}_T$-nilpotency assumptions.

THEOREM 13.1.4. *Let X be a T-local space. Then the following diagram is a homotopy pullback.*

$$
\begin{array}{ccc}
X & \longrightarrow & \hat{X}_T \\
\downarrow & & \downarrow \\
X_0 & \longrightarrow & (\hat{X}_T)_0
\end{array}
$$

THEOREM 13.1.5. *Let*

$$
\begin{array}{ccc}
P & \xrightarrow{\ \mu\ } & \prod_{p \in T} X_p \\
{\scriptstyle \nu} \downarrow & & \downarrow {\scriptstyle \phi_0} \\
Y & \xrightarrow[\ \omega\ ]{} & (\prod_{p \in T} X_p)_0
\end{array}
$$

be a homotopy pullback of connected spaces in which

(i) *the space X_p is p-complete;*
(ii) *the map ϕ_0 is a rationalization of $\prod_{p \in T} X_p$; and*
(iii) *the space Y is rational.*

Then the space P is T-local, the map μ is a completion of P at T, and the map ν is a rationalization of P. Therefore the map ω is a rationalization of the map μ.

Theorem 13.3.1 below gives a restatement that may make the comparison with the algebraic analogue, Theorem 12.3.1, more transparent.

REMARK 13.1.6. By Corollary 2.2.3, the hypothesis that P is connected in Theorem 13.1.5 is equivalent to saying that every element $z \in \pi_1((\prod_{p \in T} X_p)_0)$ is the product of an element $\phi_{0*}(x)$ and an element $\omega_*(y)$ where $x \in \pi_1(\prod X_p)$ and $y \in \pi_1(Y)$. This means that ω must satisfy the analogue of (ii) in the definition of a formal localization, Definition 8.1.5. However, we require no hypothesis analogous to (i) there. We will add such a hypothesis in §13.6, where we fix Y and assume finite generation conditions. It is remarkable and illuminating that no connection between Y and the X_p other than the condition given by requiring P to be connected is needed for the validity of Theorem 13.1.5. Of course, that condition always holds when the X_p and Y are simply connected.

REMARK 13.1.7. We shall derive Theorem 13.1.4 from Theorem 13.1.5, whereas we gave an independent proof of the analogue for localization. Since X is connected, the homotopy pullback to which Theorem 13.1.4 compares it must also be connected, as in the previous remark. The algebraic result Proposition 12.2.5 proves this, and Theorem 13.1.4 depends on that result. However, the proof of Proposition 12.2.5 referred forward to the topology. We shall carefully avoid circularity.

13.2. Global to local fracture theorems: spaces

The following general observation is elementary but important. It helps explain why one should expect to be able to reconstruct T-local spaces from their rationalizations and their completions at T.

LEMMA 13.2.1. *Let $f\colon X \longrightarrow Y$ be any map. Then $f_*\colon H_*(X; \mathbb{Z}_T) \longrightarrow H_*(Y; \mathbb{Z}_T)$ is an isomorphism if and only if $f_*\colon H_*(X; \mathbb{Q}) \longrightarrow H_*(Y; \mathbb{Q})$ and, for all $p \in T$, $f_*\colon H_*(X; \mathbb{F}_p) \longrightarrow H_*(Y; \mathbb{F}_p)$ are isomorphisms.*

PROOF. Since $\mathbb{Q}$ and $\mathbb{F}_p$ are modules over the PID $\mathbb{Z}_T$, the universal coefficient theorem gives the forward implication. For the converse, the Bockstein long exact sequences of homology groups induced by the short exact sequences

$$0 \longrightarrow \mathbb{Z}/p^{n-1} \longrightarrow \mathbb{Z}/p^n \longrightarrow \mathbb{Z}/p \longrightarrow 0$$

show that $f_*\colon H_*(X; \mathbb{Z}/p^n) \longrightarrow H_*(Y; \mathbb{Z}/p^n)$ is an isomorphism for $n \geq 1$ since it is an isomorphism when $n = 1$. Since $\mathbb{Z}/p^\infty = \mathrm{colim}\,\mathbb{Z}/p^n$ and $\mathbb{Q}/\mathbb{Z}_T \cong \oplus_{p \in T}\mathbb{Z}/p^\infty$, this implies that $f_*\colon H_*(X; \mathbb{Q}/\mathbb{Z}_T) \longrightarrow H_*(Y; \mathbb{Q}/\mathbb{Z}_T)$ is an isomorphism. Indeed, homology commutes with sums and colimits of coefficient groups since this already holds on the chain level. Now the Bockstein long exact sequence induced by the short exact sequence

$$0 \longrightarrow \mathbb{Z}_T \longrightarrow \mathbb{Q} \longrightarrow \mathbb{Q}/\mathbb{Z}_T \longrightarrow 0$$

shows that $f_*\colon H_*(X; \mathbb{Z}_T) \longrightarrow H_*(Y; \mathbb{Z}_T)$ is an isomorphism. $\qquad\square$

Since maps of T-local, rational, and T-complete spaces are equivalences if and only if they induce isomorphisms on homology with coefficients in $\mathbb{Z}_T$, $\mathbb{Q}$, and $\mathbb{F}_p$ for $p \in T$, the following result is an immediate consequence.

COROLLARY 13.2.2. *Let $f\colon X \longrightarrow Y$ be a map between T-local spaces. Then f is an equivalence if and only if its rationalization $f_0\colon X_0 \to Y_0$ and completion $\hat{f}_T\colon \hat{X}_T \to \hat{Y}_T$ at T are equivalences.*

Now the global to local fracture theorem, Theorem 13.1.4, is a direct consequence of the local to global fracture theorem, Theorem 13.1.5.

PROOF OF THEOREM 13.1.4, ASSUMING THEOREM 13.1.5. Let P be the homotopy pullback of the diagram

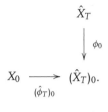

Since $\pi_1(\hat{X}_T) = \mathbb{E}_T\pi_1(X)$, Proposition 12.2.5 implies that P is connected. Therefore Theorem 13.1.5 implies that the induced map $\alpha\colon X \to P$ becomes an equivalence when rationalized or completed at T. By Corollary 13.2.2, α is an equivalence. $\qquad\square$

An alternative inductive argument is possible. Its first steps are given in the following two lemmas, which are also the first steps in the proof of Theorem 13.1.5. The following result is analogous to Lemma 8.3.3.

LEMMA 13.2.3. *Theorem 13.1.4 holds when $X = K(A, n)$, where $n \geq 1$ and A is abelian.*

PROOF. We construct the pullback P in the diagram of Theorem 13.1.4 and obtain a map $\alpha\colon X \longrightarrow P$. We check that α_0 and $\hat{\alpha}$ are equivalences using Proposition 6.2.5 for the rationalization and Proposition 11.2.5 for the completion. Corollary 13.2.2 completes the proof. This outline is a complete proof when $n \geq 2$, but it hides a subtlety when $n = 1$. To apply the cited results, we need to know that P is connected. The already proven case $n = 2$ of the present result implies the abelian group case of Proposition 12.2.5, and that result implies that P is connected. $\qquad\square$

The proof of the inductive step of the alternative proof of Theorem 13.1.4 is essentially the same as the proof of the analogous inductive step in the alternative proof of Theorem 8.1.3 given in §8.6. We need to induct up the Postnikov tower of a given nilpotent space, and the following result enables us to do so. The proof is an application of Proposition 8.6.1. Applied to the inductive construction of $K(G, 1)$ for a nilpotent group G, the argument has already been used to prove Proposition 12.2.5. Therefore a special case of this proof is used implicitly to get started with our first proof of Theorem 13.1.4.

LEMMA 13.2.4. *Suppose that Theorem 13.1.4 holds for X and let Y be the fiber of a map $k\colon X \longrightarrow K$, where K is a simply connected T-local Eilenberg-Mac Lane space. Then Theorem 13.1.4 holds for Y.*

PROOF. Consider the following homotopy commutative diagram. The homotopy pullbacks of its rows are as indicated in the column at the right since localizations and completions preserve fibers. The homotopy pullbacks of its columns are as indicated in the row at the bottom, by Lemma 13.2.3 and asssumption.

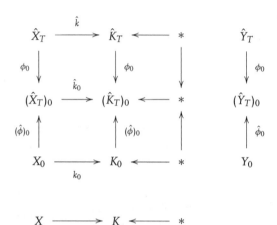

By Proposition 8.6.1, Y is equivalent to the homotopy pullback of the right column. In particular, since Y is connected, that homotopy pullback is connected. □

13.3. Local to global fracture theorems: spaces

In this section we prove the local to global result, Theorem 13.1.5. The argument is a specialization of the proof of a more general result of Dror, Dwyer, and Kan [39] in which nilpotency is relaxed to "virtual nilpotency". We abbreviate notation by letting $Z = \prod X_p$ and $Z_0 = (\prod X_p)_0$. Equivalently, Z is any T-complete space and Z_0 is its rationalization. Thus Theorem 13.1.5 can be restated as follows.

THEOREM 13.3.1. *Let*

$$
\begin{array}{ccc}
P & \xrightarrow{\mu} & Z \\
\nu \downarrow & & \downarrow \phi_0 \\
Y & \xrightarrow{\omega} & Z_0
\end{array}
$$

be a homotopy pullback of connected spaces in which

(i) the space Z is T-complete;

(ii) the map $\phi_0 \colon Z \longrightarrow Z_0$ is a rationalization of Z; and

(iii) the space Y is rational.

Then the space P is T-local, the map $\mu \colon P \longrightarrow Z$ is a completion at T, and the map $\nu \colon P \longrightarrow Y$ is a rationalization of P. Therefore ω is a rationalization of μ.

As noted in Remark 13.1.6, the assumption that P is connected means that every element of $\pi_1(Z_0)$ is a product of an element in the image of ω_* and an element in the image of ϕ_{0*}. Observe that no such hypothesis was needed in the analogous algebraic result, Theorem 12.3.1.

The proof of the theorem is based on the following lemma. It is convenient to use the language of fiber squares, which are just pullback squares in which one of the maps being pulled back is a fibration.

LEMMA 13.3.2. *For a homotopy pullback P as displayed in Theorem 13.3.1 there is an integer $n \geq 1$ and a factorization*

$$
\begin{array}{ccccccccccc}
P & \xrightarrow{\mu_n} & Z_n & \longrightarrow & \cdots & \longrightarrow & Z_i & \xrightarrow{\sigma_{i-1}} & Z_{i-1} & \longrightarrow & \cdots & \longrightarrow & Z_1 & = & Z \\
\downarrow{\scriptstyle \nu} & & \downarrow{\scriptstyle \psi_n} & & & & \downarrow{\scriptstyle \psi_i} & & \downarrow{\scriptstyle \psi_{i-1}} & & & & \downarrow{\scriptstyle \phi_0} & & \\
Y & \xrightarrow{\omega_n} & V_n & \longrightarrow & \cdots & \longrightarrow & V_i & \xrightarrow{\tau_{i-1}} & V_{i-1} & \longrightarrow & \cdots & \longrightarrow & V_1 & = & Z_0
\end{array}
$$

of the homotopy pullback square such that

(i) *each Z_i and V_i is connected and nilpotent;*

(ii) *each V_i is rational;*

(iii) *there are maps $\omega_i \colon Y \longrightarrow V_i$ and $\tau_{i-1} \colon V_i \longrightarrow V_{i-1}$ such that $\omega_1 = \omega$ and $\tau_{i-1} \circ \omega_i = \omega_{i-1}$;*

(iv) *there are maps $\mu_i \colon P \longrightarrow Z_i$ and $\sigma_{i-1} \colon Z_i \longrightarrow Z_{i-1}$ such that $\mu_1 = \mu$ and $\sigma_{i-1} \circ \mu_i = \mu_{i-1}$;*

(v) *each map $\psi_i \colon Z_i \longrightarrow V_i$ is a $\mathbb{Q}$-homology equivalence;*

(vi) *each map $\sigma_{i-1} \colon Z_i \longrightarrow Z_{i-1}$ is an $\mathbb{F}_T$-homology equivalence;*

(vii) *the following are fiber squares*

$$
\begin{array}{ccccc}
P & \xrightarrow{\mu_i} & Z_i & \xrightarrow{\sigma_{i-1}} & Z_{i-1} \\
\downarrow{\scriptstyle \nu} & & \downarrow{\scriptstyle \psi_i} & & \downarrow{\scriptstyle \psi_{i-1}} \\
Y & \xrightarrow{\omega_i} & V_i & \xrightarrow{\tau_{i-1}} & V_{i-1};
\end{array}
$$

(viii) the map $\omega_{n*}\colon \pi_1 Y \longrightarrow \pi_1 V_n$ is surjective.

PROOF OF THEOREM 13.3.1, ASSUMING LEMMA 13.3.2. Since Y and Z_0 are rational, they are T-local. Since Z is T-complete, it is T-local. By Proposition 6.2.5, P is nilpotent and T-local. With n as in Lemma 13.3.2, let F be the common fiber of the maps μ_n and ω_n and consider the following diagram.

$$
\begin{array}{ccccc}
F & \longrightarrow & P & \xrightarrow{\mu_n} & Z_n \\
\| & & \downarrow{\scriptstyle \nu} & & \downarrow{\scriptstyle \psi_n} \\
F & \longrightarrow & Y & \xrightarrow{\omega_n} & V_n
\end{array}
$$

Since Y and V_n are connected, nilpotent, and rational and $\omega_{n*}\colon \pi_1(Y) \to \pi_1(V_n)$ is a surjection, F is connected, nilpotent, and rational by Proposition 6.2.5.

By Proposition 4.4.1, $\pi_1 V_n$ and $\pi_1 Z_n$ act nilpotently on the rational homology of F. Applying Theorem 24.6.2 to the map of rational homology Serre spectral sequences induced by the displayed map of fibration sequences, we see that ν is a rational homology isomorphism since ψ_n is a rational homology isomorphism.

Since F is rational, Theorem 6.1.1 implies that $\tilde{H}_q(F;\mathbb{Z})$ is uniquely divisible for each q, hence the universal coefficient theorem implies that $\tilde{H}_i(F;\mathbb{F}_T) = 0$ for each i. Therefore the $\mathbb{F}_p$-homology Serre spectral sequence of the fibration sequence

$$
F \xrightarrow{} P \xrightarrow{\mu_n} Z_n
$$

collapses to show that $\mu_n\colon P \longrightarrow Z_n$ is an $\mathbb{F}_T$-homology equivalence. Since each of the maps σ_n is an $\mathbb{F}_T$-homology equivalence, $\mu\colon P \to Z$ is an $\mathbb{F}_T$-homology equivalence. $\qquad\square$

PROOF OF LEMMA 13.3.2. We may assume that ϕ_0 is a fibration, and we agree to arrange inductively that the map $\psi_k \colon Z_k \longrightarrow V_k$ is a fibration for each i. That is, we arrange that the squares in our factorization are fiber squares.

We proceed by induction on i. Suppose that we have constructed spaces Z_i and V_i and maps ψ_i, σ_{i-1}, τ_{i-1}, μ_i, and ω_i for $1 \leq i \leq j$ such that (i) through (vii) are satisfied. If $\omega_{j_*} \colon \pi_1 Y \longrightarrow \pi_1 V_j$ is surjective, there is nothing to prove. Otherwise let C_j be the cokernel of the map

$$\omega_{j_*} \colon H_1(Y; \mathbb{Z}) \longrightarrow H_1(V_j; \mathbb{Z}).$$

Observe that the abelian group C_j is rational.

Let F be the fiber of the map $\psi_j \colon Z_j \longrightarrow V_j$. Since

$$
\begin{array}{ccc}
P & \xrightarrow{\ \phi_j\ } & Z_j \\
{\scriptstyle \nu}\downarrow & & \downarrow{\scriptstyle \psi_j} \\
Y & \xrightarrow[\ \omega_j\ ]{} & V_j
\end{array}
$$

is a fiber square of connected spaces, for every element $x \in \pi_1(V_j)$ there are elements $y \in \pi_1(Y)$ and $z \in \pi_1(Z_j)$ such that $x = \omega_{j_*}(y)\psi_{j_*}(z)$.

Let δ_j be the composite surjection

$$\delta_j \colon \pi_1 V_j \longrightarrow H_1(V_j; \mathbb{Z}) \longrightarrow C_j$$

obtained from the Hurewicz homomorphism. Precomposing, we obtain a map

$$\varepsilon_j \colon \pi_1 Z_j \longrightarrow \pi_1 V_j \xrightarrow{\ \delta_j\ } C_j,$$

and it is also surjective since the naturality of the Hurewicz homomorphism implies that, with the notations above, $\omega_{j_*}(y)$ maps to 0 in C_j and thus x and $\psi_{j_*}(z)$ have the same image in C_j. Let D_j be the kernel of δ_j and E_j be the kernel of ε_j.

Define $\tau_j \colon V_{j+1} \longrightarrow V_j$ to be the fiber of the map $\alpha_j \colon V_j \longrightarrow K(C_j, 1)$, unique up to homotopy, that realizes the epimorphism δ_j on π_1. The image of $\pi_1(Y)$ is contained in D_j, so the map $\omega_j \colon Y \to V_j$ lifts to a map $\omega_{j+1} \colon Y \to V_{j+1}$ such that $\tau_j \circ \omega_{j+1} = \omega_j$.

Observe that $\alpha_j \circ \psi_j \colon Z_j \longrightarrow K(C_j, 1)$ is the map that realizes the epimorphism ε_j on π_1 and define $\sigma_j \colon Z_{j+1} \longrightarrow Z_j$ to be the fiber of $\alpha_j \circ \psi_j$.

Equivalently, Z_{j+1} is the pullback $Z_j \times_{V_j} V_{j+1}$ (compare Lemma 1.2.7), and this description gives a map $\psi_{j+1} \colon Z_{j+1} \longrightarrow V_{j+1}$ such that the right square commutes in the following diagram. The universal property of the pullback gives a map $\mu_{j+1} \colon P \longrightarrow Z_{j+1}$ such that $\sigma_j \circ \mu_{j+1} = \mu_j$ and the left square commutes.

$$
\begin{array}{ccccc}
P & \xrightarrow{\ \mu_{j+1}\ } & Z_{j+1} & \xrightarrow{\ \sigma_j\ } & Z_j \\
\downarrow{\scriptstyle \nu} & & \downarrow{\scriptstyle \psi_{j+1}} & & \downarrow{\scriptstyle \psi_j} \\
Y & \xrightarrow{\ \omega_{j+1}\ } & V_{j+1} & \xrightarrow{\ \tau_j\ } & V_j
\end{array}
$$

This gives the spaces and maps required at the next stage. We must verify (i)–(vii) with i replaced by $j+1$, and we must see how the construction leads inductively to (viii).

(i) Since Z_j and V_j are connected and $\varepsilon_j \colon \pi_1 Z_j \longrightarrow C_j$ and $\delta_j \colon \pi_1 V_j \longrightarrow C_j$ are surjective, the exact sequences

$$
\pi_1 Z_j \xrightarrow{\ \epsilon_j\ } C_j \longrightarrow \pi_0 Z_{j+1} \longrightarrow \pi_0 Z_j
$$

$$
\pi_1 V_j \xrightarrow{\ \delta_j\ } C_j \longrightarrow \pi_0 V_{j+1} \longrightarrow \pi_0 V_j
$$

imply that Z_{j+1} and V_{j+1} are connected. Since Z_j, V_j, and $K(C_j, 1)$ are nilpotent, Proposition 4.4.1 implies that Z_{j+1} and V_{j+1} are nilpotent.

(ii) Since C_j is rational, $K(C_j, 1)$ is rational. Since V_j and $K(C_j, 1)$ are nilpotent and rational and V_{j+1} is connected, V_{j+1} is rational.

(iii) and (iv) The required maps and relations are part of the construction.

(v) Since Z_j, V_j, and $K(C_j, 1)$ are nilpotent and Z_{j+1} and V_{j+1} are connected Proposition 4.4.1 implies that C_j acts nilpotently on $\tilde{H}_*(Z_{j+1}; \mathbb{Q})$ and $\tilde{H}_*(V_{j+1}; \mathbb{Q})$. In the map of rational homology Serre spectral sequences induced by the map of fiber sequences

$$
\begin{array}{ccccc}
Z_{j+1} & \xrightarrow{\ \sigma_j\ } & Z_j & \longrightarrow & K(C_j, 1) \\
\downarrow{\scriptstyle \psi_{j+1}} & & \downarrow{\scriptstyle \psi_j} & & \downarrow \\
V_{j+1} & \xrightarrow{\ \tau_j\ } & V_j & \longrightarrow & K(C_j, 1)
\end{array}
$$

the maps

$$\psi_{j_*} : \tilde{H}_*(Z_j; \mathbb{Q}) \to \tilde{H}_*(V_j; \mathbb{Q}) \quad \text{and} \quad \text{id} : \tilde{H}_*(K(C_j, 1); \mathbb{Q}) \to \tilde{H}_*(K(C_j, 1); \mathbb{Q})$$

are isomorphisms. Therefore Theorem 24.6.2 implies that

$$\psi_{j+1_*} : \tilde{H}_*(Z_{j+1}; \mathbb{Q}) \longrightarrow \tilde{H}_*(V_{j+1}; \mathbb{Q})$$

is an isomorphism.

(vi) Since C_j is rational, $\tilde{H}_*(K(C_j, 1); \mathbb{F}_T) = 0$ by the universal coefficient theorem. With $\mathbb{F}_T$ coefficients, the Serre spectral sequence of the fibration

$$Z_{j+1} \xrightarrow{\sigma_j} Z_j \longrightarrow K(C_j, 1)$$

collapses, and its edge homomorphism is an isomorphism

$$\sigma_{j*} : \tilde{H}_*(Z_{j+1}; \mathbb{F}_T) \longrightarrow \tilde{H}_*(Z_j; \mathbb{F}_T).$$

(vii) The right square in the diagram in (vii) is a pullback by construction. For the left square, recall that $P \cong Z_j \times_{V_j} Y$. Therefore

$$P \cong Z_j \times_{V_j} Y$$
$$\cong Z_j \times_{V_j} (V_{j+1} \times_{V_{j+1}} Y)$$
$$= Z_{j+1} \times_{V_{j+1}} Y.$$

(viii) The group C_j is the cokernel of the map

$$\pi_1 Y / [\pi_1 Y, \pi_1 Y] \to \pi_1 V_j / [\pi_1 V_j, \pi_1 V_j].$$

The short exact sequence

$$1 \to \pi_1(V_{j+1}) \to \pi_1(V_j) \xrightarrow{\delta_j} C_j \to 1$$

gives an isomorphism $\pi_1(V_{j+1}) \to \ker \delta_j$. Applied with $G_i = \pi_1 V_i$ and $H = \pi_1 Y$, Lemma 13.3.3 below implies that there is an integer $n \geq 1$ such that $\pi_1 V_n = \pi_1 Y$. □

LEMMA 13.3.3. *Let G be a nilpotent group and let $H \subset G$ be a subgroup. Let $G_1 = G$ and, inductively,*

$$G_{i+1} = \ker(G_i \longrightarrow \operatorname{coker}(H/[H, H] \longrightarrow G_i/[G_i, G_i])).$$

Then there is an integer $n \geq 1$ such that $G_j = H$ for $j \geq n$.

PROOF. Clearly $H \subset G_i$ for each i. We claim that G_i is contained in the subgroup of G generated by H and $\Gamma_i G$, the ith term of the lower central series of G. Since G is nilpotent, this gives the conclusion.

We prove the claim by induction on i, the case $i = 1$ being trivial since $\Gamma_1 G = G$. Thus suppose that G_i is contained in the subgroup of G generated by H and $\Gamma_i G$. Let $g \in G_{i+1}$. By definition, there are elements $h \in H$ and $k \in [G_i, G_i]$ such that $g = hk$. To show that g is in the subgroup generated by H and $\Gamma_{i+1} G$, it suffices to show that $[G_i, G_i]$ is contained in that subgroup. An element in $[G_i, G_i]$ is the product of elements in $[H, H]$, $[H, \Gamma_i G]$, and $[\Gamma_i G, \Gamma_i G]$. Since $\Gamma_{i+1} G = [G, \Gamma_i G]$, such elements are all in H or in $\Gamma_{i+1} G$. $\square$

13.4. The tensor product of a space and a ring

We describe an old idea in a general way. While the idea deserves further exploration, we use it only to establish an appropriate context for understanding a functor that we will use in our discussion of the complete genus of a space. We saw in §12.4 that the lack of a functorial formal completion of nilpotent groups impeded the naive analysis of the complete genus of f-nilpotent groups. This section and the next will solve the analogous problem for (simple) spaces.

The idea is to form the tensor product of a space X and a ring R to obtain a space "$X \otimes R$" such that $\pi_*(X \otimes R)$ is naturally isomorphic to $\pi_*(X) \otimes R$. Remember that our default is that $\otimes$ means $\otimes_{\mathbb{Z}}$. Of course, some restrictions must be placed on X and R. Since we would not want to try to understand R-nilpotent groups in this generality, we insist that fundamental groups be abelian. We could allow nontrivial actions of the fundamental group on higher homotopy groups, but we prefer to forego that complication. Therefore we restrict to simple spaces X.

Although not essential to the general idea, we also assume that the homotopy groups of X, or equivalently the integral homology groups of X, are finitely generated over $\mathbb{Z}_T$ for some set of primes T. In the rest of this chapter, we agree to say that a simple T-local space with this property is T-local of finite type. Our main interest is the set of all primes, when X is simple and of finite type, and the empty set of primes, when X is simple and rational of finite type. Since we want tensoring over R to be an exact functor on abelian groups, we also insist that the underlying abelian group of R be torsion free.

We have already seen two examples of the tensor product of a space and a ring. For any set of primes $S \subset T$, the localization of X at S can be thought of as $X \otimes \mathbb{Z}_S$ and the completion of X at S can be thought of as $X \otimes \hat{\mathbb{Z}}_S$. We didn't need the finite type hypothesis to conclude that $\pi_*(X_S) \cong \pi_*(X) \otimes \mathbb{Z}_S$, but we did need it to conclude that $\pi_*(\hat{X}_S) \cong \pi_*(X) \otimes \hat{\mathbb{Z}}_S$. In both cases, with our $X \otimes R$ notation, we started with $K(A, n) \otimes R = K(A \otimes R, n)$ and inducted up the Postnikov tower of X to obtain the construction. We explain how the same construction goes in our more general situation. When we localized or completed k-invariants, we exploited universal properties special to those contexts. In general, we need some substitute to allow us to tensor k-invariants with R.

Although we shall not make explicit use of it, we observe that our finite type hypothesis simplifies the task of finding such a substitute. For any finitely generated free $\mathbb{Z}_T$-module F, any $\mathbb{Z}_T$-module B, not necessarily finitely generated, and any abelian group A, the natural map

$$\alpha : \mathrm{Hom}\,(F, B) \otimes A \longrightarrow \mathrm{Hom}\,(F, B \otimes A)$$

specified by $\alpha(\phi \otimes a)(f) = \phi(f) \otimes a$ is an isomorphism. Here again, by default, Hom means $\mathrm{Hom}_{\mathbb{Z}}$, but Hom is the same as $\mathrm{Hom}_{\mathbb{Z}_T}$ when its arguments are T-local. Therefore our claimed isomorphism is a tautology when F is free on one generator, and it follows by induction when F is free on a finite number of generators. Applying this to the cellular cochains of X with coefficients in B and passing to homology gives the following result.

LEMMA 13.4.1. *If X is $f\mathbb{Z}_T$-nilpotent, B is a T-local abelian group, and A is any abelian group, then the canonical map*

$$\alpha : H^*(X; B) \otimes A \longrightarrow H^*(X; B \otimes A)$$

is an isomorphism of graded T-local abelian groups.

Now restrict to $A = R$. The unit $\mathbb{Z} \longrightarrow R$ induces a natural homomorphism $\nu : B \longrightarrow B \otimes R$ for T-local abelian groups B. Let $X_1 \otimes R = K(\pi_1(X) \otimes R, 1)$. Inducting up the Postnikov tower of X, we try to construct $X_{n+1} \otimes R$ as the fiber of an induced k-invariant k_R^{n+2} in the following diagram, where $B = \pi_{n+1}(X)$.

$$
\begin{array}{ccccccc}
K(B, n+1) & \xrightarrow{\iota} & X_{n+1} & \xrightarrow{\pi} & X_n & \xrightarrow{k^{n+2}} & K(B, n+2) \\
\downarrow{\nu} & & \downarrow{\phi_{n+1}} & & \downarrow{\phi_n} & & \downarrow{\nu} \\
K(B \otimes R, n+1) & \xrightarrow{\iota} & X_{n+1} \otimes R & \xrightarrow{\pi} & X_n \otimes R & \xrightarrow{k_R^{n+2}} & K(B \otimes R, n+2)
\end{array}
$$

We need a class $k_R^{n+2} \in H^{n+2}(X_n \otimes R; \pi_{n+1}(X) \otimes R)$ such that $\phi_n^*(k_R^{n+2}) = v_*(k^{n+2})$.

The difficulty is that, in this generality, the cohomology of the Eilenberg-Mac Lane spaces $K(B \otimes R, n)$ can be quite badly behaved. Thus we may not have enough cohomological control to start and continue the induction. If we can do so, then cocellular approximation of maps gives the functoriality of the construction. We urge the interested reader to follow up and determine conditions under which the construction can be completed. The cases $R = \mathbb{R}$ and $R = \mathbb{C}$ would be of particular interest. When T is the empty set, the realification of rational spaces case has been studied using the algebraization of rational homotopy theory [23, 35].

However, what is relevant to our work is that if one can construct a tensor product functor F by some other means, then it must in fact be constructible in the fashion just outlined. To be precise about this, suppose that we have a functor

$$F: \mathrm{Ho}_{fT}^s \mathcal{T} \longrightarrow \mathrm{Ho}_T \mathcal{T},$$

where $\mathrm{Ho}_T^s \mathcal{T}$ is the homotopy category of simple T-local spaces and $\mathrm{Ho}_{f\,T}^s \mathcal{T}$ is its full subcategory of spaces of finite T-type. Let $I: \mathrm{Ho}_{f\,T}^s \mathcal{T} \longrightarrow \mathrm{Ho}_T \mathcal{T}$ be the inclusion and suppose that we have a natural transformation $\eta: I \to F$. Suppose finally that the functor $\pi_n F$ takes values in the category of R-modules and therefore of $\mathbb{Z}_T \otimes R$-modules and that we have a natural isomorphism

$$\xi_n: \pi_n FX \longrightarrow \pi_n(X) \otimes R$$

of $\mathbb{Z}_T \otimes R$-modules such that the following diagram of $\mathbb{Z}_T$-modules commutes.

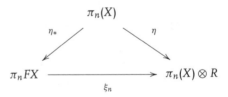

Since we are working up to homotopy, we may write $FK(B, n) = K(B \otimes R, n)$, where B is a finitely generated $\mathbb{Z}_T$-module and $n \geq 1$. Using ξ_n, we may then identify $\eta: K(B, n) \longrightarrow K(B \otimes R, n)$ with the map induced by $\eta: B \longrightarrow B \otimes R$. Inductively, we can identify $F(X_n)$ with the nth stage $(FX)_n$ of a Postnikov tower for FX. To see this consider the following diagram, where again $B = \pi_{n+1}(X)$.

$$
\begin{array}{ccccccc}
K(B, n+1) & \xrightarrow{\ \iota\ } & X_{n+1} & \xrightarrow{\ \pi\ } & X_n & \xrightarrow{\ k^{n+2}\ } & K(B, n+2) \\
\downarrow{\scriptstyle\eta} & & \downarrow{\scriptstyle\eta} & & \downarrow{\scriptstyle\eta} & & \downarrow{\scriptstyle\eta} \\
FK(B, n+1) & \xrightarrow{\ F\iota\ } & F(X_{n+1}) & \xrightarrow{\ F\pi\ } & F(X_n) & \xrightarrow{\ Fk^{n+2}\ } & FK(B, n+2) \\
\| & & \downarrow{\scriptstyle\mu_{n+1}} & & \downarrow{\scriptstyle\mu_n} & & \| \\
K(B \otimes R, n+1) & \xrightarrow{\ \iota\ } & (FX)_{n+1} & \xrightarrow{\ \pi\ } & (FX)_n & \xrightarrow{\ k_R^{n+2}\ } & K(B \otimes R, n+2)
\end{array}
$$

The top three squares commute by the naturality of η. Suppose inductively that we have an equivalence μ_n. Define k_R^{n+2} to be the composite $Fk^{n+2} \circ \mu^{-1}$ for a chosen homotopy inverse μ^{-1}. Since $Fk^{n+2} \circ F\pi \simeq *$, there is a map μ_{n+1} such that $\pi \circ \mu_{n+1} \simeq \mu_n \circ F\pi$. Since $(-) \otimes R$ is an exact functor, a five lemma argument shows that μ_{n+1} is an equivalence. Arguing the same way, we obtain an equivalence $\nu \colon FK(B, n+1) \longrightarrow K(B \otimes R, n+1)$ such that $\iota \circ \nu \simeq \mu_{n+1} \circ F\iota$. But we may as well replace the bottom arrow ι by $\iota \circ \nu$ since fibration sequences are defined up to equivalence in the homotopy category. Then the diagram commutes, as desired.

We conclude that the construction of a tensor product functor F implicitly constructs the required k-invariants k_R^{n+2}.

13.5. Sullivan's formal completion

We take $R = \hat{\mathbb{Z}} = \prod_p \hat{\mathbb{Z}}_p$ and construct a tensor product functor on the category of simple spaces equivalent to CW complexes with countably many cells, following Sullivan [133, 135]. Of course, it suffices to define the functor on countable CW complexes since we are working in homotopy categories. Since T-local spheres are constructed as countable cell complexes, any simply connected T-local space of finite type for any set of primes T will be in the domain of the construction, but since we are not insisting on T-local cells the domain also includes all simple T-local spaces of finite T-type.

DEFINITION 13.5.1. Let X be a countable CW complex. Choose a cofinal sequence of finite subcomplexes X_i and define the formal completion FX to be the telescope of the completions $\hat{X}_i$. Passage to telescopes from the completion maps $\hat{\phi} \colon X_i \longrightarrow \hat{X}_i$ induces a map $\eta \colon X \simeq \mathrm{tel}\, X_i \longrightarrow FX$. If $f \colon X \longrightarrow Y$ is a cellular map between countable CW complexes, choose a cofinal sequence $\mu(i)$ such that $f(X_i) \subset Y_{\mu(i)}$ and define $Ff \colon FX \longrightarrow FY$ by passage to telescopes from the completions $\hat{X}_i \longrightarrow \hat{Y}_{\mu(i)}$ of the restrictions of f.

It is tedious but elementary to check that different choices of cofinal sequences lead to equivalent spaces and maps. The point is that the telescopes depend up to equivalence only on choices of cofinal sequences.

PROPOSITION 13.5.2. *Let $Ho^s_{ct}\mathscr{T}$ be the full subcategory of $Ho^s\mathscr{T}$ whose objects are the simple spaces that are equivalent to countable CW complexes and let $I\colon Ho^s_{ct}\mathscr{T} \longrightarrow Ho^s\mathscr{T}$ be the inclusion. Then Definition 13.5.1 gives a functor $F\colon Ho^s_{ct}\mathscr{T} \longrightarrow Ho^s\mathscr{T}$ and a natural transformation $\eta\colon I \longrightarrow F$. The composite functor $\pi_n F$ takes values in $\hat{\mathbb{Z}}$-modules and there are natural isomorphisms*

$$\xi_n\colon \pi_n FX \longrightarrow \pi_n(X) \otimes \hat{\mathbb{Z}}$$

of $\hat{\mathbb{Z}}$-modules such that the following diagram of abelian groups commutes.

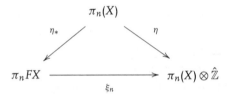

PROOF. It is routine to check that F is a functor and η is natural. It is part of Theorem 11.1.2 that the homotopy groups $\pi_n(\hat{X}_i)$ are $\hat{\mathbb{Z}}$-modules in a natural way and that they are naturally isomorphic under $\pi_n(X_i)$ to $\pi_n(X_i) \otimes \hat{\mathbb{Z}}$. Since the homotopy groups of telescopes are the colimits of the homotopy groups of their terms and tensor products commute with colimits, the last statement follows. $\square$

In effect, at least for countable complexes (and use of more general colimits could eliminate the countability restriction), we now have two different extensions of the restriction of the completion functor on finite complexes to a functor defined on more general spaces, namely the original completion functor $\hat{X}$ of Chapter 10 and the formal completion FX. The latter does not satisfy a universal property, but the previous section gives it a general conceptual home. Restricting F to T-local spaces, it gives a tensor product functor of the sort discussed there for every set of primes T. We are particularly interested in the set of all primes. Completion is trivial when restricted to rational spaces, but the functor

$$F\colon Ho^s_{f\mathbb{Q}}\mathscr{T} \longrightarrow Ho_T\mathscr{T}$$

on the homotopy category of simple rational spaces of finite type fits into the following result, in which the fact that F is already defined on all simple spaces of finite type plays a central role.

THEOREM 13.5.3. *Let X be a simple space of finite type with rationalization $\phi_0 \colon X \longrightarrow X_0$. Then the map $\eta \colon X \longrightarrow FX$ is a completion of X and the map $F(\phi_0) \colon FX \longrightarrow F(X_0)$ is a rationalization of FX. Therefore the naturality diagram*

$$
\begin{array}{ccc}
X & \xrightarrow{\ \eta\ } & FX \\
\phi_0 \downarrow & & \downarrow F(\phi_0) \\
X_0 & \xrightarrow[\ \eta\ ]{} & F(X_0)
\end{array}
$$

can be identified naturally with the homotopy pullback diagram

$$
\begin{array}{ccc}
X & \xrightarrow{\ \hat{\phi}\ } & FX \\
\phi_0 \downarrow & & \downarrow \phi_0 \\
X_0 & \xrightarrow[\ (\hat{\phi})_0\ ]{} & (FX)_0.
\end{array}
$$

In particular, $F(X_0)$ and $(FX)_0$ may be identified.

PROOF. The first statement means that $\eta \colon X \longrightarrow FX$ satisfies the universal property that characterizes $\eta \colon X \longrightarrow \hat{X}$. The last statement is intuitively obvious. One composite first tensors with $\mathbb{Q}$ and then tensors with $\hat{\mathbb{Z}}$, while the other first tensors with $\hat{\mathbb{Z}}$ and then tensors with $\mathbb{Q}$. Thus both composites amount to tensoring with $\hat{\mathbb{Q}}$. Thinking cocellularly and using the cocellular description of F given in the previous section, we see by induction up the Postnikov tower of X that the two composites are tensor products of X with $\hat{\mathbb{Q}}$ and can be specified by the same k-invariants.

We work with the cellular definition to give a formal proof. Remember that localizations and completions are defined by universal properties in the homotopy category and so are not uniquely specified. Choose a cofinal sequence of finite subcomplexes X_i of X. With any construction of $\hat{X}$ the inclusions $X_i \longrightarrow X$ induce maps $\hat{X}_i \longrightarrow \hat{X}$ under X_i, well defined up to homotopy. These maps induce a map $\alpha \colon \operatorname{tel} \hat{X}_i \longrightarrow \hat{X}$ under X which on homotopy groups induces the isomorphism $\operatorname{colim} \pi_*(X_i) \otimes \hat{Z} \cong \pi_*(X) \otimes \hat{Z}$. Therefore α is an equivalence under X, hence $\eta\colon X \longrightarrow FX$ satisfies the defining universal property of a completion of X. Similarly, since $F(X_0)$ is rational, with any construction of $(FX)_0$ its universal property gives a map $\beta\colon (FX)_0 \longrightarrow F(X_0)$ under FX that induces the isomorphism $\pi_*(X) \otimes \hat{Z} \otimes \mathbb{Q} \longrightarrow \pi_*(X) \otimes \hat{\mathbb{Q}} \otimes \hat{Z}$ on homotopy groups. Therefore β is an equivalence under FX and $F(\phi_0)$ satisfies the defining universal property of a rationalization of FX. Since the first diagram commutes, its lower arrow η satisfies the defining property of a localization of $\hat{\phi}$ in the second diagram. $\qquad\square$

13.6. Formal completions and the adèlic genus

We have defined the formal completion functor in Definition 13.5.1. We differentiate between *the* formal completion, as defined there, and *a* formal completion as specified in the following definition, which is restricted to rational spaces. Recall the algebraic notion of a formal localization of an $f\mathbb{Q}$-nilpotent π-group from Definition 12.4.2.

DEFINITION 13.6.1. Let Y be an $f\mathbb{Q}$-nilpotent space, Z be an $f\hat{\mathbb{Z}}_T$-nilpotent space, and $\phi_0\colon Z \longrightarrow Z_0$ be a rationalization. A formal completion of Y at T associated to ϕ_0 is a map $\omega\colon Y \longrightarrow Z_0$ with the following properties. Let π be the pullback of $\pi_1(Y)$ and $\pi_1(Z)$ over $\pi_1(Z_0)$.

(i) The homomorphism $\omega_*\colon \pi_1(Y) \longrightarrow \pi_1(Z_0)$ is a formal completion associated to $(\phi_0)_*\colon \pi_1(Z) \longrightarrow \pi_1(Z_0)$.

(ii) For $n \geq 2$, the homomorphism $\omega_*\colon \pi_n(Y) \longrightarrow \pi_n(Z_0)$ of π-groups is a formal completion associated to $(\phi_0)_*\colon \pi_n(Z) \longrightarrow \pi_n(Z_0)$.

When Y, Z, and therefore Z_0 are simple, the only requirement on the map ω is that for each $n \geq 1$, there must be a commutative diagram of $\mathbb{Q}$-modules

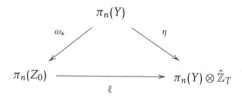

in which ξ is an isomorphism of $\hat{\mathbb{Q}}_T$-modules. No compatibility of ω with the map ϕ_0 is required.

As in algebra, this notion encodes what we see in our cocellular constructions of localizations and completions. From a global to local point of view, we have the following analogue of Lemma 6.2.1.

PROPOSITION 13.6.2. *For any* $T' \supset T$, *the rationalization* $(\hat{\phi})_0$ *of the completion* $\hat{\phi} \colon X \longrightarrow \hat{X}_T$ *of an* $f\mathbb{Z}_{T'}$-*nilpotent space* X *is a formal completion* $X_0 \longrightarrow (\hat{X}_T)_0$.

From a local to global point of view, we have the following addendum to Theorem 13.3.1.

THEOREM 13.6.3. *Let*

$$
\begin{array}{ccc}
P & \xrightarrow{\;\mu\;} & Z \\[2pt]
{\scriptstyle\nu}\big\downarrow & & \big\downarrow{\scriptstyle\phi_0} \\[2pt]
Y & \xrightarrow[\;\omega\;]{} & Z_0
\end{array}
$$

be a homotopy pullback of nilpotent spaces such that

(i) Z *is* $f\hat{\mathbb{Z}}_T$-*complete and the torsion subgroup of each* $\pi_n(Z)$ *is finite;*
(ii) $\phi_0 \colon Z \longrightarrow Z_0$ *is a rationalization of* Z;
(iii) Y *is rational; and*
(iv) ω *is a formal completion associated to* ϕ_0.

Then P *is* $f\hat{\mathbb{Z}}_T$-*nilpotent.*

PROOF. This holds by its algebraic analogue Theorem 12.4.4 and the abelian π-group analogue of that result, which admits essentially the same proof. □

REMARK 13.6.4. In view of Remark 12.4.5, some such hypotheses as in Theorem 13.6.3 are needed to construct global spaces of finite type from $f\hat{\mathbb{Z}}_T$-spaces. There are $f\hat{\mathbb{Z}}_p$-spaces Z such that there is no $f\mathbb{Z}_{(p)}$-space P whose completion at p is equivalent to Z.

We now specialize T to be the set of all primes. We again have two alternative notions of genus, in analogy with Definition 12.4.6.

DEFINITION 13.6.5. The adèlic genus of an f-nilpotent space X is the set of homotopy types of f-nilpotent spaces Y such that X_0 is equivalent to Y_0 and $\hat{X}_p$ is isomorphic to $\hat{Y}_p$ for all primes p. The complete genus of X is defined by dropping the requirement that X_0 be isomorphic to Y_0. Write $\hat{G}_0(X)$ for the adèlic genus of X and $\hat{G}(X)$ for the complete genus. Recall that $G(X)$ denotes the (local) genus of X. If two spaces are in the same local genus, they are in the same adèlic genus, and if two spaces are in the same adèlic genus they are in the same complete genus. Therefore we have inclusions

$$G(X) \subset \hat{G}_0(X) \subset \hat{G}(X).$$

We can carry out the naive analysis of the adèlic genus exactly as in the algebraic analogue discussed in §12.4. Fixing a rationalization $\phi_0 \colon \hat{X} \longrightarrow (\hat{X})_0$ the results above imply that all elements of $\hat{G}_0(X)$ can be constructed as homotopy pullbacks of formal completions $\omega \colon X \longrightarrow (\hat{X})_0$ associated to ϕ_0.

However, we can obtain a precise analysis by restricting to simple spaces, which we do henceforward. Thus we now require all spaces in a given genus to be simple. When X and therefore X_0 is simple, the argument of §13.4 shows that Sullivan's formal completion $\eta \colon X_0 \longrightarrow F(X_0)$ is a formal completion of X_0 in the sense of Definition 13.6.1. Theorem 13.5.3 leads to an easy proof of the following description of the adèlic genus $G(X)$. We agree to write FX instead of $\hat{X}$ in what follows, regarding X as the homotopy pullback displayed in the upper diagram of Theorem 13.5.3, and we agree to write FX_0 for both $F(X_0)$ and $(FX)_0$ since Theorem 13.5.3 shows that they can be identified.

Let $hAut(FX_0)$ denote the group of self-homotopy equivalences via $\hat{\mathbb{Q}}$-maps (equivalently, $\hat{\mathbb{Z}}$-maps) of FX_0, let $hAut(FX)$ denote the group of self-homotopy equivalences via $\hat{\mathbb{Z}}$-maps of FX, and let $hAut(X_0)$ denote the group of self-homotopy equivalences of X_0. Since F is a functor and η is natural, we have a homomorphism

$$F \colon hAut(X_0) \longrightarrow hAut(FX_0)$$

such that $(F\varepsilon) \circ \eta \simeq \eta \circ \varepsilon$. Similarly, by the universal property of localization we have a homomorphism $hAut(FX) \longrightarrow hAut(FX_0)$, denoted $\zeta \mapsto \tilde{\zeta}$, such that $\tilde{\zeta} \circ \phi_0 \simeq \phi_0 \circ \zeta$, where $\phi_0 = F\phi_0 \colon FX \longrightarrow FX_0$.

THEOREM 13.6.6. *For a simple space X of finite type, there is a canonical bijection between $\hat{G}_0(X)$ and the set of double cosets*

$$hAut(X_0) \backslash hAut(FX_0) / hAut(FX).$$

PROOF. We define a function Ψ from the set of double cosets to $\hat{G}_0(X)$ by sending $\xi \in hAut(FX_0)$ to the homotopy pullback of the diagram

$$X_0 \xrightarrow{\ \eta\ } FX_0 \xrightarrow{\ \xi\ } FX_0 \xleftarrow{\ \phi_0\ } FX.$$

A little diagram chasing shows that varying ξ within its double coset gives equivalent pullback diagrams and hence equivalent homotopy pullbacks.

If $\Psi(\xi)$ and $\Psi(\theta)$ are equivalent, say by an equivalence γ, then we obtain equivalences γ_0 and $F\gamma$, unique up to homotopy, such that the left and top squares are homotopy commutative in the diagram

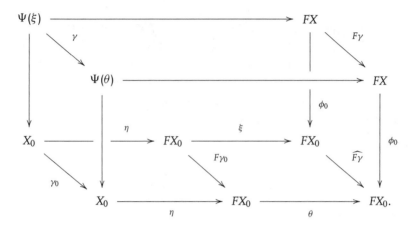

As above, these induce equivalences $F\gamma_0$ and $\widehat{F\gamma}$ such that the bottom left square and the right square are homotopy commutative. It follows that θ is homotopic to $\widehat{F\gamma} \circ \xi \circ (F\gamma_0)^{-1}$ and is thus in the same double coset as ξ. Therefore Ψ is injective.

We can construct any simple space Y of finite type as the homotopy pullback of the maps

$$Y_0 \xrightarrow{\ \eta\ } FY_0 \xleftarrow{\ \phi_0\ } FY.$$

Suppose that Y is in the same genus as X. Then we have equivalences $\alpha\colon Y_0 \longrightarrow X_0$ and $\beta\colon FY \longrightarrow FX$. Define $\xi = \tilde{\beta} \circ (F\alpha)^{-1}$. Another little diagram chase shows that Y is equivalent to $\Psi(\xi)$. Therefore Ψ is surjective. $\square$

Recall that the function space $F(X, Y)$ is nilpotent when X is finite and Y is of finite type. Taking $Y = X$, this suggests that the automorphism groups above should be nilpotent, or nearly so, when X is finite and that their analysis should be closely related to the algebraic analysis of the genus of a finitely generated nilpotent group (see Remark 12.4.9). This idea was worked out in detail by Wilkerson [144], who proved the following theorem.

THEOREM 13.6.7. *If X is a simply connected finite CW complex, then $\hat{G}(X)$ and therefore $\hat{G}_0(X)$ and $G(X)$ are finite sets.*

It is natural to ask whether these three notions of genus are genuinely different. The question was answered by Belfi and Wilkerson [10], and we merely record their answers. Say that X is π_*-finite or H_*-finite if it has only finitely many nonzero homotopy groups or only finitely many nonzero homology groups, all of them assumed to be finitely generated.

THEOREM 13.6.8. *Let X be an f-nilpotent space. If X_0 admits an H-space structure, then $\hat{G}_0(X) = \hat{G}(X)$. If, further, X is either π_*-finite or H_*-finite, then $G(X) = \hat{G}(X)$. However, there are simply connected examples such that $\hat{G}_0(X) \neq \hat{G}(X)$, and these can be chosen to be π_*-finite or H_*-finite. Similarly, there are simply connected examples such that $G(X) \neq \hat{G}_0(X)$, and these too can be chosen to be π_*-finite or H_*-finite.*

PART 4

An introduction to model category theory

14

AN INTRODUCTION TO MODEL CATEGORY THEORY

We now switch gears and consider abstract homotopy theory. Here we adopt a more categorical perspective than earlier in the book. To prove that localizations and completions exist in full generality, we make use of model category theory. Since its use pervades modern algebraic topology and since some aspects of the theory that we believe to be important are not in the existing expository works on the subject, we give a treatment here. Model category theory is due to Quillen [113]. Nice introductions are given in [43, 54], and there are two expository books on the subject [65, 66]. Given these sources and the elementary nature of most of the basic proofs, we leave some of the verifications to the reader. The literature on the subject is huge and growing. Just as Hopf algebras began in algebraic topology and then were seen to be fundamental in other subjects, homotopy theoretic methology, such as model category theory, began in algebraic topology and then was seen to be fundamental in other subjects.

In fact, by now the very term "homotopy theory" admits of two interpretations. There is the homotopy theory of topological spaces, which is the core of algebraic topology, and there is also homotopy theory as a general methodology applicable to many other subjects. In the latter sense, homotopy theory, like category theory, provides a language and a substantial body of results that are applicable throughout mathematics. The two are intertwined, so that there is a subject of categorical homotopy theory (studied by algebraic topologists) and of homotopical category theory (a closely related subject studied by category theorists). Model category theory provides a central organizational principle for this branch of mathematics. To explain properly the ideas that are involved, we outline some categorical concepts that appear wherever categories do and give just a hint of the higher categorical structures that begin to emerge in the study of model categories.

In part to emphasize this categorical perspective, we develop the definition of a model category in its modern conceptual form in §14.1 and §14.2. We focus on weak factorization systems (WFSs), since they are the conceptual key to the efficacy of the definition. We discuss homotopies and the homotopy category of a model category in §14.3 and §14.4.

We deliberately give an uninterrupted development of the theory in this and the following two chapters, reserving all discussion of examples to the two chapters that follow. The reader is encouraged to skip directly from this chapter to Chapter 17 to begin looking at the examples before seeing the rest of the theory.

14.1. Preliminary definitions and weak factorization systems

Let $\mathcal{M}$ be a category. We insist that categories have sets of morphisms between pairs of objects; category theorists would say that $\mathcal{M}$ is locally small. Similarly, we understand limits and colimits to be small, meaning that they are defined with respect to functors out of small categories $\mathcal{D}$. We assume once and for all that $\mathcal{M}$ is bicomplete. This means that $\mathcal{M}$ is complete (has all limits) and cocomplete (has all colimits). In particular, it has an initial object $\emptyset$ and a terminal object $*$ (the coproduct and product of the empty set of objects respectively).

A model structure on $\mathcal{M}$ consists of three interrelated classes of maps $(\mathcal{W}, \mathcal{C}, \mathcal{F})$, called the weak equivalences, the cofibrations, and the fibrations. The weak equivalences are the most important, since the axioms are designed to lead to a well-behaved homotopy category $\mathrm{Ho}\mathcal{M}$ that is obtained by inverting the weak equivalences. This is a localization process that is analogous to the localization of rings at multiplicatively closed subsets, and it is characterized by an analogous universal property. Formally, this means that there must be a functor $\gamma : \mathcal{M} \longrightarrow \mathrm{Ho}\mathcal{M}$ such that $\gamma(w)$ is an isomorphism if $w \in \mathcal{W}$ and γ is initial with respect to this property. We shall require in addition that the objects of $\mathrm{Ho}\mathcal{M}$ are the objects of $\mathcal{M}$ and that γ is the identity on objects.[1] That is, if $F : \mathcal{M} \longrightarrow \mathcal{H}$ is any functor such that $F(w)$ is an isomorphism for $w \in \mathcal{W}$, then there is a unique functor $\tilde{F} : \mathrm{Ho}\mathcal{M} \longrightarrow \mathcal{H}$ such that $\tilde{F} \circ \gamma = F$. It follows that $\tilde{F} = F$ on objects. We say that such a functor γ is a localization of $\mathcal{M}$ at $\mathcal{W}$. Since it is defined by a universal property, it is unique up to canonical isomorphism.

---

1. The addition is an inessential convenience. The literature is divided on this point. Some authors do and others do not insist that the objects of $\mathcal{M}$ and $\mathrm{Ho}\mathcal{M}$ coincide.

One might attempt to construct Ho$\mathcal{M}$ by means of words in the morphisms of $\mathcal{M}$ and formal inverses of the morphisms of $\mathcal{W}$, but the result of such a construction is not locally small in general. Moreover, it would be very hard to "do homotopy theory" in a category constructed so naively. The cofibrations and fibrations are extra data that allow one to do homotopy theory, and there can be many model structures on $\mathcal{M}$ with the same weak equivalences. We prefer to build up to the definition of model categories by first isolating its key categorical constituents. The three classes of maps turn out to be subcategories that contain all isomorphisms and are closed under retracts.

DEFINITION 14.1.1. A class $\mathcal{K}$ of maps in $\mathcal{M}$ is closed under retracts if, when given a commutative diagram

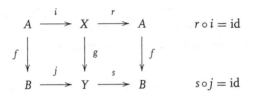

with $g \in \mathcal{K}$, it follows that $f \in \mathcal{K}$. The special case when $i \circ r = \mathrm{id}$ and $j \circ s = \mathrm{id}$ shows that every map isomorphic to a map in $\mathcal{K}$ is also in $\mathcal{K}$. It follows that if all identity maps are in $\mathcal{K}$, then so are all isomorphisms. It is conceptually helpful to think in terms of the arrow category of $\mathcal{M}$, denoted $\mathscr{A}r\mathcal{M}$, whose objects are the maps of $\mathcal{M}$ and whose morphisms $f \longrightarrow g$ are the commutative squares

The following observation is often applied to classical homotopy categories, where it shows that a retract of a homotopy equivalence is a homotopy equivalence.

LEMMA 14.1.2. *In any category, if f is a retract of g and g is an isomorphism, then f is an isomorphism.*

PROOF. Adopting the notations of the diagram in Definition 14.1.1, define $f^{-1} = rg^{-1}j$. Then f^{-1} is the inverse isomorphism of f. $\qquad\square$

REMARK 14.1.3. Observe that for any maps $f: X \longrightarrow Y$ and $g: Y \longrightarrow X$ such that $fg = \mathrm{id}$, f is tautologically a retract of gf via the diagram

$$
\begin{array}{ccccc}
X & = & X & = & X \\
f \downarrow & & \downarrow gf & & \downarrow f \\
Y & \xrightarrow{g} & X & \xrightarrow{f} & Y.
\end{array}
$$

DEFINITION 14.1.4. A subcategory $\mathcal{W}$ of $\mathcal{M}$ qualifies as a subcategory of weak equivalences if all identity maps in $\mathcal{M}$ are in $\mathcal{W}$, $\mathcal{W}$ is closed under retracts, and $\mathcal{W}$ satisfies the two out of three property: for a composite $w = v \circ u$, if any two of u, v, and w are in $\mathcal{W}$, then so is the third.

The original source of the following basic idea goes back to Lemma 1.1.1.

DEFINITION 14.1.5. Consider commutative squares

$$
\begin{array}{ccc}
A & \xrightarrow{g} & E \\
i \downarrow & {\scriptstyle \lambda} \nearrow & \downarrow p \\
X & \xrightarrow{f} & B.
\end{array}
$$

Say that (i, p) has the lifting property if for every such square there is a lift λ making the triangles commute. For a class of maps $\mathcal{L}$, we say that p satisfies the right lifting property (RLP) with respect to $\mathcal{L}$ if (i, p) has the lifting property for every $i \in \mathcal{L}$; we let $\mathcal{L}^{\boxempty}$ denote the class of all such maps p. Dually, for a class of maps $\mathcal{R}$, we say that i satisfies the left lifting property (LLP) with respect to $\mathcal{R}$ if (i, p) has the lifting property for every $p \in \mathcal{R}$; we let $^{\boxempty}\mathcal{R}$ denote the class of all such maps i. We write $\mathcal{L} \boxempty \mathcal{R}$ if (i, p) has the lifting property whenever $i \in \mathcal{L}$ and $p \in \mathcal{R}$. This means that $\mathcal{L} \subset {}^{\boxempty}\mathcal{R}$ or, equivalently, $\mathcal{R} \subset \mathcal{L}^{\boxempty}$.

Of course, these last inclusions can be proper. When they are equalities, the resulting classes have some very useful properties, which are catalogued in the following definitions and result. Here we mention transfinite colimits for the first time since §2.5. We shall make little use of them until §19.3, before which we only need sequential colimits. Transfinite colimits play a substantial role in the foundational literature of model category theory, and they are crucial to the construction of Bousfield localizations, but they play little if any direct role in the calculational applications.

DEFINITION 14.1.6. Let λ be a (nonempty) ordinal. We may regard λ as an ordered set and hence as a small category. A λ-sequence is a colimit-preserving functor $X_* \colon \lambda \longrightarrow \mathcal{M}$, so that $X_\beta \cong \mathrm{colim}_{\alpha < \beta} X_\alpha$ for $\beta < \lambda$. The transfinite composite of X is the induced map

$$X_0 \longrightarrow \mathrm{colim}_{\alpha < \lambda} X_\alpha.$$

When λ is a regular cardinal (see Definition 2.5.2) viewed as a category, it is clearly a λ-filtered category in the sense of Definition 2.5.3, and it has all colimits. The role of the colimit condition is illustrated in Proposition 2.5.4.

DEFINITION 14.1.7. Let $\mathcal{L}$ be a class of maps in $\mathcal{M}$. We say that $\mathcal{L}$ is left saturated if the following (redundant) closure properties hold.

(i) $\mathcal{L}$ is a subcategory of $\mathcal{M}$ that contains all isomorphisms.
(ii) $\mathcal{L}$ is closed under retracts.
(iii) Any coproduct of maps in $\mathcal{L}$ is in $\mathcal{L}$.
(iv) Any pushout of a map in $\mathcal{L}$ is in $\mathcal{L}$; that is, if the following diagram is a pushout and i is in $\mathcal{L}$, then j is in $\mathcal{L}$.

(v) If λ is an ordinal and X is a λ-sequence such that $X_\alpha \longrightarrow X_{\alpha+1}$ is in $\mathcal{L}$ for all $\alpha + 1 < \lambda$, then the transfinite composite of X is in $\mathcal{L}$.

The dual properties specify the notion of a right saturated class of maps. Here coproducts, pushouts, and transfinite composites must be replaced by products, pullbacks, and transfinite sequential limits.

PROPOSITION 14.1.8. *Let $\mathcal{K}$ be any class of maps in $\mathcal{M}$. Then $^{\square}\mathcal{K}$ is left saturated and $\mathcal{K}^{\square}$ is right saturated.*

PROOF. The reader is urged to carry out this categorical exercise. For (i), successive lifts show that a composite of maps in $^{\square}\mathcal{K}$ is in $^{\square}\mathcal{K}$, so that $^{\square}\mathcal{K}$ is a subcategory, and it is obvious from the definition of the lifting property that isomorphisms are in $^{\square}\mathcal{K}$. For the other parts, given a lifting problem for the relevant categorical colimit, the hypothesis gives lifts in induced lifting problems that by the universal property of the relevant colimit fit together to give a solution of the original lifting problem. It is instructive to understand why the conclusion does *not* imply that $^{\square}\mathcal{K}$ is closed under all colimits in $\mathscr{A}r\mathscr{M}$. For example, (iii) and (iv) do not imply that a pushout in $\mathscr{A}r\mathscr{M}$ of maps in $^{\square}\mathcal{K}$ is again in $^{\square}\mathcal{K}$; compare Exercise 1.1.2. □

We often use the result above together with the following evident inclusions.

LEMMA 14.1.9. *For any class* $\mathcal{K}$, $\mathcal{K} \subset {}^{\square}(\mathcal{K}^{\square})$. *If* $\mathcal{J} \subset \mathcal{K}$, *then* $\mathcal{J}^{\square} \supset \mathcal{K}^{\square}$.

DEFINITION 14.1.10. An ordered pair $(\mathscr{L}, \mathscr{R})$ of classes of morphisms of $\mathscr{M}$ factors $\mathscr{M}$ if every morphism $f : X \longrightarrow Y$ factors as a composite

$$X \xrightarrow{\ i(f)\ } Z(f) \xrightarrow{\ p(f)\ } Y$$

with $i(f) \in \mathscr{L}$ and $p(f) \in \mathscr{R}$. Let dom and cod denote the domain and codomain functors $\mathscr{A}r\mathscr{M} \longrightarrow \mathscr{M}$. The factorization is functorial[2] if i and p are functors $\mathscr{A}r\mathscr{M} \longrightarrow \mathscr{A}r\mathscr{M}$ such that

$$\text{dom} \circ i = \text{dom}, \quad \text{cod} \circ i = \text{dom} \circ p, \quad \text{cod} \circ p = \text{cod}$$

and $p(f) \circ i(f) = f$. Equivalently, a functorial factorization consists of a functor $Z : \mathscr{A}r\mathscr{M} \to \mathscr{M}$ and natural transformations i: dom $\to Z$ and p: $Z \to$ cod such that the composite natural transformation $p \circ i$: dom $\longrightarrow$ cod sends f to f.

Mapping cylinders and mapping path fibrations [93, pp. 43, 48] give the original source for the following idea, but it also arises from analogous categorical contexts.

DEFINITION 14.1.11. A weak factorization system, abbreviated WFS, in $\mathscr{M}$ is an ordered pair $(\mathscr{L}, \mathscr{R})$ of classes of morphisms of $\mathscr{M}$ that factors $\mathscr{M}$ and satisfies both

$$\mathscr{L} = {}^{\square}\mathscr{R} \quad \text{and} \quad \mathscr{R} = \mathscr{L}^{\square}.$$

The required equalities say that the maps in $\mathscr{L}$ are precisely the maps that have the LLP with respect to the maps in $\mathscr{R}$ and the maps in $\mathscr{R}$ are precisely the maps that have the RLP with respect to the maps in $\mathscr{L}$. A WFS is functorial if the factorization is functorial.

---

2. The definition given in some standard sources is not correct.

Category theorists also study strong factorization systems, for which the relevant lifts λ are required to be unique.[3] The difference is analogous to the difference between the class of fibrations and the class of covering maps as the choice of $\mathscr{R}$. Our focus on weak rather than strong factorization systems is illustrative of the difference of focus between homotopical categorical theory and classical category theory. There is an equivalent form of the definition of a WFS that is generally used in the definition of model categories. We give it in Proposition 14.1.13 below, using the following observation to prove the equivalence.

LEMMA 14.1.12 (THE RETRACT ARGUMENT). *Let $f = q \circ j \colon A \longrightarrow B$ be a factorization through an object Y. If f has the LLP with respect to q, then f is a retract of j. Dually, if f has the RLP with respect to j, then f is a retract of q.*

PROOF. A lift k in the square

$$\begin{array}{ccc} A & \xrightarrow{\ j\ } & Y \\ f\downarrow & \nearrow^{k} & \downarrow q \\ B & =\!=\!= & B \end{array}$$

gives a retraction

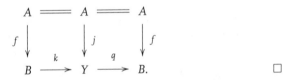

$$\begin{array}{ccccc} A & =\!=\!= & A & =\!=\!= & A \\ f\downarrow & & j\downarrow & & \downarrow f \\ B & \xrightarrow{\ k\ } & Y & \xrightarrow{\ q\ } & B. \end{array}$$

$\square$

PROPOSITION 14.1.13. *Let $(\mathscr{L}, \mathscr{R})$ factor $\mathscr{M}$. Then $(\mathscr{L}, \mathscr{R})$ is a WFS if and only if $\mathscr{L} \boxtimes \mathscr{R}$ and $\mathscr{L}$ and $\mathscr{R}$ are closed under retracts.*

PROOF. If $(\mathscr{L}, \mathscr{R})$ is a WFS, then certainly $\mathscr{L} \boxtimes \mathscr{R}$. Suppose that f is a retract of g, where $g \in \mathscr{L}$. Let $p \in \mathscr{R}$ and assume that ℓ and k make the right-hand square commute in the following diagram. We must find a lift in the right-hand square, and there is a lift λ as drawn since $g \in \mathscr{L}$.

---

3. They write $(\mathscr{E}, \mathscr{M})$, thinking of these as classes $\mathscr{E}$ of epimorphisms and $\mathscr{M}$ of monomorphisms [11, §5.5]. Reversing the order, in many categories $(\mathscr{M}, \mathscr{E})$ is a weak but not a strong factorization system. We think of cofibrations as analogous to monomorphisms and fibrations as analogous to epimorphisms.

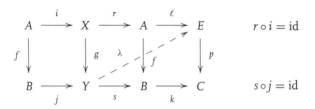

The composite $\lambda \circ j$ gives a lift in the right-hand square. Therefore $\mathcal{L}$ is closed under retracts. Dually, $\mathcal{R}$ is closed under retracts. For the converse, we have $\mathcal{L} \subset {}^{\boxtimes}\mathcal{R}$ and $\mathcal{R} \subset \mathcal{L}^{\boxtimes}$ and must show that equality holds. Let $f \in {}^{\boxtimes}\mathcal{R}$. Factor f as $q \circ j$ with $j \in \mathcal{L}$ and $q \in \mathcal{R}$. By Lemma 14.1.12, f is a retract of j and is thus in $\mathcal{L}$. Dually, $\mathcal{R} = \mathcal{L}^{\boxtimes}$. Therefore $(\mathcal{L}, \mathcal{R})$ is a WFS. □

14.2. The definition and first properties of model categories

DEFINITION 14.2.1. A model structure on $\mathcal{M}$ consists of classes $(\mathcal{W}, \mathcal{C}, \mathcal{F})$ of morphisms of $\mathcal{M}$, the weak equivalences, cofibrations, and fibrations, such that

(i) $\mathcal{W}$ has the two out of three property.
(ii) $(\mathcal{C}, \mathcal{F} \cap \mathcal{W})$ is a (functorial) weak factorization system.
(iii) $(\mathcal{C} \cap \mathcal{W}, \mathcal{F})$ is a (functorial) weak factorization system.

We emphasize that this is a general definition in which the three classes of maps need not have anything to do with the classes of maps with the same names in the classical topological setting. The parenthetical (functorial) in the definition is a matter of choice, depending on taste and convenience. Quillen's original definition did not require it, but many more recent sources do. There are interesting model categories for which the factorizations cannot be chosen to be functorial (see, for example, [71]), but they can be so chosen in the examples that are most commonly used. We will not go far enough into the theory for the difference to matter significantly. Observe that the model axioms are self-dual in the sense that the cofibrations and fibrations of $\mathcal{M}$ are the fibrations and cofibrations of a model structure on the opposite category $\mathcal{M}^{\mathrm{op}}$ that has the same weak equivalences. Therefore results about model categories come in dual pairs, of which it suffices to prove only one.

The maps in $\mathcal{F} \cap \mathcal{W}$ are called acyclic (or trivial) fibrations; those in $\mathcal{C} \cap \mathcal{W}$ are called acyclic (or trivial) cofibrations. The definition requires every map to factor both as the composite of a cofibration followed by an acyclic fibration and as an acyclic cofibration followed by a fibration; we will say a little more about that in Remark 14.2.6 below. It also requires there to be a lift in any commutative square

$$
\begin{array}{ccc}
A & \longrightarrow & E \\
i \downarrow & \nearrow & \downarrow p \\
X & \longrightarrow & B
\end{array}
$$

in which i is a cofibration, p is a fibration, and either i or p is acyclic. More precisely, it requires the following two pairs of equalities.

14.2.2 $\qquad \mathscr{C} = {}^{\boxtimes}(\mathscr{F} \cap \mathscr{W}) \quad \text{and} \quad \mathscr{F} \cap \mathscr{W} = \mathscr{C}^{\boxtimes}$

14.2.3 $\qquad \mathscr{F} = (\mathscr{C} \cap \mathscr{W})^{\boxtimes} \quad \text{and} \quad \mathscr{C} \cap \mathscr{W} = {}^{\boxtimes}\mathscr{F}$

Proposition 14.1.8 shows that $\mathscr{C}$ and $\mathscr{C} \cap \mathscr{W}$ are left saturated and $\mathscr{F}$ and $\mathscr{F} \cap \mathscr{W}$ are right saturated, which is one motivation for our original definition of a WFS.

By (14.2.2) and (14.2.3), to specify a model structure on a category with a chosen class of weak equivalences that satisfies the two out of three property, we need only specify either the cofibrations or the fibrations, not both. Moreover, by Proposition 14.1.13, the equalities (14.2.2) and (14.2.3) are equivalent to the statement that the relevant four classes are closed under retracts and satisfy

14.2.4 $\qquad \mathscr{C}\boxtimes(\mathscr{F} \cap \mathscr{W}) \quad \text{and} \quad \mathscr{F}\boxtimes(\mathscr{C} \cap \mathscr{W}).$

It is usual to define model categories by requiring (14.2.4) and requiring $\mathscr{F}$, $\mathscr{C}$, and $\mathscr{W}$ to be closed under retracts. The following observation (due to Joyal and Tierney[4]) shows that our axioms imply that $\mathscr{W}$ is closed under retracts and are therefore equivalent to the usual ones.

LEMMA 14.2.5. *The class $\mathscr{W}$ as well as the classes $\mathscr{C}$, $\mathscr{C} \cap \mathscr{W}$, $\mathscr{F}$, and $\mathscr{F} \cap \mathscr{W}$ in a model structure are subcategories that contain all isomorphisms and are closed under retracts. Therefore $\mathscr{W}$ is a subcategory of weak equivalences in the sense of Definition 14.1.4.*

PROOF. Proposition 14.1.8 implies that $\mathscr{C}$, $\mathscr{C} \cap \mathscr{W}$, $\mathscr{F}$, $\mathscr{F} \cap \mathscr{W}$ are subcategories that contain all isomorphisms and are closed under retracts. The two out of three property implies that $\mathscr{W}$ is closed under composition, and, together

---

4. It is Proposition 7.8 of A. Joyal and M. Tierney. Quasi-categories vs Segal spaces. Categories in algebra, geometry and mathematical physics, 277–326. Contemp. Math., 431. Amer. Math. Soc., Providence, 2007.

with either factorization property, it also implies that $\mathscr{W}$ coincides with the class of composites $p \circ i$ such that $i \in \mathscr{C} \cap \mathscr{W}$ and $p \in \mathscr{F} \cap \mathscr{W}$. Since all identity maps are in both $\mathscr{C} \cap \mathscr{W}$ and $\mathscr{F} \cap \mathscr{W}$, they are in $\mathscr{W}$. It remains to show that $\mathscr{W}$ is closed under retracts. Suppose given a retract diagram

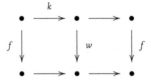

with $w \in \mathscr{W}$. First assume that $f \in \mathscr{F}$ and use either factorization and the two out of three property to factor w as $v \circ u$, where $u \in \mathscr{C} \cap \mathscr{W}$ and $v \in \mathscr{F} \cap \mathscr{W}$. Let $s = u \circ k$ in the following expansion of the previous diagram.

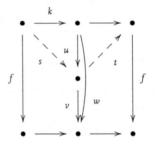

Since $f \in \mathscr{F}$ and $u \in \mathscr{C} \cap \mathscr{W}$ there is a lift t that makes the diagram commute. Then $t \circ s = $ id. Thus f is a retract of v, hence f is in $\mathscr{F} \cap \mathscr{W}$ since v is in $\mathscr{F} \cap \mathscr{W}$.

For the general case, factor f as $p \circ i$ where $i \in \mathscr{C} \cap \mathscr{W}$ and $p \in \mathscr{F}$ and construct the following expansion of our first diagram.

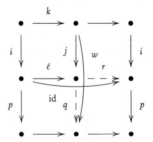

Here the top left square is a pushout and, by three applications of the universal property of pushouts, there is a map r such that the upper right square commutes and $r \circ \ell = $ id, there is a map q such that the lower left square commutes and $q \circ j = w$, and the lower right square commutes. By Proposition 14.1.8,

j is in $\mathscr{C} \cap \mathscr{W}$ since it is a pushout of a map in $\mathscr{C} \cap \mathscr{W}$ and $\mathscr{C} \cap \mathscr{W} = {}^{\boxtimes}\mathscr{F}$. Therefore, q is in $\mathscr{W}$ by the two out of three property. The diagram shows that the fibration p is a retract of q, hence p is in $\mathscr{W}$ by the first part. Since i is in $\mathscr{C} \cap \mathscr{W}$, it follows that $f = p \circ i$ is in $\mathscr{W}$. □

The following evident observation is often used but seldom made explicit.

REMARK 14.2.6. The definition of a model category implies that for any map $f : X \longrightarrow Y$, we have the following commutative solid arrow diagram in which $i(f)$ is an acyclic cofibration, $p(f)$ is a fibration, $j(f)$ is a cofibration, and $q(f)$ is an acyclic fibration. Therefore there is a lift $\xi(f)$ that makes the diagram commute.

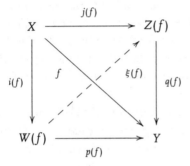

If f is a weak equivalence, then this is a diagram of weak equivalences. When the factorizations can be chosen to be functorial, one can ask whether they can be so chosen that $\xi(f)$ is natural.[5] As we explain in Proposition 15.1.11 and Remark 15.2.4, they often can be so chosen, although they usually are not so chosen.

DEFINITION 14.2.7. An object X of a model category $\mathscr{M}$ is cofibrant if the unique map $\emptyset \longrightarrow X$ is a cofibration. An acyclic fibration $q : QX \longrightarrow X$ in which QX is cofibrant is called a cofibrant approximation or cofibrant replacement[6] of X. We can obtain q by factoring $\emptyset \to X$. Dually, X is fibrant if $X \to *$ is a fibration. An acyclic cofibration $r : X \longrightarrow RX$ in which RX is fibrant is a fibrant approximation or fibrant replacement of X. We can obtain r by factoring

---

5. As far as we know, this question was first raised by Emily Riehl; it is considered in her paper [118], which studies model categories categorically.

6. The words "approximation" and "replacement" are both in common use; we usually use the former when thinking about a single object and the latter when thinking about a functorial construction.

$X \longrightarrow *$. We say that X is bifibrant[7] if it is both cofibrant and fibrant. Let $\mathcal{M}_c$, $\mathcal{M}_f$, and $\mathcal{M}_{cf}$ denote the full subcategories of cofibrant, fibrant, and bifibrant objects of $\mathcal{M}$, respectively.

Cofibrant and fibrant replacements are very important to the theory. Even when they are functorial, they are not unique, and there often are several different cofibrant or fibrant replacement functors with different good properties. Given two cofibrant replacements $q: QX \longrightarrow X$ and $q': Q'X \longrightarrow X$, the lifting property gives a weak equivalence $\xi: QX \longrightarrow Q'X$ such that $q' \circ \xi = q$. The following remark plays a role in the construction of the homotopy category $\text{Ho}\mathcal{M}$.

REMARK 14.2.8. It is central to the theory that we can replace objects by ones that are both fibrant and cofibrant. The two obvious composite ways to do this are weakly equivalent as we see from the following diagram, in which the labeled arrows are weak equivalences. Here the maps Qr and Rq are given if we have functorial factorizations Q and R. If not, we obtain them by applying lifting properties to the acyclic fibration q on the right or the acyclic cofibration r on the left; the unlabeled arrows from $\emptyset$ and to $*$ are included in the diagram to clarify that application of the lifting properties. The difference is an illustrative example of why it is often convenient but usually not essential to include the functoriality of factorizations in the definition of a model category.

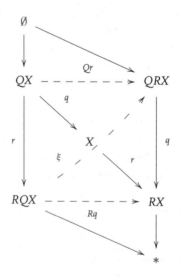

The lifting property gives a weak equivalence ξ that makes the diagram commute.[8]

In many model categories, either all objects are fibrant or all objects are cofibrant (but rarely both). For example, all objects are fibrant in the model structures that we shall define on $\mathscr{U}$, and all objects are cofibrant in the usual model structure on simplicial sets. In such cases, many results and arguments simplify. For example, the following result becomes especially helpful when all objects are cofibrant or all objects are fibrant.

LEMMA 14.2.9 (KEN BROWN'S LEMMA). *Let* $F\colon \mathscr{M} \longrightarrow \mathscr{N}$ *be a functor, where* $\mathscr{M}$ *is a model category and* $\mathscr{N}$ *is a category with a subcategory of weak equivalences. If F takes acyclic cofibrations between cofibrant objects to weak equivalences, then F takes all weak equivalences between cofibrant objects to weak equivalences. Dually, if F takes acyclic fibrations between fibrant objects to weak equivalences, then F takes all weak equivalences between fibrant objects to weak equivalences.*

PROOF. Let $f\colon X \longrightarrow Y$ be a weak equivalence between cofibrant objects of $\mathscr{M}$. The map f and the identity map of Y specify a map $X \amalg Y \longrightarrow Y$, and we factor it as the composite of a cofibration j and an acyclic fibration p to obtain the following commutative diagram in $\mathscr{M}$.

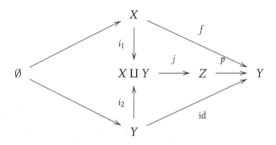

The left square is a pushout, hence i_1 and i_2 are cofibrations, and this implies that $X \amalg Y$ and Z are cofibrant. By the two out of three property in $\mathscr{M}$, ji_1 and ji_2 are weak equivalences and thus acyclic cofibrations between cofibrant objects. By hypothesis, F takes them to weak equivalences. By the two out of three property in $\mathscr{N}$, $F(p)$ is a weak equivalence since $F(p)F(ji_2) = \mathrm{id}$ and $F(f)$ is a weak equivalence since $F(f) = F(p)F(ji_1)$. □

---

8. Similarly to footnote 5, when our factorizations are functorial it is natural to ask whether ξ can be chosen to be natural. The question is answered in [118].

REMARK 14.2.10. The weak equivalences in $\mathcal{N}$ might be the isomorphisms. For example, $\mathcal{N}$ might be the homotopy category of a model category.

14.3. The notion of homotopy in a model category

As will be formalized in Addendum 16.4.10, in most examples there is a familiar and classical notion of a homotopy between maps. It is defined in terms of canonical cylinder and path objects, such as $X \times I$ and Map (I, X) in the case of spaces. Quillen [113] developed a notion of homotopy in general model categories and showed how to derive many familiar results using the model theoretic notion. However, in the examples, it turns out that the classical notion of homotopy suffices to describe the model theoretical notion. We shall make this assertion precise at the end of §16.4. Therefore, when actually working with model categories, one usually ignores the background material on the model theoretical notion of homotopy. For that reason, we just describe how the general theory goes, emphasizing the model theoretic analogue of the Whitehead theorems but leaving some detailed verifications to the reader. We largely follow [43, §4].

We consider a fixed model category $\mathcal{M}$ throughout this section. There are several variant model theoretical notions of cylinder and path objects that abstract the properties of the classical cylinder and path objects.

DEFINITION 14.3.1. A cylinder object for $X \in \mathcal{M}$ is an object Cyl X together with maps $i_0 : X \longrightarrow$ Cyl X, $i_1 : X \longrightarrow$ Cyl X, and $p : $ Cyl $(X) \longrightarrow X$ such that $p \circ i_0 = \mathrm{id} = p \circ i_1$ and p is a weak equivalence; by the two out of three property, i_0 and i_1 are also weak equivalences. A cylinder object is good if the map $i = i_0 + i_1 : X \amalg X \longrightarrow$ Cyl (X) is a cofibration. A good cylinder object Cyl (X) is very good if p is an acyclic fibration. Factorization of the folding map $X \amalg X \longrightarrow X$ shows that every X has at least one very good cylinder object. A left homotopy between maps $f, g : X \longrightarrow Y$ is a map $h : $ Cyl $(X) \longrightarrow Y$ such that $h \circ i_0 = f$ and $h \circ i_1 = g$, where Cyl (X) is any cylinder object for X; h is good or very good if Cyl (X) is good or very good. Define $\pi^\ell(X, Y)$ to be the set of equivalence classes of maps $X \longrightarrow Y$ under the equivalence relation generated by left homotopy.

LEMMA 14.3.2. *If X is cofibrant and Cyl X is a good cylinder object, then i_0 and i_1 are cofibrations and thus acyclic cofibrations.*

PROOF. The inclusions ι_0 and ι_1 of X in $X \amalg X$ are cofibrations since the following pushout diagram displays both of them as pushouts of $\emptyset \longrightarrow X$.

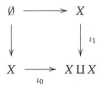

Therefore their composites i_0 and i_1 with i are cofibrations. □

LEMMA 14.3.3. *If h is a left homotopy from f to g and either f or g is a weak equivalence, then so is the other.*

PROOF. By the two out of three property, f is a weak equivalence if and only if h is a weak equivalence, and similarly for g. □

We emphasize that the definition of left homotopy allows the use of any cylinder object and that the notion of left homotopy and its good and very good variants are not equivalence relations in general. Even in some categories with canonical cylinders, such as the category of simplicial sets, homotopy is not an equivalence relation in general. If we were only interested in model category theory, we could restrict attention to very good cylinder objects. However, the canonical cylinder objects in the examples are generally not very good, so we must allow the more general versions in order to make the promised comparisons. For example, in the standard model structure on topological spaces of §17.2, $X \times I$ is a cylinder object, but it is not good unless X is cofibrant, and similarly for categories of chain complexes. We shall return to this point in Addendum 16.4.10.

DEFINITION 14.3.4. Dually, a path object[9] for X is an object $\mathrm{Cocyl}\, X$ together with maps $p_0 \colon \mathrm{Cocyl}\, X \longrightarrow X$, $p_1 \colon \mathrm{Cocyl}\, X \longrightarrow X$, and $i \colon X \longrightarrow \mathrm{Cocyl}\, X$ such that $p_0 \circ i = \mathrm{id} = p_1 \circ i$ and i and hence p_0 and p_1 are weak equivalences. A path object is good if the map $p = (p_0, p_1) \colon \mathrm{Cocyl}\, X \longrightarrow X \times X$ is a fibration. A good path object is very good if i is an acyclic cofibration. Factorization of the diagonal map $X \longrightarrow X \times X$ show that every X has at least one very good path object. There are evident dual definitions of right homotopies, good right homotopies and very good right homotopies. Define $\pi^r(X, Y)$ to be the set of equivalence classes of maps $X \longrightarrow Y$ under the equivalence relation generated by right homotopy.

Of course, the following duals of Lemmas 14.3.2 and 14.3.3 hold.

---

9. The term "cocylinder" is also used and, inconsistently, we use notation that reflects that term.

LEMMA 14.3.5. *If X is fibrant and Cocyl X is a good path object, then p_0 and p_1 are fibrations and thus acyclic fibrations.*

LEMMA 14.3.6. *If h is a right homotopy from f to g and either f or g is a weak equivalence, then so is the other.*

Lemmas 14.3.5 and 14.3.2 have the following immediate consequences.

LEMMA 14.3.7 (**HEP**). *Let i: A $\longrightarrow$ X be a cofibration and Y be a fibrant object. Then i satisfies the right homotopy extension property with respect to Y. That is, for any good path object Cocyl Y and any maps f and h that make the following square commute, there is a lift $\tilde{h}$ that makes the triangles commute.*

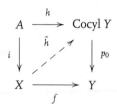

LEMMA 14.3.8 (**CHP**). *Let p: E $\longrightarrow$ B be a fibration and X be a cofibrant object. Then p satisfies the left covering homotopy property with respect to X. That is, for any good cylinder object Cyl X and any maps f and h that make the following square commute, there is a lift $\tilde{h}$ that makes the triangles commute.*

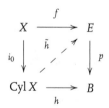

We record several further easily proven observations about these notions.

PROPOSITION 14.3.9. *The notion of left homotopy satisfies the following properties. The notion of right homotopy satisfies the dual properties. Consider maps $f, g: X \longrightarrow Y$ and, for (iii), e: W $\longrightarrow$ X.*

   (i) There is a left homotopy between f and g if and only if there is a good left homotopy between f and g.

   (ii) If Y is fibrant, there is a good left homotopy between f and g if and only if there is a very good left homotopy between f and g.

(iii) If Y is fibrant and f is left homotopic to g, then f ∘ e is left homotopic to g ∘ e.

(iv) If X is cofibrant, then left homotopy is an equivalence relation on $\mathcal{M}(X, Y)$.

PROOF. We give the reader the essential ideas in the following sketch proofs.

(i) Factor $i\colon X \amalg X \longrightarrow \operatorname{Cyl} X$ to obtain a good cylinder $\operatorname{Cyl}' X$ with an acyclic fibration to $\operatorname{Cyl} X$; composition with a homotopy gives a good homotopy.

(ii) Factor the weak equivalence $p\colon \operatorname{Cyl} X \longrightarrow X$ to obtain a very good cylinder $\operatorname{Cyl}' X$ together with an acyclic cofibration $j\colon \operatorname{Cyl} X \longrightarrow \operatorname{Cyl}' X$. Since Y is fibrant, a left homotopy h defined on $\operatorname{Cyl} X$ in the diagram

$$
\begin{array}{ccc}
\operatorname{Cyl} X & \xrightarrow{\ h\ } & Y \\
{\scriptstyle j}\downarrow & \nearrow{\scriptstyle h'} & \downarrow \\
\operatorname{Cyl}' X & \longrightarrow & *
\end{array}
$$

lifts to a left homotopy h' defined on $\operatorname{Cyl}' X$.

(iii) Use a very good cylinder $\operatorname{Cyl} X$ to define a homotopy $h\colon f \simeq g$ and choose a good cylinder $\operatorname{Cyl} W$. Use the lifting property to obtain λ making the following diagram commute. Then $h \circ \lambda$ gives the required homotopy $f \circ e \simeq g \circ e$.

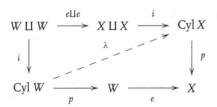

(iv) f is left homotopic to f since X itself gives a cylinder for X. If f is left homotopic to g, use of the interchange map on $X \amalg X$ shows that g is left homotopic to f. For transitivity, observe that the pushout of a pair of good cylinders $\operatorname{Cyl} X$ and $\operatorname{Cyl}' X$ along the cofibrations i_1 and i_0 in the diagram

$$
\operatorname{Cyl} X \xleftarrow{\ i_1\ } X \xrightarrow{\ i_0\ } \operatorname{Cyl}' X
$$

gives another good cylinder $\operatorname{Cyl}'' X$. Given left homotopies $f \simeq g$ and $g \simeq h$ defined on $\operatorname{Cyl} X$ and $\operatorname{Cyl}' X$, use the universal property of pushouts to obtain a homotopy $f \simeq h$ defined on $\operatorname{Cyl}'' X$. $\qquad\square$

COROLLARY 14.3.10. *If Y is fibrant, then composition in $\mathcal{M}$ induces composition*

$$\pi^\ell(X, Y) \times \pi^\ell(W, X) \longrightarrow \pi^\ell(W, Y).$$

If W is cofibrant, then composition in $\mathcal{M}$ induces composition

$$\pi^r(X, Y) \times \pi^r(W, X) \longrightarrow \pi^r(W, Y).$$

PROOF. If $e, e' \colon W \longrightarrow X$ and $f, f' \colon X \longrightarrow Y$ are left homotopic, then $f \circ e$ and $f' \circ e$ are left homotopic by (iii) and $f \circ e$ and $f \circ e'$ are left homotopic by composing a homotopy Cyl $W \longrightarrow X$ with f. This implies the first part. The second is dual. ☐

The previous results give properties of left homotopies and of right homotopies, thought of separately. Perhaps the real force of Quillen's approach to homotopies is the comparison between left and right homotopies.

PROPOSITION 14.3.11. *Consider maps $f, g \colon X \longrightarrow Y$.*

(i) *If X is cofibrant and f is left homotopic to g, then f is right homotopic to g.*
(ii) *If Y is fibrant and f is right homotopic to g, then f is left homotopic to g.*

PROOF. For (i), there is a left homotopy $h \colon \text{Cyl} X \longrightarrow Y$ defined on some good cylinder object Cyl X. Choose any fixed good path object Cocyl Y. Since i_0 is an acyclic cofibration, by Lemma 14.3.2, and p is a fibration, there is a lift λ in the following diagram.

The composite λi_1 is a right homotopy from f to g. The proof of (ii) is dual. ☐

DEFINITION 14.3.12. When X is cofibrant and Y is fibrant, we say that f is homotopic to g, written $f \simeq g$, if f is left or, equivalently, right homotopic to g. We then write $\pi(X, Y)$ for the set of homotopy classes of maps $X \longrightarrow Y$.

The previous proof has the following consequence, which is the key to comparing classical homotopies with model theoretic homotopies.

COROLLARY 14.3.13. *Let X be cofibrant and Y be fibrant. Fix a good cylinder object* Cyl X *and a good path object* Cocyl Y. *If $f \simeq g$, then f is left homotopic to g via a homotopy defined on* Cyl X *and f is right homotopic to g via a homotopy mapping to* Cocyl Y.

The following result can be viewed as giving weak model theoretic analogues of the dual Whitehead theorems. Another variant is given in Theorem 14.4.8 below.

THEOREM 14.3.14. *The following versions of the dual Whitehead theorems hold.*

(i) *If X is cofibrant and $p\colon Z \longrightarrow Y$ is an acyclic fibration, then the function*
$p_*\colon \pi^\ell(X, Z) \longrightarrow \pi^\ell(X, Y)$ *is a bijection.*

(ii) *If Y is fibrant and $i\colon W \longrightarrow X$ is an acyclic cofibration, then the function*
$i^*\colon \pi^r(X, Y) \longrightarrow \pi^r(W, Y)$ *is a bijection.*

PROOF. For a map $f\colon X \longrightarrow Y$ and for a left homotopy h between maps $k, \ell\colon X \longrightarrow Z$ that is defined on a good cylinder Cyl X, lifts in the diagrams

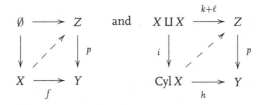

show that p_* is surjective and injective, respectively. Here we have used that $p \circ (k + \ell)$ restricts to pk and $p\ell$ on the two copies of X in $X \amalg X$. $\square$

The topological analogue of (i) does not require a fibration hypothesis and therefore has one of the implications in the following result as a formal consequence [93, pp. 73–74]. In the model theoretical version, neither implication is obvious.

THEOREM 14.3.15. *Let $f\colon X \longrightarrow Y$ be a map between bifibrant objects X and Y. Then f is a homotopy equivalence if and only if f is a weak equivalence.*

PROOF. Factor f as the composite of an acyclic cofibration $i\colon X \longrightarrow Z$ and a fibration $p\colon Z \longrightarrow Y$ and observe that Z is also bifibrant. By Theorem 14.3.14(ii), the functions

$$i^*\colon \pi(Z, X) \longrightarrow \pi(X, X) \quad \text{and} \quad i^*\colon \pi(Z, Z) \longrightarrow \pi(X, Z)$$

are bijections. Choose $j\colon Z \longrightarrow X$ such that $i^*(j) = \mathrm{id}_X$, so that $ji \simeq \mathrm{id}_X$. Then $i^*(ij) = iji \simeq i$ and therefore $ij \simeq \mathrm{id}_Z$. Thus i is a homotopy equivalence with homotopy inverse j. If f is a weak equivalence, then p is an acyclic fibration and the dual argument gives that p is a homotopy equivalence with a homotopy inverse $q\colon Y \longrightarrow Z$. The composite $g = jq$ is then a homotopy inverse to f.

Conversely, assume that f is a homotopy equivalence with homotopy inverse g. Since i is a weak equivalence, it suffices to prove that p is a weak equivalence to deduce that f is a weak equivalence. Let $h\colon \mathrm{Cyl}\,Y \longrightarrow Y$ be a good left homotopy from $fg = pig$ to the identity map of Y. Choose a lift k in the diagram

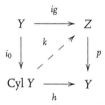

and let $q = ki_1\colon Y \longrightarrow Z$. Then k is a good homotopy from ig to q such that $pq = \mathrm{id}_Y$. Moreover, $p \simeq pij = fj$ and therefore

$$qp \simeq igp \simeq igfj \simeq \mathrm{id}_Z.$$

By Lemma 14.3.3, this implies that qp is a weak equivalence. By Remark 14.1.3, p is a retract of qp since $pq = \mathrm{id}$. Therefore p is also a weak equivalence. □

REMARK 14.3.16. We have used HELP and coHELP in several places, notably §3.3. The first author has long viewed them to be a central organizational convenience in classical homotopy theory. Implicitly and explicitly, we shall again use HELP in developing the q-model structures on spaces in §17.2 and on chain complexes in §18.4. These generalizations of the HEP and CHP in classical homotopy theory are themselves specializations of dual model theoretic generalizations of HELP and coHELP that were introduced and given the names left and right HELP by Vogt [139]. He used them to give a characterization of the weak equivalences in any model cateogory in terms of lifting properties.

14.4. The homotopy category of a model category

To begin with, we reconsider the cofibrant and fibrant replacements of Definition 14.2.7 from a homotopical point of view. In the definition of a model category, we built in the choice of having functorial factorizations. When we have them, we have cofibrant and fibrant replacement functors Q and R. In general, we have functors up to homotopy. To make this precise, observe that the results of the previous section, in particular Corollary 14.3.10 and Proposition 14.3.11, validate the following definitions of homotopy categories.

DEFINITION 14.4.1. Consider the full categories $\mathcal{M}_c$, $\mathcal{M}_f$, and $\mathcal{M}_{cf}$ of cofibrant, fibrant, and bifibrant objects of $\mathcal{M}$, respectively. Define their homotopy categories $h\mathcal{M}_c$, $h\mathcal{M}_f$, and $h\mathcal{M}_{cf}$ to be the categories with the same objects and with morphisms the equivalence classes of maps with respect to right homotopy, left homotopy, and homotopy, respectively. In the first two cases, we understand equivalence classes under the equivalence relation generated by right or left homotopy.

Consider a map $f\colon X \to Y$. Choose cofibrant replacements $q\colon QX \to X$ and $q\colon QY \to Y$ and fibrant replacements $r\colon X \to RX$ and $r\colon Y \to RY$. Then we can obtain lifts $Qf\colon QX \longrightarrow QY$ and $Rf\colon RX \longrightarrow RY$ in the diagrams

14.4.2

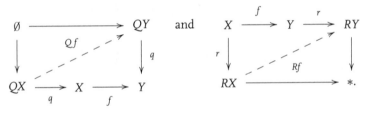

Thus we have a kind of point set level naturality of q and r even when we do not have functors Q and R. These constructions enjoy the following properties.

LEMMA 14.4.3. *Consider a map $f\colon X \longrightarrow Y$.*

(i) f is a weak equivalence if and only if Qf is a weak equivalence.

(ii) The left (and hence right) homotopy classes of Qf depend only on the left homotopy class of the composite fq.

(iii) If Y is fibrant, the right homotopy class of Qf depends only on the right homotopy class of f.

(iv) f is a weak equivalence if and only if Rf is a weak equivalence.

(v) The right (and hence left) homotopy classes of Rf depend only on the right homotopy class of the composite rf.

(vi) *If X is cofibrant, the left homotopy class of Rf depends only on the left homotopy class of f.*

PROOF. Statements (iv)–(vi) are dual to statements (i)–(iii). Parts (i) and (iv) hold by the two out of three property. Parts (ii) and (v) hold by Theorem 14.3.14 and Proposition 14.3.11, applied with X replaced by the cofibrant object QX or Y replaced by the fibrant object RY. If Y is fibrant, then so is QY, and if X is cofibrant, then so is RX. Parts (iii) and (vi) hold by two more applications of Theorem 14.3.14. □

Lemma 14.4.3 and Remark 14.2.8, the latter elaborated to show that ξ becomes natural on passage to the homotopy category $h\mathcal{M}_{cf}$, imply the following statement.

PROPOSITION 14.4.4. *Cofibrant and fibrant replacement induce functors*

$$Q \colon \mathcal{M} \longrightarrow h\mathcal{M}_c \quad \text{and} \quad R \colon \mathcal{M} \longrightarrow h\mathcal{M}_f.$$

When restricted to fibrant and cofibrant objects, respectively, these functors factor through homotopy categories to induce functors

$$hQ \colon h\mathcal{M}_f \longrightarrow h\mathcal{M}_{cf} \quad \text{and} \quad hR \colon h\mathcal{M}_c \longrightarrow h\mathcal{M}_{cf}.$$

Moreover, bifibrant replacement RQ and QR induce naturally equivalent functors

$$hR \circ Q, \ hQ \circ R \colon \mathcal{M} \longrightarrow h\mathcal{M}_{cf}.$$

The notations hQ, hR, and analogues are generally abbreviated to Q and R, by abuse of notation, and we agree to write RQ for the functor $hR \circ Q \colon \mathcal{M} \to h\mathcal{M}_{cf}$ induced by chosen objectwise cofibrant and fibrant replacements, as above, or by chosen functorial replacements if we have them. By a similar abuse of notation, we write RQf for either a map in $\mathcal{M}_{cf}$ obtained by successive lifts in (14.4.2) or for its homotopy class (which is well-defined), letting the context determine the meaning.

DEFINITION 14.4.5. Define the homotopy category $\mathrm{Ho}\mathcal{M}$ to have objects the objects of $\mathcal{M}$ and morphism sets

$$\mathrm{Ho}\mathcal{M}(X, Y) = h\mathcal{M}_{cf}(RQX, RQY) = \pi(RQX, RQY),$$

with the evident composition. Define $\gamma : \mathcal{M} \longrightarrow \mathrm{Ho}\mathcal{M}(X, Y)$ to be the identity on objects and to send a map f to RQf. Observe that $\mathrm{Ho}\mathcal{M}$ is equivalent to $h\mathcal{M}_{cf}$ via the functor that sends X to RQX and f to RQf.

PROPOSITION 14.4.6. *The class of maps f such that $\gamma(f)$ is an isomorphism is precisely $\mathscr{W}$, and every map in $\mathrm{Ho}\mathcal{M}$ is a composite of morphisms in $\gamma(\mathcal{M})$ and inverses of morphisms in $\gamma(\mathscr{W})$.*

PROOF. By Lemma 14.4.3, if $f : X \longrightarrow Y$ is a weak equivalence, then so is RQf. By Theorem 14.3.15, RQf is then a homotopy equivalence and hence an isomorphism in $h\mathcal{M}_{cf}$. Conversely, if RQf is an isomorphism in $h\mathcal{M}_{cf}$, then RQf is a homotopy equivalence and therefore a weak equivalence, again by Theorem 14.3.15. This implies that f is a weak equivalence.

For the second part, note that for any X we have weak equivalences

$$X \xleftarrow{\quad q \quad} QX \xrightarrow{\quad r \quad} RQX,$$

and these induce an isomorphism $\xi_X = \gamma(r)\gamma(q)^{-1} : \gamma(X) \to \gamma(RQX)$ in $\mathrm{Ho}\mathcal{M}$. When $X \in \mathcal{M}_{cf}$, q and r are homotopy equivalences and thus ξ_X is a map in $\mathcal{M}_{cf}$. If Y is also in $\mathcal{M}_{cf}$, the maps ξ_X and ξ_Y in $\mathcal{M}_{cf}$ induce an isomorphism of $\pi(X, Y)$ with $\pi(RQX, RQY)$. For any X and Y, this identifies $\mathrm{Ho}\mathcal{M}(RQX, RQY) = \pi(RQRQX, RQRQY)$ with $\mathrm{Ho}\mathcal{M}(X, Y) = \pi(RQX, RQY)$. Since passage to homotopy classes of maps is a surjection

$$\mathcal{M}(RQX, RQY) \longrightarrow \pi(RQX, RQY) = \mathrm{Ho}\mathcal{M}(X, Y),$$

every map $f : X \longrightarrow Y$ in $\mathrm{Ho}\mathcal{M}(X, Y)$ is represented by a composite $\xi_Y^{-1} g \xi_X$ for some map $g : RQX \longrightarrow RQY$ in $\mathcal{M}$. $\qquad\square$

THEOREM 14.4.7. *The functor $\gamma : \mathcal{M} \longrightarrow \mathrm{Ho}\mathcal{M}$ is a localization of $\mathcal{M}$ at $\mathscr{W}$.*

PROOF. Let $F : \mathcal{M} \longrightarrow \mathscr{H}$ be a functor that sends weak equivalences to isomorphisms. We must construct $\tilde{F} : \mathrm{Ho}\mathcal{M} \longrightarrow \mathscr{H}$ such that $\tilde{F} \circ \gamma = F$. We let $\tilde{F} = F$ on objects. With the notations of the proof of Proposition 14.4.6, we can and must define $\tilde{F}$ on morphisms by sending a map ξ_X to $F(r)F(q)^{-1}$ and sending a map $f \in \mathrm{Ho}\mathcal{M}(X, Y)$ represented by $\xi_Y^{-1} g \xi_X$ to $F(\xi_Y)^{-1} F(g) F(\xi_X)$. $\qquad\square$

Alternative versions of the dual Whitehead theorems drop out formally from the construction of $\mathrm{Ho}\mathcal{M}$.

THEOREM 14.4.8 (WHITEHEAD). *The following versions of the dual Whitehead theorems hold.*

(i) *A map* $p\colon Z \longrightarrow Y$ *between fibrant objects is a weak equivalence if and only if* $p_*\colon \pi(X, Z) \longrightarrow \pi(X, Y)$ *is a bijection for all cofibrant objects X.*

(ii) *A map* $i\colon W \longrightarrow X$ *betweeen cofibrant objects is a weak equivalence if and only if* $i^*\colon \pi(X, Y) \longrightarrow \pi(W, Y)$ *is a bijection for all fibrant objects Y.*

PROOF. In any category $\mathscr{C}$, a map $f\colon A \longrightarrow B$ is an isomorphism if and only if either $f_*\colon \mathscr{C}(C, A) \longrightarrow \mathscr{C}(C, B)$ or $f^*\colon \mathscr{C}(B, C) \longrightarrow \mathscr{C}(A, C)$ is an isomorphism for one C in each isomorphism class of objects in $\mathscr{C}$. In fact, we need only test on objects C and D that are isomorphic to A and B. Recalling that a map f in $\mathscr{M}$ is a weak equivalence if and only if $\gamma(f)$ is an isomorphism in Ho$\mathscr{M}$, we apply this categorical triviality in the homotopy category Ho$\mathscr{M}$. Here every object is isomorphic to a cofibrant object and Ho$\mathscr{M}(X, Y)$ can be identified with $\pi(X, Y)$ when X is cofibrant and Y is fibrant. For (i), it suffices to use cofibrant approximations of Z and Y as test objects and for (ii), it suffices to use fibrant approximations of W and X. □

It is important to understand when functors defined on $\mathscr{M}$ are homotopy invariant, in the sense that they take homotopic maps to the same map. There are three results along this line, the most obvious of which is the least useful.

LEMMA 14.4.9. *Any functor* $F\colon \mathscr{M} \longrightarrow \mathscr{H}$ *that takes weak equivalences to isomorphisms identifies left or right homotopic maps.*

PROOF. For a cylinder Cyl X, $Fi_0 = Fi_1$ since both are inverse to Fp. Therefore, for a homotopy $h\colon \mathrm{Cyl}\, X \longrightarrow Y$ from f to g,

$$Ff = F(hi_0) = FhFi_0 = FhFi_1 = F(hi_1) = Fg.$$

The proof for right homotopies is dual. □

However, the hypothesis on F here is too strong and rarely holds in practice. The following dual pair of lemmas often do apply. Ken Brown's lemma (14.2.9) is relevant to these results and to their applications in the next chapter.

LEMMA 14.4.10. *Any functor* $F\colon \mathscr{M}_c \longrightarrow \mathscr{H}$ *that takes acyclic cofibrations to isomorphisms identifies right homotopic maps.*

PROOF. By (i) and (ii) of the dual of Proposition 14.3.9, if f, g are right homotopic maps $X \longrightarrow Y$ where X and Y are cofibrant, then there is a very good homotopy $h: X \longrightarrow \text{Cocyl } Y$ between them. Then $i: Y \longrightarrow \text{Cocyl } Y$ is an acyclic cofibration, hence Cocyl Y is also cofibrant. Therefore Fi is defined and is an isomorphism, hence so are Fp_0 and Fp_1. The conclusion follows as in Lemma 14.4.9. $\square$

LEMMA 14.4.11. *Any functor* $F: \mathcal{M}_f \longrightarrow \mathcal{H}$ *that takes acyclic fibrations to isomorphisms identifies left homotopic maps.*

15

COFIBRANTLY GENERATED AND PROPER
MODEL CATEGORIES

This chapter develops several disparate basic features of model category theory. There is a standard construction of WFSs and model categories, which is based on Quillen's "small object argument". The latter is a general method for starting with a set, $\mathcal{I}$ say, of maps of $\mathcal{M}$ and constructing from $\mathcal{I}$ a functorial WFS $(^\boxempty(\mathcal{I}^\boxempty), \mathcal{I}^\boxempty)$. We explain this construction of WFSs in §15.1.

The method has the attractive feature that $^\boxempty(\mathcal{I}^\boxempty)$ is constructed from $\mathcal{I}$ in a concrete cellular fashion. The reader should have in mind the set $\mathcal{I}$ of inclusions $S^n \longrightarrow D^{n+1}$ used to construct cell complexes of spaces. The method in general involves transfinite colimits, although Quillen [113] originally considered only sequential colimits. The transfinite version of the argument was implicit in Bousfield [16], but was only later codified in the notion of a cofibrantly generated model category. That notion offers a very convenient packaging of sufficient conditions to verify the model axioms, as we explain in §15.2.

The small object argument is often repeated in the model category literature, but it admits a useful variant that we feel has not been sufficiently emphasized in print.[1] We call the variant the compact object argument. In many basic examples, such as topological spaces, chain complexes, and simplicial sets, only sequential colimits are required. When this is the case, we obtain a more concrete type of cofibrantly generated model category called a compactly generated model category. In such cases we are free to ignore transfinite cell complexes. Compactly generated model categories are attractive to us since the relevant cell theory is much closer to classical cell theory in algebraic topology (e.g., [93]) and in homological algebra (e.g., [77]) than the transfinite version. Appreciation of the naturality of the more general notion can best be obtained by reading §19.3, where we construct Bousfield

---

1. It is discussed in [97], but that is not a book for those new to the subject.

localizations of spaces at homology theories. The examples of cofibrantly generated model categories that we construct before that section are compactly generated.

We feel that the general case of cofibrantly generated model categories is overemphasized in the model category literature, and we urge the reader not to get bogged down in the details of the requisite smallness condition. At least on a first reading, it suffices to focus on the simpler compactly generated case. As we observe at the end of §15.2, there is an attractive intermediate notion that arises when cofibrations are constructed using sequential colimits and acyclic cofibrations are constructed using transfinite colimits. It is that kind of cofibrantly generated model category that appears in §19.3.

We describe model structures in over and under "slice categories" in §15.3. This gives a frequently used illustration of how one creates new model structures from given ones. We then describe left and right proper model structures in §15.4. These conditions on a model structure are central to the applications of model category theory, and they play an important role in the development of Bousfield localization in Chapter 19. We illustrate their use in the relatively technical §15.5, which is best skipped on a first reading. It uses properness to relate lifting properties to hom sets in homotopy categories.

15.1. The small object argument for the construction of WFSs

The essential starting point is to define $\mathcal{I}$-cell complexes. When $\mathcal{I}$ is the set $\{S^n \longrightarrow D^{n+1}\}$ of standard cell inclusions, all CW complexes will be examples of $\mathcal{I}$-cell complexes as we define them. However, for reasons we will explain, that fails with the usual model theoretic definition of an $\mathcal{I}$-cell complex. Recall Definition 14.1.6.

DEFINITION 15.1.1. Let $\mathcal{I}$ be a set of maps in $\mathcal{M}$. For an object $X \in \mathcal{M}$ and an ordinal λ, a relative $\mathcal{I}$-cell λ-complex under X is a map $g : X \longrightarrow Z$ that is a transfinite composite of a λ-sequence Z_* such that $Z_0 = X$ and, for a successor ordinal $\alpha + 1 < \lambda$, $Z_{\alpha+1}$ is obtained as the pushout in a diagram

$$
\begin{array}{ccc}
\coprod A_q & \xrightarrow{\ j\ } & Z_\alpha \\
\coprod i_q \downarrow & & \downarrow \\
\coprod B_q & \xrightarrow[\ k\ ]{} & Z_{\alpha+1},
\end{array}
$$

where the $i_q \colon A_q \longrightarrow B_q$ run through some set S_α of maps in $\mathcal{I}$. The restrictions of j to the A_q are called attaching maps, and the restrictions of k to the B_q are called cells. We say that f is a simple relative $\mathcal{I}$-cell λ-complex under X if the cardinality of each S_α is one, so that we adjoin a single cell at each stage. Define $\mathscr{C}(\mathcal{I})$ to be the class of retracts of relative $\mathcal{I}$-cell complexes in $\mathcal{M}$. An object $X \in \mathcal{M}$ is an $\mathcal{I}$-cell complex if there is a relative $\mathcal{I}$-cell complex $\emptyset \longrightarrow X$.

DEFINITION 15.1.2. A relative $\mathcal{I}$-cell ω-complex is called a sequential, or classical, relative $\mathcal{I}$-cell complex. Here the indices α just run over the natural numbers, so that $Z = \operatorname{colim} Z_n$ as in the classical definition of CW complexes.

In the model category literature, relative $\mathcal{I}$-cell complexes are generally defined to be simple, so that they are transfinite composites of pushouts of maps in $\mathcal{I}$. Coproducts are then not mentioned in the definition, but they appear in its applications. Using coproducts in the definition keeps us closer to classical cell theory, minimizes the need for set theoretic arguments, and prescribes $\mathcal{I}$-cell complexes in the form that they actually appear in all versions of the small object argument. Note that we have placed no restriction on the cardinality of the sets S_α. Such a restriction is necessary if we want to refine a given $\mathcal{I}$-cell complex to a simple $\mathcal{I}$-cell λ-complex for some prescribed value of λ, but we shall avoid use of such refinements. We digress to say just a bit about how the comparison of simple and general $\mathcal{I}$-cell complexes works, leaving the details to the literature [65, Ch. 10], but we will make no use of simple cell complexes and therefore no use of the comparison.

PROPOSITION 15.1.3. *By ordering the elements of coproducts along which pushouts are taken, a relative $\mathcal{I}$-cell λ-complex $f \colon X \longrightarrow Z$ can be reinterpreted as a simple relative $\mathcal{I}$-cell complex. That is, we can reinterpret the maps $Z_\alpha \longrightarrow Z_{\alpha+1}$ as simple relative cell complexes, obtained by attaching one cell at a time, and then reindex so as to interpolate these simple cell complexes into the original cell complex to obtain a simple cell κ-complex for some ordinal $\kappa \geq \lambda$.*

We can determine the cardinality of κ in terms of the cardinalities of λ and the sets S_α. Recall the definition of a regular cardinal from Definition 2.5.2.

DEFINITION 15.1.4. A relative $\mathcal{I}$-cell λ-complex is regular if λ is a regular cardinal and the indexing sets S_α for the attaching maps all have cardinality less than that of λ.

There is no real loss of generality in indexing $\mathcal{I}$-cell complexes on regular cardinals since we can add in identity maps to reindex a relative $\mathcal{I}$-cell λ-complex to a relative $\mathcal{I}$-cell κ-complex that is indexed on a regular cardinal $\kappa \geq \lambda$. There is real loss of generality in restricting the cardinalities of the sets S_α. The point of the restriction is that we can now apply Proposition 15.1.3 without increasing the cardinality of the ordinal λ that we start with.

COROLLARY 15.1.5. *A regular $\mathcal{I}$-cell λ-complex can be reinterpreted as a simple relative $\mathcal{I}$-cell λ-complex.*

That ends the digression. We now offer two parallel sets of details. One focuses solely on sequential colimits, that is, on $\mathcal{I}$-cell ω-complexes. The other makes use of $\mathcal{I}$-cell λ-complexes for larger ordinals λ. We argue that the latter should be used when necessary, but that their use when unnecessary only makes arguments unaesthetically complicated. As already noted, we shall arrange our work so that we have no need to make explicit use of ordinals larger than ω until Chapter 19.

DEFINITION 15.1.6. An object A of $\mathcal{M}$ is compact with respect to $\mathcal{I}$ if for every relative $\mathcal{I}$-cell ω-complex $f : X \longrightarrow Z = \mathrm{colim}_n Z_n$, the canonical map

$$\mathrm{colim}_n \mathcal{M}(A, Z_n) \longrightarrow \mathcal{M}(A, Z)$$

is a bijection. The set $\mathcal{I}$ is compact, or permits the compact object argument, if every domain object A of a map in $\mathcal{I}$ is compact with respect to $\mathcal{I}$. When $\mathcal{I}$ is compact, we interpret $\mathscr{C}(I)$ to mean the class of retracts of relative $\mathcal{I}$-cell ω-complexes, excluding the relative $\mathcal{I}$-cell λ-complexes for $\lambda > \omega$.

DEFINITION 15.1.7. Let κ be a cardinal. An object A of $\mathcal{M}$ is κ-small with respect to $\mathcal{I}$ if for every cardinal $\lambda \geq \kappa$ and every relative $\mathcal{I}$-cell λ-complex $f : X \longrightarrow Z = \mathrm{colim}_{\beta < \lambda} Z_\beta$, the canonical map

$$\mathrm{colim}_{\beta < \lambda} \mathcal{M}(A, Z_\beta) \longrightarrow \mathcal{M}(A, Z)$$

is a bijection. An object A is small with respect to $\mathcal{I}$ if it is κ_A-small for some κ_A. The set $\mathcal{I}$ is small, or permits the small object argument, if every domain object A of a map in $\mathcal{I}$ is small with respect to $\mathcal{I}$.

LEMMA 15.1.8. *If $\mathcal{I}$ is small, there is a regular cardinal λ such that every domain object A of a map in $\mathcal{I}$ is λ-small.*

PROOF. Take λ to be any regular cardinal that contains all κ_A. □

REMARK 15.1.9. In Definition 15.1.6, we do not restrict to regular $\mathcal{I}$-cell ω-complexes. The cardinalities of the sets S_n of attaching maps are unrestricted, just as in the usual definition of CW complexes. The regular $\mathcal{I}$-cell ω-complexes are the complexes with finite skeleta, and we can order their cells to make them simple ω-complexes. We cannot so order general ω-complexes without increasing the cardinality. Observe that "compact" and "ω-small" are very different notions: "compact" nowhere mentions cardinalities greater than ω, whereas ω-small refers to all cardinalities greater than ω. In marked contrast, Proposition 15.1.3 implies that the notion of A being κ-small with respect to $\mathcal{I}$ is unchanged if we restrict attention to simple relative $\mathcal{I}$-cell complexes in Definition 15.1.7, as is generally done in the literature [65, 66].

For clarity, we isolate the central construction of the small object argument.

CONSTRUCTION 15.1.10. Let $f : X \longrightarrow Y$ be a map in $\mathcal{M}$. Let S be the set of all commutative squares

$$
\begin{array}{ccc}
A_q & \xrightarrow{\ k_q\ } & X \\
{\scriptstyle i_q}\big\downarrow & & \big\downarrow{\scriptstyle f} \\
B_q & \xrightarrow[\ j_q\ ]{} & Y,
\end{array}
$$

where i_q is a map in $\mathcal{I}$. We construct the single step factorization diagram for f:

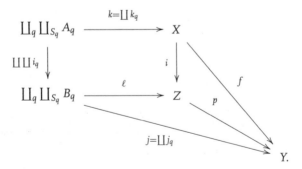

The square is a pushout diagram that defines Z, i, and ℓ, and the map p is given by the universal property of pushouts. The diagram displays a factorization of f as pi, where i is a one step relative $\mathcal{I}$-cell complex and the map ℓ can be viewed as solving the lifting problem for $(\amalg i_q, p)$ defined by the maps ik and j.

PROPOSITION 15.1.11 (THE SMALL OBJECT ARGUMENT). *Assume that a set $\mathcal{I}$ of maps in $\mathcal{M}$ is either compact or small. Then there is a functorial WFS $(\mathscr{C}(\mathcal{I}), \mathcal{I}^{\boxtimes})$. Moreover, the construction is functorial with respect to inclusions of subsets $\mathcal{J} \subset \mathcal{I}$.*

PROOF. Let $\lambda = \omega$ in the compact case and let λ be the regular cardinal of Lemma 15.1.8 in the small case. Let $f : X \longrightarrow Y$ be a map in $\mathcal{M}$. We shall construct a functorial factorization $X \xrightarrow{i} Z \xrightarrow{p} Y$ in which i is a relative $\mathcal{I}$-cell λ-complex and p is in $\mathcal{I}^{\boxtimes}$. With the language of Definition 14.1.6, we construct a λ-sequence Z_* of objects over Y, and we then define $i : X \to Z$ and $p : Z \to Y$ by taking the transfinite composite of this λ-sequence. We set $Z_0 = X$ and let $i_0 = \mathrm{id} : X \longrightarrow Z_0$ and $p_0 = f : Z_0 \longrightarrow Y$. Inductively, suppose that we have completed the construction up through Z_α, so that we have a relative $\mathcal{I}$-cell complex $i_\alpha : X \longrightarrow Z_\alpha$ and a map $p_\alpha : Z_\alpha \longrightarrow Y$ such that $p_\alpha \circ i_\alpha = f$. We construct $Z_{\alpha+1}$ together with a factorization $Z_\alpha \longrightarrow Z_{\alpha+1} \xrightarrow{p_{\alpha+1}} Y$ of p_α by applying the single step factorization diagram to the map p_α. The composite $X \xrightarrow{i_\alpha} Z_\alpha \longrightarrow Z_{\alpha+1}$ is a relative $\mathcal{I}$-cell complex $i_{\alpha+1}$, and $p_{\alpha+1}$ solves a lifting problem as specified in Construction 15.1.10. We define Z_β, i_β, and p_β on limit ordinals β by passage to colimits. With $\beta = \lambda$, this completes the construction of the factorization.

To see that p is in $\mathcal{I}^{\boxtimes}$, consider the following diagram, in which $\iota \in \mathcal{I}$ and maps k and j are given such that the outer square commutes.

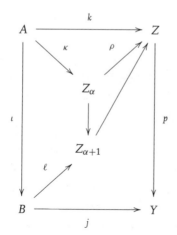

Since $\mathcal{I}$ is compact or small, k factors through some Z_α, giving maps κ and ρ making the top triangle commute, and then the arrows ι, j, κ, and $p\rho$ display one of the squares used in the single step construction of $Z_{\alpha+1}$ from Z_α. The single step construction gives the map ℓ, and the diagonal composite gives the required lift.

To see that we have a WFS $(\mathscr{C}(\mathcal{I}), \mathcal{I}^{\boxtimes})$, we must show that

$$\mathscr{C}(\mathcal{I}) = {}^{\boxtimes}(\mathcal{I}^{\boxtimes}) \quad \text{and} \quad \mathscr{C}(\mathcal{I})^{\boxtimes} = \mathcal{I}^{\boxtimes}.$$

If $f: X \longrightarrow Y$ is in ${}^{\boxtimes}(\mathcal{I}^{\boxtimes})$ and we factor f as above, then the retract argument of Lemma 14.1.12 shows that f is a retract of i and is therefore in $\mathscr{C}(\mathcal{I})$. Conversely, since $\mathcal{I}$ is contained in the left saturated class ${}^{\boxtimes}(\mathcal{I}^{\boxtimes})$ (see Definition 14.1.7, Proposition 14.1.8, and Lemma 14.1.9), we see that $\mathscr{C}(\mathcal{I}) \subset {}^{\boxtimes}(\mathcal{I}^{\boxtimes})$. By Lemma 14.1.9, $\mathcal{I}^{\boxtimes} \supset \mathscr{C}(\mathcal{I})^{\boxtimes}$ and $\mathscr{C}(\mathcal{I})^{\boxtimes} \supset ({}^{\boxtimes}(\mathcal{I}^{\boxtimes}))^{\boxtimes} = \mathcal{I}^{\boxtimes}$.

To see the functoriality of the factorization, consider a commutative square

$$
\begin{array}{ccc}
X & \xrightarrow{f} & Y \\
{\scriptstyle r}\downarrow & & \downarrow{\scriptstyle s} \\
X' & \xrightarrow{f'} & Y'.
\end{array}
$$

Construct a factorization $X' \xrightarrow{i} Z' \xrightarrow{p} Y'$ in the same way as above. Inductively, assume that we have obtained a map $t_\alpha : Z_\alpha \longrightarrow Z'_\alpha$ that makes the following diagram commute.

$$
\begin{array}{ccccc}
X & \xrightarrow{i_\alpha} & Z_\alpha & \xrightarrow{p_\alpha} & Y \\
{\scriptstyle r}\downarrow & & {\scriptstyle t_\alpha}\downarrow & & \downarrow{\scriptstyle s} \\
X' & \xrightarrow{i_\alpha} & Z'_\alpha & \xrightarrow{p_\alpha} & Y'
\end{array}
$$

For the next stage, the composite with s of a square used to construct $Z_{\alpha+1}$ from Z_α gives one of the squares used to construct $Z'_{\alpha+1}$ from Z'_α. The map $t: Z \longrightarrow Z'$ obtained by passage to colimits satisfies $ti = ir$ and $sp = pt$, verifying the claimed functoriality of the factorization.

Finally, let $\mathcal{J} \subset \mathcal{I}$. An ordinal λ such that the domains of maps in $\mathcal{I}$ are λ-small with respect to relative $\mathcal{I}$-cell complexes has the same property for $\mathcal{J}$.

We may use λ-complexes to construct both WFSs, and we write superscripts $\mathcal{I}$ and $\mathcal{J}$ to distinguish them. Inductively, assume that we have obtained a map $t_\alpha \colon Z_\alpha^{\mathcal{J}} \longrightarrow Z_\alpha^{\mathcal{I}}$ that makes the following diagram commute.

$$
\begin{array}{ccccc}
X & \xrightarrow{\ i_\alpha\ } & Z_\alpha^{\mathcal{J}} & \xrightarrow{\ p_\alpha\ } & Y \\
\Big\| & & \Big\downarrow{\scriptstyle t_\alpha} & & \Big\| \\
X & \xrightarrow[\ i_\alpha\ ]{} & Z_\alpha^{\mathcal{I}} & \xrightarrow[\ p_\alpha\ ]{} & Y
\end{array}
$$

The composite of t_α with an attaching map used to construct $Z_{\alpha+1}^{\mathcal{J}}$ from $Z_\alpha^{\mathcal{J}}$ gives one of the attaching maps used to construct $Z_{\alpha+1}^{\mathcal{I}}$ from $Z_\alpha^{\mathcal{I}}$, and the universal property of pushouts gives the next map $t_{\alpha+1}$. Passage to colimits gives a map $t \colon Z^{\mathcal{J}} \longrightarrow Z^{\mathcal{I}}$ under X and over Y. For $\mathcal{K} \subset \mathcal{J}$, the composite of this construction for $\mathcal{K} \subset \mathcal{J}$ and $\mathcal{J} \subset \mathcal{I}$ gives the construction for $\mathcal{K} \subset \mathcal{I}$. $\qquad\square$

REMARK 15.1.12. There is a general property of a category, called local presentability [12, §5.2], that ensures that any set $\mathcal{I}$ permits the small object argument. In such categories, this leads to a more uniform and aesthetically satisfactory treatment of the small object argument. It is satisfied by most algebraically defined categories. It is not satisfied by the category of compactly generated topological spaces. However, if instead of using all compact Hausdorff spaces in the definition of compactly generated spaces [93, p. 37], one only uses standard simplices, one obtains the locally presentable category of "combinatorial spaces". It appears that one can redo all of algebraic topology with combinatorial spaces replacing compactly generated spaces. We will not say more about that point of view. The insight is due to Jeff Smith (unpublished); published sources are [9, 47].

15.2. Compactly and cofibrantly generated model categories

The model categories in most common use, although by no means the only interesting ones, are obtained by use of the small object argument. Their cofibrations and acyclic cofibrations admit cellular descriptions.

DEFINITION 15.2.1. A model structure $(\mathcal{W}, \mathcal{C}, \mathcal{F})$ on $\mathcal{M}$ is generated by sets $\mathcal{I}$ and $\mathcal{J}$ if $\mathcal{F} = \mathcal{J}^{\boxtimes}$ and $\mathcal{F} \cap \mathcal{W} = \mathcal{I}^{\boxtimes}$. Since $\mathcal{I}^{\boxtimes} = \mathcal{C}(\mathcal{I})^{\boxtimes}$ and $\mathcal{J}^{\boxtimes} = \mathcal{C}(\mathcal{J})^{\boxtimes}$, it is equivalent that $\mathcal{C} = \mathcal{C}(\mathcal{I})$ and $\mathcal{C} \cap \mathcal{W} = \mathcal{C}(\mathcal{J})$. The sets $\mathcal{I}$ and $\mathcal{J}$ are called the generating cofibrations and generating acyclic cofibrations. We say that the model structure is cofibrantly generated if the sets $\mathcal{I}$ and $\mathcal{J}$ are small. We say that the model structure is compactly generated if the sets $\mathcal{I}$ and $\mathcal{J}$ are compact, in which case only sequential cell complexes are used to define $\mathscr{C}(\mathcal{I})$ and $\mathscr{C}(\mathcal{J})$.

It is possible for a model category to be generated by sets $\mathcal{I}$ and $\mathcal{J}$ that are neither small nor compact, although this possibility is seldom encountered and never mentioned in the literature.

REMARK 15.2.2. A combinatorial model category is a locally presentable category that is also a cofibrantly generated model category [9].

There are several variant formulations of the following criterion for detecting cofibrantly generated model categories. The version we give is [97, Thm. 4.5.6]. Recall Definition 14.1.4.

THEOREM 15.2.3. *Let $\mathcal{M}$ be a bicomplete category with a given subcategory $\mathcal{W}$ of weak equivalences and given sets $\mathcal{I}$ and $\mathcal{J}$ of maps. Assume that $\mathcal{I}$ and $\mathcal{J}$ are compact or small. Then $\mathcal{M}$ is a compactly generated or cofibrantly generated model category with generating cofibrations $\mathcal{I}$ and generating acyclic cofibrations $\mathcal{J}$ if and only if the following two conditions hold.*

(i) (Acyclicity condition) Every relative $\mathcal{J}$-cell complex is a weak equivalence.
(ii) (Compatibility condition) $\mathcal{I}^{\boxtimes} = \mathcal{J}^{\boxtimes} \cap \mathcal{W}$.

PROOF. The necessity of the conditions is obvious. Thus assume that (i) and (ii) hold. Define $\mathscr{C} = {}^{\boxtimes}(\mathcal{I}^{\boxtimes})$ and $\mathscr{F} = \mathcal{J}^{\boxtimes}$. We must show that $(\mathscr{C}, \mathscr{F} \cap \mathcal{W})$ and $(\mathscr{C} \cap \mathcal{W}, \mathscr{F})$ are WFSs. In view of the definitions of $\mathscr{F}$ and $\mathscr{C}$, the small object argument gives WFSs

$$(\mathscr{C}, \mathcal{I}^{\boxtimes}) \quad \text{and} \quad (\mathscr{C}(\mathcal{J}), \mathscr{F}).$$

Since we have assumed in (ii) that $\mathcal{I}^{\boxtimes} = \mathscr{F} \cap \mathcal{W}$, it only remains to prove that $\mathscr{C}(\mathcal{J}) = \mathscr{C} \cap W$. Since $\mathcal{W}$ is a category of weak equivalences, it is closed under retracts, hence (i) gives that $\mathscr{C}(\mathcal{J}) \subset \mathcal{W}$. Moreover,

$$\mathcal{J} \subset {}^{\boxtimes}(\mathcal{J}^{\boxtimes}) \subset {}^{\boxtimes}(\mathcal{J}^{\boxtimes} \cap \mathcal{W}) = {}^{\boxtimes}(\mathscr{F} \cap \mathcal{W}) = {}^{\boxtimes}(\mathcal{I}^{\boxtimes}) = \mathscr{C}.$$

It follows by left saturation that $\mathscr{C}(\mathcal{J}) \subset \mathscr{C}$ and thus $\mathscr{C}(\mathcal{J}) \subset \mathscr{C} \cap \mathcal{W}$. Conversely, let $f \in \mathscr{C} \cap \mathcal{W}$ and factor f as a composite $q \circ j$, where $j \in \mathscr{C}(\mathcal{J})$ and

$q \in \mathcal{F}$. Then j is in $\mathcal{W}$, hence q is in $\mathcal{W}$ by the two out of three property. Thus q is in $\mathcal{F} \cap \mathcal{W} = \mathcal{I}^{\boxempty}$. Since f is in $\mathcal{C}$, it has the LLP with respect to q. By the retract argument, Lemma 14.1.12, f is a retract of j. Since j is in $\mathcal{C}(\mathcal{J})$, f is in $\mathcal{C}(\mathcal{J})$. □

In practice, the compatibility condition is often easily verified by formal arguments. For example, Theorem 15.2.3 is often used to transport a model structure across an adjunction, as we formalize in Theorem 16.2.5 below, and then only the acyclicity condition need be verified. The acyclicity condition clearly holds if $\mathcal{J} \subset \mathcal{W}$ and $\mathcal{W}$ is a left saturated class of maps in the sense of Definition 14.1.7. This rarely applies since pushouts of weak equivalences are generally not weak equivalences. However, pushouts of coproducts of maps in $\mathcal{J}$ often are weak equivalences, and verifying that is usually the key step in verifying the acyclicity condition.

The following two remarks pull in opposite directions. The first suggests that it may sometimes be useful to expand $\mathcal{I}$, whereas the second suggests that it is worthwhile to keep $\mathcal{I}$ as small as possible.

REMARK 15.2.4. The sets $\mathcal{I}$ and $\mathcal{J}$ are not uniquely determined. In practice, with the standard choices, the maps in $\mathcal{J}$ are relative $\mathcal{I}$-cell complexes but are not themselves in $\mathcal{I}$. However, we are free to replace $\mathcal{I}$ by $\mathcal{I} \cup \mathcal{J}$. The hypotheses of Theorem 15.2.3 still hold. This gives a new set $\mathcal{I}$ of generating cofibrations that contains $\mathcal{J}$. Now we have not only factorization functors $Z^{\mathcal{I}}(f)$ and $Z^{\mathcal{J}}(f)$ on maps $f : X \longrightarrow Y$, but we also have a natural comparison map $Z^{\mathcal{J}}(f) \longrightarrow Z^{\mathcal{I}}(f)$ under X and over Y. This places us in a categorical context of maps of WFSs [118].

REMARK 15.2.5. Notice that use of the small object argument applies independently to $\mathcal{I}$ and $\mathcal{J}$ to construct the two required WFSs. The assumption that $\mathcal{I}$ and $\mathcal{J}$ be compact or small in Theorem 15.2.3 should be interpreted as allowing $\mathcal{I}$ to be compact and $\mathcal{J}$ to be small. This combination appears in the most common examples, such as Bousfield localizations, and it allows one to model the cofibrations using only sequential $\mathcal{I}$-cell complexes.

15.3. Over and under model structures

There are several elementary constructions on model categories that lead to new model categories, often cofibrantly generated if the original one was. We record a useful elementary example that plays an important role in parametrized homotopy theory [97] and has many other applications. It shows in particular how to construct model structures on based spaces from model structures on unbased spaces.

DEFINITION 15.3.1. Let $\mathcal{M}$ be a category and B be an object of $\mathcal{M}$. The slice category $B/\mathcal{M}$ of objects under B has objects the maps $i = i_X \colon B \longrightarrow X$ in $\mathcal{M}$ and morphisms the maps $f \colon X \longrightarrow Y$ such that $f \circ i_X = i_Y$. Forgetting the maps i_X gives a functor $U \colon B/\mathcal{M} \longrightarrow \mathcal{M}$. Dually, we have the slice category $\mathcal{M}/B$ of objects over B and a functor $U \colon \mathcal{M}/B \longrightarrow \mathcal{M}$. Combining these, we have the category $\mathcal{M}_B$ of sectioned objects over B; these have projections $p \colon X \longrightarrow B$ and sections $s \colon B \longrightarrow X$ such that $p \circ s = \mathrm{id}$, and we have forgetful functors U from $\mathcal{M}_B$ to both $B/\mathcal{M}$ and $\mathcal{M}/B$. Thinking of B as the identity map $B \longrightarrow B$, we may identify $\mathcal{M}_B$ with either $B/(\mathcal{M}/B)$ or $(B/\mathcal{M})/B$. For $X \in \mathcal{M}$, the canonical inclusion makes $X \amalg B$ an object under B and the canonical projection makes $X \times B$ an object over B. For $Y \in \mathcal{M}/B$, $Y \amalg B$ is an object under B via the canonical inclusion and over B via the given map $Y \longrightarrow B$ and the identity map of B. We have adjunctions

15.3.2
$$\mathcal{M}(UY, X) \cong (\mathcal{M}/B)(Y, B \times X),$$

15.3.3
$$\mathcal{M}(X, UY) \cong (B/\mathcal{M})(X \amalg B, Y),$$

and

15.3.4
$$(\mathcal{M}/B)(Y, UZ) \cong \mathcal{M}_B(Y \amalg B, Z).$$

For a pair of objects A and B, we have the category $A/\mathcal{M}/B$ of objects under A and over B. It has forgetful functors, again denoted U, to $A/\mathcal{M}$ and $\mathcal{M}/B$. All forgetful functors in sight are faithful. The categories $B/\mathcal{M}$, $\mathcal{M}/B$, and $\mathcal{M}_B$ are bicomplete. The category $A/\mathcal{M}/B$ is not, but it is a coproduct of bicomplete categories. For each fixed map $f \colon A \longrightarrow B$, let $(A/\mathcal{M}/B)_f$ be the subcategory of diagrams $A \xrightarrow{s} X \xrightarrow{p} B$ such that $p \circ s = f$. In particular, $\mathcal{M}_B = (B/\mathcal{M}/B)_{\mathrm{id}}$. The map f is an object of both $\mathcal{M}/B$ and $A/\mathcal{M}$, and $(A/\mathcal{M}/B)_f$ may be identified with either $f/(\mathcal{M}/B)$ or $(A/\mathcal{M})/f$. Therefore $(A/\mathcal{M}/B)_f$ is bicomplete, and $A/\mathcal{M}/B$ is the disjoint union over f of its subcategories $(A/\mathcal{M}/B)_f$.

DEFINITION 15.3.5. Consider a functor $U \colon \mathcal{N} \longrightarrow \mathcal{M}$.

(i) We say that U reflects a property if a map f in $\mathcal{M}$ has the property whenever Uf does; it creates the property if f has the property if and only if Uf has the property, that is, if f preserves and reflects the property.

(ii) If we have a class $\mathscr{L}$ of maps in $\mathscr{M}$, we define a corresponding class $U^{-1}\mathscr{L}$ of maps in $\mathscr{N}$ by letting f be in $U^{-1}\mathscr{L}$ if and only if Uf is in $\mathscr{L}$; if we think of $\mathscr{L}$ as specifying a property, then we think of U as creating the class of maps in $\mathscr{N}$ that satisfies the corresponding property.

(iii) We say that a model structure on $\mathscr{M}$ creates a model structure on $\mathscr{N}$ if $\mathscr{N}$ is a model category with the classes of weak equivalences and fibrations created by the functor U. The cofibrations in $\mathscr{N}$ must then be the maps that satisfy the LLP with respect to the acyclic fibrations. The dual notion, with cofibrations and fibrations reversed, is much less frequently encountered.

(iv) We say that a model structure on $\mathscr{M}$ strongly creates a model structure on $\mathscr{N}$ if $\mathscr{N}$ is a model category with the classes of weak equivalences, cofibrations, and fibrations all created by the functor U.

It rarely happens that $\mathscr{M}$ strongly creates a model structure on $\mathscr{N}$, but it does happen in the present context of over and under categories. The definition of left and right proper model categories will be given shortly, in Definition 15.4.1 below.

THEOREM 15.3.6. *Let $\mathscr{M}$ be a model category, let A and B be objects of $\mathscr{M}$, and let $f : A \longrightarrow B$ be a map. Then the forgetful functors U strongly create model structures on $B/\mathscr{M}$, $\mathscr{M}/B$, $\mathscr{M}_B$, and $(A/\mathscr{M}/B)_f$. If $\mathscr{M}$ is left or right proper, then so are $B/\mathscr{M}$, $\mathscr{M}/B$, $\mathscr{M}_B$, and $(A/\mathscr{M}/B)_f$. If $\mathscr{M}$ has generating sets $\mathcal{I}$ and $\mathcal{J}$, then $B/\mathscr{M}$, $\mathscr{M}/B$, $\mathscr{M}_B$, and $(A/\mathscr{M}/B)_f$ have generating sets suitably constructed from $\mathcal{I}$ and $\mathcal{J}$.*

PROOF. The model axioms are inherited from $\mathscr{M}$, the point being that the WFSs of $\mathscr{M}$ directly imply WFSs for the slice categories since lifting maps and factorizations in $\mathscr{M}$ between maps in slice categories provide lifting maps and factorizations in slice categories. Properness is a similar formal check.

For the last statement, we do not use the small object argument. Rather, we appeal to the notion of a model category that is generated but not necessarily compactly or cofibrantly generated given in Definition 15.2.1. Define $\mathcal{I}/B$ to be the set of maps in $\mathscr{M}/B$ such that Ui is a map in $\mathcal{I}$. Define $B/\mathcal{I}$ to be the set of maps in $B/\mathscr{M}$ obtained by applying the functor $(-) \amalg B$ to the maps in $\mathcal{I}$. Define $\mathcal{I}_B$ to be the set of maps $i \amalg B$ in $\mathscr{M}_B$, where i is a map in $\mathcal{I}/B$. Generalizing the last, define $\mathcal{I}_f$ to be the set of maps

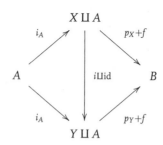

in $(A/\mathscr{M}/B)_f$, where i_A is the inclusion onto the summand A and i is a map in $\mathcal{I}/B$. The last two definitions can be viewed as specializations of the first two to appropriate slice categories of arrow categories. Define $B/\mathcal{J}$, $\mathcal{J}/B$, $\mathcal{J}_B$, and $\mathcal{J}_f$ similarly. In all cases, since $\mathscr{F} = \mathcal{J}^{\boxslash}$ and $\mathscr{F} \cap \mathscr{W} = \mathcal{I}^{\boxslash}$ in $\mathscr{M}$ and the forgetful functors to $\mathscr{M}$ create the model structures, formal arguments give the corresponding equalities in our slice categories. For $\mathscr{M}/B$, it is clear from the definitions that $p \in (\mathcal{I}/B)^{\boxslash}$ if and only if $U_p \in \mathcal{I}$ and similarly for $\mathcal{J}$. For $B/\mathscr{M}$, it is clear from the adjunction (15.3.3) and the definitions that $p \in (B/\mathcal{I})^{\boxslash}$ if and only if $U_p \in \mathcal{I}$ and similarly for $\mathcal{J}$. The analogues for $\mathscr{M}_B$ and $(A/\mathscr{M}/B)_f$ follow formally. $\qquad\square$

REMARK 15.3.7. The forgetful functor $U : \mathscr{M}/B \longrightarrow \mathscr{M}$ clearly carries relative cell complexes to relative cell complexes (both defined starting from either $\mathcal{I}$ or $\mathcal{J}$). It follows that if $\mathcal{I}$ and $\mathcal{J}$ are compact or small, then so are $\mathcal{I}/B$ and $\mathcal{J}/B$. Therefore, if $\mathscr{M}$ is compactly or cofibrantly generated, then so is $\mathscr{M}/B$. Provided that B itself is a cell complex, $U : B/\mathscr{M} \longrightarrow \mathscr{M}$ carries relative cell complexes to relative cell complexes. In this case, we see using the adjunction (15.3.3) that if $\mathcal{I}$ and $\mathcal{J}$ are compact or small, then so are $\mathcal{I}/B$ and $\mathcal{J}/B$. Without the proviso on B, this seems to be false. It follows that if B is both an $\mathcal{I}$-cell complex and a $\mathcal{J}$-cell complex and $\mathscr{M}$ is compactly or cofibrantly generated, then so is $B/\mathscr{M}$. This criterion is uninteresting in general, since it implies that B is contractible, but it does apply to categories of based objects.

We shall define Quillen adjunctions below, in Definition 16.2.1, and the following result will be immediate from the definition.

COROLLARY 15.3.8. *The adjunctions (15.3.2), (15.3.3), and (15.3.4) are Quillen adjunctions.*

15.4. Left and right proper model categories

The following concept will play a significant role in our discussion of Bousfield localization in Chapter 19, and it is important throughout model category theory. We will give a conceptual reinterpretation of the definition in Proposition 16.2.4.

DEFINITION 15.4.1. A model category $\mathcal{M}$ is left proper if the pushout of a weak equivalence along a cofibration is a weak equivalence. This means that if i is a cofibration and f is a weak equivalence in a pushout diagram

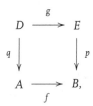

then g is a weak equivalence. It is right proper if the pullback of a weak equivalence along a fibration is a weak equivalence. This means that if p is a fibration and f is a weak equivalence in a pullback diagram

$$
\begin{array}{ccc}
D & \xrightarrow{\;g\;} & E \\
{\scriptstyle q}\downarrow & & \downarrow{\scriptstyle p} \\
A & \xrightarrow{\;f\;} & B,
\end{array}
$$

then g is a weak equivalence. It is proper if it is both left and right proper.

Over and under model structures often play a helpful technical role in proofs, as is well illustrated by their use in the proof of the following result. One point is that appropriate maps can be viewed as cofibrant or fibrant objects in model categories, allowing us to apply results about such objects to maps.

PROPOSITION 15.4.2. *Let $\mathcal{M}$ be a model category. Then any pushout of a weak equivalence between cofibrant objects along a cofibration is a weak equivalence, hence $\mathcal{M}$ is left proper if every object of $\mathcal{M}$ is cofibrant. Dually, any pullback of a weak equivalence between fibrant objects along a fibration is a weak equivalence, hence $\mathcal{M}$ is right proper if every object of $\mathcal{M}$ is fibrant.*

PROOF. Consider a pushout diagram

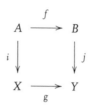

in which i is a cofibration, f is a weak equivalence, and A and B are cofibrant. We must show that g is a weak equivalence. Observe that X and Y are also cofibrant. By Theorem 14.4.8, it suffices to show that $g^*: \pi(Y, Z) \longrightarrow \pi(X, Z)$ is a bijection for all fibrant objects Z. For a map $t: X \longrightarrow Z$, we see by applying Theorem 14.4.8(ii) to f that there is a map $s: B \longrightarrow Z$ such that $s \circ f \simeq t \circ i$. Applying HEP, in the form given in Lemma 14.3.7, to the cofibration i, we can homotope t to a map t' such that $s \circ f = t' \circ i$. Then s and t' define a map $r: Y \longrightarrow Z$ such that $r \circ g = t'$. Thus g^* is surjective. To see that g^* is injective, let u and v be maps $Y \longrightarrow Z$ such that $u \circ g \simeq v \circ g$ via a good right homotopy $h: X \longrightarrow \text{Cocyl } Z$. Working in the model category $\mathcal{M}/(Z \times Z)$ of Theorem 15.3.6, in which the good cocylinder $(p_0, p_1): \text{Cocyl } Z \longrightarrow Z \times Z$ is a fibrant object, we apply Theorem 14.4.8(ii) again to obtain a map $k: B \longrightarrow \text{Cocyl } Z$ over $Z \times Z$ such that $k \circ f \simeq h \circ i$. Applying Lemma 14.3.7 again, we can homotope h over $Z \times Z$ to a map h' such that $k \circ f = h' \circ i$. Then h' and k define a right homotopy $Y \longrightarrow \text{Cocyl } Z$ from u to v. □

There is another condition that one can ask of a model category that appears to be much stronger but in fact is equivalent to its being left proper. Roughly, the condition states that pushouts preserve weak equivalences. The statement has often been verified in special situations and is used over and over again in applications. We have already seen instances of it in Lemma 2.1.3 and its dual Lemma 2.2.4.

DEFINITION 15.4.3. A model category $\mathcal{M}$ satisfies the left gluing lemma if, for any commutative diagram

$$
\begin{array}{ccccc}
A & \xleftarrow{\;i\;} & C & \xrightarrow{\;k\;} & B \\
{\scriptstyle f}\downarrow & & {\scriptstyle g}\downarrow & & \downarrow{\scriptstyle h} \\
A' & \xleftarrow{\;j\;} & C' & \xrightarrow{\;\ell\;} & B'
\end{array}
$$

in which i and j are cofibrations and f, g, and h are weak equivalences, the induced map of pushouts

$$A \cup_C B \longrightarrow A' \cup_{C'} B'$$

is a weak equivalence. The right gluing lemma is stated dually.

PROPOSITION 15.4.4. *A model category $\mathcal{M}$ is left or right proper if and only if it satisfies the left or right gluing lemma.*

PROOF. We prove the left case. If the left gluing lemma holds, then we see that $\mathcal{M}$ is left proper by taking f, g, and k to be identity maps, so that $i = j$. The conclusion then shows that the pushout $A \cong A \cup_B B \longrightarrow A \cup_B B'$ of the weak equivalence h along the cofibration i is a weak equivalence. Thus assume that $\mathcal{M}$ is left proper. We proceed in three steps.

Step 1. If k and ℓ are both weak equivalences, then by left properness so are the horizontal arrows in the commutative diagram

$$
\begin{array}{ccc}
A & \longrightarrow & A \cup_C B \\
f \downarrow & & \downarrow \\
A' & \longrightarrow & A' \cup_{C'} B'.
\end{array}
$$

Since f is a weak equivalence, the right vertical arrow is a weak equivalence by the two out of three property of weak equivalences.

Step 2. If k and ℓ are both cofibrations, consider the commutative diagram

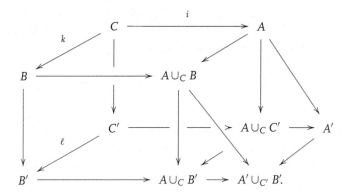

The back, front, top, and two bottom squares are pushouts, and the middle composite $C' \longrightarrow A'$ is j. Since k and ℓ are cofibrations, so are the remaining three arrows from the back to the front. Similarly, i and its pushouts are cofibrations. Since $C \longrightarrow C'$, $A \longrightarrow A'$, and $B \longrightarrow B'$ are weak equivalences, left properness and the two out of three property imply that $A \longrightarrow A \cup_C C'$, $A \cup_C C' \longrightarrow A'$, $A \cup_C B \longrightarrow A \cup_C B'$, and $A \cup_C B' \longrightarrow A' \cup_{C'} B'$ are weak equivalences. Composing the last two, $A \cup_C B \longrightarrow A' \cup_{C'} B'$ is a weak equivalence.

Step 3. To prove the general case, construct the following commutative diagram.

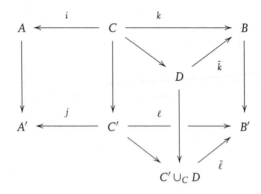

Here we first factor k as the composite of a cofibration and a weak equivalence $\bar{k}$ and then define a map $\bar{\ell}$ by the universal property of pushouts. By left properness, $D \longrightarrow C' \cup_C D$ is a weak equivalence, and by the two out of three property, so is $\bar{\ell}$. By the second step,

$$A \cup_C D \longrightarrow A' \cup_{C'} (C' \cup_C D) \cong A' \cup_C D$$

is a weak equivalence and by the first case, so is

$$A \cup_C B \cong (A \cup_C D) \cup_D B \longrightarrow (A' \cup_C D) \cup_{(C' \cup_C D)} B' \cong A' \cup_{C'} B'. \qquad \square$$

In our examples, we sometimes prove the gluing lemma directly, because it is no more difficult. However, in deeper examples it is often easier to check that the model structure is left proper and to use the previous result to deduce that the left gluing lemma holds.

15.5. Left properness, lifting properties, and the sets [X, Y]

We work in a fixed model category $\mathcal{M}$ in this section. Let $i\colon A \longrightarrow X$ be a map in $\mathcal{M}$ and Z be an object of $\mathcal{M}$. Consider the induced map

15.5.1
$$i^*\colon [X, Z] \longrightarrow [A, Z]$$

of hom sets in Ho$\mathcal{M}$. We describe how we can sometimes deduce that i^* is a bijection directly from lifting properties and, conversely, how we can sometimes deduce lifting properties when i^* is a bijection. While i will be a cofibration and Z will be fibrant, the force of these results comes from the fact that the relevant lifting properties concern pairs of maps, neither of which need be a weak equivalence. This should be a standard part of model category theory, although we have not found exactly what we want in the literature. The work in this section will be needed in our discussion of Bousfield localization, and it well illustrates how important it is to know whether a model structure is left or right proper. The results here all have evident duals, with cofibrations and fibrations reversed.

We defined homotopy pushouts of spaces in Definition 2.1.1. We now generalize to arbitrary model categories and then specialize to a particular case of interest.

DEFINITION 15.5.2. Define the homotopy pushout (or double mapping cylinder) $M(f, g)$ of a pair of maps $f\colon A \longrightarrow X$ and $g\colon A \longrightarrow Y$ to be the pushout of $f \amalg g\colon A \amalg A \longrightarrow X \amalg Y$ along the cofibration $i_0 + i_1\colon A \amalg A \longrightarrow \mathrm{Cyl}\, A$ of a good cylinder object $\mathrm{Cyl}\, A$. Thus a map $h\colon M(f, g) \longrightarrow Z$ specifies a good left homotopy $h i_0 f \simeq h i_0 g$. When $X = Y$ and $f = g$, we call the homotopy pushout $M(f, f)$ a spool[2] of f and denote it by $\mathrm{Spl}\, f$. We define homotopy pullbacks dually.

We have the following pair of results, the first of which is just an observation.

LEMMA 15.5.3. *If* $i\colon A \longrightarrow X$ *and the canonical map* $k\colon \mathrm{Spl}\, i \longrightarrow \mathrm{Cyl}\, X$ *are cofibrations between cofibrant objects and both satisfy the LLP with respect to a fibration* $Z \longrightarrow *$, *then* $i^*\colon [X, Z] \longrightarrow [A, Z]$ *is a bijection.*

---

2. This well-chosen term is due to Bousfield [16].

PROOF. Since $[X, Z] = \pi(X, Z)$ and similarly for $[A, Z]$, the LLP for i gives that i^* is a surjection and the LLP for k gives that i^* is an injection. ☐

LEMMA 15.5.4. *Assume that $\mathcal{M}$ is left proper. If $i: A \longrightarrow X$ is a cofibration and Z is a fibrant object such that*

$$i^*: [X, Z] \longrightarrow [A, Z]$$

is a bijection, then $Z \longrightarrow *$ satisfies the RLP with respect to i.

The proof uses the notion of a cofibrant approximation of a map.

DEFINITION 15.5.5. A cofibrant approximation of a map $f: X \longrightarrow Y$ is a commutative diagram

$$
\begin{array}{ccc}
QX & \xrightarrow{\ f'\ } & QY \\
{\scriptstyle q_X}\downarrow & & \downarrow{\scriptstyle q_Y} \\
X & \xrightarrow{\ f\ } & Y
\end{array}
$$

in which QX and QY are cofibrant, f' is a cofibration, and q_X and q_Y are weak equivalences.

LEMMA 15.5.6. *Any map $f: X \longrightarrow Y$ has a cofibrant approximation.*

PROOF. Let $q_X: QX \longrightarrow X$ be a cofibrant approximation of X and factor fq_X as $q_Y f'$, where $f': QX \longrightarrow QY$ is a cofibration and $q_Y: QY \longrightarrow Y$ is an acyclic fibration. ☐

Observe that cofibrant approximation of maps gives a way to arrange that the hypothesis on i in Lemma 15.5.3 is satisfied. We will consider analogues for the map k in Lemma 15.5.3 at the end of the section. Lemma 15.5.4 is proven by concatenating the following two results, using that i^* is isomorphic to j^* if j is a cofibrant approximation of i.

LEMMA 15.5.7. *Assume that $\mathcal{M}$ is left proper. Let $i: A \longrightarrow X$ be a cofibration, let $j: B \longrightarrow Y$ be a cofibrant approximation of i, and let $p: E \longrightarrow B$ be a fibration. If p satisfies the RLP with respect to j, then p satisfies the RLP with respect to i.*

PROOF. Consider the following diagram.

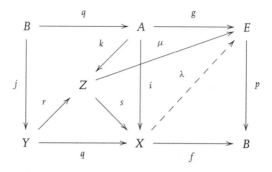

Here $i\colon A \longrightarrow X$ is a cofibration, f and g are given such that the right square commutes, and we seek a lift λ. As constructed in Lemma 15.5.6, the two maps q are cofibrant approximations and j is a cofibration. The upper left square is a pushout, so that k is a cofibration and, since $\mathcal{M}$ is left proper, r is a weak equivalence. The map s is given by the universal property of pushouts and is a weak equivalence since r and q are so. There is a lift $Y \longrightarrow E$ by hypothesis, and the lift μ is then given by the universal property of pushouts. We may regard X, Z, and E as objects in the category $(A/\mathcal{M}/B)_{pg}$. Since i and k are cofibrations in $\mathcal{M}$, X and Z are cofibrant objects of the model category $(A/\mathcal{M}/B)_{pg}$, and s is a weak equivalence between them. Similarly E is a fibrant object of $(A/\mathcal{M}/B)_{pg}$. Therefore the induced map

$$s^*\colon \pi(X, E)_{pg} \longrightarrow \pi(Z, E)_{pg}$$

of homotopy classes of maps in $(A/\mathcal{M}/B)_{pg}$ is a bijection. This implies that there is a lift $\lambda\colon X \longrightarrow E$ such that $\lambda s \simeq \mu$ under A and over B. The two triangles in the right square commute since λ is a map under A and over B. $\square$

LEMMA 15.5.8. *If $j\colon B \longrightarrow Y$ is a cofibration between cofibrant objects and Z is a fibrant object such that*

$$j^*\colon [Y, Z] \longrightarrow [B, Z]$$

is a bijection, then $Z \longrightarrow *$ satisfies the RLP with respect to j.

PROOF. $[Y, Z] = \pi(Y, Z)$ and similarly for B, so for any map $g\colon B \longrightarrow Z$ there is a map $\lambda'\colon Y \longrightarrow Z$ and a homotopy $h\colon B \longrightarrow \operatorname{Cocyl} Z$ from $\lambda'j$ to g. By HEP, Lemma 14.3.7, h extends to a homotopy $\tilde{h}\colon Y \longrightarrow \operatorname{Cocyl} Z$ such that $p_0\tilde{h} = \lambda'$ and $\tilde{h}j = h$. The map $\lambda = p_1\tilde{h}$ satisfies $\lambda j = g$. $\square$

Returning to Lemma 15.5.3, we consider its hypothesis about $\operatorname{Spl} i$. We therefore assume that $i\colon A \longrightarrow X$ is a cofibration between cofibrant objects.

A map $h: \operatorname{Spl} i \longrightarrow Z$ specifies two maps $X \longrightarrow Z$ and a good left homotopy between their restrictions to A. When $\operatorname{Cyl} X$ is very good, we can form a commutive diagram

15.5.9

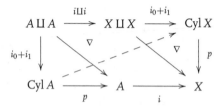

Passage to pushouts displays the map $k: \operatorname{Spl} i \longrightarrow \operatorname{Cyl} X$ as a cofibration between cofibrant objects. Here the map $\operatorname{Cyl} i$ is obtained by noticing that the solid arrow part of the following diagram commutes, so that we can obtain the dotted lift.

There is an alternative way to verify the hypothesis on k, but it presupposes familiarity with the notion of an enriched model structure that will be discussed in §16.4.

REMARK 15.5.10. When we have a good cylinder functor in the sense of Addendum 16.4.10, we automatically have a map $\operatorname{Cyl} i$ such that (15.5.9) commutes. The following diagrams display the idea behind the two ways of getting the map k; the symbol $\odot$ indicates tensors, as defined in §16.3.

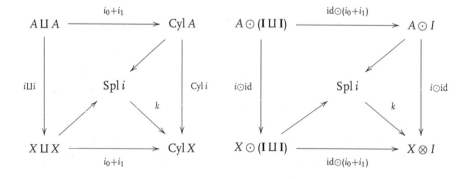

In the first, we are working in a general model category and k: Spl $i \longrightarrow$ Cyl X is obtained by passage to pushouts from the right square of (15.5.9). If $\mathcal{M}$ is left proper, then k is a weak equivalence if i is a weak equivalence. In the second, we are working in a $\mathcal{V}$-model category where the unit object $\mathbf{I}$ of the monoidal model category $\mathcal{V}$ has a good cylinder object I. The second diagram is a pushout product, hence k there is a cofibration and is a weak equivalence if i is a weak equivalence.

Thus the hypotheses of Lemma 15.5.3 hold quite generally. The discussion leads to the following definition, which will be used at one place in §19.3.

DEFINITION 15.5.11. A subcategory of weak equivalences $\mathcal{L}$ has good spools if for every $\mathcal{L}$-acyclic cofibration i: $A \longrightarrow X$ between cofibrant objects, there is a spool Spl i such that k: Spl $i \longrightarrow$ Cyl X is an $\mathcal{L}$-acyclic cofibration.

16

CATEGORICAL PERSPECTIVES
ON MODEL CATEGORIES

Just as we have categories, functors, and natural transformations, we have homotopy categories derived from model categories, which we call derived homotopy categories, together with derived functors and derived natural transformations. These functors and natural transformations come in two flavors, left and right, and are discussed in §16.1. Left and right Quillen adjoints are the most common source of left and right derived functors, and we discuss these in §16.2. We return to these ideas in §16.5 where, following Hovey [66, §1.4] and Shulman [127], we describe the 2-categorical way of understanding the passage to derived homotopy categories and we explain that double categories, rather than categories or 2-categories, give the appropriate conceptual framework for understanding maps between model categories. In that framework, left and right Quillen adjoints are treated symmetrically as morphisms in a single double category, rather than asymmetrically as the morphisms in either of a pair of categories.

In §16.3, we outline the theory of enriched categories. In most of mathematics, categories appear not just with sets of morphisms but with morphism objects that lie in some well-behaved category $\mathcal{V}$. For example, the objects of $\mathcal{V}$ might be abelian groups, modules over a commutative ring, topological spaces, simplicial sets, chain complexes over a commutative ring, or spectra. We return to model category theory and describe enriched model category theory in §16.4. One small point that deserves more emphasis than it receives in the literature is that there is a familiar classical notion of homotopy in the enriched categories that appear in nature, and the model categorical notion of homotopy and the classical notion of homotopy can be used interchangably in such contexts.

16.1. Derived functors and derived natural transformations

Having defined model categories, we want next to define functors and natural transformations between them in such a way that they give "derived" functors

314

and natural transformations on passage to their derived homotopy categories. To that end, we must first define what we mean by derived functors between the derived homotopy categories $\mathrm{Ho}\mathscr{M}$ and $\mathrm{Ho}\mathscr{N}$ of model categories $\mathscr{M}$ and $\mathscr{N}$. However, we focus primarily on a single model category $\mathscr{M}$ in this section.

We say that a functor $F\colon \mathscr{M} \longrightarrow \mathscr{N}$ between categories with weak equivalences is homotopical if it takes weak equivalences to weak equivalences, and we say that a functor $F\colon \mathscr{M} \longrightarrow \mathrm{Ho}\mathscr{N}$ or more generally a functor $F\colon \mathscr{M} \longrightarrow \mathscr{H}$ for any other category $\mathscr{H}$ is homotopical if it takes weak equivalences to isomorphisms. The universal property of localization then gives a functor $\tilde{F}\colon \mathrm{Ho}\mathscr{M} \longrightarrow \mathrm{Ho}\mathscr{N}$ in the first case or $\tilde{F}\colon \mathrm{Ho}\mathscr{M} \longrightarrow \mathscr{H}$ in the general case such that the first or the second of the following two diagrams commutes.

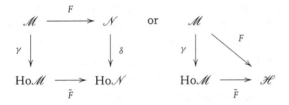

Here $\gamma\colon \mathscr{M} \longrightarrow \mathrm{Ho}\mathscr{M}$ and $\delta\colon \mathscr{N} \longrightarrow \mathrm{Ho}\mathscr{N}$ denote the localization functors.

The functor $\tilde{F}$ is a derived functor of F. However, this will not suffice for the applications. As we have already said, most functors F that we encounter are not homotopical, and then it is too much to expect that diagrams such as those above commute; rather, we often obtain diagrams like these that commute up to a natural transformation that is characterized by a universal property.

There are two kinds of derived functor in common use: left and right. We note parenthetically that there are also functors that in some sense deserve the name of a derived functor and yet are neither left nor right derived in the sense we are about to define. The theory of such functors is not well understood, but they appear in applications (e.g., [97]) and have been given a formal description in [128]. In this section, we focus attention on a functor $F\colon \mathscr{M} \longrightarrow \mathscr{H}$, where the target category $\mathscr{H}$ is arbitrary. When we return to the model category $\mathscr{N}$, we will apply the following definition with $\mathscr{H}$ taken to be $\mathrm{Ho}\mathscr{N}$ and with F replaced by $\delta \circ F$ for some functor $F\colon \mathscr{M} \longrightarrow \mathscr{N}$.

DEFINITION 16.1.1. A left derived functor of a functor $F\colon \mathscr{M} \longrightarrow \mathscr{H}$ is a functor $\mathbb{L}F\colon \mathrm{Ho}\mathscr{M} \to \mathscr{H}$ together with a natural transformation $\mu\colon \mathbb{L}F \circ \gamma \to F$ such that for any functor $K\colon \mathrm{Ho}\mathscr{M} \longrightarrow \mathscr{H}$ and natural transformation $\xi\colon K \circ \gamma \longrightarrow F$, there is a unique natural transformation $\sigma\colon K \longrightarrow \mathbb{L}F$ such that the composite $\mu \circ (\sigma \cdot \gamma): K \circ \gamma \longrightarrow \mathbb{L}F \circ \gamma \longrightarrow F$ coincides with ξ. That is, $(\mathbb{L}F, \mu)$ is terminal among pairs (K, ξ).

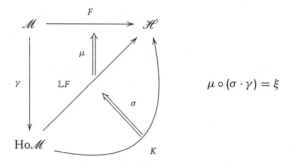

$$\mu \circ (\sigma \cdot \gamma) = \xi$$

It is a categorical convention to write arrows for functors and double arrows for natural transformations, as in the diagram above. We use ∘ to denote the composite of two functors or the composite of two natural transformations. We use · to denote the composite of a natural transformation and a functor; categorically, that is often called whiskering. For example, $\sigma \cdot \gamma$ is defined by applying σ to objects in the image of γ. Since left derived functors are characterized by a universal property, they are unique up to canonical isomorphism if they exist. Confusingly, they are examples of what are known categorically as right Kan extensions; we shall ignore that categorical perspective. Of course, we also have a dual definition.

DEFINITION 16.1.2. A right derived functor of a functor $F: \mathcal{M} \longrightarrow \mathcal{H}$ is a functor $\mathbb{R}F: \text{Ho}\mathcal{M} \to \mathcal{H}$ together with a natural transformation $\nu: F \to \mathbb{R}F \circ \gamma$ such that for any functor $P: \mathcal{M} \longrightarrow \text{Ho}\mathcal{M}$ and natural transformation $\zeta: F \longrightarrow P \circ \gamma$, there is a unique natural transformation $\rho: \mathbb{R}F \longrightarrow P$ such that the composite $(\rho \cdot \gamma) \circ \nu: F \longrightarrow \mathbb{R}F \circ \gamma \longrightarrow P \circ \gamma$ coincides with ζ. That is, $(\mathbb{R}F, \nu)$ is initial among pairs (P, ζ).

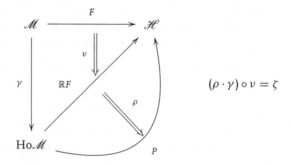

$$(\rho \cdot \gamma) \circ \nu = \zeta$$

It is important to realize that these definitions depend only on the functor F and the localization $\gamma : \mathcal{M} \longrightarrow \mathrm{Ho}\mathcal{M}$, not on any possible model structure that might be present on $\mathcal{M}$ or any extra structure on $\mathcal{H}$. We have assumed that $\mathcal{M}$ has a model structure since that structure leads to a convenient general way to construct left and right derived functors of suitably well-behaved functors.

PROPOSITION 16.1.3. *If $F : \mathcal{M} \longrightarrow \mathcal{H}$ takes acyclic cofibrations between cofibrant objects to isomorphisms, then the left derived functor $(\mathbb{L}F, \mu)$ exists. Moreover, for any cofibrant object X of $\mathcal{M}$, $\mu : \mathbb{L}FX \longrightarrow FX$ is an isomorphism.*

PROOF. By Ken Brown's lemma, 14.2.9, and Remark 14.2.10, F carries all weak equivalences between cofibrant objects to isomorphisms. By Proposition 14.4.4, cofibrant replacement induces a functor $Q : \mathcal{M} \longrightarrow h\mathcal{M}_c$. By Lemma 14.4.10, F passes to right homotopy classes to induce a functor $F : h\mathcal{M}_c \longrightarrow \mathcal{H}$. The functor $F \circ Q$ carries weak equivalences in $\mathcal{M}$ to isomorphisms. Define $\mathbb{L}F : \mathrm{Ho}\mathcal{M} \longrightarrow \mathcal{H}$ to be the functor induced from $F \circ Q$ by the universal property of γ and define $\mu : \mathbb{L}F \circ \gamma \longrightarrow F$ at an object X to be the map $Fq : FQX \longrightarrow FX$ in $\mathcal{H}$. If X is cofibrant, then q is a weak equivalence between cofibrant objects and Fq is an isomorphism. It is an easy exercise to verify that $(\mathbb{L}F, \mu)$ satisfies the required universal property. $\square$

PROPOSITION 16.1.4. *If $F : \mathcal{M} \longrightarrow \mathcal{H}$ takes acyclic fibrations between fibrant objects to isomorphisms, then the right derived functor $(\mathbb{R}F, \nu)$ exists. Moreover, for any fibrant object Y of $\mathcal{M}$, $\nu : FY \longrightarrow \mathbb{R}FY$ is an isomorphism.*

Now return to the pair of model categories $\mathcal{M}$ and $\mathcal{N}$. The following language is standard. Its purpose is just to distinguish between the two choices of target category, $\mathcal{N}$ and $\mathrm{Ho}\mathcal{N}$. We will later omit the word "total".

DEFINITION 16.1.5. *Let $F : \mathcal{M} \longrightarrow \mathcal{N}$ be a functor. A total left derived functor of F is a functor $\mathbb{L}F : \mathrm{Ho}\mathcal{M} \longrightarrow \mathrm{Ho}\mathcal{N}$ such that $\mathbb{L}F$ is a derived functor of $\delta \circ F$. A total right derived functor is defined dually.*

COROLLARY 16.1.6. *If $F : \mathcal{M} \longrightarrow \mathcal{N}$ takes acyclic cofibrations between cofibrant objects to weak equivalences, then the total left derived functor $(\mathbb{L}F, \mu)$ exists.*

COROLLARY 16.1.7. *If $F : \mathcal{M} \longrightarrow \mathcal{N}$ takes acyclic fibrations between fibrant objects to weak equivalences, then the total right derived functor $(\mathbb{R}F, \mu)$ exists.*

Having defined and shown the existence of derived functors, the obvious next step is to define and show the existence of derived natural transformations.

DEFINITION 16.1.8. Let $\alpha \colon F \longrightarrow F'$ be a natural transformation between functors $F, F' \colon \mathcal{M} \longrightarrow \mathcal{H}$ with left derived functors $(\mathbb{L}F, \mu)$ and $(\mathbb{L}F', \mu')$. A derived natural transformation of α is a natural transformation $\mathbb{L}\alpha \colon \mathbb{L}F \to \mathbb{L}F'$ such that the following diagram of natural transformations commutes and is terminal among commuting squares of the same form.

$$
\begin{array}{ccc}
& \mathbb{L}\alpha & \\
\mathbb{L}F & \longrightarrow & \mathbb{L}F' \\
\mu \downarrow & & \downarrow \mu' \\
F & \longrightarrow & F' \\
& \alpha &
\end{array}
$$

The terminality condition is displayed schematically in the commutative diagram of functors and natural transformations

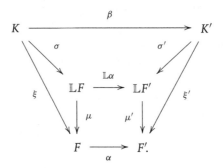

Given β, ξ, and ξ' that make the outer trapezoid commute, there must exist natural transformations σ and σ' that make the upper trapezoid commute. Derived natural transformations between right derived functors are defined dually.

REMARK 16.1.9. In the context of total derived functors of a pair of functors $F, F' \colon \mathcal{M} \longrightarrow \mathcal{N}$, the bottom arrow $\alpha \colon F \longrightarrow F'$ must be replaced by the arrow $\delta \circ \alpha \colon \delta \circ F \longrightarrow \delta \circ F'$ in the diagrams in the preceding definition.

In the context of Corollary 16.1.6, the existence is obvious. In effect, we define $\mathbb{L}\alpha \colon \mathbb{L}FX \longrightarrow \mathbb{L}F'X$ to be $\alpha \colon FQX \longrightarrow F'QX$. The verification of terminality is straightforward. The dual existence statement is similar.

LEMMA 16.1.10. *If $F, F' : \mathcal{M} \longrightarrow \mathcal{N}$ are functors that take acyclic cofibrations between cofibrant objects to weak equivalences and $\alpha : F \longrightarrow F'$ is a natural transformation, then $\mathbb{L}\alpha$ exists.*

LEMMA 16.1.11. *If $F, F' : \mathcal{M} \longrightarrow \mathcal{N}$ are functors that take acyclic fibrations between fibrant objects to weak equivalences and $\alpha : F \longrightarrow F'$ is a natural transformation, then $\mathbb{R}\alpha$ exists.*

REMARK 16.1.12. In practice, the functors that have left derived functors are often left adjoints and the functors that have right derived functors are often right adjoints. Virtually every functor mentioned in this book is a left or right adjoint with a left or right derived functor. However, a given functor F might be both a left and a right adjoint and it might have both a left derived functor $\mathbb{L}F$ and a right derived functor $\mathbb{R}F$. There will then be a natural map $\mathbb{L}F \longrightarrow \mathbb{R}F$, but it need not be an isomorphism in general [128].

16.2. Quillen adjunctions and Quillen equivalences

Corollaries 16.1.6 and 16.1.7 apply in particular to the left and right adjoints of a special kind of adjunction between model categories. Such adjunctions suffice to give most of the derived functors that one needs in the applications. We again assume that $\mathcal{M}$ and $\mathcal{N}$ are model categories.

DEFINITION 16.2.1. Let $F : \mathcal{M} \longrightarrow \mathcal{N}$ and $U : \mathcal{N} \longrightarrow \mathcal{M}$ be left and right adjoint. The pair (F, U) is a Quillen adjunction if the following equivalent conditions are satisfied.

(i) F preserves cofibrations and U preserves fibrations.
(ii) F preserves cofibrations and acyclic cofibrations.
(iii) U preserves fibrations and acyclic fibrations.
(iv) F preserves acyclic cofibrations and U preserves acyclic fibrations.

The Quillen adjunction (F, U) is a Quillen equivalence if for any map $f : FX \longrightarrow Y$ with adjoint $g : X \longrightarrow UY$, where X is cofibrant and Y is fibrant, f is a weak equivalence in $\mathcal{N}$ if and only if g is a weak equivalence in $\mathcal{M}$.

The verification that the conditions listed in the lemma are in fact equivalent is an exercise in the definition of an adjoint functor and the meaning of the lifting properties; we leave it to the reader. The letters F and U are meant to suggest "free" and "underlying". Many of the applications concern adjunctions

(F, U) where F specifies free structured objects (such as monoids, algebras, etc.) and U forgets the structure on these objects.

PROPOSITION 16.2.2. *If (F, U) is a Quillen adjunction, then the total derived functors $\mathbb{L}F$ and $\mathbb{R}U$ exist and form an adjoint pair. If (F, U) is a Quillen equivalence, then $(\mathbb{L}F, \mathbb{R}U)$ is an adjoint equivalence between $\mathrm{Ho}\mathcal{M}$ and $\mathrm{Ho}\mathcal{N}$.*

PROOF. Corollaries 16.1.6 and 16.1.7 give that $\mathbb{L}F$ and $\mathbb{R}U$ exist. If X is cofibrant in $\mathcal{M}$ and Y is fibrant in $\mathcal{N}$, then the adjunction $\mathcal{N}(FX, Y) \cong \mathcal{M}(X, UY)$ passes to homotopy classes of maps to give

$$h\mathcal{N}(FX, Y) \cong h\mathcal{M}(X, UY).$$

Now let X and Y be general objects of $\mathcal{M}$ and $\mathcal{N}$, respectively. Since $\mathbb{L}F = FQ$ and $\mathbb{R}U = UR$ we have

$$\mathrm{Ho}\mathcal{N}(\mathbb{L}FX, Y) \cong \mathrm{Ho}\mathcal{N}(FQX, RY) \cong \mathrm{Ho}\mathcal{M}(QX, URY) \cong \mathrm{Ho}\mathcal{M}(X, \mathbb{R}UY).$$

The weak equivalences $r\colon Y \longrightarrow RY$ and $q\colon QX \longrightarrow X$ induce the first and last isomorphisms. The middle isomorphism is an instance of the isomorphism on the level of homotopy classes of maps.

For the last statement, we must show that the unit and counit of the derived adjunction are isomorphisms. When X is cofibrant, as we may assume, the unit of this adjunction is induced by the composite of the unit $\eta\colon X \longrightarrow UFX$ of the adjunction (F, U) and the map $UrF\colon UFX \longrightarrow URFX$. By the definition of a Quillen equivalence, this composite is a weak equivalence since its adjoint is the weak equivalence $r\colon FX \longrightarrow RFX$. The dual argument applies to the counit. $\square$

The converse of the last statement also holds, and gives one among several equivalent conditions that we record in the following result. We leave the verification to the reader (or see [66, I.3.13, I.3.16]). Recall Definition 15.3.5.

PROPOSITION 16.2.3. *Let (F, U) be a Quillen adjunction between $\mathcal{M}$ and $\mathcal{N}$. Then the following statements are equivalent.*

(i) (F, U) is a Quillen equivalence.

(ii) $(\mathbb{L}F, \mathbb{R}U)$ is an adjoint equivalence of categories.

(iii) F reflects weak equivalences between cofibrant objects and the composite

$$\varepsilon \circ FqU\colon FQUY \longrightarrow FUY \longrightarrow Y$$

is a weak equivalence for all fibrant Y.

(iv) U reflects weak equivalences between fibrant objects and the composite

$$UrF \circ \eta \colon X \longrightarrow UFX \longrightarrow URFX$$

is a weak equivalence for all cofibrant X.

If U creates the weak equivalences in $\mathscr{N}$, the following statement can be added.

(v) $\eta \colon X \longrightarrow UFX$ is a weak equivalence for all cofibrant X.

If F creates the weak equivalences in $\mathscr{M}$, the following statement can be added.

(vi) $\varepsilon \colon FUY \longrightarrow Y$ is a weak equivalence for all fibrant Y.

We illustrate these notions with a reinterpretation of what it means for a model category $\mathscr{M}$ to be left or right proper. Let $f \colon A \longrightarrow B$ be a map in $\mathscr{M}$. We have functors $f^* \colon B/\mathscr{M} \longrightarrow A/\mathscr{M}$ and $f_* \colon \mathscr{M}/A \longrightarrow \mathscr{M}/B$ given by precomposition or postcomposition with f. The functor f^* has the left adjoint $f_!$ given on objects $i \colon A \longrightarrow X$ by letting $f_!(i)$ be the pushout $B \longrightarrow X \cup_A B$. Dually, the functor f_* has the right adjoint $f^!$ given on objects $p \colon E \longrightarrow B$ by letting $f^!(p)$ be the pullback $A \times_B E \longrightarrow A$.

PROPOSITION 16.2.4. A model category $\mathscr{M}$ is left proper if and only if $(f_!, f^*)$ is a Quillen equivalence between the under model categories $A/\mathscr{M}$ and $B/\mathscr{M}$. It is right proper if and only if $(f_*, f^!)$ is a Quillen equivalence between the over model categories $\mathscr{M}/A$ and $\mathscr{M}/B$.

PROOF. Recall Definition 15.3.5. Since the forgetful functors U from the under and over categories to $\mathscr{M}$ strongly create their model structures, by Theorem 15.3.6, and since $Uf^* = U$ on $B/\mathscr{M}$ and $Uf_* = U$ on $\mathscr{M}/A$, it is evident that f^* is a Quillen right adjoint that creates the weak equivalences in $B/\mathscr{M}$ and that f_* is a Quillen left adjoint that creates the weak equivalences in $A/\mathscr{M}$. A cofibrant object of $A/\mathscr{M}$ is a cofibration $i \colon A \longrightarrow X$, and the unit η on the object i is precisely the map $X \longrightarrow X \cup_A B$ that the definition of left proper asserts to be a weak equivalence when f is a weak equivalence. Therefore, Proposition 16.2.3(v) gives the claimed characterization of left proper. Dually, Proposition 16.2.3(vi) gives the claimed characterization of right proper. □

The criterion of Proposition 16.2.3(v) is especially useful since a standard way to build a model category structure on $\mathscr{N}$ is to use an adjunction (F, U) to create it from a model structure on $\mathscr{M}$, setting $\mathscr{W}_{\mathscr{N}} = U^{-1}(\mathscr{W}_{\mathscr{M}})$ and $\mathscr{F}_{\mathscr{N}} = U^{-1}(\mathscr{F}_{\mathscr{M}})$. The following result, which is [65, 11.3.2], is frequently used for this purpose.

THEOREM 16.2.5. *Let (F, U) be an adjunction between $\mathscr{M}$ and $\mathscr{N}$, where $\mathscr{M}$ is a cofibrantly or compactly generated model category with sets $\mathcal{I}$ and $\mathcal{J}$ of generating cofibrations and generating acyclic cofibrations. Let $F\mathcal{I}$ and $F\mathcal{J}$ be the sets of maps in $\mathscr{N}$ obtained by applying F to the maps in $\mathcal{I}$ and $\mathcal{J}$. Define weak equivalences and fibrations in $\mathscr{N}$ by requiring the functor U to create them. Then $\mathscr{N}$ is a cofibrantly or compactly generated model category with generating cofibrations $F\mathcal{I}$ and generating acyclic cofibrations $F\mathcal{J}$ if the following conditions are satisfied.*

(i) $F\mathcal{I}$ and $F\mathcal{J}$ are small or compact.
(ii) Every relative $F\mathcal{J}$-cell complex is a weak equivalence.

Moreover, (F, U) is then a Quillen adjunction.

PROOF. Since functors preserve composition and retracts, the weak equivalences in $\mathscr{N}$ form a subcategory of weak equivalences. Now the conclusion follows from Theorem 15.2.3 since exercises in the use of adjunctions show that the compatibility condition $F\mathcal{I}^{\boxtimes} = F\mathcal{J}^{\boxtimes} \cap \mathscr{W}$ in $\mathscr{N}$ follows formally from the compatibility condition $\mathcal{I}^{\boxtimes} = \mathcal{J}^{\boxtimes} \cap \mathscr{W}$ in $\mathscr{M}$. Since U preserves fibrations and weak equivalences, the last statement is clear. $\square$

REMARK 16.2.6. With the notations of Definition 15.1.6 or 15.1.7, the verification of (i) reduces by adjunction to consideration of maps $A \longrightarrow U\mathrm{colim}_{\beta < \lambda} Z_{\beta}$, where A is small in $\mathscr{M}$ and Z is a relative cell complex in $\mathscr{N}$ (defined using sequences in the compact case). Often U commutes with the colimits relevant to the small object argument and the smallness of A in $\mathscr{M}$ implies the smallness of FA in $\mathscr{N}$. The verification of (ii) concerns the preservation of weak equivalences under colimits. Since the weak equivalences are created by the right adjoint U, this is not formal.

If (F, U) and (G, V) are adjoint pairs, where the target category of F is the source category of G, then the composites (GF, UV) also form an adjoint pair. If (F, U) and (G, V) are Quillen adjunctions, then so is (GF, UV). It is standard in the model category literature to define the category of model categories to have objects the model categories and morphisms the left Quillen functors or, alternatively, the right Quillen functors. Obviously the asymmetry is unaesthetic.

More deeply, it masks one of the greatest difficulties of model category theory and of the theory of derived functors in general. Good properties of left adjoints are preserved under composition of left adjoints. However, it is in practice very often necessary to compose left adjoints with right adjoints.

Results about such composites are hard to come by. They are often truly deep mathematics. We shall say a little bit more about this in §16.5.

However, the categorical considerations of greatest relevance to the applications of model category theory concern enrichment from hom sets to hom objects. The model axioms do not refer to any enrichment that might be present, but there are additional axioms that do relate enrichment to the model structure. We shall turn to those in §16.4 after summarizing the categorical background in the next section.

16.3. Symmetric monoidal categories and enriched categories

In practice, categories come in nature with more structure than just sets of morphisms. This extra structure is central to all of category theory, homotopical or not. While every mathematician who makes use of categories should understand enrichment, this is not the place for a full exposition. The most thorough source is Kelly's book [75], and an introduction can be found in Borceux [12, Ch. 6]. We outline what is most relevant to model categories in this section.

A monoidal structure on a category $\mathscr{V}$ is a product, $\otimes$ say, and a unit object I such that the product is associative and unital up to coherent natural isomorphisms; $\mathscr{V}$ is symmetric if $\otimes$ is also commutative up to coherent natural isomorphism. Informally, coherence means that diagrams that intuitively should commute do in fact commute. (The symmetry coherence admits a weakening that gives braided monoidal categories, but those will not concern us.) A symmetric monoidal category $\mathscr{V}$ is closed if it has internal hom objects $\underline{\mathscr{V}}(X, Y)$ in $\mathscr{V}$ together with adjunction isomorphisms

$$\mathscr{V}(X \otimes Y, Z) \cong \mathscr{V}(X, \underline{\mathscr{V}}(Y, Z)).$$

These isomorphisms of hom sets imply isomorphisms of internal hom objects in $\mathscr{V}$

$$\underline{\mathscr{V}}(X \otimes Y, Z) \cong \underline{\mathscr{V}}(X, \underline{\mathscr{V}}(Y, Z)).$$

The proof is an exercise in the use of the Yoneda lemma: these two objects represent isomorphic functors of three variables.

From now on, we let $\mathscr{V}$ be a bicomplete closed symmetric monoidal category. Such categories appear so often in nature that category theorists have invented a name for them: such a category is often called a "cosmos". We will require our cosmos $\mathscr{V}$ to be a model category in the next section, but we ignore model category theory for the moment. When $\otimes$ is the cartesian product, we say that $\mathscr{V}$ is cartesian closed, but the same category $\mathscr{V}$ can admit other symmetric monoidal structures.

EXAMPLES 16.3.1. We give examples of cosmoi $\mathscr{V}$.

(i) The category $\mathscr{S}et$ of sets is closed cartesian monoidal.

(ii) The category $\mathscr{U}$ of (compactly generated) spaces is cartesian closed. The space $\underline{\mathscr{U}}(X, Y)$ is the function space of maps $X \longrightarrow Y$ with the k-ification of the compact open topology.

(iii) The category $\mathscr{U}_*$ of based spaces is closed symmetric monoidal under the smash product. The smash product would not be associative if we used just spaces, rather than compactly generated spaces [97, §1.7]. The based space $\underline{\mathscr{U}_*}(X, Y)$ is the function space $F(X, Y)$ of based maps $X \longrightarrow Y$.

(iv) The category $s\mathscr{S}et$ of simplicial sets is cartesian closed. A simplicial set is a contravariant functor $\Delta \longrightarrow \mathscr{S}et$, where Δ is the category of sets $n = \{0, 1, \cdots, n\}$ and monotonic maps. There are n-simplices $\Delta[n]$ in $s\mathscr{S}et$, and $n \mapsto \Delta[n]$ gives a covariant functor $\Delta \longrightarrow s\mathscr{S}et$. The internal hom [92, §I.6] in $s\mathscr{S}et$ is specified by

$$\underline{s\mathscr{S}et}(X, Y)_n = s\mathscr{S}et(X \times \Delta[n], Y).$$

(v) For a commutative ring R, the category $\mathscr{M}_R$ of R-modules is closed symmetric monoidal under the functors $\otimes_R$ and Hom_R; in particular, the category $\mathscr{A}b$ of abelian groups is closed symmetric monoidal.

(vi) For a commutative ring R, the category Ch_R of $\mathbb{Z}$-graded chain complexes of R-modules (with differential lowering degree) is closed symmetric monoidal under the graded tensor product and hom functors

$$(X \otimes_R Y)_n = \Sigma_{p+q=n} X_p \otimes_R Y_q; \quad d(x \otimes y) = d(x) \otimes y + (-1)^p x \otimes d(y)$$

$$\text{Hom}_R(X, Y)_n = \Pi_i \text{Hom}_R(X_i, Y_{i+n}); \quad d(f)_i = d \circ f_i - (-1)^n f_{i-1} \circ d.$$

Here the symmetry $\gamma: X \otimes Y \longrightarrow Y \otimes X$ is defined with a sign,

$$\gamma(x \otimes y) = (-1)^{pq} y \otimes x \text{ for } x \in X_p \text{ and } y \in Y_q.$$

(vii) The category $\mathscr{C}at$ of small categories is cartesian closed.

Example (iii) generalizes from $\mathscr{U}$ to an arbitrary cartesian closed category $\mathscr{V}$.

EXAMPLE 16.3.2. Let $\mathscr{V}$ be cartesian closed. Its unit object $*$ is a terminal object, so there is a unique map $t: V \longrightarrow *$ for any $V \in \mathscr{V}$. Let $\mathscr{V}_* = */\mathscr{V}$ be the category of based objects in $\mathcal{V}$, with base maps denoted $i: * \longrightarrow V$. For $V, W \in \mathcal{V}_*$, define the smash product $V \wedge W$ and the function object $F(V, W)$ to be the pushout and pullback in $\mathcal{V}$ displayed in the diagrams

$$
\begin{array}{ccc}
V \amalg W & \xrightarrow{\ j\ } & V \times W \\
\downarrow & & \downarrow \\
* & \longrightarrow & V \wedge W
\end{array}
\qquad \text{and} \qquad
\begin{array}{ccc}
F(V, W) & \longrightarrow & \underline{\mathcal{V}}(V, W) \\
\downarrow & & \downarrow{\scriptstyle i^*} \\
\underline{\mathcal{V}}(*, *) & \xrightarrow[\ i_*\ ]{} & \underline{\mathcal{V}}(*, W).
\end{array}
$$

Here j has coordinates (id, i) on $V \cong V \times *$ and (i, id) on $W \cong * \times W$, and the base map $i: * \longrightarrow F(V, W)$ is induced by the canonical isomorphism $* \longrightarrow \underline{\mathcal{V}}(*, *)$ and the map $\mathcal{V}(t, i): * \cong \mathcal{V}(*, *) \longrightarrow \underline{\mathcal{V}}(V, W)$. The unit $S^0 = *_+$ in $\mathcal{V}_*$ is the coproduct of two copies of $*$, with one of them giving the base map $i: * \longrightarrow S^0$.

There are two ways of thinking about enriched categories. One can think of "enriched" as an adjective, in which case one thinks of enrichment as additional structure on a preassigned ordinary category. Alternatively, one can think of "enriched category" as a noun, in which case one thinks of a self-contained definition of a new kind of object. From that point of view, one constructs an ordinary category from an enriched category. Thinking from the two points of view simultaneously, it is essential that the constructed ordinary category be isomorphic to the ordinary category that one started out with. Either way, there is a conflict of notation between that preferred by category theorists and that in common use by "working mathematicians" (to whom [80] is addressed). We give the definition in its formulation as a noun, but we use notation that implicitly takes the working mathematician's point of view that we are starting with a preassigned category $\mathcal{M}$.

DEFINITION 16.3.3. Let $\mathcal{V}$ be a symmetric monoidal category. A $\mathcal{V}$-category $\mathcal{M}$, or a category $\mathcal{M}$ enriched in $\mathcal{V}$, consists of

(i) a class of objects, with typical objects denoted X, Y, Z;
(ii) for each pair of objects (X, Y), a hom object $\underline{\mathcal{M}}(X, Y)$ in $\mathcal{V}$;
(iii) for each object X, a unit map $\mathrm{id}_X: \mathbf{I} \longrightarrow \underline{\mathcal{M}}(X, X)$ in $\mathcal{V}$; and
(iv) for each triple of objects (X, Y, Z), a composition morphism in $\mathcal{V}$

$$
\underline{\mathcal{M}}(Y, Z) \otimes \underline{\mathcal{M}}(X, Y) \longrightarrow \underline{\mathcal{M}}(X, Z).
$$

The evident associativity and unity diagrams are required to commute.

$$\underline{\mathcal{M}}(Y,Z) \otimes \underline{\mathcal{M}}(X,Y) \otimes \underline{\mathcal{M}}(W,X) \longrightarrow \underline{\mathcal{M}}(Y,Z) \otimes \underline{\mathcal{M}}(W,Y)$$

$$\downarrow \qquad\qquad\qquad\qquad\qquad\qquad\qquad\qquad \downarrow$$

$$\underline{\mathcal{M}}(X,Z) \otimes \underline{\mathcal{M}}(W,X) \longrightarrow \underline{\mathcal{M}}(W,Z)$$

$$\mathbf{I} \otimes \underline{\mathcal{M}}(X,Y) \overset{\cong}{\longleftarrow} \underline{\mathcal{M}}(X,Y) \overset{\cong}{\longrightarrow} \underline{\mathcal{M}}(X,Y) \otimes \mathbf{I}$$

$$\downarrow \qquad\qquad\qquad\qquad \downarrow \text{id} \qquad\qquad\qquad\qquad \downarrow$$

$$\underline{\mathcal{M}}(Y,Y) \otimes \underline{\mathcal{M}}(X,Y) \longrightarrow \underline{\mathcal{M}}(X,Y) \longleftarrow \underline{\mathcal{M}}(X,Y) \otimes \underline{\mathcal{M}}(X,X)$$

The underlying category of the enriched category has the same objects and has morphism sets specified by

16.3.4 $$\mathcal{M}(X,Y) = \mathscr{V}(\mathbf{I}, \underline{\mathcal{M}}(X,Y)).$$

The unit element of $\mathcal{M}(X,X)$ is id_X. The composition is the evident composite

$$\mathscr{V}(\mathbf{I}, \underline{\mathcal{M}}(Y,Z)) \times \mathscr{V}(\mathbf{I}, \underline{\mathcal{M}}(X,Y))$$

$$\downarrow \otimes$$

$$\mathscr{V}(\mathbf{I} \otimes \mathbf{I}, \underline{\mathcal{M}}(Y,Z) \otimes \underline{\mathcal{M}}(X,Y))$$

$$\downarrow$$

$$\mathscr{V}(\mathbf{I}, \underline{\mathcal{M}}(X,Z)),$$

where we have used the unit isomorphism $\mathbf{I} \otimes \mathbf{I} \cong \mathbf{I}$.

As said, we have given the definition in its "noun" form. In its "adjectival" form, one starts with a preassigned ordinary category $\mathcal{M}$, prescribes the appropriate enrichment, and requires a canonical isomorphism between the original category $\mathcal{M}$ and the underlying category of the prescribed enriched category. Rigorously, equality must be replaced by isomorphism in (16.3.4), but one generally regards that canonical isomorphism as an identification. Less formally, we start with an ordinary category $\mathcal{M}$, construct the hom objects $\underline{\mathcal{M}}(X,Y)$ in $\mathscr{V}$, and check that we have the identification (16.3.4). For example, any cosmos is naturally enriched over itself via its internal hom objects. The reader is urged to think through the identifications (16.3.4) in Examples 16.3.1.

EXAMPLES 16.3.5. When $\mathcal{V}$ is one of the cosmoi specified in Examples 16.3.1, categories enriched in $\mathcal{V}$, or $\mathcal{V}$-categories, have standard names.

(i) Categories as usually defined are categories enriched in $\mathcal{S}et$.
(ii) Categories enriched in $\mathcal{A}b$ are called $\mathcal{A}b$-categories. They are called additive categories if they have zero objects and biproducts [80, p. 196]. They are called abelian categories if, further, all maps have kernels and cokernels, every monomorphism is a kernel, and every epimorphism is a cokernel [80, p. 198].
(iii) Categories enriched in $\mathcal{U}$ are called topological categories.
(iv) Categories enriched in $s\mathcal{S}et$ are called simplicial categories.
(v) Categories enriched in Ch_R for some R are called DG-categories.
(vi) Categories enriched in $\mathcal{C}at$ are called (strict) 2-categories and, inductively, categories enriched in the Cartesian monoidal category of $(n-1)$-categories are called (strict) n-categories.

Examples of all six sorts are ubiquitous. For any ring R, not necessarily commutative, the category of left R-modules is abelian. Many categories of structured spaces, such as the categories of topological monoids and of topological groups, are topological categories. The letters DG stand for "differential graded". We shall return to the last example in §16.5.

Most of the model category literature focuses on simplicial categories. Although there are technical reasons for this preference, we prefer to work with naturally occurring enrichments wherever possible, and these may or may not be simplicial. In our examples, we focus on $\mathcal{V} = \mathcal{U}$ and $\mathcal{V} = Ch_R$. These have features in common that are absent when $\mathcal{V} = s\mathcal{S}et$.

Of course, the definition of a $\mathcal{V}$-category is accompanied by the notions of a $\mathcal{V}$-functor $F\colon \mathcal{M} \longrightarrow \mathcal{N}$ and a $\mathcal{V}$-natural transformation $\eta\colon F \longrightarrow G$ between two $\mathcal{V}$-functors $\mathcal{M} \longrightarrow \mathcal{N}$. For the former, we require maps

$$F\colon \underline{\mathcal{M}}(X, Y) \longrightarrow \underline{\mathcal{N}}(FX, FY)$$

in $\mathcal{V}$ that preserve composition and units. For the latter, we require maps $\eta\colon FX \longrightarrow GX$ in $\mathcal{N}$ such that the following naturality diagrams commute in $\mathcal{V}$ for all objects $X, Y \in \mathcal{M}$.

$$\begin{array}{ccc}
\underline{\mathscr{M}}(Y,X) & \xrightarrow{\ \ F\ \ } & \underline{\mathscr{N}}(FY,FX) \\
{\scriptstyle G}\downarrow & & \downarrow{\scriptstyle \eta_*} \\
\underline{\mathscr{N}}(GY,GX) & \xrightarrow[\ \eta^*\ ]{} & \underline{\mathscr{N}}(FY,GX)
\end{array}$$

Here we have used that maps $f\colon X' \longrightarrow X$ and $g\colon Y \longrightarrow Y'$ in $\mathscr{M}$ induce maps

$$f^*\colon \underline{\mathscr{M}}(X,Y) \longrightarrow \underline{\mathscr{M}}(X',Y) \quad\text{and}\quad g_*\colon \underline{\mathscr{M}}(X,Y) \longrightarrow \underline{\mathscr{M}}(X,Y')$$

in $\mathscr{V}$. Indeed, f is an element of the set $\mathscr{V}(\mathrm{I}, \underline{\mathscr{M}}(X,X'))$, and f^* is the composite

$$\underline{\mathscr{M}}(X',Y) \cong \underline{\mathscr{M}}(X',Y) \otimes \mathrm{I} \xrightarrow{\ \mathrm{id}\otimes f\ } \underline{\mathscr{M}}(X',Y) \otimes \underline{\mathscr{M}}(X,X') \xrightarrow{\ \circ\ } \underline{\mathscr{M}}(X,Y).$$

The general idea is that one first expresses categorical notions diagrammatically on hom sets, and one then sees how to reinterpret the notions in the enriched sense.

However, there are important enriched categorical notions that take account of the extra structure given by the enrichment and are not just reinterpretations of ordinary categorical notions. In particular, there are weighted (or indexed) colimits and limits. The most important of these (in nonstandard notation) are tensors $X \odot V$ (sometimes called copowers) and cotensors (sometimes called powers) $\Phi(V,X)$ in $\mathscr{M}$ for objects $X \in \mathscr{M}$ and $V \in \mathscr{V}$. These are characterized by natural isomorphisms

16.3.6 $\qquad \mathscr{M}(X \odot V, Y) \cong \mathscr{V}(V, \underline{\mathscr{M}}(X,Y)) \cong \mathscr{M}(X, \Phi(V,Y))$

of hom sets. Taking X to be the initial object $\varnothing$ or Y to be the terminal object $*$ and using that initial and terminal objects are unique up to isomorphism, we see that

16.3.7 $\qquad\qquad\qquad \varnothing \odot V \cong \varnothing \quad\text{and}\quad \Phi(V,*) \cong *$

for all objects $V \in \mathscr{V}$. There are natural maps

16.3.8

$$X \odot (V \otimes W) \longrightarrow (X \odot V) \odot W \quad\text{and}\quad \Phi(V, \Phi(W,X)) \longrightarrow \Phi(V \otimes W, X)$$

and we require these to be isomorphisms. The first of these is the adjoint of the composite

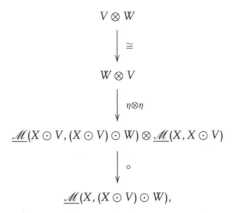

where η is the unit of the first adjunction in (16.3.6), and the second is defined similarly. By a check of represented functors and the Yoneda lemma, (16.3.6) and the isomorphisms (16.3.8) imply natural isomorphisms of objects in $\mathscr{V}$

16.3.9 $\underline{\mathscr{M}}(X \odot V, Y) \cong \underline{\mathscr{V}}(V, \underline{\mathscr{M}}(X, Y)) \cong \underline{\mathscr{M}}(X, \Phi(V, Y)).$

Again, examples are ubiquitous. If R is a ring, X and Y are left R-modules, and V is an abelian group, then $X \otimes V$ and $\mathrm{Hom}\,(V, Y)$ are left R-modules that give tensors and cotensors in the abelian category of left R-modules. This works equally well if X and Y are chain complexes of R-modules and V is a chain complex of abelian groups. We shall return to this example in Chapter 18.

We say that the $\mathscr{V}$-category $\mathscr{M}$ is $\mathscr{V}$-bicomplete if it has all weighted colimits and limits. We dodge the definition of these limits by noting that $\mathscr{M}$ is $\mathscr{V}$-bicomplete if it is bicomplete in the ordinary sense and has all tensors and cotensors [75, Thm. 3.73]. The category $\mathscr{V}$ is itself a $\mathscr{V}$-bicomplete $\mathscr{V}$-category. Its tensors and cotensors are given by its product $\otimes$ and internal hom functor $\underline{\mathscr{V}}$.

16.4. Symmetric monoidal and enriched model categories

Let $\mathscr{M}$ be a $\mathscr{V}$-bicomplete $\mathscr{V}$-category, where $\mathscr{V}$ is a cosmos (bicomplete closed symmetric monoidal category). The reader may wish to focus on the case $\mathscr{M} = \mathscr{V}$. Before turning to model category theory, we consider some constructions on the arrow category of $\mathscr{M}$.

CONSTRUCTION 16.4.1. Let $i: A \longrightarrow X$ and $p: E \longrightarrow B$ be maps in $\mathscr{M}$. Define $\underline{\mathscr{M}}_{\square}[i, p]$ to be the pullback in $\mathscr{V}$ of the bottom and right arrows in the following diagram, and define $\underline{\mathscr{M}}[i, p]$ to be the map in $\mathscr{V}$ given by the universal property of pullbacks.

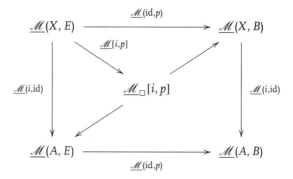

The functor $\mathcal{V}(\mathbf{I}, -)$ is a right adjoint and therefore preserves limits, such as pullbacks; applying this functor and changing notation from $\underline{\mathcal{M}}$ to $\mathcal{M}$ gives the analogous commutative diagram of sets for the underlying category $\mathcal{M}$. The relevance to model category theory is clear from the following observation.

LEMMA 16.4.2. *The pair (i, p) has the lifting property if and only if the function*

$$\mathcal{M}[i, p]: \mathcal{M}(X, E) \longrightarrow \mathcal{M}_\square[i, p] = \mathcal{M}(A, E) \times_{\mathcal{M}(A, B)} \mathcal{M}(X, B)$$

is surjective.

CONSTRUCTION 16.4.3. Let $k: V \longrightarrow W$ be a map in $\mathcal{V}$ and let $f: X \longrightarrow Y$ be a map in $\mathcal{M}$.

(i) Define the pushout product $f \,\square\, k$ by the following diagram, in which $f \boxtimes k$ denotes the pushout of the top and left pair of maps and $f \,\square\, k$ is given by the universal property of pushouts.

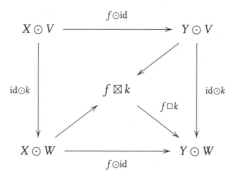

(ii) Define $\Phi[k,f]$ by the following diagram, in which $\Phi_\square[k,f]$ denotes the pullback of the bottom and right pair of maps and $\Phi[k,f]$ is given by the universal property of pullbacks.

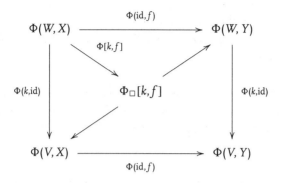

By (16.3.7), we have the following special cases.

LEMMA 16.4.4. *If $i\colon \emptyset \longrightarrow X$ is the unique map, then $i\square k$ can be identified with*

$$\mathrm{id} \odot k\colon X \otimes V \longrightarrow X \otimes W.$$

If $p\colon X \longrightarrow *$ is the unique map, the $\Phi_\square[k,p]$ can be identified with

$$\Phi(k,\mathrm{id})\colon \Phi(W,X) \longrightarrow \Phi(V,X).$$

In the rest of the section, we assume further that $\mathscr{V}$ and $\mathscr{M}$ are model categories.

LEMMA 16.4.5. *Consider (generic) maps $i\colon A \longrightarrow X$ and $p\colon E \longrightarrow B$ in $\mathscr{M}$ and $k\colon V \longrightarrow W$ in $\mathscr{V}$. The following statements are equivalent.*

  (i) *The map $\mathscr{M}[i,p]$ in $\mathscr{V}$ is a fibration if i is a cofibration and p is a fibration, and it is acyclic if in addition either i or p is acyclic.*
  (ii) *The map $i\square k$ in $\mathscr{M}$ is a cofibration if i and k are cofibrations, and it is acyclic if in addition either i or k is acyclic.*
  (iii) *The map $\Phi[k,p]$ in $\mathscr{M}$ is a fibration if k is a cofibration and p is a fibration, and it is acyclic if in addition either k or p is acyclic.*

PROOF. We show that (i) and (ii) are equivalent. A dual argument shows that (i) and (iii) are equivalent. By the first adjunction of (16.3.6) and a diagram chase that we leave to the reader, the pair $(k, \mathscr{M}[i,p])$ has the lifting property if and only if the pair $(i\square k, p)$ has the lifting property. By the model axioms, this lifting property holds for all cofibrations k if and only if $\mathscr{M}[i, p]$ is an acyclic fibration, and it holds for all acyclic cofibrations k if and only if $\mathscr{M}[i, p]$ is a fibration. Assume that i is a cofibration. If (i) holds and k is a cofibration, then $i \square k$ has the LLP with respect to all acyclic fibrations p and is therefore a cofibration. If, further, i or k is acyclic, then $i \square k$ has the LLP with respect to all fibrations p and is therefore an acyclic cofibration. Thus (ii) holds. Similarly, (ii) implies (i). $\qquad\square$

The following observation admits several variants. It concerns the verification of the acyclicity part of Lemma 16.4.5(ii), and there are analogous observations concerning the verification of the acyclicity parts of the other two statements.

REMARK 16.4.6. Consider the pushout-product diagram of Construction 16.4.3(i). If k is an acyclic cofibration and the functors $X \odot (-)$ and $Y \odot (-)$ preserve acyclic cofibrations, then, since pushouts preserve acyclic cofibrations, we can conclude by the two out of three property that $f \square k$ is a weak equivalence. Similarly, if the left map id $\odot k$ is a cofibration, the map f is a weak equivalence, the functors $(-) \odot V$ and $(-) \odot W$ preserve weak equivalences, and $\mathscr{M}$ is left proper, we can conclude that $f \square k$ is a weak equivalence.

DEFINITION 16.4.7. The model structure on the cosmos $\mathscr{V}$ is monoidal if

(i) the equivalent conditions of Lemma 16.4.5 hold for $\mathscr{M} = \mathscr{V}$; and
(ii) for some (and hence any) cofibrant replacement $q\colon Q\mathbf{I} \longrightarrow \mathbf{I}$, the induced map id $\otimes q\colon X \otimes Q\mathbf{I} \longrightarrow X \otimes \mathbf{I} \cong X$ is a weak equivalence for all cofibrant $X \in \mathscr{V}$.

Assume that this holds. Then $\mathscr{M}$ is said to be a $\mathscr{V}$-model category if

(i) the equivalent conditions of Lemma 16.4.5 hold for $\mathscr{M}$; and
(ii) for some (and hence any) cofibrant replacement $q\colon Q\mathbf{I} \longrightarrow \mathbf{I}$, the induced map id $\odot q\colon X \odot Q\mathbf{I} \longrightarrow X \odot \mathbf{I} \cong X$ is a weak equivalence for all cofibrant $X \in \mathscr{M}$.

By Lemma 16.4.4, conditions (ii) and (iii) of Lemma 16.4.5 imply preservation properties of the functors $X \odot (-)$ and $\Phi(-, X)$ as special cases. The case $\mathscr{M} = \mathscr{V}$ is of particular interest.

LEMMA 16.4.8. *Assume that $\mathscr{M}$ is a $\mathscr{V}$-model structure. If $X \in \mathscr{M}$ is cofibrant, then the functor $X \odot (-)$ preserves cofibrations and acyclic cofibrations. If X is* *fibrant, then the functor* $\Phi(-, X)$ *converts cofibrations and acyclic cofibrations in* $\mathscr{V}$ *to fibrations and acyclic fibrations in* $\mathscr{M}$.

Of course, these statements can be rephrased in terms of Quillen adjunctions. In all of the examples that we shall encounter in this book, the unit object $\mathbf{I} \in \mathscr{V}$ is cofibrant and therefore the unit conditions in Definition 16.4.7 hold trivially. The reader, like many authors, may prefer to assume once and for all that $\mathbf{I}$ is cofibrant, but there are interesting examples where that fails (e.g., in [46]). We agree to assume that the unit $*$ of any given cartesian closed model category is cofibrant. Then Theorem 15.3.6 and Example 16.3.2 lead to the following observation.

LEMMA 16.4.9. *Let* $\mathscr{V}$ *be a cartesian closed monoidal model category and give* $\mathscr{V}_* = */\mathscr{V}$ *its induced model structure as the category of objects under* $*$ *from Theorem 15.3.6. Then* $\mathscr{V}_*$ *is a monoidal model category under the smash product.*

PROOF. The weak equivalences, cofibrations, and fibrations $(\mathscr{W}_*, \mathscr{C}_*, \mathscr{F}_*)$ of $\mathscr{V}_*$ are created by the forgetful functor $U \colon \mathscr{V}_* \longrightarrow \mathscr{V}$, and U has left adjoint $(-)_+$, the addition of a disjoint base point. The unit object $S^0 = *_+$ is cofibrant in $\mathscr{V}_*$ since $*$ is cofibrant in $\mathscr{V}$. We must prove that $\mathscr{C}_* \square \mathscr{C}_* \subset \mathscr{C}_*$ and $(\mathscr{W}_* \cap \mathscr{C}_*) \square \mathscr{C}_* \subset (\mathscr{W}_* \cap \mathscr{C}_*)$. We prove the first of these. The proof of the second is similar.

Observe that the functor $(-)_+$ is strong symmetric monoidal in the sense that $V_+ \wedge W_+$ is naturally isomorphic to $(V \times W)_+$. This implies that if i and j are maps in $\mathscr{V}$, then $i_+ \square j_+$ is isomorphic to $(i \square j)_+$. Let $\mathscr{C}_+$ denote the class of maps i_+ in $\mathscr{V}_*$, where $i \in \mathscr{C}$. We saw in Corollary 15.3.8 that $\mathscr{C}_* = {}^{\boxtimes}((\mathscr{C}_+)^{\boxtimes})$. Since $\mathscr{V}$ is monoidal, $\mathscr{C}_+ \square \mathscr{C}_+ \subset \mathscr{C}_+ \subset \mathscr{C}_*$. Formal adjointness arguments from the closed structure on $\mathscr{V}_*$ imply that

$$ {}^{\boxtimes}((\mathscr{C}_+)^{\boxtimes}) \square {}^{\boxtimes}((\mathscr{C}_+)^{\boxtimes}) \subset {}^{\boxtimes}((\mathscr{C}_+)^{\boxtimes}), $$

which says that $\mathscr{C}_* \square \mathscr{C}_* \subset \mathscr{C}_*$. $\qquad\square$

The definitions so far are standard, but we require an important addendum. It connects up naturally occurring homotopies with model theoretic homotopies. Although it is often used implicitly, as far as we know it has never been made explicit. It applies to all of the examples in this book, but it does not always apply.

ADDENDUM 16.4.10. We require of a monoidal model category $\mathcal{V}$ that it contain a fixed chosen good cylinder object $I = \mathrm{Cyl}\,\mathbf{I}$ such that the maps

$$X \otimes I \longrightarrow X \otimes \mathbf{I} \cong X \quad \text{and} \quad Y \cong \underline{\mathcal{V}}(\mathbf{I}, Y) \longrightarrow \underline{\mathcal{V}}(I, Y)$$

induced by the weak equivalence $p \colon I \longrightarrow \mathbf{I}$ are weak equivalences for all cofibrant objects X and fibrant objects Y of $\mathcal{V}$. Similarly, we require of a $\mathcal{V}$-model category $\mathcal{M}$ that the maps

$$X \odot I \longrightarrow X \quad \text{and} \quad Y \longrightarrow \Phi(I, Y)$$

are weak equivalences for all cofibrant objects X and fibrant objects Y of $\mathcal{M}$. For general objects X and Y of $\mathcal{M}$, we define classical homotopies to be maps $X \odot I \longrightarrow Y$. By adjunction, these are the same as maps $X \longrightarrow \Phi(I, Y)$, so that there is no distinction between left and right classical homotopies.

REMARK 16.4.11. In practice, the functors $X \otimes -$ on $\mathcal{V}$ and $X \odot -$ from $\mathcal{V}$ to $\mathcal{M}$ preserve weak equivalences for all cofibrant objects X in $\mathcal{V}$ or $\mathcal{M}$, and dually for the functors $\underline{\mathcal{V}}(-, Y)$ and $\Phi(-, Y)$. Then the addendum holds as a special case. Proposition 14.3.9 implies that all classical homotopies are model theoretic homotopies. When X is cofibrant and Y is fibrant, Lemma 16.4.8 implies that $X \odot I$ is a good cylinder object and $\Phi(I, Y)$ is a good path object. Since Corollary 14.3.13 shows that we can then use any fixed good cylinder object $\mathrm{Cyl}\,X$ and any fixed good path object $\mathrm{Cocyl}\,Y$ to define left and right homotopies, we are entitled to choose the classical cylinders $X \odot I$ and path objects $\Phi(I, Y)$. We conclude in particular that the model theoretic set $\pi(X, Y)$ of homotopy classes of maps coincides with the classical set $\pi(X, Y)$ of classical homotopy classes of maps $X \longrightarrow Y$.

The remark has the following consequence, which shows that our addendum to the standard definitions reconciles classical homotopy theory with model theoretic homotopy theory.

THEOREM 16.4.12. *Let $\mathcal{M}$ be a $\mathcal{V}$-model category that satisfies Addendum 16.4.10. Then $\mathrm{Ho}\,\mathcal{M}$ is equivalent to the homotopy category $\mathrm{h}\mathcal{M}_{cf}$ of bifibrant objects and classical homotopy classes of maps between them.*

REMARK 16.4.13. We shall not go into the details, but in fact $\mathrm{Ho}\,\mathcal{M}$ is enriched over $\mathrm{Ho}\,\mathcal{V}$, and the equivalence of the theorem is an equivalence of $\mathrm{Ho}\,\mathcal{V}$-categories. The reader is invited to verify these assertions.

16.5. A glimpse at higher categorical structures

In this section we give an informal introduction to the application of some of the more elementary parts of higher category theory to model category theory. The relevant categorical notions give the right formal context for the study of left and right derived functors. We shall ignore issues of size in this informal discussion, allowing classes of objects and categories that are not locally small.

We have defined categories enriched in a symmetric monoidal category $\mathcal{V}$, and we have observed that one such $\mathcal{V}$ is the Cartesian monoidal category $\underline{\mathcal{C}at}$, the category of categories and functors. We have also observed that a category enriched in $\underline{\mathcal{C}at}$ is called a 2-category. An example is $\underline{\mathcal{C}at}$ itself. The internal hom $\underline{\mathcal{C}at}(\mathcal{C}, \mathcal{D})$ is the category of functors $\mathcal{C} \longrightarrow \mathcal{D}$ and natural transformations between them. We have composition functors

$$\underline{\mathcal{C}at}(\mathcal{D}, \mathcal{E}) \times \underline{\mathcal{C}at}(\mathcal{C}, \mathcal{D}) \longrightarrow \underline{\mathcal{C}at}(\mathcal{C}, \mathcal{E})$$

between these internal homs given by composition of functors and natural transformations. For the latter, if $\alpha : F \longrightarrow F'$ and $\beta : G \longrightarrow G'$ are natural transformations between composable pairs of functors, then $\beta \circ \alpha$ is the common composite in the commutative diagram

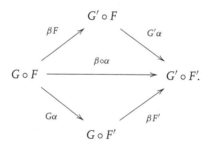

We think of functors as morphisms between categories and natural transformations as morphisms between functors. We also have two 2-categories $\mathcal{M}od\mathcal{C}at_\ell$ and $\mathcal{M}od\mathcal{C}at_r$ of model categories. Their objects (or 0-cells) are model categories, their morphisms (or 1-cells) are Quillen left adjoints and Quillen right adjoints, respectively, and their morphisms between morphisms (or 2-cells) are natural transformations in both cases.

The reason for introducing this language is that there is a sensible notion of a pseudo-functor between 2-categories, and it provides the right language to describe what exactly $\mathbb{L}$ and $\mathbb{R}$ are. A pseudo-functor $\mathbb{F} : \mathcal{C} \longrightarrow \mathcal{D}$ between 2-categories assigns 0-cells, 1-cells, and 2-cells in $\mathcal{D}$ to 0-cells, 1-cells, and 2-cells in $\mathcal{C}$. For each fixed pair X, Y of 0-cells, $\mathbb{F}$ specifies a functor

$\mathscr{C}(X, Y) \longrightarrow \mathscr{D}(\mathbb{F}X, \mathbb{F}Y)$. However, for 0-cells X, $\mathbb{F}(\mathrm{Id}_X)$ need not equal $\mathrm{Id}_{\mathbb{F}X}$ and, for composable 1-cells F and G, $\mathbb{F}(G \circ F)$ need not equal $\mathbb{F}G \circ \mathbb{F}F$. Rather, there are 2-cell isomorphisms connecting these. These isomorphisms are subject to coherence axioms asserting that certain associativity and left and right unity diagrams commute; see, for example, [76]. Given the precise definition, the following result becomes clear.

PROPOSITION 16.5.1. *Passage from model categories to their derived homotopy categories* Ho$\mathscr{M}$, *derived functors, and derived natural transformations specify pseudo-functors*

$$\mathbb{L}: \mathscr{M}\!od\mathscr{C}at_\ell \longrightarrow \mathscr{C}at \quad \text{and} \quad \mathbb{R}: \mathscr{M}\!od\mathscr{C}at_r \longrightarrow \mathscr{C}at.$$

Obviously, it is unsatisfactory to have $\mathbb{L}$ and $\mathbb{R}$ part of distinct structures. After all, left and right Quillen adjoints are directly related, and one sometimes must compose them. The proper framework is given by viewing model categories as forming not a pair of 2-categories, but rather as a single double category. We sketch the idea, referring the interested reader to [127] for details.

Just as we can define a category enriched in any symmetric monoidal category $\mathscr{V}$, we can define an internal category in any complete category $\mathscr{V}$. It has object and morphism objects $\mathscr{O}b$ and $\mathscr{M}or$ in $\mathscr{V}$ together with maps

$$S, T: \mathscr{M}or \longrightarrow \mathscr{O}b, \quad I: \mathscr{O}b \longrightarrow \mathscr{M}or, \quad \text{and} \quad C: \mathscr{M}or \times_{\mathscr{O}b} \mathscr{M}or \longrightarrow \mathscr{M}or$$

in $\mathscr{V}$, called source, target, identity, and composition. These must satisfy the usual unit and associativity laws for a category, expressed diagrammatically. An ordinary category is an internal category in the category of sets. Internal categories in the category $\mathscr{U}$ of topological spaces appear frequently. A double category is just an internal category in $\mathscr{C}at$. A 2-category can be viewed as a double category whose object category is discrete, meaning that it has only identity morphisms.

The definition just given is the quickest possible, but it obscures the essential symmetry of the notion of a double category $\mathscr{D}$. It has 0-cells, namely the objects of the category $\mathscr{O}b$, it has both "vertical" and "horizontal" 1-cells, namely the morphisms of $\mathscr{O}b$ and the objects of $\mathscr{M}or$, and it has 2-cells, namely the morphisms of $\mathscr{M}or$. When one writes out the category axioms for the functors S, T, I, and C in the original definition, one finds that they are completely symmetric with respect to the vertical and horizontal 1-cells. We cannot compose horizontal and vertical 1-cells in general, but we nevertheless think of 2-cells as fillers in diagrams

16.5.2

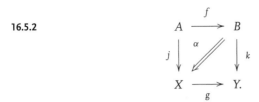

Here A, B, X, and Y are 0-cells, j and k are vertical 1-cells, f and g are horizontal 1-cells, and α is a 2-cell. With the original definition, α is a morphism from f to g in $\mathcal{M}or$, and the fact that $\mathcal{M}or$ is a category leads to vertical composition of 2-cells. The composition functor C leads to horizontal composition of 2-cells, and the fact that C is a functor expresses a symmetric interchange law between these two composition laws for 2-cells. A double category has as part of its structure vertical and horizontal 2-categories $\mathcal{C}_v$ and $\mathcal{C}_h$ of 0-cells, vertical or horizontal 1-cells, and vertical or horizontal 2-cells. Here a 2-cell α as pictured above is vertical if f and g are identity horizontal 1-cells (objects of $\mathcal{M}or$ in the image of I) and is horizontal if j and k are identity vertical 1-cells (identity morphisms in the category $\mathcal{O}b$).

An example that is relevant to the study of a single model category may clarify the idea. Thus let $\mathcal{M}$ be a category with a subcategory $\mathcal{W}$ of weak equivalences. We form a double category $\mathcal{D}(\mathcal{M}, \mathcal{W})$ whose objects are the objects of $\mathcal{M}$, whose horizontal and vertical 1-cells are the morphisms of $\mathcal{M}$ and of $\mathcal{W}$, and whose 2-cells are commutative diagrams

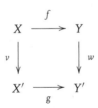

in which $v, w \in \mathcal{W}$. Thinking in terms of the arrow category of $\mathcal{M}$, we can view this square as a morphism $v \longrightarrow w$ between vertical arrows or as a morphism $f \longrightarrow g$ between horizontal arrows.

Any 2-category $\mathcal{C}$ determines a double category $\mathcal{D}(\mathcal{C})$.[1] Its vertical and horizontal 1-cells are both the 1-cells of $\mathcal{C}$. Its 2-cells α in (16.5.2) are the 2-cells $\alpha: kf \longrightarrow gj$ in $\mathcal{C}$. In particular, we have the double category $\mathcal{D}(\mathcal{C}at)$ whose

---

1. Category theorists often call $\mathcal{D}(\mathcal{C})$ the double category of quintets in $\mathcal{C}$ since its 2-cells can be viewed as quintets (f, g, j, k, α).

1-cells are functors and whose 2-cells (16.5.2) are natural transformations $\alpha \colon kf \longrightarrow gj$.

The reason for introducing this language is that it gives a way of expressing $\mathbb{L}$ and $\mathbb{R}$ as part of a single kind of functor, namely a double pseudo-functor $\mathbb{F} \colon \mathscr{D} \longrightarrow \mathscr{E}$ between double categories. Such an $\mathbb{F}$ assigns 0-cells, vertical 1-cells, horizontal 1-cells, and 2-cells in $\mathscr{D}$ to the corresponding kinds of cells in $\mathscr{E}$. These assignments must come with coherence isomorphism 2-cells that give restrictions of $\mathbb{F}$ to pseudo-functors $\mathbb{F}_v \colon \mathscr{D}_v \longrightarrow \mathscr{E}_v$ and $\mathbb{F}_h \colon \mathscr{D}_h \longrightarrow \mathscr{E}_h$. The coherence isomorphisms must be doubly natural, in a sense that is made precise in [127, §6]. Once the definition has been made precise, diagram chasing proves the following generalization of Proposition 16.5.1; see [127, Thm. 7.6].

THEOREM 16.5.3. *Model categories are the 0-cells of a double category $\mathscr{M}od$ whose vertical and horizontal 1-cells are the Quillen left and right adjoints and whose 2-cells are the natural transformations. There is a double pseudo-functor $\mathbb{F} \colon \mathscr{M}od \longrightarrow \mathscr{Q}(\mathscr{C}at)$ such that $\mathbb{F}_v = \mathbb{L}$ and $\mathbb{F}_h = \mathbb{R}$.*

This result encodes many relationships between left and right Quillen adjoints in a form that is familiar in other categorical contexts. The Quillen adjunctions (F, U) are examples of the correct categorical notion of an adjunction in a double category, called a conjunction in [127, §5]. For example, the theorem has the following direct categorical corollary [127, Cor. 7.8], which should be contrasted with Remark 16.1.12.

COROLLARY 16.5.4. *If F is both a Quillen left adjoint and a Quillen right adjoint, then $\mathbb{L}F \cong \mathbb{R}F$.*

The real force of Theorem 16.5.3 concerns the comparisons it induces between composites of left and right derived adjoint base change functors [127, §9], but it would take us too far afield to say anything about that here.

17

MODEL STRUCTURES ON THE CATEGORY OF SPACES

We give an idiosyncratic introduction to examples, focusing on the kinds of model categories encountered in classical algebraic topology and classical homological algebra. We treat these examples in parallel, discussing model structures in topology here and model structures in homological algebra in the next chapter.

The central point we want to make is that there are three intertwined model structures on the relevant categories, and that the least familiar, which is called the mixed model structure and was introduced relatively recently by Michael Cole [33], is in some respects the most convenient. In fact, we shall argue that algebraic topologists, from the very beginning of the subject, have by preference worked implicitly in the mixed model structure on topological spaces. A general treatment of topological model categories from our point of view is given in [97, Part II], and the treatment here is largely extracted from that source.[1] We offer a new perspective on the philosophy in §19.1, where we show that the mixed model structure is fundamental conceptually as well as pragmatically.

In contrast to our conventions in the first half of the book, spaces are no longer assumed to be of the homotopy types of CW complexes, although we do still assume that they are compactly generated. We also focus on unbased rather than based spaces. This is reasonable in view of Theorem 15.3.6, which shows how to construct model structures on the category of based spaces from model structures on the category of unbased spaces.

---

1. While that treatment focused on more advanced examples, its discussion of the general philosophy may nevertheless be helpful to the reader.

17.1. The Hurewicz or *h*-model structure on spaces

The most obvious homotopical notion of a weak equivalence is an actual homotopy equivalence. We call such maps *h*-equivalences for short. It is natural to expect there to be a model category structure on the category $\mathscr{U}$ of compactly generated spaces in which the weak equivalences are the *h*-equivalences.

In [93] and in the first part of this book (see especially §1.1 and §1.3), the words "cofibration" and "fibration" are used only in their classical topological sense. The cofibrations are the maps $i: A \longrightarrow X$ that satisfy the homotopy extension property (HEP). This means that for all spaces B they satisfy the LLP with respect to the map $p_0: B^I \longrightarrow B$ given by evaluation at 0. The fibrations are the maps $p: E \longrightarrow B$ that satisfy the covering homotopy property (CHP). This means that for all spaces A they satisfy the RLP with respect to the inclusion $i_0: A \longrightarrow A \times I$. This notion of fibration was first defined by Hurewicz [70]. Since we are now doing model category theory and will have varying notions of cofibration and fibration, we call these Hurewicz cofibrations and Hurewicz fibrations, conveniently abbreviated to *h*-cofibrations and *h*-fibrations.

These cofibrations and fibrations were not considered by Quillen in his paper introducing model categories [113], but Strøm [132] later proved a version of the following result. Technically, he worked in the category of all spaces, not just the compactly generated ones, and in that category the model theoretic cofibrations are the Hurewicz cofibrations that are closed inclusions. However, as we left as an exercise in [93, p. 46], the Hurewicz cofibrations in $\mathscr{U}$ are closed inclusions. A thorough discussion of variant *h*-type model structures on topological categories is given in [97, Ch. 4].

THEOREM 17.1.1 (*h*-MODEL STRUCTURE). *The category $\mathscr{U}$ is a monoidal model category whose weak equivalences, cofibrations, and fibrations, denoted $(\mathscr{W}_h, \mathscr{C}_h, \mathscr{F}_h)$, are the h-equivalences, h-cofibrations, and h-fibrations. All spaces are both h-fibrant and h-cofibrant, hence this model structure is proper.*

Before turning to the proof, we record the following corollary, which is immediate from Theorem 15.3.6, Corollary 15.3.8, and Lemma 16.4.9.

COROLLARY 17.1.2. *The category $\mathscr{U}_*$ of based spaces in $\mathscr{U}$ is a proper model category whose weak equivalences, cofibrations, and fibrations are the based maps that are h-equivalences, h-cofibrations, and h-fibrations when regarded as maps in $\mathscr{U}$.*

The pair (T, U), where T is given by adjoining disjoint basepoints and U is the forgetful functor, is a Quillen adjunction relating $\mathscr{U}$ to $\mathscr{U}_*$. Moreover, $\mathscr{U}_*$ is a monoidal model category with respect to the smash product.

REMARK 17.1.3. In the first half of the book, we worked thoughout in the category $\mathscr{T}$ of *nondegenerately* based spaces in $\mathscr{U}$. These are precisely the h-cofibrant objects in the h-model category of all based spaces. There is an elementary cofibrant approximation functor, called whiskering. For a based space X, let wX be the wedge $X \vee I$, where I is given the basepoint 0 when constructing the wedge, and give wX the basepoint $1 \in I$. The map $q \colon wX \longrightarrow X$ that shaves the whisker I is an h-equivalence, and it is a based h-equivalence when X is in $\mathscr{T}$. While $\mathscr{T}$ cannot be a model category, since it is not cocomplete, it is far more convenient than $\mathscr{U}_*$. The h-classes of maps specified in Corollary 17.1.2 are not the ones natural to the based context, where one wants all homotopies to be based. However, when restricted to $\mathscr{T}$, the h-classes of maps do coincide with the based homotopy equivalences, based cofibrations, and based fibrations by [93, p. 44] and Lemmas 1.3.3 and 1.3.4.

To begin the proof of Theorem 17.1.1, we return to an unfinished piece of business from Chapter 1, namely the proof of Lemma 1.1.1. The lifting axioms needed to prove Theorem 17.1.1 are the same as the unbased version of that result.

PROPOSITION 17.1.4. *Consider a commutative diagram of spaces*

17.1.5

$$
\begin{array}{ccc}
A & \xrightarrow{\ g\ } & E \\
{\scriptstyle i}\downarrow & {\raise2pt\hbox{$\scriptstyle\lambda$}}\nearrow & \downarrow{\scriptstyle p} \\
X & \xrightarrow[\ f\ ]{} & B
\end{array}
$$

in which i is an h-cofibration and p is an h-fibration. If either i or p is an h-equivalence, then there exists a lift λ.

PROOF. First, assume that i is a homotopy equivalence. A result in [93, p. 44] shows that A is a deformation retract of X. Thus there is a retraction $r \colon X \longrightarrow A$ and a homotopy $h \colon ir \simeq \mathrm{id}$ relative to A. Since (X, A) is an NDR pair, there is also a map $u \colon X \longrightarrow I$ such that $A = u^{-1}(0)$ [93, p. 43]. We deform the homotopy h to a more convenient homotopy $j \colon ir \simeq \mathrm{id}$ by setting

$$j(x, t) = \begin{cases} h(x, t/u(x)) & \text{if } t < u(x) \\ h(x, 1) & \text{if } t \geq u(x). \end{cases}$$

Since $fji_0 = fir = pgr$ and p is an h-fibration, there is a lift v in the diagram

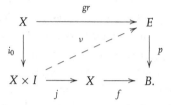

We obtain a lift λ in (17.1.5) by setting $\lambda(x) = v(x, u(x))$. Since $u(A) = 0$ and $ri = $ id, $\lambda i = vi_0 i = gri = g$, while $p\lambda(x) = pv(x, u(x)) = fj(x, u(x)) = f(x)$.

Second, assume that p is a homotopy equivalence. A result in [93, p. 50] shows that there is a section $s\colon B \longrightarrow E$ and a homotopy $h\colon s \circ p \simeq$ id over B. Let $Mi = X \times 0 \cup A \times I$ be the mapping cylinder and construct a diagram

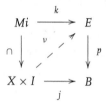

by letting $k(x, 0) = s(f(x))$, $k(a, t) = h(g(a), t)$, and $j(x, t) = f(x)$. The left arrow is the inclusion of a deformation retract [93, p. 43], hence is an h-acyclic h-cofibration. By the first part, there is a lift v. Setting $\lambda(x) = v(x, 1)$, we obtain a lift λ in (17.1.5). $\qquad\square$

Note that $\mathcal{W}_h$ is closed under retracts by Lemma 14.1.2 and obviously satisfies the two out of three property. To complete the proof of the model axioms, it only remains to prove the factorization axioms. We need the following saturation property of the h-acyclic h-cofibrations.

LEMMA 17.1.6. *A pushout of an h-acyclic h-cofibration is an h-acyclic h-cofibration and the inclusion of the initial term in a colimit of a sequence of h-acyclic h-cofibrations is an h-acyclic h-cofibration.*

PROOF. This follows easily from the already cited result, [93, p. 44], that an h-acyclic h-cofibration is the inclusion of a deformation retraction. $\qquad\square$

It is standard that any map $f: X \longrightarrow Y$ factors as composites

$$X \xrightarrow{\;j\;} Mf \xrightarrow{\;r\;} Y \quad \text{and} \quad X \xrightarrow{\;\nu\;} Nf \xrightarrow{\;\rho\;} Y,$$

where j is an h-cofibration, r and ν are h-equivalences, and ρ is an h-fibration [93, pp. 43, 48]. We can modify the first construction to arrange that r is also an h-fibration or modify the second construction to arrange that ν is also an h-cofibration. It suffices to do the second of these since if we then replace r by a composite $\rho \circ \nu$ of an h-acyclic h-cofibration ν and an h-fibration ρ, then ρ will be acyclic by the two out of three property for $\mathscr{W}_h$ and therefore f will be the composite $\rho \circ (\nu \circ j)$ of an h-acyclic h-fibration and an h-cofibration.

LEMMA 17.1.7. *Any map $f: X \longrightarrow Y$ factors as the composite of an h-acyclic h-cofibration and an h-fibration, hence f also factors as the composite of an h-cofibration and an h-acyclic h-fibration.*

PROOF. Let $Z_0 = X$ and $\rho_0 = f$. Inductively, assume that we have constructed a map $\rho_n: Z_n \longrightarrow Y$. By the definition of $N\rho_n$ as $Z_n \times_Y Y^I$, we have projections $N\rho_n \to Z_n$ and $N\rho_n \to Y^I$, and the latter has an adjoint map $N\rho_n \times I \to Y$. Construct the following diagram, in which Z_{n+1} is the displayed pushout and ν_n and λ_n are the canonical maps.

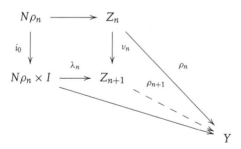

The universal property of pushouts gives an induced map ρ_{n+1}. By Lemma 17.1.6, ν_n is an h-acyclic h-cofibration since it is a pushout of such a map. Let Z be the colimit of the Z_n and let $\nu: X \longrightarrow Z$ and $\rho: Z \longrightarrow Y$ be the colimits of the maps ν_n and ρ_n. Certainly $f = \rho \circ \nu$ and, again by Lemma 17.1.6, ν is an h-acyclic h-cofibration. By an exercise in point-set topology, since we are working with compactly generated spaces the canonical continuous bijection $\operatorname{colim} N\rho_n \longrightarrow N\rho$ is a homeomorphism. The left adjoint $(-) \times I$ commutes with colimits, and the colimit of the maps λ_n gives a lift λ in the canonical example

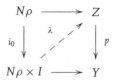

of (17.1.5). The adjoint of λ is a path-lifting function $N\rho \longrightarrow Z^I$. As observed in [93, p. 47], this implies that ρ is an h-fibration. □

We have completed the proof that $\mathcal{U}$ is a model category. It is monoidal since a standard lemma [93, p. 43] on the product of NDR-pairs verifies Lemma 16.4.5(ii). Since h-acyclic h-cofibrations and h-acyclic h-fibrations are inclusions and projections of deformation retractions, by [93, pp. 44, 50], every object is both h-cofibrant and h-fibrant by a glance at the relevant lifting properties. By Proposition 15.4.2, this implies that the model structure is proper.

This completes the proof of Theorem 17.1.1.

17.2. The Quillen or q-model structure on spaces

In the first half of the book, we assumed that all spaces had the homotopy types of CW complexes. Therefore there was no distinction between a homotopy equivalence and a weak homotopy equivalence. By contrast, we now define a q-equivalence to be a weak homotopy equivalence, namely a map that induces a bijection on path components and an isomorphism on homotopy groups for all choices of basepoints.

LEMMA 17.2.1. *The subcategory $\mathcal{W}_q$ of weak homotopy equivalences in $\mathcal{U}$ is a subcategory of weak equivalences.*

PROOF. We first show that the collection of maps $\mathcal{W}_q$ satisfies the two out of three property. Let $h = g \circ f$, $f \colon X \longrightarrow Y$ and $g \colon Y \longrightarrow Z$. If f and g or if g and h are weak equivalences, then clearly so is h or f. To see that if f and h are weak equivalences, then so is g, note that a point $y \in Y$ may not be of the form $f(x)$ but is nevertheless a choice of basepoint for which we must check that $g_* \colon \pi_*(Y, y) \longrightarrow \pi_*(Z, g(y))$ is an isomorphism. Since f induces a bijection on path components, there is a point $f(x)$ and a path α from $f(x)$

to y. Conjugating with α and $g \circ \alpha$ gives vertical isomorphisms such that the following diagram commutes, and the bottom arrow g_* is an isomorphism since f_* and $h_* = g_* \circ f_*$ are isomorphisms.

$$
\begin{array}{ccc}
\pi_*(Y, y) & \xrightarrow{\ g_*\ } & \pi_*(Z, g(y)) \\
\downarrow & & \downarrow \\
\pi_*(X, x) \xrightarrow{\ f_*\ } \pi_*(Y, f(x)) & \xrightarrow{\ g_*\ } & \pi_*(Z, g(f(x)))
\end{array}
$$

Therefore the upper arrow g_* is an isomorphism. This verifies the two out of three property, and it is clear that $\mathscr{W}_q$ is a subcategory that contains all isomorphisms and is closed under retracts. $\qquad\square$

We define a q-fibration to be a Serre fibration, which is a map that satisfies the RLP with respect to $\mathcal{I}$, where $\mathcal{I}$ denotes the set of inclusions $S^{n-1} \longrightarrow D^n$, $n \geq 0$. Here S^{-1} is empty. We let $\mathcal{J}$ denote the set of maps $i_0 \colon D^n \longrightarrow D^n \times I$, $n \geq 0$. We define a q-cofibration to be a map that satisfies the LLP with respect to the q-acyclic q-fibrations.

THEOREM 17.2.2 (q-MODEL STRUCTURE). *The category $\mathscr{U}$ is a compactly generated, proper, and monoidal model category whose weak equivalences, cofibrations, and fibrations, denoted $(\mathscr{W}_q, \mathscr{C}_q, \mathscr{F}_q)$, are the q-equivalences, q-cofibrations, and q-fibrations. The sets $\mathcal{I}$ and $\mathcal{J}$ are generating sets for the q-cofibrations and the q-acyclic q-cofibrations. Every space is q-fibrant.*

PROOF. By Proposition 2.5.4, any compact space K is compact in the sense of Definition 15.1.6. Therefore $\mathcal{I}$ and $\mathcal{J}$ are compact, and we understand $\mathscr{C}(\mathcal{I})$ and $\mathscr{C}(\mathcal{J})$ to mean the retracts of the sequential relative cell complexes of Definition 15.1.1. By the compact object argument, Proposition 15.1.11, we have functorial WFSs $(\mathscr{C}(\mathcal{I}), \mathcal{I}^{\boxtimes})$ and $(\mathscr{C}(\mathcal{J}), \mathcal{J}^{\boxtimes})$. To verify the model axioms, we need only verify the acylicity and compatibility conditions of Theorem 15.2.3. For the acyclicity, if $i \colon A = X_0 \longrightarrow \operatorname{colim} X_q = X$ is a relative $\mathcal{J}$-cell complex, then, by inspection, each map $X_q \longrightarrow X_{q+1}$ of the colimit system is the inclusion of a deformation retraction and therefore i is a q-equivalence (in fact an h-equivalence).

For the compatibility, we must show that $\mathcal{I}^{\boxtimes} = \mathcal{J}^{\boxtimes} \cap \mathscr{W}_q$. The maps in $\mathcal{J}$ are relative CW complexes, so they are in $\mathscr{C}(\mathcal{I})$, and this implies that $\mathcal{I}^{\boxtimes} \subset \mathcal{J}^{\boxtimes}$. To show that $\mathcal{I}^{\boxtimes} \subset \mathscr{W}_q$, observe that the inclusion of a basepoint, $* \longrightarrow S^n$, and the inclusion of the two bases of the reduced cylinder, $S^n \vee S^n \longrightarrow S^n_+ \wedge I$, are relative CW complexes and are thus in $\mathscr{C}(\mathcal{I})$. If $p: E \longrightarrow B$ is in $\mathcal{I}^{\boxempty}$, then liftings with respect to $* \longrightarrow S^n$ show that $p_*: \pi_n(E, x) \longrightarrow \pi_n(B, p(x))$ is surjective and liftings with respect to $S^n \vee S^n \longrightarrow S^n_+ \wedge I$ show that p_* is injective. Conversely, suppose that $p: E \longrightarrow B$ is in $\mathcal{J}^{\boxempty} \cap \mathscr{W}_q$ and consider a lifting problem

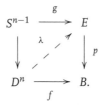

We use the square to construct the solid arrow portion of the following diagram, in which j is the constant homotopy at g, $j(x, t) = g(x)$, and $h = p \circ j$.

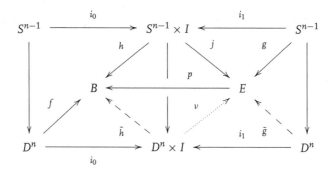

By the key lemma of [93, p. 68], since p is a weak equivalence there are dashed arrows $\tilde{g}$ and $\tilde{h}$ making the dashed and solid arrow part of the diagram commute. Since the pair $(D^n \times I, D^n \times \{0\})$ is homeomorphic to the pair $(D^n \times I, D^n \times \{1\} \cup S^{n-1} \times I)$ and $p \in \mathcal{J}^{\boxempty}$, p satisfies the RLP with respect to the latter pair. Therefore, there is a lift v such that $p \circ v = \tilde{h}$, $v \circ i_1 = \tilde{g}$, and v restricts to j on $S^{n-1} \times I$. The composite $\lambda = v \circ i_0$ is the desired lift in our original diagram.

To see that the q-model structure is monoidal, note that the product of cells is a cell, in the sense that if i and j are the inclusions $S^{m-1} \longrightarrow D^m$ and $S^{n-1} \longrightarrow D^n$, then $i \square j$ is homeomorphic to the inclusion $S^{m+n-1} \longrightarrow D^{m+n}$. Similarly, if j here is instead $i_0: D^n \longrightarrow D^n \times I$, then $i \square j$ is homeomorphic to $i_0: D^{m+n} \longrightarrow D^{m+n} \times I$. It follows inductively that if i and j are relative $\mathcal{I}$-cell complexes, then $i \square j$ is a relative $\mathcal{I}$-cell complex, and that if i is a relative $\mathcal{I}$-cell complex and j is a relative $\mathcal{J}$-cell complex, then $i \square j$ is a relative $\mathcal{J}$-cell complex. Moreover, the functor $i \square (-)$ on the arrow category preserves retracts. Therefore $i \square j$ is a q-cofibration if i or j is so and is a q-acyclic q-cofibration if, further, either i or j is so.

Since every space is q-fibrant, the q-model structure is right proper. Rather than prove directly that it is left proper, we prove the gluing lemma and use Proposition 15.4.4. □

LEMMA 17.2.3 (THE GLUING LEMMA). *Assume that i and j are q-cofibrations and f, g, and h are q-equivalences in the following commutative diagram.*

$$
\begin{array}{ccccc}
A & \xleftarrow{\ i\ } & C & \xrightarrow{\ k\ } & B \\
{\scriptstyle f}\downarrow & & {\scriptstyle g}\downarrow & & \downarrow{\scriptstyle h} \\
A' & \xleftarrow{\ j\ } & C' & \xrightarrow{\ \ell\ } & B'
\end{array}
$$

Then the induced map of pushouts

$$ X = A \cup_C B \longrightarrow A' \cup_{C'} B' = X' $$

is a q-equivalence.

PROOF. Since a q-cofibration is an h-cofibration, a lemma of [93, p. 78] shows that the natural maps $M(i, k) \longrightarrow X$ and $M(j, \ell) \longrightarrow X'$ from the double mapping cylinders to the pushouts are h-equivalences and thus q-equivalences. Now, breaking the double mapping cylinders into overlapping unions of single mapping cylinders, as in [93, p. 78], we see that the conclusion will hold in general if it holds for a map $(X; A, B) \longrightarrow (X'; A', B')$ of excisive triads, with $C = A \cap B$ and $C' = A' \cap B'$. This case is a theorem proven in [93, pp. 78–80]. □

Again by Theorem 15.3.6, Corollary 15.3.8, and Lemma 16.4.9, Theorem 17.2.2 has the following consequence.

COROLLARY 17.2.4. *The category $\mathcal{U}_*$ of based spaces in $\mathcal{U}$ is a compactly generated proper model category whose weak equivalences, cofibrations, and fibrations are the based maps that are q-equivalences, q-cofibrations, and q-fibrations in $\mathcal{U}$. The sets of generating based q-cofibrations and q-acyclic based q-cofibrations, $\mathcal{I}_+$ and $\mathcal{J}_+$, are obtained from $\mathcal{I}$ and $\mathcal{J}$ by adjoining disjoint basepoints to the source and target of all maps. The pair (T, U), where T is given by adjoining disjoint basepoints, is a*

Quillen adjunction relating $\mathscr{U}$ to $\mathscr{U}_*$. Moreover, $\mathscr{U}_*$ is a monoidal model structure with respect to the smash product.

REMARK 17.2.5. The q-cofibrant based spaces are those for which $* \longrightarrow X$ is a retract of a relative $\mathcal{I}$-cell complex, so they are retracts of cell complexes with $*$ as a vertex. Note that the most natural kind of based cell complexes would start with based cells $S^n \longrightarrow D^{n+1}$ for chosen basepoints of spheres and would have based attaching maps. While such cell complexes are useful, they can only model connected based spaces.

17.3. Mixed model structures in general

Every h-equivalence is a q-equivalence, every h-fibration is a q-fibration, and therefore every q-cofibration is an h-cofibration. This situation occurs often when one first builds a homotopy category by identifying homotopic maps and then constructs a "derived" homotopy category by inverting weak equivalences. Following Cole [33], we explain how to mix model structures in such a situation. Throughout this section, we work in a bicomplete category $\mathscr{M}$ with two model structures,

$$(\mathscr{W}_h, \mathscr{C}_h, \mathscr{F}_h) \quad \text{and} \quad (\mathscr{W}_q, \mathscr{C}_q, \mathscr{F}_q),$$

such that

$$\mathscr{W}_h \subset \mathscr{W}_q, \quad \mathscr{F}_h \subset \mathscr{F}_q, \quad \text{and therefore} \quad \mathscr{C}_q \subset \mathscr{C}_h.$$

There are dual results with the roles of cofibrations and fibrations reversed, but our applications will not use them. We give the basic theorems and several elaborations, all adapted from [33]. Since his short paper gives complete proofs, we shall be just a little sketchy. The details are all elementary. What is deep is the insight that these results should be true.

THEOREM 17.3.1 (THE MIXED MODEL STRUCTURE). *Define* $\mathscr{W}_m = \mathscr{W}_q$, $\mathscr{F}_m = \mathscr{F}_h$, *and*

$$\mathscr{C}_m = {}^{\boxtimes}(\mathscr{F}_h \cap \mathscr{W}_q) = {}^{\boxtimes}(\mathscr{F}_m \cap \mathscr{W}_m).$$

Then $(\mathscr{W}_m, \mathscr{C}_m, \mathscr{F}_m)$ *is a model structure on* $\mathscr{M}$.

PROOF. Clearly $\mathscr{W}_m$, $\mathscr{F}_m$, and $\mathscr{C}_m$ are subcategories of $\mathscr{M}$ that are closed under retracts and $\mathscr{W}_m$ satisfies the two out of three property. One of the lifting properties holds by definition. Since $\mathscr{F}_h = \mathscr{F}_m$, to see that

$$\mathscr{F}_m = (\mathscr{C}_m \cap \mathscr{W}_m)^{\boxtimes}$$

it suffices to show that

17.3.2
$$\mathscr{C}_m \cap \mathscr{W}_m = \mathscr{C}_h \cap \mathscr{W}_h.$$

A map in $\mathscr{C}_h \cap \mathscr{W}_h$ is clearly in $\mathscr{W}_m$, and it has the LLP with respect to $\mathscr{F}_h$ and thus with respect to $\mathscr{F}_h \cap \mathscr{W}_q$, hence is also in $\mathscr{C}_m$. For the opposite inclusion, if i is in $\mathscr{C}_m \cap \mathscr{W}_m$ and we factor i as $p \circ j$, where j is in $\mathscr{C}_h \cap \mathscr{W}_h$ and p is in $\mathscr{F}_h$, then p is in $\mathscr{W}_q$ by the two out of three property. Therefore i satisfies the LLP with respect to p. Thus there is a section s of p such that $s \circ i = j$. By the retract argument, Lemma 14.1.12, i is a retract of j and hence is in $\mathscr{C}_h \cap \mathscr{W}_h$.

For the factorization axioms, let $f : X \longrightarrow Y$ be any map. By (17.3.2), factoring f as $p \circ i$ where i is in $\mathscr{C}_h \cap \mathscr{W}_h$ and p is in $\mathscr{F}_h$ gives one of the required factorizations. For the other, factor f as $p \circ i$ where i is in $\mathscr{C}_q$ and p is in $\mathscr{F}_q \cap \mathscr{W}_q$. Then factor p as $q \circ j$ where j is in $\mathscr{C}_h \cap \mathscr{W}_h$ and q is in $\mathscr{F}_h$. Then $f = q \circ (j \circ i)$. Clearly $\mathscr{C}_q \subset \mathscr{C}_m$, hence i is in $\mathscr{C}_m$. By (17.3.2), j is also in $\mathscr{C}_m$, hence so is $j \circ i$. By the two out of three property, q is in $\mathscr{W}_q$ and thus in $\mathscr{F}_m \cap \mathscr{W}_m$. $\qquad\square$

The mixed model structure relates well to Quillen adjunctions. Let $\mathscr{N}$ be another category that, like $\mathscr{M}$, has h- and q-model structures such that $\mathscr{W}_h \subset \mathscr{W}_q$ and $\mathscr{F}_h \subset \mathscr{F}_q$.

PROPOSITION 17.3.3. *Let (F, U) be an adjunction relating $\mathscr{M}$ and $\mathscr{N}$. If (F, U) is a Quillen adjunction with respect to both the h- and the q-model structures, then (F, U) is a Quillen adjunction with respect to the m-model structures. If, further, (F, U) is a Quillen equivalence with respect to the q-model structures, then (F, U) is a Quillen equivalence with respect to the m-model structures.*

PROOF. Since m-fibrations are h-fibrations, U preserves m-fibrations. Since the m-acyclic m-fibrations are the maps in $\mathscr{F}_h \cap \mathscr{W}_q = \mathscr{F}_h \cap (\mathscr{W}_q \cap \mathscr{F}_q)$, U preserves them too. This proves the first statement. The second statement is clear since F and U induce an adjoint equivalence between the homotopy categories $\mathrm{Ho}\mathscr{M}$ and $\mathrm{Ho}\mathscr{N}$ defined with respect to $\mathscr{W}_q = \mathscr{W}_m$; see Proposition 16.2.3. $\qquad\square$

The main value of the m-model structure comes from an analysis of the m-cofibrations and the m-cofibrant objects, which we give next, using the following result in the proof. A key point is that m-cofibrations satisfy properties that show how to relate the weak equivalences and cofibrations in the h- and q-model structures. When we specialize to spaces, the statements in the rest of the section give model theoretic refinements of classical results, and the reader may wish to skip to the next section, referring back to this one as needed. For example, the first part of the following result refines the Whitehead theorem that a weak equivalence between CW complexes is a homotopy equivalence.

PROPOSITION 17.3.4. *Let i and j be m-cofibrations in the commutative diagram*

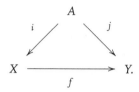

(i) *If f is a q-equivalence, then f is an h-equivalence. In particular, a q-equivalence between m-cofibrant objects is an h-equivalence.*

(ii) *If f is an h-cofibration, then f is an m-cofibration. In particular, an h-cofibration between m-cofibrant objects is an m-cofibration.*

PROOF. The proof of (i) is analogous to the proof of Ken Brown's lemma (14.2.9), and uses a similar commutative diagram and factorization:

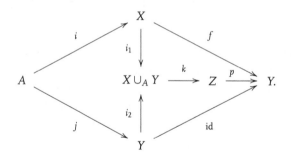

The left square is a pushout, hence i_1 and i_2 are m-cofibrations. Factor the induced map $X \cup_A Y \longrightarrow Y$ as $p \circ k$, where $k \in \mathscr{C}_m$ and $p \in \mathscr{F}_m \cap \mathscr{W}_m$. By the two out of three property, $k \circ i_1$ and $k \circ i_2$ are in $\mathscr{W}_m$. By (17.3.2), they are in $\mathscr{C}_h \cap \mathscr{W}_h \subset \mathscr{W}_h$. Since $p \circ (k \circ i_2) = \mathrm{id}$, p is in $\mathscr{W}_h$ by the two out of three property. But then $f = p \circ (k \circ i_1)$ is also in $\mathscr{W}_h$.

For (ii), factor f as $p \circ k$ where $k \colon X \longrightarrow Z$ is in $\mathscr{C}_m$ and $p \colon Z \longrightarrow Y$ is in $\mathscr{F}_m \cap \mathscr{W}_m$. Then $k \circ i$ is in $\mathscr{C}_m$ and we can apply (i) to the diagram

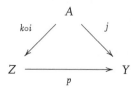

to see that p is in $\mathscr{W}_h$ and thus in $\mathscr{F}_h \cap \mathscr{W}_h$. Therefore f has the LLP with respect to p. By the retract argument, f is a retract of k and is thus an m-cofibration. □

THEOREM 17.3.5. *A map $j\colon A \longrightarrow X$ is an m-cofibration if and only if j is an h-cofibration that factors as a composite $f \circ i$, where i is a q-cofibration and f is an h-equivalence. An object X is m-cofibrant if and only if it is h-cofibrant and has the h-homotopy type of a q-cofibrant object.*

PROOF. If j is in $\mathscr{C}_m$, then it is certainly in $\mathscr{C}_h$. We can factor it as $j = f \circ i$, where i is in $\mathscr{C}_q$ and f is in $\mathscr{F}_q \cap \mathscr{W}_q$. Since i and j are both in $\mathscr{C}_m$, Proposition 17.3.4(i) shows that the q-equivalence f must be an h-equivalence.

For the converse, we are given a factorization $j = f \circ i$, where $i\colon A \longrightarrow Y$ is in $\mathscr{C}_q$ and $f\colon Y \longrightarrow X$ is in $\mathscr{W}_h$. Factor j as $p \circ k$, where $k\colon A \longrightarrow E$ is in $\mathscr{C}_m$ and $p\colon E \longrightarrow X$ is in $\mathscr{F}_m \cap \mathscr{W}_m$. Since $i \in \mathscr{C}_q \subset \mathscr{C}_m$, there is a lift in the diagram

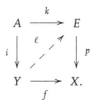

Since f and p are in $\mathscr{W}_q$, so is ℓ. Since i and k are in $\mathscr{C}_m$, Proposition 17.3.4(i) shows that ℓ is in $\mathscr{W}_h$. But then p is in $\mathscr{W}_h$ by the two out of three property and is thus in $\mathscr{F}_h \cap \mathscr{W}_h$. Since j is in $\mathscr{C}_h$, it has the LLP with respect to p. By the retract argument, j is a retract of k and is thus in $\mathscr{C}_m$.

For the second statement, applying the first part with $A = \emptyset$ gives the forward implication. For the converse, let Y be a q-cofibrant object that is isomorphic to X in the homotopy category $h\mathscr{M}$ of the h-model structure. Then Y is h-cofibrant and the isomorphism must be given by an h-homotopy equivalence $Y \longrightarrow X$. The first statement applies with $A = \emptyset$ to show that X is m-cofibrant. □

The following analogue is not an alternative characterization but rather a convenient factorization up to retract property of m-cofibrations.

LEMMA 17.3.6. *Any m-cofibration $j\colon A \longrightarrow X$ is a retract of an m-cofibration $ki\colon A \longrightarrow Z$ such that i is a q-cofibration and k is an h-acyclic h-cofibration.*

PROOF. Factor j as $p \circ i$, where $i \in \mathscr{C}_q$ and $p \in \mathscr{F}_q \cap \mathscr{W}_q$. Then factor p as $q \circ k$, where $k \in \mathscr{C}_h \cap \mathscr{W}_h$ and $q \in \mathscr{F}_h$. Then $q \in \mathscr{W}_q$ by the two out of three property and j has the LLP with respect to q. By the retract argument, j is a retract of $k \circ i$. □

We use this to compare properness in the q- and m-model structures.

PROPOSITION 17.3.7. *If $\mathcal{M}$ is right q-proper, then $\mathcal{M}$ is right m-proper; $\mathcal{M}$ is left m-proper if and only if $\mathcal{M}$ is left q-proper.*

PROOF. Since $\mathscr{F}_m \subset \mathscr{F}_q$, the first statement is clear from the definition of right proper. Similarly, since $\mathscr{C}_q \subset \mathscr{C}_m$, the forward implication of the second statement is clear from the definition of left proper. Assume that $\mathcal{M}$ is left q-proper and consider a pushout diagram

$$
\begin{array}{ccc}
A & \xrightarrow{\ f\ } & B \\
{\scriptstyle j}\downarrow & & \downarrow \\
X & \xrightarrow[\ g\ ]{} & X \cup_A B
\end{array}
$$

in which j is an m-cofibration and f is a q-equivalence. We must show that g is a q-equivalence. As in Lemma 17.3.6, let j be a retract of an m-cofibration $ki\colon A \longrightarrow Z$ where $i\colon A \longrightarrow A'$ is a q-cofibration and $k\colon A' \longrightarrow Z$ is an h-acyclic h-cofibration. Then g is a retract of the pushout $h\colon Z \longrightarrow Z \cup_A B$ of f along ki. Since $\mathscr{W}_m = \mathscr{W}_q$ is closed under retracts, it suffices to prove that h is in $\mathscr{W}_q$. The diagram

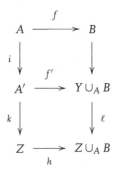

displays a composite of pushout diagrams. Since i is in $\mathscr{C}_q$ and $\mathcal{M}$ is left q-proper, f' is in $\mathscr{W}_q$. Since k is in $\mathscr{C}_h \cap \mathscr{W}_h$, so is its pushout ℓ. Since $\mathscr{W}_h \subset \mathscr{W}_q$, h is in $\mathscr{W}_q$ by the two out of three property. □

We also use Lemma 17.3.6 to relate monoidal properties of the three model structures. Thus let $\mathcal{M}$ be a $\mathscr{V}$-bicomplete $\mathscr{V}$-category, where, like

$\mathcal{M}$, the cosmos $\mathcal{V}$ has h- and q-model structures such that $\mathcal{W}_h \subset \mathcal{W}_q$ and $\mathcal{F}_h \subset \mathcal{F}_q$.

THEOREM 17.3.8. *Assume that $\mathcal{V}$ is h-monoidal and q-monoidal. Then it is m-monoidal. If $\mathcal{M}$ is a $\mathcal{V}$-model category with respect to both the h- and q-model structures, then $\mathcal{M}$ is a $\mathcal{V}$-model category with respect to the m-model structure.*

PROOF. We must show that $\otimes \colon \mathcal{V} \times \mathcal{V} \longrightarrow \mathcal{V}$ and $\odot \colon \mathcal{M} \times \mathcal{V} \longrightarrow \mathcal{M}$ take a pair (f, g) of maps in $\mathcal{C}_m$ to a map in $\mathcal{C}_m$ that is m-acyclic if either f or g is so. The acyclicity part follows easily from (17.3.2). For the cofibration part, we use Lemma 17.3.6 to break the verification into steps that follow from statements about the h- or q-model structures separately and from consideration of composites. The unit conditions (ii) of Definition 16.4.7 are fussy and make use of factorizations

$$\emptyset \to \Gamma_q I \to \Gamma_m I \to \Gamma_h I \to I$$

of cofibrant approximations in the three model categories. Since I is q-cofibrant and therefore m- and h-cofibrant in all examples encountered in this book, we refer the reader to [33, 6.6] for details. $\square$

Note that we can vary the situation here by, for example, taking the h- and q-model structures to be the same on $\mathcal{V}$, while using different h- and q-model structures on $\mathcal{M}$. In particular, it is sensible to make this choice when studying simplicial model categories since the category of simplicial sets does not have an h-model as opposed to q-model structure.

Turning to more technical results, we show that Proposition 17.3.4 admits more elaborate analogues. These results help make it easy to recognize h-equivalences and m-cofibrations when we see them.

PROPOSITION 17.3.9. *Consider a commutative diagram*

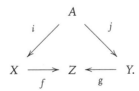

(i) *If i and j are m-cofibrations, f is a q-equivalence, and g is an h-equivalence, then f is an h-equivalence.*

(ii) *If i is an h-cofibration, j is an m-cofibration, and f and g are h-equivalences, then i is an m-cofibration.*

PROOF. For (i), factor g as $p \circ k$, where $k \colon Y \longrightarrow W$ is in $\mathscr{C}_m$ and $p \colon W \longrightarrow Z$ is in $\mathscr{F}_m \cap \mathscr{W}_m$. Since $i \in \mathscr{C}_m$, i has the LLP with respect to p and there is a lift ℓ in the diagram

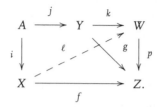

Then $k \in \mathscr{W}_m$ by the two out of three property and, by (17.3.2), k is in $\mathscr{C}_h \cap \mathscr{W}_h \subset \mathscr{W}_h$. Again by the two out of three property, p is in $\mathscr{W}_h$. Since f and p are in $\mathscr{W}_q$, so is ℓ. Since i and $k \circ j$ are in $\mathscr{C}_m$, ℓ is in $\mathscr{W}_h$ by Proposition 17.3.4(i), hence $f = p \circ \ell$ is in $\mathscr{W}_h$.

For (ii), factor i as $p \circ k$ where $k \colon A \longrightarrow W$ is in $\mathscr{C}_m$ and $p \colon W \longrightarrow X$ is in $\mathscr{F}_m \cap \mathscr{W}_m$. We see that $f \circ p$ is in $\mathscr{W}_h$ by applying (i) to the diagram

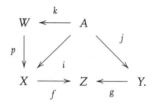

Since f is in $\mathscr{W}_h$, so is p. Thus p is in $\mathscr{F}_h \cap \mathscr{W}_h$ and i has the LLP with respect to p. By the retract argument, i is a retract of k and is thus in $\mathscr{C}_m$. $\square$

PROPOSITION 17.3.10. *Assume that $\mathscr{M}$ is left h-proper and consider a commutative diagram (not necessarily a pushout) in which f is an h-equivalence and i and j are h-cofibrations.*

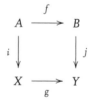

(i) *If i and j are m-cofibrations and g is a q-equivalence, then g is an h-equivalence.*

(ii) *If g is an h-equivalence and i or j is an m-cofibration, then so is the other.*

(iii) *If i and j are m-cofibrations and g is an h-cofibration, then g is an m-cofibration.*

PROOF. Factor the given square through a pushout P to obtain

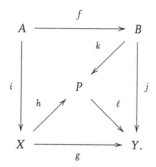

Since $\mathcal{M}$ is left h-proper, i is in $\mathscr{C}_h$, and f is in $\mathscr{W}_h$, we see that h is in $\mathscr{W}_h$. Therefore, by the two out of three property, if g is in either $\mathscr{W}_h$ or $\mathscr{W}_q$, then so is ℓ. Moreover, if i is in $\mathscr{C}_m$, then so is its pushout k.

For (i), we have that g is in $\mathscr{W}_q$, by hypothesis, and thus ℓ is in $\mathscr{W}_q$. But then, by Proposition 17.3.4(i), ℓ is in $\mathscr{W}_h$ and therefore so is $g = \ell \circ h$.

For (ii), assume first that i is in $\mathscr{C}_m$. Then k is in $\mathscr{C}_m$ and Proposition 17.3.9(ii) shows that j is in $\mathscr{C}_m$. Assume next that j is in $\mathscr{C}_m$. Factor i as $p \circ m$, where $m \colon A \longrightarrow Z$ is in $\mathscr{C}_m$ and $p \colon Z \longrightarrow Y$ is in $\mathscr{F}_m \cap \mathscr{W}_m$. In the square

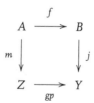

gp is in $\mathscr{W}_q$ and (i) applies to show that $g \circ p$ is in $\mathscr{W}_h$ and thus p is in $\mathscr{F}_h \cap \mathscr{W}_h$. Since i is in $\mathscr{C}_h$, it has the LLP with respect to p. By the retract argument, i is a retract of m and is thus in $\mathscr{C}_m$.

For (iii), factor g as $p \circ k$, where $k \colon X \longrightarrow Z$ is in $\mathscr{C}_m$ and $p \colon Z \longrightarrow Y$ is in $\mathscr{F}_m \cap \mathscr{W}_m$. We may apply (i) to the square

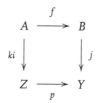

to see that p is in $\mathcal{W}_h$. Then g has the LLP with respect to p. By the retract argument, g is a retract of k and is thus in $\mathcal{C}_m$. $\qquad\square$

17.4. The mixed model structure on spaces

Returning to the case $\mathcal{M} = \mathcal{U}$, we advertise how well the mixed model structure captures the familiar viewpoint of classical algebraic topology. We first summarize the properties of the m-model structure on spaces. The following definition is dictated by Theorem 17.3.5.

DEFINITION 17.4.1. A map $j \colon A \longrightarrow X$ in $\mathcal{U}$ is an m-cofibration if j is an h-cofibration that factors as a composite $f \circ i$, where i is a q-cofibration and f is an h-equivalence.

THEOREM 17.4.2 (m-MODEL STRUCTURE). *The category $\mathcal{U}$ is a proper monoidal model category whose weak equivalences, cofibrations, and fibrations, denoted either $(\mathcal{W}_m, \mathcal{C}_m, \mathcal{F}_m)$ or $(\mathcal{W}_q, \mathcal{C}_m, \mathcal{F}_h)$, are the q-equivalences, m-cofibrations, and h-fibrations. Every space is m-fibrant. A space is m-cofibrant if and only if it has the homotopy type of a CW complex. The identity functor on $\mathcal{U}$ is a right Quillen equivalence from the m-model structure to the q-model structure and therefore is a left Quillen equivalence from the q-model structure to the m-model structure.*

PROOF. Only the characterization of m-cofibrant spaces requires comment. Recall that every space is h-cofibrant. As we shall explain shortly, every cell complex is homotopy equivalent to a CW complex. Moreover, any retract up to homotopy of a CW complex is homotopy equivalent to a CW complex by [93, p. 80, #3]. These facts imply the stated characterization. $\qquad\square$

COROLLARY 17.4.3. *The category $\mathcal{U}_*$ of based spaces in $\mathcal{U}$ is a proper model category whose weak equivalences, cofibrations, and fibrations are the based maps that are q-equivalences, m-cofibrations, and h-fibrations in $\mathcal{U}$. The pair (T, U), where T is given by adjoining disjoint basepoints, is a Quillen adjunction relating $\mathcal{U}$ to $\mathcal{U}_*$. Moreover, $\mathcal{U}_*$ is a monoidal model structure with respect to the smash product. A based space is m-cofibrant if and only if it is h-cofibrant (nondegenerately based) and of the homotopy type of a CW complex.*

PROOF. For the last statement, the inclusion of the basepoint in an m-cofibrant based space is an h-cofibration; compare Remark 17.1.3. $\qquad\square$

Therefore the category of CW homotopy types in $\mathscr{T}$ used in the first half of this book is the full subcategory of the category $\mathscr{U}_*$ whose objects are the m-cofibrant based spaces. As promised in the introduction, this gives a model theoretic justification for our original choice of a convenient category in which to work.

It has long been accepted that the q-equivalences give the definitively right weak equivalences for classical homotopy theory. This dictates $\mathscr{W}_q$ as our preferred subcategory of weak equivalences. The fibrations most frequently used in practice are the Hurewicz fibrations rather than the Serre fibrations. There are good reasons for this. For example, Hurewicz fibrations, but not Serre fibrations, are determined locally, in the sense of the following result (e.g., [88, 3.8]). It generalizes a theorem of Hurewicz [70].

THEOREM 17.4.4. *Let $p\colon E \longrightarrow B$ be a surjective map and assume that B has a numerable open cover $\{U\}$ such that each restriction $p^{-1}U \longrightarrow U$ is an h-fibration. Then p is an h-fibration.*

The proof uses the characterization of h-fibrations in terms of path-lifting functions [93, §7.2], and that characterization itself often gives an easy way of checking that a map is an h-fibration. Paradoxically, this means that it is often easiest to prove that a map is a q-fibration by proving the stronger statement that it is an h-fibration. These considerations argue for the h-fibrations, $\mathscr{F}_h$, as our preferred subcategory of fibrations.

With the mixed model structure, we combine these preferences, which is just what the working algebraic topologist does in practice, and has done for the past half century. The mixed model structure has all of the good formal properties of the q-model structure. It is almost certainly not cofibrantly generated, but that is irrelevant to the applications.

The results of the previous section imply that the class $\mathscr{C}_m$ of m-cofibrations is also very well-behaved. Any map with the name cofibration should at least be a classical Hurewicz cofibration, and an m-cofibration is an h-cofibration that is a q-cofibration up to homotopy equivalence. Using [93, §6.5], we see that an m-cofibration $A \longrightarrow X$ is an h-cofibration that is cofiber homotopy equivalent under A to a retract of a relative cell complex and thus, by a relative generalization of an argument to follow, of a relative CW complex.

Proposition 17.3.4(ii) gives a weak two out of three property that makes it easy to recognize when a map is an m-cofibration. Since m-cofibrations are more general than q-cofibrations, Proposition 17.3.4(i) generalizes the relative version of the Whitehead theorem that a weak equivalence between cell complexes is a homotopy equivalence. Since $\mathcal{U}$ is h-proper, Proposition 17.3.10 gives interesting generalizations of these results.

Cell complexes are more general than CW complexes, but any cell complex is homotopy equivalent to a CW complex. Therefore, in the mixed model structure, we can use cell and CW complexes interchangeably. That too conforms with historical preference. There are two quite different ways to see this. One is to first approximate any space X by a weakly equivalent CW complex, (e.g., [93, §10.5]), and then use the Whitehead theorem (e.g., [93, §10.3]) to show that any cell approximation of X is homotopy equivalent to the constructed CW approximation. The other is to inductively deform any given cell complex to a homotopy equivalent CW complex by cellular approximation of its attaching maps. The geometric realization of the total singular complex of X gives a particularly nice functorial CW approximation (e.g., [93, §16.2]), and it is a functorial cofibrant approximation in both the q- and the m-model structures.

A central reason for preferring the m- to the q-model structure is that it is generally quite hard to check whether a given space is actually homeomorphic to a cell or CW complex, whereas there are powerful classical theorems that characterize spaces of the homotopy types of CW complexes. For an elementary example of this, there are many contractible spaces that cannot be homeomorphic to cell complexes; cones on badly behaved spaces give examples. In particular, Milnor [103] gives a characterization that has the following consequence, among many others. By an n-ad, we understand a space X together with $n - 1$ closed subspaces X_i. It is a CW n-ad if X is a CW complex and the X_i are subcomplexes. It is an m-cofibrant n-ad if it is homotopy equivalent to a CW n-ad.

THEOREM 17.4.5 (MILNOR). *If X is m-cofibrant and C is compact, then the function space X^C of maps $C \longrightarrow X$ is m-cofibrant. If $(X; X_1, \cdots, X_{n-1})$ is an m-cofibrant n-ad and $(C; C_1, \cdots, C_{n-1})$ is a compact n-ad, then the function space n-ad $(X^C; X_1^{C_1}, \cdots, X_{n-1}^{C_{n-1}})$ is an m-cofibrant n-ad.*

Moreover, the m-cofibrant objects mesh naturally with the h-fibrations [121, 129]. The following result again requires Hurewicz rather than Serre fibrations.

THEOREM 17.4.6 (STASHEFF, SCHÖN). *If $p \colon E \longrightarrow B$ is an h-fibration and B is m-cofibrant, then E is m-cofibrant if and only if each fiber F is m-cofibrant.*

It is to be emphasized that the same hierarchichal picture of h-, m-, and q-model structures is present in all of the myriad other examples of topological model categories, although it has usually not been made explicit. Examples appear in [46, 82, 83] and many other places.

17.5. The model structure on simplicial sets

We have mentioned that most of the model theoretic literature focuses on simplicial sets. Although the standard model structure on simplicial sets is very convenient and useful, the proofs that it is indeed a model structure are notoriously involved. The senior author has long believed that simpler proofs should be possible. Correspondence between him and Pete Bousfield in the course of writing this book have led to several variant proofs, primarily due to Bousfield, that are simpler than those to be found in the literature. We explain the most concise variant. We give the statement and relevant background material in this section and turn to the details of proof in the next.

We refer to the 1967 book [92], as reprinted in this series, for much of the background and details of definitions, although we could instead refer to the 1999 book of Goerss and Jardine [53]. A very good source for our purposes is the 1990 book of Fritsch and Piccinini [49], which gives complete and elementary proofs of the key results we need. The crux of our proof comes from results in two wonderful 1957 papers, by Kan [74] and Milnor [102].[2] In retrospect, each gives a fibrant replacement functor on simplicial sets with unusually nice properties: both of these functors preserve fibrations and finite limits. Our proof of the model axioms requires use of these properties, but it is irrelevant which of the functors we use.

Our starting point is the following definition, part (iii) of which is not but ought to be the standard definition of a weak equivalence of simplicial sets. As explained, for example, in [92, 6.11], the naive notion of homotopy between maps that is obtained by use of the simplicial 1-simplex $I = \Delta[1]$ gives an equivalence relation when the target simplicial set is a Kan complex, and then the set $\pi(X, Z)$ of homotopy classes of maps $X \longrightarrow Z$ makes good sense.

DEFINITION 17.5.1. Let $f : X \longrightarrow Y$ be a map of simplicial sets.

(i) f is a cofibration if the map $f_q : X_q \longrightarrow Y_q$ of q-simplices is an injection for each q; equivalently, f is a categorical monomorphism. In particular, every simplicial set is cofibrant.

---

2. Regrettably, the material of Kan's paper was not included in [92].

(ii) f is a fibration if it is a Kan fibration. This means that f satisfies the RLP with respect to all horns $\Lambda^k[n] \longrightarrow \Delta[n]$. In particular, the fibrant objects are the Kan complexes.

(iii) f is a weak equivalence if

$$f^*: \pi(Y, Z) \longrightarrow \pi(X, Z)$$

is a bijection for all Kan complexes Z.

Here $\Lambda^k[n]$ is the subcomplex of $\Delta[n]$ generated by all $(n-1)$-dimensional faces of the standard n-simplex except the kth. Let $\mathscr{C}$, $\mathscr{F}$, and $\mathscr{W}$ denote the three classes of maps of simplicial sets just defined. Let $\mathcal{I}$ denote the set of inclusions $\partial\Delta[n] \longrightarrow \Delta[n]$ and let $\mathcal{J}$ denote the set of inclusions $\Lambda^k[n] \longrightarrow \Delta[n]$.

We abbreviate our previous notation by writing $\mathscr{S} = s\mathscr{S}et$ for the category of simplicial sets. Let (T, S) denote the adjoint pair between $\mathscr{S}$ and $\mathscr{U}$ given by geometric realization and the total singular complex; TX is usually denoted $|X|$, but we follow the notation in [92]. We shall prove the following theorem.

THEOREM 17.5.2. *The classes $(\mathscr{W}, \mathscr{C}, \mathscr{F})$ give the cosmos $\mathscr{S}$ a structure of compactly generated, proper, monoidal model category with generating sets $\mathcal{I}$ of cofibrations and $\mathcal{J}$ of acyclic cofibrations. Moreover (T, S) is a Quillen equivalence between $\mathscr{S}$ and $\mathscr{U}$ with its q-model structure.*

The proof uses several equivalent characterizations of the weak equivalences. Note that in contrast with spaces, homotopy groups do not enter into the definition of weak equivalences. One reason is that they are not easily defined for general simplicial sets. However, they admit a direct combinatorial definition for Kan complexes, in which case they are studied in detail in [92, Ch. I]. This suggests the following definition and theorem.

DEFINITION 17.5.3. A map $f: X \longrightarrow Y$ of Kan complexes is a combinatorial weak equivalence if $f_*: \pi_n(X, x) \longrightarrow \pi_n(Y, f(x))$ is an isomorphism for all $n \geq 0$ and all base vertices $x \in X$.

The class of combinatorial weak equivalences forms a subcategory of weak equivalences between Kan complexes, in the sense of Definition 14.1.4, and the following characterizations hold.

THEOREM 17.5.4. *Let X and Y be Kan complexes. The following conditions on a map $f: X \longrightarrow Y$ are equivalent.*

(i) f is a combinatorial weak equivalence.

(ii) f is a homotopy equivalence.

(iii) f is a weak equivalence.

The equivalence of (i) and (ii) is [92, 12.5], and we will not repeat the proof. The proof there uses the combinatorial theory of minimal Kan complexes, but other proofs are possible. The equivalence of (ii) and (iii) is formal. Since the Kan complexes are the bifibrant objects in $\mathscr{S}$, this result should be viewed as the simplicial analogue of the Whitehead theorem that a weak equivalence between CW complexes is a homotopy equivalence.

We need a few standard results about the functors S and T.

LEMMA 17.5.5. *The functor S takes spaces to Kan complexes, takes Serre fibrations to Kan fibrations, and preserves limits.*

PROOF. The first two statements follow directly from the adjunction and observations about topological simplices. The third holds since S is a right adjoint. □

LEMMA 17.5.6. *The functor T preserves finite limits, hence so does the composite functor ST.*

PROOF. T preserves finite products by [92, 14.3].[3] Recall from [92, Ch. III] or [93, Ch. 16] that each point of TX can be written uniquely in the form $|x, u|$, where x is a nondegenerate simplex in some X_q and u is an interior point of the topological n-simplex Δ_n [92, 14.2], so that TX is a CW complex with one q-cell for each nondegenerate q-simplex. Using that and the proof that T preserves finite products, one can easily check that the natural map

$$T(Y \times_X Z) \longrightarrow TY \times_{TX} TZ$$

is a continuous bijection between subcomplexes of $T(X \times Y) \cong TX \times TY$. Consideration of cell structures shows that the bijection is a homeomorphism.[4] □

We do not have to use the following result, but the shape of the theory is clarified by stating it.

---

3. The cited result assumes that $T(X) \times T(Y)$ is a CW complex, which of course is automatic using compactly generated spaces.

4. The similar argument using equalizers is given in detail in Gabriel and Zisman [50, p. 51].

THEOREM 17.5.7. *The functor T takes Kan fibrations to Serre fibrations, in fact to Hurewicz fibrations, hence the functor ST preserves Kan fibrations.*

REMARK 17.5.8. With Serre fibrations in the conclusion, this result is due to Quillen [114]. All known proofs use the combinatorial theory of minimal fibrations. Quillen used the result of Gabriel and Zisman [50] that T takes minimal fibrations to Serre fibrations. With Hurewicz fibrations in the conclusion, the result is due to Fritsch and Piccinini [49, Thm. 4.5.25].

The adjunction (T, S) relates well to homotopies and homotopy groups.

LEMMA 17.5.9. *Let X be a simplicial set and Y be a space. Then*

$$\pi(TX, Y) \cong \pi(X, SY)$$

and

$$\pi_n(Y, y) \cong \pi_n(SY, y)$$

for any $y \in Y$ (regarded as a 0-simplex of SY).

PROOF. Homotopies $X \times \Delta[1] \longrightarrow SY$ correspond by adjunction to maps $T(X \times \Delta[1]) \longrightarrow Y$. Since T preserves products and $T\Delta[1]$ is homeomorphic to the interval I, these adjoint maps are in bijective correspondence with homotopies $TX \times I \longrightarrow Y$. This proves the first statement. The argument extends to pairs, and the statement about homotopy groups is an immediate comparison of definitions, using that T takes $(\Delta[n], \partial\Delta[n])$ to its evident topological analogue. $\qquad\square$

The following result is due to Milnor [102]; see [49, Cor. 4.5.31] or [92, Thm. 16.6(i)]. Actually, one can either first prove the theorem and then deduce the corollary or first prove the corollary and then deduce the theorem.

THEOREM 17.5.10. *For a Kan complex X, the unit $\eta: X \longrightarrow STX$ of the adjunction is a combinatorial weak equivalence and therefore a homotopy equivalence.*

COROLLARY 17.5.11. *For a space Y, the counit $\varepsilon: TSY \longrightarrow Y$ of the adjunction is a q-equivalence.*

PROOF. As with any adjunction, the composite

$$SY \xrightarrow{\ \eta\ } STSY \xrightarrow{\ S\varepsilon\ } SY$$

is the identity map. By the two out of three property and Theorem 17.5.10, $S\varepsilon$ is a combinatorial weak equivalence. By Lemma 17.5.9, this implies that ε induces an isomorphism on all homotopy groups and is thus a q-equivalence. □

We advertised the significance of the CW approximation functor TS and gave an intuitive sketch of a more direct proof of Corollary 17.5.11 in [93, pp. 122–124]. As noted there, the cellular chains of the CW complex TSY are naturally isomorphic to the singular chains of Y. Now the following result recovers the usual definition of weak equivalences of simplicial sets.

PROPOSITION 17.5.12. *A map* $f: X \longrightarrow Y$ *of simplicial sets is a weak equivalence if and only if* $Tf: TX \longrightarrow TY$ *is a q-equivalence and therefore a homotopy equivalence.*

PROOF. If f is a weak equivalence, then Lemma 17.5.9 and the fact that SW is a Kan complex for any space W imply that

$$(Tf)^*: \pi(TY, W) \longrightarrow \pi(TX, W)$$

is a bijection for any W. Therefore Tf is a homotopy equivalence. Conversely, if Tf is a homotopy equivalence, then Lemma 17.5.9 implies that

$$f^*: \pi(Y, SW) \longrightarrow \pi(X, SW)$$

is a bijection for any space W. In particular, this holds when $W = TZ$ for a Kan complex Z. By Theorem 17.5.10, Z is homotopy equivalent to STZ. Therefore

$$f^*: \pi(Y, Z) \longrightarrow \pi(X, Z)$$

is a bijection for all Kan complexes Z. □

COROLLARY 17.5.13. *For any simplicial set X, the unit* $\eta: X \longrightarrow STX$ *is a weak equivalence.*

PROOF. The composite

$$TX \xrightarrow{\ T\eta\ } TSTX \xrightarrow{\ \varepsilon\ } TX$$

is the identity map and ε is a homotopy equivalence, hence $T\eta$ is a homotopy equivalence. $\square$

Proposition 17.5.12 means that the left adjoint T creates the weak equivalences in $\mathscr{S}$. In most other adjoint pair situations, it is the right adjoint that creates the weak equivalences. The difference is central to the relative difficulty in proving the model category axioms in $\mathscr{S}$.

The simplicial simplices $\Delta[n]$ admit barycentric subdivisions $Sd\Delta[n]$ and these can be used to construct a subdivision functor $Sd: \mathscr{S} \longrightarrow \mathscr{S}$. The functor Sd has a right adjoint Ex. For a simplicial set X, the set of n-simplices of $Ex(X)$ is the set of maps of simplicial sets $Sd\Delta[n] \longrightarrow X$. There are "last vertex" maps $Sd\Delta[n] \longrightarrow \Delta[n]$, and these maps induce a natural injection $e: X \longrightarrow Ex(X)$ of simplicial sets. Applying the functor Ex and the map e iteratively and passing to colimits, there results a functor $Ex^\infty: \mathscr{S} \longrightarrow \mathscr{S}$ and a natural map $\eta: X \longrightarrow Ex^\infty(X)$.

LEMMA 17.5.14. *The functor Ex^∞ preserves finite limits.*

PROOF. This holds since Ex^∞ is the colimit of the right adjoints Ex^n along the injections $e: Ex^n \longrightarrow Ex^{n+1}$. $\square$

Kan [74] proved the following two results. The first is straightforward. See, for example, [49, Lem. 4.6.18, Prop. 4.6.19]. Kan's original proof of the second was quite difficult, but a straightforward argument directly analogous to Milnor's proof of Theorem 17.5.10 is given in [49, Lem. 4.6.20, Cor. 4.6.22].

THEOREM 17.5.15. *The functor Ex^∞ takes simplicial sets to Kan complexes and takes fibrations to fibrations.*

THEOREM 17.5.16. *The map $Te: TX \longrightarrow TEx(X)$ is a homotopy equivalence. Therefore the map $T\eta: TX \longrightarrow TEx^\infty(X)$ is a homotopy equivalence.*

PROPOSITION 17.5.17. *A map $f: X \longrightarrow Y$ is a weak equivalence if and only if the map $Ex^\infty(f): Ex^\infty(X) \longrightarrow Ex^\infty(Y)$ is a weak equivalence and therefore a homotopy equivalence.*

PROOF. In view of Proposition 17.5.12 and Theorem 17.5.16, this is immediate from the naturality of η. $\square$

We summarize the results collected above in the following omnibus theorem, which we state in model categorical language. It makes sense to ask for such data in any model category, but such data rarely exists.

THEOREM 17.5.18. *Let* $R: \mathscr{S} \longrightarrow \mathscr{S}$ *denote either of the functors* ST *or* Ex^{∞} *and let* $\eta: X \longrightarrow RX$ *be the natural map. Then* R *and* η *satisfy the following properties.*

(i) $\eta: X \longrightarrow RX$ *is a weak equivalence.*

(ii) *RX is fibrant.*

(iii) *If* $p: E \longrightarrow B$ *is a fibration, then so is* $Rp: RE \longrightarrow RB$.

(iv) *The functor R preserves finite limits.*

Moreover, a map $f: X \longrightarrow Y$ *is in* $\mathscr{W}$ *if and only if* Rf *is a homotopy equivalence.*

The most difficult of these statements is (iii) in the case $R = ST$, and use of $R = Ex^{\infty}$ instead of $R = ST$ circumvents the need for that. However, the essential conceptual point is that our proof of the model axioms uses the listed properties, and we are free to use either choice of R.

17.6. The proof of the model axioms

We use the criterion of Theorem 15.2.3 to complete the proof of Theorem 17.5.2, and we give full details starting from the results recorded in the previous section. Since $\mathcal{I}$ and $\mathcal{J}$ consist of maps between simplicial sets with only finitely many nondegenerate simplices, any map from one of them to a sequential colimit factors through a finite stage. Therefore the small-object argument applies in its compact form to both $\mathcal{I}$ and $\mathcal{J}$. Thus we have WFSs $(\mathscr{C}(\mathcal{I}), \mathcal{I}^{\boxempty})$ and $(\mathscr{C}(\mathcal{J}), \mathcal{J}^{\boxempty})$. By Definition 17.5.1, the cofibrations are the injections and the fibrations are the Kan fibrations, which means that $\mathcal{J}^{\boxempty} = \mathscr{F}$. We begin with the following observation.

LEMMA 17.6.1. *Every cofibration* $i: K \longrightarrow L$ *is isomorphic to a relative* $\mathcal{I}$-cell *complex. Therefore* $\mathcal{I}^{\boxempty} = \mathscr{C}^{\boxempty}$ *and* $\mathcal{I}^{\boxempty} \subset \mathscr{F}$.

PROOF. Let $L_0 = K$ and L_1 be the union of K and the subcomplex of L generated by the vertices of L not in K. Inductively, suppose we have constructed a relative cell complex $K \longrightarrow L_n$ and an inclusion $L_n \longrightarrow L$ that is a bijection on q-simplices for $q < n$. Let J_n be the set of n-simplices in L but not in L_n. They must be nondegenerate and their boundaries must consist of simplices in L_n. We may regard an n-simplex as a map $\Delta[n] \longrightarrow L$, and we form the pushout diagram

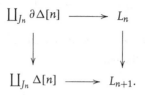

The universal property of pushouts gives a map $L_{n+1} \longrightarrow L$, and the induced map $\operatorname{colim} L_n \longrightarrow L$ is evidently a bijection. The first part of the last statement is immediate and implies the second part since the maps in $\mathcal{J}$ are injections. □

We must prove the acyclicity and compatibility conditions of Theorem 15.2.3. As a left adjoint, T preserves all colimits and, for $j \in \mathcal{J}$, Tj is homeomorphic to an inclusion $i_0 : D^n \longrightarrow D^n \times I$. Therefore, up to isomorphism, T carries a relative $\mathcal{J}$-cell complex in the category of simplicial sets to a relative $\mathcal{J}$-complex in the category of topological spaces, hence acyclicity is immediate from the corresponding property for the q-model structure on $\mathcal{U}$. It remains to prove the compatibility condition $\mathcal{I}^{\boxempty} = \mathcal{J}^{\boxempty} \cap \mathcal{W}$. For that purpose we need two standard and elementary combinatorial lemmas. The first gives the CHEP in $\mathcal{S}$.

LEMMA 17.6.2 (CHEP). *Let $i: A \longrightarrow X$ be a cofibration and $p: E \longrightarrow B$ be a fibration. Let $j: Mi = X \times \{0\} \cup A \times I \longrightarrow X \times I$ be the inclusion of the mapping cylinder Mi in the cylinder $X \times I$. For any commutative square*

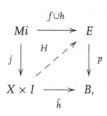

there is a homotopy H that makes the diagram commute. The analogue with the vertex $\{0\}$ replaced by the vertex $\{1\}$ in the definition of Mi also holds.

PROOF. As in Lemma 1.3.2, $h: A \times I \longrightarrow E$ is a homotopy of the restriction of $f: X \longrightarrow E$ to A, and $\bar{h}$ is a homotopy of pf whose restriction to A is covered by h. By adjunction, the statement is equivalent to the assertion that i has the LLP with respect to the canonical map $E^I \longrightarrow Np$, where $Np = B^I \times_B E$ is defined using either $p_0\colon B^I \longrightarrow B$ or $p_1\colon B^I \longrightarrow B$. Lemma 17.6.1 implies that this will hold for all $i\colon A \to X$ if it holds for $i \in \mathcal{I}$, that is, for $i\colon \partial\Delta[n] \to \Delta[n]$.

We claim that in this case the map $j\colon Mi \longrightarrow \Delta[n] \times I$ is a relative $\mathcal{J}$-cell complex with $n+1$ cells, all of them of dimension $n+1$. Since p has the RLP with respect to $\mathcal{J}$, this implies the conclusion. The notation is marginally simpler if we start with the vertex 1 rather than 0 in I. More precisely, starting with $K_0 = \Delta[n] \times \{1\} \cup \partial\Delta[n] \times I$ and ending with $K_{n+1} = \Delta[n] \times I$, we claim that there are $n+1$ nondegenerate simplices $v_m\colon \Delta[n+1] \longrightarrow \Delta[n] \times I$, $0 \leq m \leq n$, that are in $\Delta[n] \times I$ and are not in Mi and that $\Delta[n] \times I$ is constructed from Mi by inductively attaching these simplices along attaching maps $\alpha_m\colon \Lambda^{m+1}[n+1] \longrightarrow K_m$. We display the $(m+1)^{st}$ step of the construction in the commutative diagram

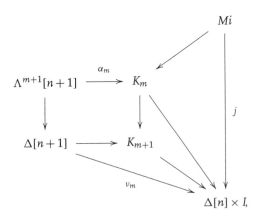

where the left square is a pushout.

Recall that $\Delta[n]_q$ is the set of order-preserving functions $\underline{q} \longrightarrow \underline{n}$, where $\underline{n} = \{0, 1, \cdots, n\}$. Thus $(\Delta[n] \times I)_q$ consists of pairs of order-preserving functions. We can order these lexicographically and determine which are nondegenerate. An inspection that is implicit or explicit in any proof that realization preserves products shows that the nondegenerate $(n+1)$-simplices of $\Delta[n] \times I$ can be identified with the $n+1$ order-preserving injections $v_m\colon \underline{n+1} \longrightarrow \underline{n} \times \underline{1}$ whose images are the ordered chains

$$\{(0,0), (1,0), \cdots, (m,0), (m,1), (m+1,1), \cdots, (n,1)\},$$

where $0 \leq m \leq n$. The displayed pushouts build in these simplices inductively. The faces $d_i v_m$ are obtained by deleting the ith entry in the chain, and these are all in $\partial \Delta[n] \times I$ except for the cases $d_0 v_0 \in \Delta[n] \times \{1\}$, $d_{m+1} v_m = d_{m+1} v_{m+1}$, which has image

$$\{(0,0), (1,0), \cdots, (m,0), (m+1,1), \cdots, (n,1)\},$$

and $d_{n+1} v_n \in \Delta[n] \times \{0\}$. We regard v_m as the map $\Delta[n+1] \longrightarrow \Delta[n] \times I$ that takes the $n+1$ simplex ι_{n+1} given by the identity map of $\underline{n+1}$ to the simplex v_m. To define the attaching map α_m, we must specify the images in K_m of the faces $d_i \iota_{n+1}$ for $i \neq m+1$. These images are forced by commutativity of the diagram, since they are given by faces of $K_m \subset \Delta[n] \times I$ that are in the image of v_m. The argument with $\{1\}$ replaced by $\{0\}$ is similar, but working downward rather than upward on m. $\square$

The second combinatorial lemma gives that each $\Delta[n]$ is contractible.

LEMMA 17.6.3. *Let* $r \colon \Delta[n] \longrightarrow *$ *be the trivial map. The last vertex n of* $\Delta[n]$ *gives a map* $n \colon * \longrightarrow \Delta[n]$, *and there is a homotopy* $h \colon \mathrm{id} \simeq nr$.

PROOF. We must construct $h \colon \Delta[n] \times I \longrightarrow \Delta[n]$ that restricts to the identity map on $\Delta[n] \times \{0\}$ and restricts to nr on $\Delta[n] \times \{1\}$. Identifying the q-simplices of $\Delta[n] \times I$ with lexicographically ordered pairs of order-preserving functions as in the previous proof, we see that the order-preserving function $\underline{n} \times \underline{1} \longrightarrow \underline{n}$ that sends $(m,0)$ to m and $(m,1)$ to n determines h. Observe the asymmetry: homotopy is not an equivalence relation here since $\Delta[n]$ is not a Kan complex, and there is no analogous homotopy $nr \simeq \mathrm{id}$. $\square$

The following result gives the special case of the compatibility condition that applies to fibrations with trivial base space. We separate it out for clarity.

LEMMA 17.6.4. *The following conditions on a Kan complex F are equivalent.*

(i) $\pi_n(F, v) = *$ *for all* $n \geq 0$ *and all base vertices* v.

(ii) F *is contractible.*

(iii) *The trivial map* $r_F \colon F \longrightarrow *$ *is in* $\mathcal{I}^{\boxempty}$.

PROOF. Statements (i) and (ii) are equivalent by Theorem 17.5.4, applied to r_F. We could check that (iii) implies (i) directly from the definition of homotopy groups in [92, p. 7], but we instead observe that (iii) implies (ii) by specialization of the first part of the proof of the next result. To see that (ii) implies (iii), let $i: A \longrightarrow X$ be a map in $\mathcal{I}$ (or, more generally, any injection) and let $g: A \longrightarrow F$ be a map. We must show that g extends to a map $\tilde{g}: X \longrightarrow F$ such that $\tilde{g}i = g$. Since F is contractible, we can choose a base vertex $v: * \longrightarrow F$ and a homotopy $h: vr_F \simeq \text{id}$. Let $r_X: X \longrightarrow *$ be the trivial map. By the CHEP, we can extend

$$vr_X \cup h(g \times \text{id}): Mi = X \times \{0\} \cup A \times I \longrightarrow F$$

to a homotopy $H: X \times I \longrightarrow F$ such that $Hj = vr_X \cup h(g \times \text{id}): Mi \longrightarrow F$. Then the map $\tilde{g} = H_1$ satisfies $\tilde{g}i = g$. □

PROPOSITION 17.6.5 (COMPATIBILITY). $\mathcal{I}^\boxtimes = \mathcal{J}^\boxtimes \cap \mathcal{W}$.

PROOF. Let $p: E \longrightarrow B$ be in $\mathcal{I}^\boxtimes$. By Lemma 17.6.1, p has the RLP with respect to all injections and in particular is a fibration. We can construct a section s of p and a homotopy $h: sp \simeq \text{id}$ as lifts in the diagrams

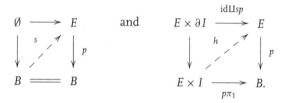

Therefore p is a homotopy equivalence and is thus in $\mathcal{W}$.

Conversely, let $p: E \longrightarrow B$ be in $\mathcal{J}^\boxtimes \cap \mathcal{W}$, so that p is an acyclic Kan fibration. The fiber F_b over b is the pullback of p along $b: \Delta[0] \longrightarrow B$. Applying $R = ST$ or $R = Ex^\infty$, we obtain a Kan fibration between Kan complexes with fiber RF_b over $\eta(b)$. Since p is in $\mathcal{W}$, Rp is a homotopy equivalence. It induces isomorphisms on all homotopy groups by Theorem 17.5.4. Fibrations between Kan complexes have long exact sequences of homotopy groups [92, Thm. 7.6], and the long exact sequence for Rp shows that $\pi_n(RF_b, \eta(b)) = *$ for all n. By Lemma 17.6.4, F_b is contractible and $r_F: F \longrightarrow *$ is in $\mathcal{I}^\boxtimes$.

Consider a lifting problem

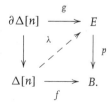

By Lemma 17.6.3, there is a homotopy $h\colon \Delta[n] \times I \longrightarrow B$ from f to the constant map c_b at b. We make two applications of the CHEP.

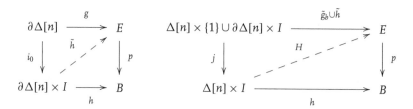

The lift $\tilde{h}$ is a homotopy from g to a map $g_b = \tilde{h}_1\colon \partial\Delta[n] \longrightarrow F_b$ that covers the restriction of h to a homotopy $\partial\Delta[n] \times I \longrightarrow B$. Since r_F is in $\mathcal{I}^{\boxtimes}$, there is a lift $\tilde{g}_b\colon \Delta[n] \longrightarrow F_b$ of g_b. The lift H is a homotopy from a lift $\lambda = H_0$ in our original diagram to the map $\tilde{g}_b$. $\qquad\square$

This completes the proof of the model axioms. The rest of the proof of Theorem 17.5.2 is straightforward. The functor T converts inclusions of simplicial sets to inclusions of subcomplexes in CW complexes, and it preserves weak equivalences by Proposition 17.5.12. Therefore T is a Quillen left adjoint, and (T, S) is a Quillen equivalence by Theorem 17.5.10 and Corollary 17.5.11.

Since every object of $\mathcal{S}$ is cofibrant, $\mathcal{S}$ is left proper by Proposition 15.4.2. Since we have fibrant replacement functors R that preserve pullbacks and fibrations, to check that $\mathcal{S}$ is right proper it suffices to consider pullback squares in which all objects are fibrant. Therefore Proposition 15.4.2 also implies that $\mathcal{S}$ is right proper.

To see that $\mathcal{S}$ is monoidal, let $i\colon A \longrightarrow B$ and $j\colon X \longrightarrow Y$ be cofibrations. The pushout product $i \boxtimes j$ is readily verified to be an inclusion and thus a cofibration. We must show that it is acyclic if either i or j is acylic. Since the functor T preserves products and pushouts, it preserves pushout-products. Since $\mathcal{U}$ is a monoidal model category, the conclusion follows.

REMARK 17.6.6. The last statement is usually proven by the combinatorial theory of anodyne extensions, which are essentially just the maps in $\mathcal{C}(\mathcal{J})$, and the conclusions of that theory are then used in the proof of the model axioms. Our argument uses the CHEP, which is the starting point for the theory of anodyne extensions, but nothing more. The rest of the conclusions of that theory follow from their topological analogues, once the model structure on $\mathscr{S}$ is in place.

REMARK 17.6.7. There is no h-model structure on simplicial sets in the literature, and it does not seem sensible to try to define one. One point is just that the unit interval simplicial set $I = \Delta[1]$ is asymmetric and the obvious notion of homotopy is not an equivalence relation.

18

MODEL STRUCTURES ON CATEGORIES OF CHAIN COMPLEXES

In this chapter, we mimic the previous chapter algebraically, describing the h-, q-, and m-model structures on the category of R-chain complexes. Here R is a ring, not necessarily commutative, fixed throughout the chapter and R-modules are taken to be left R-modules; of course, either left or right works. Again, as in topology, we believe that the m-model structure is central and deserves much more attention than it has received in the literature. We shall explain why it is conceptually fundamental shortly, in §19.1.

18.1. The algebraic framework and the analogy with topology

Let Ch_R be the category of $\mathbb{Z}$-graded R-chain complexes and R-chain maps between them. Differentials lower degree, $d\colon X_n \longrightarrow X_{n-1}$. The category Ch_R is bicomplete. Limits and colimits in Ch_R are just limits and colimits of the underlying R-modules, constructed degreewise, with the naturally induced differentials. Here we use the term "R-module" for a graded R-module, without differentials (or with differential identically zero).

We can shift to cohomological grading, $X^i = X_{-i}$, without changing the mathematics. The differential would then raise degree. Homological grading emphasizes the analogy with topology. In the topological literature, chain complexes are often assumed to be bounded below, following Quillen's original treatment [113]. In cohomological grading, that corresponds to cochain complexes bounded above, which is not a commonly occurring framework. We prefer to make no boundedness assumption. While we have chosen not to deal with spectra in this book, the analogy with model structures on categories of spectra is much closer if we work in the unbounded context.

In this chapter, $\otimes$ and Hom mean $\otimes_{\mathbb{Z}}$ and $\mathrm{Hom}_{\mathbb{Z}}$ unless otherwise specified and chain complexes mean $\mathbb{Z}$-chain complexes. Recall from §16.3 that a cosmos is a bicomplete closed symmetric monoidal category. The category $Ch_{\mathbb{Z}}$ is a cosmos under $\otimes$ and Hom. Recall too that

$$(X \otimes Y)_n = \sum_{i+j=n} X_i \otimes Y_j \quad \text{and} \quad \text{Hom}\,(X, Y)_n = \prod_i \text{Hom}\,(X_i, Y_{i+n})$$

with differentials given by

$$d(x \otimes y) = d(x) \otimes y + (-1)^{\deg x} x \otimes d(y) \text{ and } (df)(x) = d(f(x)) - (-1)^n f(d(x)).$$

The category Ch_R is enriched, tensored, and cotensored over $Ch_{\mathbb{Z}}$. The chain complex of morphisms $X \longrightarrow Y$ is $\text{Hom}_R\,(X, Y)$, where $\text{Hom}_R\,(X, Y)$ is the subcomplex of $\text{Hom}\,(X, Y)$ consisting of those maps f that are maps of underlying R-modules. The reader may want to check that $\text{Hom}_R\,(X, Y)$ is closed under the differential in $\text{Hom}\,(X, Y)$.

We used the notations $\odot$ and Φ for tensors and cotensors earlier, but we use $\otimes$ and Hom here, where these again mean $\otimes_{\mathbb{Z}}$ and $\text{Hom}_{\mathbb{Z}}$. However, for clarity and brevity, we generally abbreviate notation by setting $X^K = \text{Hom}\,(K, X)$. For $X \in Ch_R$ and $K \in Ch_{\mathbb{Z}}$, the chain complexes $X \otimes K$ and X^K are R-chain complexes with $r(x \otimes k) = (rx) \otimes k$ and $(rf)(k) = rf(k)$ for $r \in R, x \in X, k \in K$, and $f \in \text{Hom}\,(K, X)$. We leave it to the reader to verify the required adjunctions

$$\text{Hom}_R\,(X \otimes K, Y) \cong \text{Hom}\,(K, \text{Hom}_R\,(X, Y)) \cong \text{Hom}_R\,(X, Y^K).$$

To emphasize the analogy with topology, we give algebraic objects topological names. Observe that since the zero module 0 is both an initial and terminal object in Ch_R, the analogy to make is with based rather than unbased spaces. For $n \in \mathbb{Z}$, we define S^n, the n-sphere chain complex, to be $\mathbb{Z}$ concentrated in degree n with zero differential. For any integer n, we define the n-fold suspension $\Sigma^n X$ of an R-chain complex X to be $X \otimes S^n$. Thus $(\Sigma^n X)_{n+q} \cong X_q$. The notation is motivated by the observation that if we define $\pi_n(X)$ to be the abelian group of chain homotopy classes of maps $S^n \longrightarrow X$ (ignoring the R-module structure), then $\pi_n(X) = H_n(X)$. The motto is that homology is a special case of homotopy and that homological algebra is a special case of homotopical algebra.

Analogously, we define D^{n+1} to be the $(n+1)$-disk chain complex. It is $\mathbb{Z}$ in degrees n and $n+1$ and zero in all other degrees. There is only one differential that can be nonzero, and we choose that differential to be the identity map $\mathbb{Z} \longrightarrow \mathbb{Z}$. The copy of $\mathbb{Z}$ in degree n is identified with S^n and is quite literally the boundary of D^{n+1}. We agree to write $S^n_R = R \otimes S^n$ and $D^{n+1}_R = R \otimes D^{n+1}$.

We define I to be the cellular chains of the unit interval. It is the chain complex with one basis element $[I]$ in degree 1, two basis elements $[0]$ and

[1] in degrees 0, and differential $d([I]) = [0] - [1]$. We define a homotopy $f \simeq g$ between maps of R-chain complexes $X \longrightarrow Y$ to be a map of R-chain complexes $h : X \otimes I \longrightarrow Y$ that restricts to f and g on $X \otimes [0]$ and $X \otimes [1]$. As the reader can check from the differential on the tensor product (or see [93, p. 90]),

$$s(x) = (-1)^{\deg x} h(x \otimes [I])$$

then specifies a chain homotopy satisfying the usual formula $ds + sd = f - g$. We record the following definition now but say more about it later. In Ch_R, coproducts are direct sums and the pushout of maps $f : A \to X$ and $g : A \to Y$ is the "difference cokernel" $(X \oplus Y)/\mathrm{Im}(f - g)$.

DEFINITION 18.1.1. Let $f : X \longrightarrow Y$ be a map of R-chain complexes. Define the mapping cylinder Mf to be the pushout $Y \cup_f (X \otimes I)$ of the diagram

$$Y \xleftarrow{\;\;f\;\;} X \xrightarrow{\;\;i_0\;\;} X \otimes I.$$

Define the mapping cocylinder Nf to be the pullback $X \times_f \mathrm{Hom}(I, Y)$ of the diagram

$$X \xrightarrow{\;\;f\;\;} Y \xleftarrow{\;\;p_0\;\;} \mathrm{Hom}(I, Y).$$

We have two natural categories of weak equivalences in Ch_R. The h-equivalences are the homotopy equivalences of R-chain complexes, and the q-equivalences are the quasi-isomorphisms, namely those maps of R-chain complexes that induce an isomorphism on passage to the homology of the underlying chain complexes. We call the subcategories consisting of these classes of weak equivalences $\mathscr{W}_h$ and $\mathscr{W}_q$. Since chain homotopic maps induce the same map on homology, $\mathscr{W}_h \subset \mathscr{W}_q$. It is easily checked that both categories are closed under retracts and satisfy the two out of three property and are thus subcategories of weak equivalences as defined in Definition 14.1.4. Similarly, it will be evident that the classes of cofibrations and fibrations that we define in this chapter are subcategories closed under retracts, and we take that for granted in our proofs of the model axioms.

We let hCh_R denote the ordinary homotopy category of Ch_R and call it the classical homotopy category of Ch_R. It is obtained from Ch_R by passing to homotopy classes of maps or, equivalently, by inverting the homotopy equivalences. It is often denoted $\mathscr{K}_R$ in the literature. We let $HoCh_R$ denote the category obtained from Ch_R, or equivalently from hCh_R, by formally inverting the quasi-isomorphisms. It is called the *derived category* of Ch_R, and the alternative notation $\mathscr{D}_R$ is standard. Just as in topology, we shall describe three interrelated model structures on Ch_R, which we name as follows.

The classical model structure is denoted by

18.1.2
$$(\mathscr{W}_h, \mathscr{C}_h, \mathscr{F}_h).$$

The Quillen, or projective, model structure is denoted by

18.1.3
$$(\mathscr{W}_q, \mathscr{C}_q, \mathscr{F}_q).$$

The mixed model structure is denoted by

18.1.4
$$(\mathscr{W}_m, \mathscr{C}_m, \mathscr{F}_m) = (\mathscr{W}_q, \mathscr{C}_m, \mathscr{F}_h).$$

After some preliminaries in §18.2 on *h*-cofibrations and *h*-fibrations, we describe these model structures successively in the rest of the chapter. The one in common use is the *q*-module structure, but, just as in topology, we argue that the *m*-model structure is probably more convenient. It well represents how one actually works in derived categories, and we shall see in the next chapter that it gives a new conceptual perspective on their construction.

REMARK 18.1.5. The *h*-model structure, which was long folklore, is due independently to Cole [32], Schwänzl and Vogt [122], and Christensen and Hovey [29, 66]. The Quillen model structure is of course due to Quillen [113]. The *q*-model structure has an injective analogue that we define in passing but do not discuss in detail; see, for example, [66, Thm. 2.3.13]. This model structure is of less interest to us since, with it, $Ch_{\mathbb{Z}}$ is not a monoidal model category [66, p. 111]. However, there are closely related contexts, such as categories of chain complexes of sheaves, where the *q*-model structure (alias the projective model structure) does not exist, but the injective model structure does. Moreover, there is a "flat" model structure that is monoidal in such contexts [51, 52, 67]. The mixed model structure is due to Cole [33].

18.2. *h*-cofibrations and *h*-fibrations in Ch_R

In this preliminary section, we mimic the topological theory of *h*-cofibrations and *h*-fibrations, replacing $\mathscr{U}$ (or $\mathscr{T}$) by Ch_R and replacing $\times$ (or $\wedge$) by $\otimes$.

DEFINITION 18.2.1. An *h*-cofibration is a map $i \colon A \longrightarrow X$ in Ch_R that satisfies the homotopy extension property (HEP). That is, for all $B \in Ch_R$, i satisfies the LLP with respect to the map $p_0 \colon B^I \longrightarrow B$ given by evaluation at the zero cycle

[0]. An h-fibration is a map $p: E \longrightarrow B$ that satisfies the covering homotopy property (CHP). That is, for all R-chain complexes A, p satisfies the RLP with respect to the map $i_0: A \longrightarrow A \otimes I$. Let $\mathscr{C}_h$ and $\mathscr{F}_h$ denote the classes of h-cofibrations and h-fibrations.

Remember that the h-equivalences are the homotopy equivalences of R-chain complexes. We say that a R-chain complex is contractible if it is homotopy equivalent to the R-chain complex 0. The following triviality is helpful.

LEMMA 18.2.2. *Let C be a contractible R-chain complex. Then $0 \longrightarrow C$ is a retract of $i_0: C \longrightarrow C \otimes I$ and $C \longrightarrow 0$ is a retract of $p_0: C^I \longrightarrow C$.*

PROOF. Here of course we mean retracts in the arrow category. A contracting homotopy $h: 0 \simeq \mathrm{id}_C$ may be viewed as either a map $C \otimes I \longrightarrow C$ or a map $C \longrightarrow C^I$. The following diagrams commute:

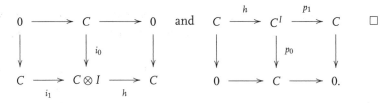

LEMMA 18.2.3. *If $i: A \longrightarrow X$ is an h-cofibration, then i is a monomorphism. If $p: E \longrightarrow B$ is an h-fibration, then p is an epimorphism.*

PROOF. Just as for spaces [93, p. 42], the mapping cylinder $Mi = X \cup_i (A \otimes I)$ must be a retract of $X \otimes I$, which is impossible if i has a nonzero kernel. The dual argument applies to p. □

The following results are direct analogues of standard results in topology concerning cofiber and fiber homotopy equivalence [93, pp. 44, 50]. Consider the category A/Ch_R of chain complexes under a chain complex R. Let X and Y be R-chain complexes under A, with given maps $i: A \to X$ and $j: A \to Y$. Two maps $f, g: X \longrightarrow Y$ under A are said to be homotopic under A, or relative to A, if there is a homotopy $h: X \otimes I \longrightarrow Y$ between them such that $h(a \otimes [I]) = 0$ for $a \in A$. That is, h restricts on $A \otimes I$ to the algebraic version of the constant homotopy at j. A cofiber homotopy equivalence is a homotopy equivalence under A. Working in the category Ch_R/B of chain complexes over a chain complex B, the notion of fiber homotopy equivalence is defined dually.

We note that the composite of homotopies $j, h \colon X \otimes I \to Y$, where $h \colon e \simeq f$ and $j \colon f \simeq g$, is the homotopy $k \colon e \simeq g$ given by e and g on $X \otimes [0]$ and $X \otimes [1]$ and by $j + h$ on $X \otimes [I]$. This composite of algebraic homotopies substitutes for the composite of topological homotopies that is obtained by cutting the unit interval in half and rescaling the homotopies.[1]

PROPOSITION 18.2.4. *Let $i \colon A \longrightarrow X$ and $j \colon A \longrightarrow Y$ be h-cofibrations and let $f \colon X \longrightarrow Y$ be a map under A. If f is a homotopy equivalence, then f is a cofiber homotopy equivalence.*

PROOF. The conclusion says that there is a map $g \colon Y \longrightarrow X$ under A and homotopies $fg \simeq \mathrm{id}$ and $gf \simeq \mathrm{id}$ under A. The proof is formally identical to the proof of the topological analogue in [93, p. 44], but with $\times$ replaced by $\otimes$. The cited proof displays composites of homotopies explicitly in terms of the unit interval, but translating in terms of composites of algebraic homotopies as just described is straightforward. □

COROLLARY 18.2.5. *Let $i \colon A \longrightarrow X$ be an h-acyclic h-cofibration. Then X/A is contractible and i is isomorphic under A to the inclusion $A \longrightarrow A \oplus X/A$.*

PROOF. By h-acyclicity and Proposition 18.2.4, there is a deformation retraction $r \colon X \longrightarrow A$. If $h \colon X \otimes I \longrightarrow X$ is a homotopy $\mathrm{id} \simeq i \circ r$ under A, then h induces a contracting homotopy on X/A. □

PROPOSITION 18.2.6. *Let $p \colon E \longrightarrow B$ and $q \colon F \longrightarrow B$ be h-fibrations and let $f \colon E \longrightarrow F$ be a map over p. If f is a homotopy equivalence, then f is a fiber homotopy equivalence.*

COROLLARY 18.2.7. *Let $p \colon E \longrightarrow B$ be an h-acyclic h-fibration. Then $\ker(p)$ is contractible and p is isomorphic over B to the projection $B \oplus \ker(p) \longrightarrow B$.*

A companion to these results relates homotopy equivalences to contractibility. Observe that for any R-chain complexes X and Y, the abelian group of 0-cycles in $\mathrm{Hom}_R(X, Y)_0$ is the group $Ch_R(X, Y)$ of maps of R-chain complexes $X \longrightarrow Y$, and $H_0(\mathrm{Hom}_R(X, Y))$ is the abelian group $hCh_R(X, Y)$ of homotopy classes of maps $X \longrightarrow Y$ of R-chain complexes.

---

1. If one wants, one can make addition of homotopies look formally the same in the two contexts by observing that the sum of homotopies is defined using a map $I \longrightarrow I \cup_{S^0} I$ in both contexts.

LEMMA 18.2.8. *Let*

$$0 \longrightarrow X \xrightarrow{\ f\ } Y \xrightarrow{\ g\ } Z \longrightarrow 0$$

be an exact sequence of R-chain complexes whose underlying exact sequence of R-modules splits degreewise. Then f is a homotopy equivalence if and only if Z is contractible and g is a homotopy equivalence if and only if X is contractible.

PROOF. If f is an h-equivalence, then $f_* \colon \mathrm{Hom}_R(A, X) \longrightarrow \mathrm{Hom}_R(A, Y)$ is an h-equivalence and thus a quasi-isomorphism for any A. Since the functor $\mathrm{Hom}_R(A, -)$ preserves finite direct sums of underlying graded R-modules, the splitting hypothesis ensures that the functor $\mathrm{Hom}_R(A, -)$ takes the given short exact sequence to a short exact sequence, and the resulting long exact sequence of homology groups shows that the homology of $\mathrm{Hom}_R(A, Z)$ is zero. Taking $A = Z$, the identity map of Z is a 0-cycle and hence a boundary, which means that Z is contractible. Conversely, if Z is contractible, then $\mathrm{Hom}_R(A, Z)$ is a contractible chain complex for any A, hence f_* is a quasi-isomorphism for any A. Since $hCh_R(A, X) \cong H_0(\mathrm{Hom}_R(A, X))$, it follows formally from the cases $A = Y$ and $A = X$ that f is an h-equivalence. A symmetric argument applies with f and Z replaced by g and X. $\qquad\square$

It is now easy to verify the algebraic analogue of Proposition 17.1.4.

PROPOSITION 18.2.9. *Consider a commutative diagram of R-chain complexes*

$$
\begin{array}{ccc}
A & \xrightarrow{\ g\ } & E \\
{\scriptstyle i}\downarrow & {\scriptstyle \lambda}\nearrow & \downarrow{\scriptstyle p} \\
X & \xrightarrow[\ f\ ]{} & B
\end{array}
$$

in which i is an h-cofibration and p is an h-fibration. If either i or p is an h-equivalence, then there exists a lift λ.

PROOF. If p is an h-acyclic h-fibration, then Corollary 18.2.7 shows that p is isomorphic to a projection $B \oplus C \longrightarrow B$, where C is contractible. Thus p is the sum of id_B and $C \longrightarrow 0$, which by Lemma 18.2.2 is a retract of $p_0 \colon C^I \longrightarrow C$. Therefore i satisfies the LLP with respect to p since it satisfies the LLP with respect to id_D and p_0. The proof when i is an h-acyclic h-cofibration is dual. $\square$

18.3. The *h*-model structure on Ch_R

We shall prove the following theorem. Recall the definition of a monoidal model category $\mathcal{V}$ and of a $\mathcal{V}$-model category from Definition 16.4.7.

THEOREM 18.3.1. *The subcategories $(\mathcal{W}_h, \mathcal{C}_h, \mathcal{F}_h)$ define a model category structure on Ch_R, called the h-model structure. Every object is h-cofibrant and h-fibrant, hence the h-model structure is proper. If R is commutative, the cosmos Ch_R is a monoidal model category under $\otimes$. In general, Ch_R is a $Ch_{\mathbb{Z}}$-model category.*

For the model structure, it remains only to prove the factorization axioms. This could be done directly, but we prefer to take a less elegant and more informative approach. In topology, a map has the HEP if and only if it is the inclusion of an NDR-pair [93, p. 43] and a map has the CHP if and only if it has this property locally [93, p. 49]. These criteria allow us to recognize such maps when we see them. Similarly, in algebra, it is not at all obvious how to recognize maps that satisfy the HEP or CHP when we see them. We shall rectify this by giving new definitions of r-cofibrations and r-fibrations and proving that these maps are precisely the maps that satisfy the HEP or CHP. The new notions are much more algebraically intuitive. To go along with this, we define an r-equivalence to be an h-equivalence.

REMARK 18.3.2. What is really going on here is that we have two model structures, the h-model structure and the r-model structure, that happen to coincide. The r stands for "relative", and the r-module structure is a starting point for relative homological algebra. In more sophisticated algebraic situations, there are h-, r-, and q-model structures, and they are all different. This happens, for example, if R is a commutative ring, A is a DG R-algebra, and we consider model structures on the category of differential graded A-modules.[2] In this situation, A need not be projective as an R-module, and then functors such as $\otimes_A$ and Hom_A rarely preserve exact sequences. Relative homological algebra rectifies this by restricting the underlying notion of an exact sequence of A-modules to sequences that are degreewise split exact as sequences of graded R-modules.

DEFINITION 18.3.3. A map $f \colon X \longrightarrow Y$ of R-chain complexes is an r-cofibration if it is a degreewise-split monomorphism; it is an r-fibration if

---

2. Details will appear in a paper by the first author.

it is a degreewise-split epimorphism. We use the term "R-split" for degreewise split from now on.

Of course, such splittings are given by maps of underlying graded R-modules that need not be maps of chain complexes. The following result (due to Cole [32]) shows that the splittings can be deformed to chain maps if the given R-split maps are homotopy equivalences. It implies the r-analogues of Corollaries 18.2.5 and 18.2.7.

PROPOSITION 18.3.4. *Let*

$$0 \longrightarrow X \overset{f}{\longrightarrow} Y \overset{g}{\longrightarrow} Z \longrightarrow 0$$

be an exact sequence of R-chain complexes whose underlying exact sequence of graded R-modules splits. If f or g is a homotopy equivalence, then the sequence is isomorphic under X and over Z to the canonical exact sequence of R-chain complexes

$$0 \longrightarrow X \longrightarrow X \oplus Z \longrightarrow Z \longrightarrow 0.$$

PROOF. We may choose a map of R-modules $r \colon Y \longrightarrow X$ such that $rf = \mathrm{id}_X$. As usual, there results a map of R-modules $i \colon Z \longrightarrow Y$ such that $gi = \mathrm{id}_Z$, $ri = 0$, and $fr + ig = \mathrm{id}_Y$.

Assume that f is a homotopy equivalence. By Lemma 18.2.8, Z is contractible. Let s be a contracting chain homotopy, so that $ds + sd = \mathrm{id}_Z$. Define

$$r' = r - r \circ d \circ i \circ s \circ g \colon Y \longrightarrow X.$$

Since $g \circ f = 0$, $r' \circ f = r \circ f = \mathrm{id}_X$, so that r' is also a splitting map. But now r' is a chain map, as we see by a tedious chain of equalities:

$$\begin{aligned}
dr' &= dr - drdisg \\
&= rfdr - rfdrdisg && \text{since } rf = \mathrm{id} \\
&= rdfr - rdfrdisg && \text{since } df = fd \\
&= rd(\mathrm{id} - ig) - rd(\mathrm{id} - ig)disg && \text{since } fr + ig = \mathrm{id} \\
&= rd - rdig - rddisg + rdigdisg \\
&= rd - rdig + rdidgisg && \text{since } dd = 0 \text{ and } dg = gd \\
&= rd - rdig + rdidsg && \text{since } gi = \mathrm{id} \\
&= rd - rdi(\mathrm{id} - ds)g \\
&= rd - rdisdg && \text{since } ds + sd = \mathrm{id} \\
&= rd - rdisgd = r'd && \text{since } dg = gd.
\end{aligned}$$

Therefore r' induces an isomorphism of chain complexes $Y \cong X \oplus Z$ under X and over Z. The argument when g is a homotopy equivalence is dual, in the sense illustrated in the following proof of the r-analogue of Proposition 18.2.9. □

PROPOSITION 18.3.5. *Consider a commutative diagram of R-chain complexes*

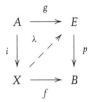

in which i is an r-cofibration and p is an r-fibration. If either i or p is an h-equivalence, then there exists a lift λ.

PROOF. If i is an h-equivalence, then i is isomorphic to a canonical inclusion $i \colon A \longrightarrow A \oplus C$ where C is contractible, and if p is an h-equivalence, then p is isomorphic to a canonical projection $p \colon B \oplus C \longrightarrow B$ where C is contractible. The constructions of lifts in the two cases are dual. Since the reader may not be comfortable with this kind of duality, we give the details in both cases. Let C be a contractible R-chain complex with contracting homotopy s, so that $ds + sd = \mathrm{id}_C$.

First, assume that p is an h-equivalence and take $E = B \oplus C$. Write $g = (g_1, g_2)$, $g_1 \colon A \longrightarrow B$ and $g_2 \colon A \longrightarrow C$, and write $\lambda = (\lambda_1, \lambda_2)$ similarly. To ensure that $p\lambda = f$, we can and must define $\lambda_1 = f$. To define λ_2, choose a retraction $r \colon X \longrightarrow A$ of underlying R-modules, so that $ri = \mathrm{id}_A$. Then define

$$\lambda_2 = dsg_2 r + sg_2 rd.$$

Since $d^2 = 0$, we see immediately that $d\lambda_2 = dsg_2 rd = \lambda_2 d$, and we have

$$
\begin{aligned}
\lambda_2 i &= (dsg_2 r + sg_2 rd)i \\
&= dsg_2 + sg_2 d && \text{since } ri = \mathrm{id} \text{ and } di = \mathrm{id} \\
&= dsg_2 + sdg_2 && \text{since } dg_2 = g_2 d \\
&= g_2 && \text{since } ds + sd = \mathrm{id}.
\end{aligned}
$$

Next assume that i is an h-equivalence and take $X = A \oplus C$. Write $f = (f_1, f_2)$, $f_1 \colon A \longrightarrow B$ and $f_2 \colon C \longrightarrow B$, and write $\lambda = (\lambda_1, \lambda_2)$ similarly. To ensure that $\lambda i = g$, we can and must define $\lambda_1 = g$. To define λ_2, choose a section $j \colon B \longrightarrow E$

of underlying R-modules, so that $pj = \mathrm{id}_B$. Then define

$$\lambda_2 = jf_2sd + djf_2s.$$

Since $d^2 = 0$, we see immediately that $d\lambda_2 = djf_2sd = \lambda_2d$, and we have

$$
\begin{aligned}
p\lambda_2 &= p(jf_2sd + djf_2s) \\
&= f_2sd + df_2s && \text{since } pj = \mathrm{id} \text{ and } pd = dp \\
&= f_2sd + f_2ds && \text{since } df_2 = f_2d \\
&= f_2 && \text{since } ds + sd = \mathrm{id}. \qquad \square
\end{aligned}
$$

PROPOSITION 18.3.6. *Let $f: X \longrightarrow Y$ be a map of R-chain complexes. Then f is an h-cofibration if and only if it is an r-cofibration and f is an h-fibration if and only if it is an r-fibration.*

PROOF. Changing notation, let $i: A \longrightarrow X$ be an h-cofibration. Then the mapping cylinder $Mi = X \cup_i A \otimes I$ is a retract of $X \otimes I$. As an R-module, Mi is the direct sum $X \oplus A \oplus \Sigma A$, where A is $A \otimes R \cdot [1]$ and ΣA is $A \otimes R \cdot [I]$; $A = A \otimes [0]$ is identified with $i(A) \subset X$. Clearly Mi retracts to the summand A. The composite retraction restricts on $X \otimes R \cdot [1]$ to a splitting $X \longrightarrow A$ of i. Conversely, let i be an r-cofibration. Since $p_0: B^I \longrightarrow B$ is an h-acylic r-fibration for any B, Proposition 18.3.5 shows that i satisfies the HEP and is thus an h-cofibration. A dual argument shows that the h-fibrations coincide with the r-fibrations. $\qquad \square$

It is now very easy to prove the factorization axioms. Just as in topology, any map $f: X \longrightarrow Y$ factors as composites

18.3.7 $$X \xrightarrow{\ j\ } Mf \xrightarrow{\ r\ } Y \quad \text{and} \quad X \xrightarrow{\ \nu\ } Nf \xrightarrow{\ \rho\ } Y,$$

where r and ν are h-equivalences. Since the topological proofs of these equivalences do not transcribe directly to algebra, we indicate quick proofs; formal arguments are also possible. Here $j(x) = x \otimes [1]$, $r(y) = y$, $r(x \otimes [1]) = f(x)$, and $r(x \otimes [I]) = 0$. Define $i: Y \longrightarrow Mf$ by $i(y) = y$. Then $ri = \mathrm{id}_Y$. A homotopy $h: Mf \otimes I \longrightarrow Mf$ from ir to id_{Mf} is given by

$$
h(z \otimes [I]) =
\begin{cases}
0 & \text{if } z \in Y \text{ (or } z = x \otimes [0]) \\
x \otimes [I] & \text{if } z = x \otimes [1] \\
0 & \text{if } z = x \otimes [I].
\end{cases}
$$

A small check, taking care with signs, shows that this works. The definitions of v and ρ are dual to those of j and r respectively, and a dual proof shows that v is an h-equivalence. An easy inspection shows that j and v are R-split monomorphisms and r and ρ are R-split epimorphisms. Therefore these elementary factorizations are model theoretic factorizations and the proof of the model axioms is complete.

It remains to prove that the h-model structure is monoidal when R is commutative and that Ch_R is an $Ch_{\mathbb{Z}}$-model category in general. Let $i\colon A \longrightarrow X$ and $j\colon Y \longrightarrow Z$ be h-cofibrations (in either situation, with $\otimes$ understood to be $\otimes_R$ for the first statement). We must prove that $i \square j\colon i \boxtimes j \longrightarrow X \otimes Z$ is an h-cofibration that is h-acyclic if i or j is h-acyclic. If $r\colon X \longrightarrow A$ and $s\colon Z \longrightarrow Y$ split i and j, they induce a splitting $r \square s$ as displayed in the following comparison of coequalizer diagrams.

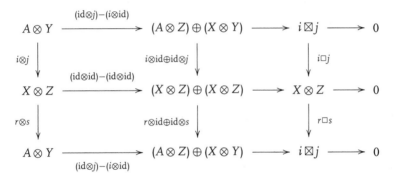

Now suppose that i is a homotopy equivalence. By Corollary 18.2.5, it is isomorphic under A to an inclusion $A \longrightarrow A \oplus C$, where C is contractible. This implies that $i \square j$ is isomorphic to the map

$$(C \otimes Y) \oplus (A \otimes Z) \xrightarrow{\text{id} \otimes j \oplus \text{id}} (C \otimes Z) \oplus (A \otimes Z).$$

This is a homotopy equivalence since a contracting homotopy for $C \simeq \{0\}$ induces a contracting homotopy for $C \otimes Y$ and $C \otimes Z$.

18.4. The q-model structure on Ch_R

We defined h-fibrations in Ch_R to be the precise algebraic analogues of Hurewicz fibrations in $\mathcal{U}$, and we define q-fibrations to be the precise analogues of Serre fibrations. Recall that $\mathcal{W}_q$ is the subcategory of quasi-isomorphisms.

DEFINITION 18.4.1. Let $\mathcal{I}$ denote the set of inclusions $S_R^{n-1} \longrightarrow D_R^n$ for all $n \in \mathbb{Z}$ and let $\mathcal{J}$ denote the set of maps $i_0 \colon D_R^n \longrightarrow D_R^n \otimes I$ for all $n \in \mathbb{Z}$. A map p in Ch_R is a q-fibration if it satifies the RLP with respect to $\mathcal{J}$. A map is a q-cofibration if it satisfies the LLP with respect to all q-acyclic q-fibrations. Let $\mathscr{C}_q$ and $\mathscr{F}_q$ denote the subcategories of q-cofibrations and q-fibrations.

The proof of the following result is almost identical to that of its topological analogue Theorem 17.1.1. There are alternative, more algebraically focused proofs, and we say a little about the steps as we go along.

THEOREM 18.4.2. *The subcategories $(\mathscr{W}_q, \mathscr{C}_q, \mathscr{F}_q)$ define a compactly generated model category structure on Ch_R, called the q-model structure. The sets $\mathcal{I}$ and $\mathcal{J}$ are generating sets for the q-cofibrations and the q-acyclic q-cofibrations. Every object is q-fibrant and the q-model structure is proper. If R is commutative, the cosmos Ch_R is a monoidal model category under $\otimes$. In general, Ch_R is an $Ch_\mathbb{Z}$-model category.*

It is easy to characterize the q-fibrations and the q-acyclic q-fibrations directly from the definitions.

PROPOSITION 18.4.3. *A map $p \colon E \longrightarrow B$ is a q-fibration if and only if it is a degreewise epimorphism.*

PROOF. By definition, p is a q-fibration if and only if p has the RLP with respect to all maps $i_0 \colon D_R^n \longrightarrow D_R^n \otimes I$. Elaborating on Lemma 18.2.2, it is easy to check that i_0 is isomorphic to the direct sum of id $\colon D_R^n \longrightarrow D_R^n$, $0 \longrightarrow D_R^n$, and $0 \longrightarrow D_R^{n+1}$. This implies that p is a q-cofibration if and only p has the RLP with respect to all maps $0 \longrightarrow D_R^n$. Since a map $D_R^n \longrightarrow B$ is a choice of a chain of B_n, we can find a lift in any diagram of the form

if and only if p is a degreewise epimorphism. $\qquad\square$

PROPOSITION 18.4.4. *A map $p \colon E \longrightarrow B$ is a q-acyclic q-fibration if and only if p is in $\mathcal{I}^{\boxtimes}$.*

PROOF. It is an exercise to check this by using that a test diagram

$$\begin{array}{ccc} S_R^{n-1} & \longrightarrow & E \\ \downarrow & & \downarrow \\ D_R^n & \longrightarrow & B \end{array}$$

consists of a cycle $e \in E_{n-1}$ and a chain $b \in B_n$ such that $d(b) = p(e)$. $\qquad\square$

In fact, there is no need to give such a direct proof of Proposition 18.4.4 since the conclusion will drop out from the verification of the model axioms. The following lemma is the exact algebraic analogue of the key lemma used in the proof of HELP (the homotopy extension and lifting property) in [93]. As in topology, it helps us organize the proof of the compatibility condition required in Theorem 15.2.3. We prove it and record the algebraic version of HELP before returning to the proof of Theorem 18.4.2 and characterizing the q-cofibrations and q-acyclic q-cofibrations, which we do in the next section.

LEMMA 18.4.5. *Let $e \colon Y \longrightarrow Z$ be a map in Ch_R. Then the induced map e_* on homology is a monomorphism in degree $n-1$ and an epimorphism in degree n if and only if whenever given maps $f \colon D_R^n \to Z$, $g \colon S_R^{n-1} \to Y$, and $h \colon S_R^{n-1} \otimes I \to Z$ such that $f | S_R^{n-1} = h \circ i_0$ and $e \circ g = h \circ i_1$ in the following diagram, there are maps $\tilde{g}$ and $\tilde{h}$ that make the entire diagram commute.*

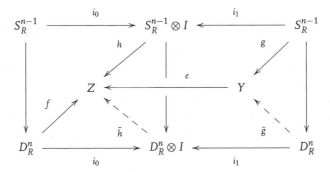

PROOF. Let i_{n-1} be the canonical basis element of S_R^{n-1} and j_n be the basis element of D_R^n with $d(j_n) = i_{n-1}$. The map g is given by a cycle $y = g(i_{n-1}) \in Y_{n-1}$ and the map f is given by a boundary relation

$$z = f(i_{n-1}) = df(j_n) \in Z_{n-1}.$$

Letting $c = (-1)^{n-1}h(i_{n-1} \otimes [I])$, we see that the homotopy h is determined by y, z, and a chain c such that $d(c) = z - e(y)$. Thus e_* sends the homology class $[y]$ to 0. If $H_{n-1}(e)$ is a monomorphism, there is a chain $k \in Y_n$ such that $d(k) = y$. Then $e(k) + c - f(j_n)$ is a cycle. If $H_n(e)$ is an epimorphism, there must be a cycle $y' \in Y_n$ and a chain $\ell \in Z_{n+1}$ such that

$$d(\ell) = e(y') - (e(k) + c - f(j_n)).$$

We then define $\tilde{g}(j_n) = k - y'$ and $\tilde{h}(j_n \otimes [I]) = (-1)^n \ell$; the definitions of these maps on other basis elements are forced by commutativity of the diagram. A little check shows that $\tilde{h}$ is in fact a chain map. For the converse, we see that $H_{n-1}(e)$ is a monomorphism by the case $h(i_{n-1} \otimes [I]) = 0$ of the diagram. Then

$$dz = h(i_{n-1} \otimes [0]) = h(i_{n-1} \otimes [1]) = ey,$$

so that f displays $eg(i_{n-1})$ as a boundary. The map $\tilde{g}$ shows that $g(i_{n-1})$ must be a boundary. We see that $H_n(e)$ is an epimorphism by the case $g = 0$ and $h = 0$ of the diagram. Then z is a cycle, and the maps $\tilde{g}$ and $\tilde{h}$ display a cycle of Y whose image under e is homologous to z. □

We shall be using the compact object argument, hence we are only interested in sequential cell complexes, like the classical cell complexes in topology and like projective resolutions in algebra. We define $\mathscr{C}(\mathcal{I})$ and $\mathscr{C}(\mathcal{J})$ to be the retracts of the relative cell complexes, as in Definition 15.1.1.

REMARK 18.4.6. An $\mathcal{I}$-cell complex X has an increasing filtration given by its successive terms, which we shall here denote by F_qX rather than X_q in order to avoid confusion with the R-module X_q of elements of degree q. The subcomplex F_qX can have elements of arbitrary degree. Let us write $X_{\leq q}$ for the elements of X of degree $\leq q$. This gives X a second filtration that corresponds to the skeletal filtration of CW complexes in topology. In topology, cellular approximation of maps allows us to replace cell complexes by equivalent CW complexes. In algebra, the comparison is much simpler and the difference is negligible since an "attaching map" $S_R^n \longrightarrow F_qX$ for a cell D_R^{n+1} necessarily has image in the elements $(F_qX)_n$ of degree n. When we restrict attention to those X such that $X_q = 0$ for $q < 0$, as in classical projective resolutions, we may as well also restrict attention to those $\mathcal{I}$-cell complexes such that $F_qX = X_{\leq q}$ (starting with $F_{-1}X = 0$ and allowing F_0X to be nonzero).

THEOREM 18.4.7 (HELP). *Let* $A \longrightarrow X$ *be a relative* $\mathcal{I}$-*cell complex and let* $e : Y \longrightarrow Z$ *be a q-equivalence. Given maps* $f : X \longrightarrow Z$, $g : A \longrightarrow Y$, *and* $h : A \otimes I \longrightarrow Z$ *such that* $f|A = h \circ i_0$ *and* $e \circ g = h \circ i_1$ *in the following diagram, there are maps* $\tilde{g}$ *and* $\tilde{h}$ *that make the entire diagram commute.*

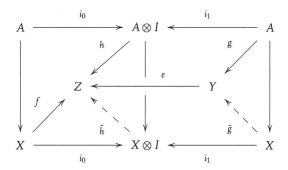

PROOF. We proceed by induction over the cellular filtration, and then by passage to unions, proceeding one layer of cells at a time. The $(q+1)^{\text{st}}$ term $F_{q+1}X$ of the filtration of X is constructed by attaching cells D^n along attaching maps $S_R^{n-1} \longrightarrow F_q X$. Starting with $F_0 X = A$, we obtain the conclusion by applying the case $S_R^{n-1} \longrightarrow D_R^n$ one cell at a time to the cells of $F_{q+1}X$ not in $F_q X$. $\qquad \square$

18.5. Proofs and the characterization of q-cofibrations

PROOF OF THEOREM 18.4.2. With $\lambda = \omega$, it is obvious that $\mathcal{I}$ and $\mathcal{J}$ are compact in the sense of Definition 15.1.6. Indeed, for any finite cell complex A and relative $\mathcal{I}$-cell complex $X \longrightarrow Z = \text{colim } F_q Z$, the canonical map

$$\text{colim}_n \, Ch_R(A, F_q Z) \longrightarrow Ch_R(A, Z)$$

is an isomorphism since a map $A \longrightarrow Z$ is determined by the images of its finitely many R-basis elements. By the compact object argument, Proposition 15.1.11, we have functorial WFSs $(\mathscr{C}(\mathcal{I}), \mathcal{I}^{\boxtimes})$ and $(\mathscr{C}(\mathcal{J}), \mathcal{J}^{\boxtimes})$. To verify the model axioms, we need only verify the acylicity and compatibility conditions of Theorem 15.2.3. For the acylicity, let $i : A = F_0 X \longrightarrow \text{colim } F_q X = X$ be a relative $\mathcal{J}$-cell complex. Arguing one cell at a time, we see that each map $F_q X \longrightarrow F_{q+1}X$ of the colimit system is the inclusion of a deformation retraction and therefore i is a q-equivalence (in fact, an h-equivalence).

For the compatibility, we must show that $\mathcal{I}^{\boxempty} = \mathcal{J}^{\boxempty} \cap \mathcal{W}_q$. The maps in $\mathcal{J}$ are $\mathcal{I}$-cell complexes, so they are in $\mathscr{C}(\mathcal{I})$, and this implies that $\mathcal{I}^{\boxempty} \subset \mathcal{J}^{\boxempty}$. To show that $\mathcal{I}^{\boxempty} \subset \mathcal{W}_q$, observe that the inclusions $0 \longrightarrow S_R^n$ and $i_0 + i_1: S_R^n \oplus S_R^n \longrightarrow S_R^n \otimes I$ are relative $\mathcal{I}$-cell complexes and are thus in $\mathscr{C}(\mathcal{I})$. If $p: E \longrightarrow B$ is in $\mathcal{I}^{\boxempty}$, then liftings with respect to $0 \longrightarrow S_R^n$ show that $p_*: H_n(E) \longrightarrow H_n(B)$ is surjective and liftings with respect to $S_R^n \oplus S_R^n \longrightarrow S_R^n \otimes I$ show that p_* is injective. Conversely, suppose that $p: E \longrightarrow B$ is in $\mathcal{J}^{\boxempty} \cap \mathcal{W}_q$ and consider a lifting problem

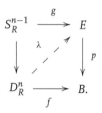

We use the square to construct the solid arrow portion of the following diagram. The map h is the composite of f and the map $S_R^{n-1} \otimes I \longrightarrow D_R^n$ that sends $i_{n-1} \otimes [I]$ to 0 and sends both $i_{n-1} \otimes [0]$ and $i_{n-1} \otimes [1]$ to i_{n-1}.

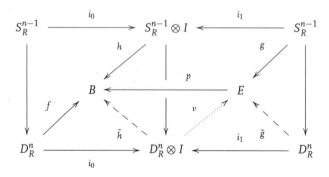

By Lemma 18.4.5, since p is a quasi-isomorphism there are dashed arrows $\tilde{g}$ and $\tilde{h}$ making the dashed and solid arrow parts of the diagram commute. Since p has the RLP with respect to i_0, it has the RLP with respect to i_1 and we obtain a lift ν such that $p \circ \nu = \tilde{h}$ and $\nu \circ i_1 = \tilde{g}$. The composite $\lambda = \nu \circ i_0$ is the desired lift in our original diagram. This completes the proof of the model axioms.

To see that the q-model structure is monoidal when R is commutative, consider $i \square j$, where i and j are the cells $i: S_R^{m-1} \longrightarrow D_R^m$ and $j: S_R^{n-1} \longrightarrow D_R^n$. We obtain a split inclusion f of relative cell complexes

$$
\begin{array}{ccc}
S_R^{m+n-2} & \longrightarrow & D_R^{m+n-1} \\
f \downarrow & & \downarrow f \\
i \boxtimes j & \xrightarrow{\ i \square j\ } & D_R^m \otimes_R D_R^n
\end{array}
$$

by setting $f(i_{m+n-2}) = i_{m-1} \otimes i_{n-1}$ and $f(j_{m+n-1}) = j_m \otimes i_{n-1}$. The quotient of $i \square j$ by f is isomorphic to the cell

$$
S_R^{m+n-1} \longrightarrow D_R^{m+n}.
$$

That is, $i \square j$ is the direct sum of two cells and so is a relative cell complex. If we replace j here by $i_0 \colon D_R^n \longrightarrow D_R^n \otimes I$, then $i \square j$ is a homotopy equivalence. Rather than prove that, we observe that the proof of Proposition 18.4.3 shows that we could have used the alternative set of generating q-acyclic q-cofibrations consisting of the maps $j \colon 0 \longrightarrow D_R^n$ and then $i \square j$ is just $i \otimes_R \mathrm{id}_{D^n}$, which is a relative $\mathcal{J}$-cell complex. It follows inductively that if i and j are relative $\mathcal{I}$-cell complexes, then $i \square j$ is a relative $\mathcal{I}$-cell complex, and that if i is a relative $\mathcal{I}$-cell complex and j is a relative $\mathcal{J}$-cell complex, then $i \square j$ is a relative $\mathcal{J}$-cell complex. Moreover, the functor $i \square (-)$ on the arrow category preserves retracts. Therefore $i \square j$ is a q-cofibration if i or j is so and is a q-acyclic q-cofibration if, further, either i or j is so. The proof that Ch_R is an $Ch_{\mathbb{Z}}$-model category for any ring R is similar.

Since every R-chain complex is q-fibrant, the q-model structure is right proper. As in topology, we prove the gluing lemma and conclude that the q-model structure is left proper by Proposition 15.4.4. $\qquad\square$

LEMMA 18.5.1 (THE GLUING LEMMA). *Assume that i and j are q-cofibrations and f, g, and h are q-equivalences in the following commutative diagram in Ch_R.*

$$
\begin{array}{ccccc}
A & \xleftarrow{\ i\ } & C & \xrightarrow{\ k\ } & B \\
f \downarrow & & \downarrow g & & \downarrow h \\
A' & \xleftarrow{\ j\ } & C' & \xrightarrow{\ \ell\ } & B'
\end{array}
$$

Then the induced map of pushouts

$$
X = A \cup_C B \longrightarrow A' \cup_{C'} B' = X'
$$

is a q-equivalence.

PROOF. The pushout $A \cup_C B$ is constructed by an exact sequence

$$0 \longrightarrow C \xrightarrow{(i,k)} A \oplus B \xrightarrow{\mathrm{id}_A - \mathrm{id}_B} A \cup_C B \longrightarrow 0.$$

Here (i, k) is a monomorphism since i is a monomorphism. We have a similar exact sequence for $A' \cup_{C'} B'$. The induced map is given by a map of exact sequences in Ch_R, and the conclusion follows by the five lemma applied to the resulting map of long exact sequences. □

Of course, one characterization of the q-cofibrations and q-acyclic q-cofibrations is that they are retracts of relative $\mathcal{I}$-cell complexes and relative $\mathcal{J}$-cell complexes. We want something more explicit. The following results should be compared with the characterization of h-cofibrations as the degreewise-split monomorphisms.

PROPOSITION 18.5.2. *Let C be an object of Ch_R.*

(i) *$0 \longrightarrow C$ is a q-acyclic q-cofibration if and only if C is a projective object of the category Ch_R.*

(ii) *If C is q-cofibrant, then C is degreewise projective.*

(iii) *If C is bounded below and degreewise projective, then C is q-cofibrant.*

PROOF. Proposition 18.4.3 implies that a map is a q-fibration if and only if it is an epimorphism in Ch_R, so (i) is a direct reinterpretation of the LLP for $0 \longrightarrow C$. For (ii), any $\mathcal{I}$-cell complex is degreewise free, hence any retract of an $\mathcal{I}$-cell complex is degreewise projective. The proof of (iii) is just like the standard construction of maps from complexes of projectives to resolutions in classical homological algebra. Given a q-acyclic q-fibration $p: E \longrightarrow B$ and a map $f: C \longrightarrow B$, we must construct a lift $\lambda: C \longrightarrow E$ such that $p\lambda = f$, and we proceed by degreewise induction, starting with 0 in degrees below the minimal degree in which C is nonzero. Suppose given $\lambda_n: C_n \longrightarrow E_n$ such that $p\lambda_n = f_n$ and $d\lambda_n = \lambda_{n-1}d$. Since p_{n+1} is an epimorphism and C_{n+1} is projective, there is a map $\mu: C_{n+1} \longrightarrow E_{n+1}$ such that $p\mu = f$. Let $\nu = d\mu - \lambda_n d: C_{n+1} \longrightarrow E_n$. Then $p\nu = 0$ and $d\nu = 0$, so that ν is a map into the cycles of the q-acyclic complex $\ker(p)$. The cycles are equal to the boundaries, so, again using that C_{n+1} is projective, there is a map $\tau: C_{n+1} \longrightarrow \ker(p)_{n+1}$ such that $d\tau = \nu$. Setting $\lambda_{n+1} = \mu - \tau$, we find that $p\lambda_{n+1} = f_{n+1}$ and $d\lambda_{n+1} = \lambda_n d$. □

PROPOSITION 18.5.3. *A map $i: A \longrightarrow X$ is a q-cofibration if and only if it is a degreewise-split monomorphism such that the cokernel $C = X/A$ is q-cofibrant.*

PROOF. Assume first that i is a q-cofibration. Then i is a retract of a relative $\mathcal{I}$-cell complex. Since a retract of a degreewise-split monomorphism is also a degreewise-split monomorphism, i is a degreewise-split monomorphim. Since C is the pushout of $A \longrightarrow 0$ along i, C is q-cofibrant. For the converse, observe that $\pi(C, Y) = 0$ if Y is q-acyclic by the model theoretic Whitehead theorem, Theorem 14.4.8. This implies that the chain complex $\mathrm{Hom}_R(C, Y)$ is q-acyclic. Consider a lifting problem

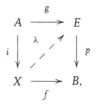

where p is a q-acyclic q-fibration. Let $Y = \ker(p)$ and observe that Y is q-acyclic. Since C is degreewise projective, we can write $X = A \oplus C$ as graded R-modules. Since the inclusion $A \longrightarrow X$ and projection $X \longrightarrow C$ are chain maps, we can write the differential on X in the form

$$d(a, c) = (d(a) + t(c), d(c)),$$

where t is a degree -1 map of graded R-modules such that $dt + td = 0$. This formula is forced by $d^2 = 0$. We write $f = f_1 + f_2$, $f_1: A \longrightarrow B$ and $f_2: C \longrightarrow B$, and we write $\lambda = \lambda_1 + \lambda_2$ similarly. We can and must define $\lambda_1 = g$ to ensure that $g = \lambda i$. We want $p\lambda_2(c) = f_2(c)$ and

$$d\lambda_2(c) = \lambda d(0, c) = \lambda(t(c), d(c)) = gt(c) + \lambda_2 d(c).$$

Since C is degreewise projective, there is a map $\tilde{f_2}: C \longrightarrow E$ of graded R-modules such that $p\tilde{f_2} = f_2$. The map $\tilde{f_2}$ is a first approximation to the required map λ_2. Define $k: C \longrightarrow E$ by

$$k = d\tilde{f_2} - \tilde{f_2}d - gt.$$

We claim that $pk = 0$, so that k may be viewed as a map $C \longrightarrow Y$ of degree -1. To see this, note that $df = fd$ implies $df_2 = f_1 t + f_2 d$. Since $pd = dp$,

$$pk = dp\tilde{f_2} - p\tilde{f_2}d - pgt = df_2 - f_2d - f_1t = 0.$$

Moreover,

$$dk + kd = -d\tilde{f_2}d - dgt + d\tilde{f_2}d - gtd = 0,$$

so that k is a cycle of degree -1 in $\mathrm{Hom}_R(C, Y)$. It must be a boundary, so there is a degree 0 map of graded R-modules $\ell : C \longrightarrow Y \subset E$ such that $d\ell - \ell d = k$. Define $\lambda_2 = \tilde{f_2} - \ell$. Certainly $p\lambda_2 = p\tilde{f_2}$, and

$$d\lambda_2 = d\tilde{f_2} - d\ell = \tilde{f_2}d + gt + k - k - \ell d = gt + \lambda_2 d. \qquad \square$$

Of course, regarding an ungraded R-module M as a DG R-module concentrated in degree 0, a q-cofibrant approximation of M is exactly a projective resolution of M. There is a dual model structure that encodes injective resolutions [66, Thm. 2.3.13].

THEOREM 18.5.4. *There is a cofibrantly generated injective model structure* $(\mathcal{W}_q, \mathcal{C}_i, \mathcal{F}_i)$ *on* Ch_R, *where the maps in* $\mathcal{C}_i$ *are the monomorphisms. The maps in* $\mathcal{F}_i$ *are the degreewise-split epimorphisms with i-fibrant kernel, and*

  (i) $D \longrightarrow 0$ *is a q-acyclic i-fibration if and only if D is an injective object of the category* Ch_R.
  (ii) *If D is i-fibrant, then D is degreewise injective.*
  (iii) *If D is bounded above and degreewise injective, then D is i-fibrant.*

The identity functor is a Quillen equivalence from the q-model structure to the i-model structure on Ch_R.

REMARK 18.5.5. Hovey [68] has studied abelian categories with model structures in which the cofibrations are the monomorphisms with cofibrant cokernel and the fibrations are the epimorphisms with fibrant kernel.

18.6. The *m*-model structure on Ch_R

We briefly describe the mixed model structure and how it relates to the familiar viewpoint of classical homological algebra. The homotopy category hCh_R and derived category $HoCh_R$ have been used in tandem since the beginnings of the subject. The mixed model structure allows more direct use of the homotopy theory of hCh_R in the study of $HoCh_R$.

We obviously have $\mathscr{W}_h \subset \mathscr{W}_q$, and $\mathscr{F}_h \subset \mathscr{F}_q$ since the h-fibrations are the R-split epimorphisms and the q-fibrations are all epimorphisms. All of the work needed to define and describe the mixed model structure

$$(\mathscr{W}_m, \mathscr{C}_m, \mathscr{F}_m) = (\mathscr{W}_q, \mathscr{C}_m, \mathscr{F}_h)$$

has already been done in §17.3. The mixed model structure has all of the good formal properties of the q-model structure. It is proper, it is monoidal when R is commutative, and it is a $Ch_{\mathbb{Z}}$-model structure in general. The identity functor on Ch_R is a right Quillen equivalence from the m-model structure to the q-model structure and therefore a left Quillen equivalence from the q-model structure to the m-model structure.

The class $\mathscr{C}_m$ of m-cofibrations is very well behaved. The h-cofibrations are the R-split monomorphisms, and Theorem 17.3.5 says that an m-cofibration is an h-cofibration that is a q-cofibration up to homotopy equivalence. More precisely, using Proposition 18.2.4, we see that an m-cofibration $A \longrightarrow X$ is an R-split monomorphism that is cofiber homotopy equivalent under A to a relative cell complex. Proposition 17.3.4(ii) gives a weak two out of three property that makes it easy to recognize when a map is an m-cofibration. Since m-cofibrations are more general than q-cofibrations, Proposition 17.3.4(i) generalizes the relative version of the Whitehead theorem that a weak equivalence between cell complexes is a homotopy equivalence. Since Ch_R is h-proper, Proposition 17.3.10 significantly generalizes these results.

Specializing, Theorem 17.3.5 implies that the m-cofibrant objects are precisely the objects of Ch_R that are of the homotopy types of q-cofibrant objects. This implies that they are homotopy equivalent to complexes of projective R-modules, and the converse holds when we restrict attention to complexes that are bounded below. Homotopy invariant constructions that start with complexes of projective R-modules automatically give m-cofibrant objects, but not necessarily degreewise-projective objects. In particular, it is very often useful to study the perfect complexes, namely the objects of Ch_R that are homotopy equivalent to bounded complexes of finitely generated projective R-modules. These are the dualizable objects of $HoCh_R$ in the sense, for example, of [95]. It would take us too far afield to go into the details of this, but the m-model structure is exactly right for studying this subcategory of Ch_R.

From the point of view of classical homological algebra, it is interesting to think of projective resolutions homotopically as analogues of approximation by CW complexes, well-defined only up to homotopy equivalence. From that point of view there is no need to restrict attention to degreewise projectives since only the homotopy type is relevant. Of course, using either a q-cofibrant or

m-cofibrant approximation P of an R-module M, regarded as a chain complex concentrated in degree 0, we have

$$\mathrm{Tor}^R_*(N, M) = H_*(N \otimes_R P) \quad \text{and} \quad \mathrm{Ext}^*_R(M, N) = H^* \mathrm{Hom}_R(P, N),$$

the latter regraded cohomologically. We can think of these as obtained by first applying the derived functors of $N \otimes_R (-)$ and $\mathrm{Hom}_R(-, N)$ and then taking homology groups or, equivalently, thinking in terms of spheres S^n_R, homotopy groups.

19

RESOLUTION AND LOCALIZATION
MODEL STRUCTURES

In §19.1, we present a new perspective on the three model structures that we described on spaces and chain complexes in the previous two chapters. Indeed, this gives a new perspective on the construction of the homotopy category Ho$\mathcal{U}$ of spaces from the naive homotopy category $h\mathcal{U}$ obtained by identifying homotopic maps and of the derived homotopy category HoCh_R of chain complexes from the naive homotopy category hCh_R. The perspective applies to many other contexts in topology and algebra where we have both a classical and a derived homotopy category. It focuses on what we call resolution model structures. This is a nonstandard name. These model structures are usually called colocalization model structures since they are dual to the more familiar localization model structures, to which we turn next. The essential point is that the mixed model structures on $\mathcal{U}$ and on Ch_R are actually examples of resolution model categories and therefore play an intrinsic conceptual role independent of the q-model structures. This theory brings into focus the conceptual unity of our three model structures.

Localization model structures codify Bousfield localization, which vastly generalizes the constructions in the first half of this book. There we concentrated on arithmetic localizations and completions that are closely related to standard algebraic constructions. We restricted our constructions to nilpotent spaces since that is the natural range of applicability of our elementary methodology and since most applications focus on such spaces. There are several different ways to generalize to non-nilpotent spaces, none of them well understood calculationally. Bousfield localization gives a general conceptual understanding of the most widely used of these, and it gives the proper perspective for generalizations to categories other than the category of spaces and indeed to other fields of mathematics.

In modern algebraic topology, especially in stable homotopy theory, Bousfield localizations at generalized homology theories play a fundamental role.

These have been less studied on the space level than on the spectrum level, where they are central to the structural study of the stable homotopy category. The model theoretic method of construction of such localizations works equally well for spaces and spectra, and it specializes to extend our arithmetic constructions to non-nilpotent spaces. We explain the general idea in §19.2. Although we start work in a general model theoretic context, we switch gears in §19.3 and give a geodesic approach to the construction of localizations of spaces at generalized homology theories. Except that we work topologically rather than simplicially, our exposition is based primarily on the original paper of Bousfield [16].

We return to the general theory in §19.4, where we place the localization of spaces at a homology theory in a wider context of localization at a map, or at a class of maps. We then relate localization to enrichment in the context of $\mathcal{V}$-model categories in §19.5. The enrichment is essential to the construction of Bousfield localization in full generality.

19.1. Resolution and mixed model structures

Our initial definition in this section requires only a category, but we prefer to start with a model category $\mathcal{M}$. In principle, the model structure can be perfectly general, but we use the notation $(\mathcal{H}, \mathcal{C}, \mathcal{F})$, choosing the letter $\mathcal{H}$ since we are thinking of actual homotopy equivalences. We write $[X, Y]_{\mathcal{H}}$ for the set of morphisms $X \longrightarrow Y$ in $\mathrm{Ho}\mathcal{M}$. We are thinking of contexts in which $[X, Y]_{\mathcal{H}} = \pi(X, Y)$. Let $\mathcal{W} \subset \mathcal{M}$ be any subcategory of weak equivalences (in the sense of Definition 14.1.4) that contains $\mathcal{H}$. We are thinking of the weak homotopy equivalences in $\mathcal{U}$ and the quasi-isomorphisms in Ch_R. Since we have the two categories $\mathcal{H}$ and $\mathcal{W}$ of weak equivalences in sight, we sometimes say that a map is $\mathcal{W}$-acyclic if it is in $\mathcal{W}$.

DEFINITION 19.1.1. An object C of $\mathcal{M}$ is said to be $\mathcal{W}$-resolvant (or $\mathcal{W}$-colocal) abbreviated to resolvant when $\mathcal{W}$ is understood, if it is cofibrant and for every map $f : X \longrightarrow Y$ in $\mathcal{W}$, the induced function

$$f_* : [C, X]_{\mathcal{H}} \longrightarrow [C, Y]_{\mathcal{H}}$$

is a bijection. A map $\gamma : \Gamma X \longrightarrow X$ in $\mathcal{W}$ from a resolvant object ΓX to X is called a resolution (or colocalization) of X.

As will become clear, the resolvant objects in $\mathcal{U}$ are the spaces of the homotopy types of CW complexes, and this codifies the Whitehead theorem for such spaces. Similarly, the (bounded below) resolvant objects in Ch_R are the chain complexes that are of the homotopy type of chain complexes of projective modules, and this codifies the analogous Whitehead theorem for such chain complexes.

The definition of a resolution prescribes a universal property. Writing "up to homotopy" to mean "in Ho$\mathcal{M}$", if $f: D \longrightarrow X$ is any map from a resolvant object D to X, there is a map $\tilde{f}$, unique up to homotopy, such that the following diagram commutes up to homotopy.

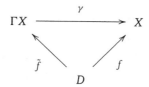

Therefore a resolution of X is unique up to homotopy if it exists. There is a natural model theoretic way to try to construct resolutions.

DEFINITION 19.1.2. A map $i: A \longrightarrow X$ in $\mathcal{M}$ is a $\mathcal{W}$-cofibration if it satisfies the LLP with respect to $\mathcal{W} \cap \mathcal{F}$. An object C is $\mathcal{W}$-cofibrant if $\emptyset \longrightarrow C$ is a $\mathcal{W}$-cofibration. Let $\mathscr{C}_{\mathcal{W}}$ denote the class of $\mathcal{W}$-cofibrations in $\mathcal{M}$.

Even without knowing that the definitions above give a model structure, we can relate $\mathcal{W}$-resolvant objects to $\mathcal{W}$-cofibrant objects.

PROPOSITION 19.1.3. *Let $\mathcal{W}$ be a subcategory of weak equivalences in $\mathcal{M}$. If $\mathcal{M}$ is right proper, then every $\mathcal{W}$-resolvant object is $\mathcal{W}$-cofibrant.*

Under mild hypotheses, we can also prove the converse. The dual of both Proposition 19.1.3 and its converse are given in Proposition 19.2.5 below, and the proofs there dualize directly.

DEFINITION 19.1.4. If the classes $(\mathcal{W}, \mathscr{C}_{\mathcal{W}}, \mathcal{F})$ specify a model structure on $\mathcal{M}$, we call it the $\mathcal{W}$-resolution model structure. When this holds, we say that the $\mathcal{W}$-resolution model structure exists.

We give some implications of the existence of the $\mathcal{W}$-model structure before turning to examples. Write $[X, Y]_{\mathcal{W}}$ for the morphism sets of the homotopy category Ho$(\mathcal{M}, \mathcal{W})$ of the $\mathcal{W}$-resolution model structure. Categorically, Ho$(\mathcal{M}, \mathcal{W})$ is obtained from $\mathcal{M}$, or from Ho$\mathcal{M}$, by formally inverting the morphisms in $\mathcal{W}$. Observe that $\mathscr{C}_{\mathcal{W}} \subset \mathscr{C}$ since $\mathscr{H} \subset \mathcal{W}$.

PROPOSITION 19.1.5. *If* $(\mathscr{W}, \mathscr{C}_{\mathscr{W}}, \mathscr{F})$ *is a model structure, then the identity functor on* $\mathscr{M}$ *is a Quillen right adjoint from the original model structure on* $\mathscr{M}$ *to the* $\mathscr{W}$-*resolution structure. If C is* $\mathscr{W}$-*cofibrant, then C is cofibrant in the original model structure on* $\mathscr{M}$ *and*

19.1.6 $$[C, X]_{\mathscr{H}} \cong [C, X]_{\mathscr{W}}.$$

Therefore C is $\mathscr{W}$-*resolvant.*

PROOF. The first statement is clear since the $\mathscr{W}$-resolution model structure has more weak equivalences and the same fibrations as the original model structure. Since $\mathscr{C}_{\mathscr{W}} \subset \mathscr{C}$, (19.1.6) is immediate from the adjunction between homotopy categories. The last statement follows by the definition of a $\mathscr{W}$-resolvant object. $\qquad\square$

COROLLARY 19.1.7. *If* $(\mathscr{W}, \mathscr{C}_{\mathscr{W}}, \mathscr{F})$ *is a model structure on* $\mathscr{M}$, *then a* $\mathscr{W}$-*cofibrant approximation* $\gamma : \Gamma X \longrightarrow X$ *of X, obtained by factoring* $\emptyset \longrightarrow X$, *is a* $\mathscr{W}$-*resolution of X.*

Of course, this construction of resolutions necessarily gives a functor Γ on the homotopy category together with a natural transformation $\gamma : \Gamma \longrightarrow \mathrm{Id}$. The first part of the following result holds by Propositions 19.1.3 and 19.1.5. The duals of the other two parts are proven in the next section, and the dual proofs work equally well.

PROPOSITION 19.1.8. *Asssume that* $\mathscr{M}$ *is right proper and* $(\mathscr{W}, \mathscr{C}_{\mathscr{W}}, \mathscr{F})$ *is a model structure. Then the following conclusions hold.*

(i) *An object C is* $\mathscr{W}$-*resolvant if and only if it is* $\mathscr{W}$-*cofibrant.*
(ii) $\mathscr{W}$ *is the class of all maps* $f : X \longrightarrow Y$ *such that*

$$f_* : [C, X]_{\mathscr{W}} \longrightarrow [C, Y]_{\mathscr{W}}$$

is a bijection for all $\mathscr{W}$-*resolvant objects C.*
(iii) *The* $\mathscr{W}$-*resolution model structure is right proper.*

These ideas give a general and conceptually pleasing way to construct resolutions, but of course the real work lies in the proof that the $\mathscr{W}$-resolution model structure exists. In the two most classical cases, we have already done that work.

PROPOSITION 19.1.9. *Start with the h-model structure on $\mathcal{U}$ and let $\mathcal{W} = \mathcal{W}_q$ be the subcategory of weak homotopy equivalences. Then the $\mathcal{W}$-resolution model structure exists since it coincides with the m-model structure.*

PROPOSITION 19.1.10. *Start with the h-model structure on Ch_R and let $\mathcal{W} = \mathcal{W}_q$ be the subcategory of quasi-isomorphisms. Then the $\mathcal{W}$-resolution model structure exists since it coincides with the m-model structure.*

These results are immediate from the definitions since in both cases the fibrations and weak equivalences of the two model structures are identical. Observe that there is no direct mention of the q-model structure in this development. Its role is to show that the $\mathcal{W}$-localization model structure exists and to help describe its properties, as summarized in our discussion of the m-model structures. Dualizing some of the discussion in the rest of the chapter leads to alternative ways to prove the model axioms in other situations where the general definitions apply.

Observe that Proposition 19.1.8 gives a direct conceptual route to the characterization of spaces of the homotopy types of CW complexes as those spaces C such that $f_* : \pi(C, X) \longrightarrow \pi(C, Y)$ is a bijection for all weak equivalences $f : X \longrightarrow Y$, and similarly for chain complexes.

19.2. The general context of Bousfield localization

In the rest of the chapter, except when we focus on spaces, we let $\mathcal{M}$ be a bicomplete category with a model structure $(\mathcal{W}, \mathcal{C}, \mathcal{F})$ and associated homotopy category Ho$\mathcal{M}$. We denote its morphism sets by $[X, Y]$ rather than $[X, Y]_{\mathcal{W}}$ for brevity and consistency with the first half of the book. One often wants to localize $\mathcal{M}$ so as to invert more weak equivalences than just those in $\mathcal{W}$, and there is a general procedure for trying to do so that is dual to the idea of the previous section. As the choice of the letter $\mathcal{W}$ is meant to suggest, we are thinking of $\mathcal{M}$ as the q- or m-model structure, but the definition is general. There are two variants, which focus attention on two slightly different points of view. Either we can start with a subcategory $\mathcal{L}$ of weak equivalences, in the sense of Definition 14.1.4, that contains $\mathcal{W}$, or we can start with an arbitrary class $\mathcal{K}$ of maps in $\mathcal{M}$. We make the former choice in this section and the latter choice when we return to the general theory in §19.4. Starting with $\mathcal{L}$, we have definitions that are precisely dual to those of the previous section.

DEFINITION 19.2.1. An object Z of $\mathcal{M}$ is said to be $\mathcal{L}$-local, abbreviated to local when $\mathcal{L}$ is understood, if it is fibrant and for every map $\xi : X \longrightarrow Y$ in $\mathcal{L}$, the induced function

$$\xi^*\colon [Y, Z] \longrightarrow [X, Z]$$

is a bijection. A map $\phi\colon X \longrightarrow LX$ in $\mathscr{L}$ to a local object LX is called a localization of X at $\mathscr{L}$, or an $\mathscr{L}$-localization of X.

Here is the example most relevant to this book.

DEFINITION 19.2.2. For spaces X and a homology theory E_* on spaces, say that $\xi\colon X \longrightarrow Y$ is an E-equivalence if it induces an isomorphism $E_*(X) \longrightarrow E_*(Y)$ and let $\mathscr{L}_E$ denote the class of E-equivalences. Say that a space Z is E-local if it is $\mathscr{L}_E$-local. Say that a map $\phi\colon X \longrightarrow X_E$ from X into an E-local space X_E is a localization at E if ϕ is an E-equivalence.

EXAMPLE 19.2.3. When $E_*(X) = H_*(X; \mathbb{Z}_T)$, these definitions agree up to nomenclature with our definition of localization at T in §5.2. Similarly, when $E_*(X) = H_*(X; \mathbb{F}_T)$, where $\mathbb{F}_T = \times_{p \in T} \mathbb{F}_p$, these definitions agree up to nomenclature with our definition of completion at T in §10.2.

Observe that the names "localization" and "completion" that are carried over from algebra in the earlier parts of the book are really the names of two examples of a generalized notion of localization. Dually to §19.1, the essential idea of Bousfield localization is to try to construct a model structure on $\mathscr{M}$ of the form $(\mathscr{L}, \mathscr{C}, \mathscr{F}_{\mathscr{L}})$.

DEFINITION 19.2.4. A map $p\colon E \longrightarrow B$ in $\mathscr{M}$ is an $\mathscr{L}$-fibration if it satisfies the RLP with respect to $\mathscr{L} \cap \mathscr{C}$. An object Z is $\mathscr{L}$-fibrant if $Z \longrightarrow *$ is an $\mathscr{L}$-fibration. Let $\mathscr{F}_{\mathscr{L}}$ denote the class of $\mathscr{L}$-fibrations in $\mathscr{M}$.

We shall give more detail of the general theory than we gave in §19.1, but many of the results and their proofs, such as the following one, dualize to prove analogous results in that context. Even without knowing that the notions above give a model structure, we can relate $\mathscr{L}$-local objects to $\mathscr{L}$-fibrant objects.

PROPOSITION 19.2.5. *Let $\mathscr{L}$ be a subcategory of weak equivalences in $\mathscr{M}$.*

(i) *If $\mathscr{M}$ is left proper, then every $\mathscr{L}$-local object is $\mathscr{L}$-fibrant.*
(ii) *If $\mathscr{L}$ has good spools, then every $\mathscr{L}$-fibrant object is $\mathscr{L}$-local.*

PROOF. For (i), if Z is $\mathscr{L}$-local and $i\colon A \longrightarrow X$ is in $\mathscr{C} \cap \mathscr{L}$, then the function $i^*\colon [X, Z] \longrightarrow [A, Z]$ is a bijection. Therefore, $Z \longrightarrow *$ satisfies the RLP with respect to i by Lemma 15.5.4. With the definition of good spools given in Definition 15.5.11, (ii) follows from Lemma 15.5.3. □

DEFINITION 19.2.6. If the classes $(\mathscr{L}, \mathscr{C}, \mathscr{F}_{\mathscr{L}})$ specify a model structure on $\mathscr{M}$, we call it the $\mathscr{L}$-localization model structure. When this holds, we say that the $\mathscr{L}$-localization model structure exists.

We give some implications of the existence of the $\mathscr{L}$-localization model structure before turning to examples. Write $[X, Y]_{\mathscr{L}}$ for the morphism sets of the homotopy category $\mathrm{Ho}(\mathscr{M}, \mathscr{L})$ of the $\mathscr{L}$-localization model structure. Again, categorically, $\mathrm{Ho}(\mathscr{M}, \mathscr{L})$ is obtained from $\mathscr{M}$ or from $\mathrm{Ho}\mathscr{M}$ by formally inverting the morphisms in $\mathscr{L}$. The following result is dual to Proposition 19.1.5 and admits a dual proof.

PROPOSITION 19.2.7. *If $(\mathscr{L}, \mathscr{C}, \mathscr{F}_{\mathscr{L}})$ is a model structure, then the identity functor on $\mathscr{M}$ is a Quillen left adjoint from the original model structure on $\mathscr{M}$ to the $\mathscr{L}$-localization structure. If Z is $\mathscr{L}$-fibrant, then Z is fibrant and*

19.2.8
$$[X, Z] \cong [X, Z]_{\mathscr{L}}.$$

Therefore Z is $\mathscr{L}$-local.

Observe that the "good spools" hypothesis of Proposition 19.2.5(ii) is not needed to prove that $\mathscr{L}$-fibrant objects are $\mathscr{L}$-local when $(\mathscr{L}, \mathscr{C}, \mathscr{F}_{\mathscr{L}})$ is a model structure.

COROLLARY 19.2.9. *If $(\mathscr{L}, \mathscr{C}, \mathscr{F}_{\mathscr{L}})$ is a model structure on $\mathscr{M}$, then an $\mathscr{L}$-fibrant approximation $\phi \colon X \longrightarrow LX$ of X, obtained by factoring $X \longrightarrow *$, is an $\mathscr{L}$-localization of X.*

This construction of localizations necessarily gives a functor L on the homotopy category together with a natural transformation $\phi \colon \mathrm{Id} \longrightarrow L$. In this general context, there is a meta-theorem that reads as follows.

THEOREM 19.2.10. *Under suitable hypotheses, $(\mathscr{L}, \mathscr{C}, \mathscr{F}_{\mathscr{L}})$ is a model structure on $\mathscr{M}$, called the $\mathscr{L}$-localization model structure.*

This is a general and conceptually pleasing way to construct localizations, and it has myriads of applications in algebraic topology and algebraic geometry.

However, in practice the logic is circular, since the essential step in the proof of Theorem 19.2.10 is the following factorization result, which already constructs the desired localizations as a special case.

THEOREM 19.2.11. *Under suitable hypotheses, every map $f : X \longrightarrow Y$ factors as an $\mathscr{L}$-acyclic cofibration $i : X \longrightarrow E$ followed by an $\mathscr{L}$-fibration $p : E \longrightarrow Y$.*

In fact, Theorem 19.2.11 directly implies Theorem 19.2.10. Indeed, one of the lifting properties is given by the definition of $\mathscr{F}_{\mathscr{L}}$, and the following lemma implies that the other lifting property and the other factorization are already given by the WFS $(\mathscr{C}, \mathscr{F} \cap \mathscr{W})$ of the original model structure on $\mathscr{M}$.

LEMMA 19.2.12. *A map $f : X \longrightarrow Y$ is in $\mathscr{F} \cap \mathscr{W}$ if and only if it is in $\mathscr{F}_{\mathscr{L}} \cap \mathscr{L}$.*

PROOF. If f is in $\mathscr{F} \cap \mathscr{W}$, then f is in $\mathscr{L}$ since $\mathscr{W} \subset \mathscr{L}$ and f is in $\mathscr{F}_{\mathscr{L}}$ since f satisfies the RLP with respect to $\mathscr{C}$ and therefore with respect to $\mathscr{C} \cap \mathscr{L}$. Conversely, assume that f is in $\mathscr{F}_{\mathscr{L}} \cap \mathscr{L}$. Factor f in our original model structure on $\mathscr{M}$ as a composite of a cofibration $i : X \longrightarrow E$ and an acyclic fibration $p : E \longrightarrow Y$. Then i is in $\mathscr{L}$ by the two out of three property, hence i has the LLP with respect to f. Therefore there is a lift $\lambda : E \longrightarrow X$ in the diagram

This implies that f is a retract of p and is thus in $\mathscr{F} \cap \mathscr{W}$. □

One can ask for minimal hypotheses under which Theorems 19.2.10 and 19.2.11 hold. It is usual to assume that $\mathscr{M}$ is left proper. This assumption is very natural in view of the following result.

PROPOSITION 19.2.13. *Assume that $\mathscr{M}$ is left proper and $(\mathscr{L}, \mathscr{C}, \mathscr{F}_{\mathscr{L}})$ is a model structure. Then the following conclusions hold.*

(i) An object Z is $\mathscr{L}$-local if and only if it is $\mathscr{L}$-fibrant.

(ii) $\mathscr{L}$ is the class of all maps $f : X \longrightarrow Y$ such that

$$f^* : [Y, Z]_{\mathscr{L}} \longrightarrow [X, Z]_{\mathscr{L}}$$

is a bijection for all $\mathscr{L}$-local objects Z.

(iii) The $\mathscr{L}$-localization model structure is left proper.

PROOF. Part (i) is immediate from Propositions 19.2.5(i) and 19.2.7. For (ii), f^* is certainly an isomorphism if f is in $\mathscr{L}$. For the converse, recall that the defining property of an $\mathscr{L}$-local object refers to $[-, -]$ rather than $[-, -]_{\mathscr{L}}$. Let Z be $\mathscr{L}$-local and consider the commutative diagram

$$
\begin{array}{ccc}
[LY, Z]_{\mathscr{L}} & \xrightarrow{(Lf)^*} & [LX, Z]_{\mathscr{L}} \\
\phi^* \downarrow & & \downarrow \phi^* \\
[Y, Z]_{\mathscr{L}} & \xrightarrow{f^*} & [X, Z]_{\mathscr{L}}.
\end{array}
$$

The vertical arrows are bijections, hence f^* is a bijection if and only if $(Lf)^*$ is a bijection. Since Z is $\mathscr{L}$-local, it is $\mathscr{L}$-fibrant, hence by Proposition 19.2.7 we can identify $(Lf)^*$ with

$$(Lf)^* : [LY, Z] \longrightarrow [LX, Z].$$

If this is a bijection for all $\mathscr{L}$-fibrant Z, such as $Z = LX$, then the image of Lf in Ho$\mathscr{M}$ is an isomorphism, hence Lf is in $\mathscr{W}$ and thus in $\mathscr{L}$. Since $Lf \circ \phi = \phi \circ f$ in Ho$(\mathscr{M}, \mathscr{L})$, the image of f in Ho$(\mathscr{M}, \mathscr{L})$ is an isomorphism, hence f is in $\mathscr{L}$. This is essentially an application of the two out of three property, but that does not quite apply since we are not assuming that ϕ is functorial and L is natural before passage to homotopy categories.

For (iii), assume that f is in $\mathscr{L}$ and i is a cofibration in the pushout diagram

$$
\begin{array}{ccc}
A & \xrightarrow{f} & B \\
i \downarrow & & \downarrow j \\
X & \xrightarrow{g} & X \cup_A B.
\end{array}
$$

We must show that g is in $\mathscr{L}$. Construct the following commutative diagram.

$$
\begin{array}{ccccc}
QX & \xleftarrow{i'} & QA & \xrightarrow{f'} & QB \\
q_X \downarrow & & q_A \downarrow & & \downarrow q_B \\
X & \xleftarrow{i} & A & \xrightarrow{f} & B
\end{array}
$$

The squares display cofibrant approximations of the maps i and f, as constructed in Lemma 15.5.6. By the gluing lemma (Proposition 15.4.4), g is weakly equivalent (in $\mathscr{A}r\mathscr{M}$) to the pushout $g' \colon QX \longrightarrow QX \cup_{QA} QB$ of f' along i'. Since $(\mathscr{L}, \mathscr{C}, \mathscr{F}_{\mathscr{L}})$ is a model structure, g' is in $\mathscr{L}$ by Proposition 15.4.2. Therefore g is in $\mathscr{L}$. $\qquad\square$

In enriched contexts, to which we shall return in §19.5, Proposition 19.2.13 implies that $\mathscr{L}$-localizations of $\mathscr{V}$-model structures are generally $\mathscr{V}$-model structures.

PROPOSITION 19.2.14. *Assume that $\mathscr{M}$ is a left proper $\mathscr{V}$-model category for some monoidal model category $\mathscr{V}$ and assume further that either*

(i) *the functors $(-) \odot V$ preserve $\mathscr{L}$-acyclic cofibrations or*

(ii) *the functors $(-) \odot V$ preserve $\mathscr{L}$ and the functors $X \odot (-)$ preserve cofibrations.*

Then if $(\mathscr{L}, \mathscr{C}, \mathscr{F}_{\mathscr{L}})$ is a model structure, it is a $\mathscr{V}$-model structure.

PROOF. The enrichment, tensors, and cotensors are given, and we know that if $i \colon A \longrightarrow X$ is a cofibration in $\mathscr{M}$ and $k \colon V \longrightarrow W$ is a cofibration in $\mathscr{V}$, then the pushout product $i\square k \colon i \boxtimes k \longrightarrow X \odot W$ is a cofibration that is a weak equivalence if either i or k is a weak equivalence. We must show that if i is in $\mathscr{L}$, then so is $i\square k$. This follows from the hypotheses, the previous result, and Remark 16.4.6. $\qquad\square$

To prove Theorem 19.2.11 and therefore Theorem 19.2.10, it is also usual to assume that $\mathscr{M}$ is cofibrantly generated. That is convenient and probably essential for a general axiomatic approach. However, it plays a relatively minor conceptual role. Any proof of Theorem 19.2.11 is likely to use transfinite induction. As we shall see in §19.3, where we prove Theorem 19.2.11 for $\mathscr{L} = \mathscr{L}_E$, the argument will also prove that the $\mathscr{L}$-model structure is cofibrantly generated when the original one is, albeit with a large and inexplicit set of generators for the acyclic cofibrations. The proof in §19.3 will start with the q-model structure on $\mathscr{U}$ and not the m-model structure. Whether or not that is essential, it is certainly convenient technically.

Nevertheless, the cofibrant generation of the $\mathscr{L}$-model structure seems to be more of a pleasant added bonus than an essential conceptual feature. To avoid obscuring the ideas, we delay turning to this feature until the end of the next section. As an historical aside, Bousfield's original treatment [16] first introduced the methods of transfinite induction into model category theory, and only later were his ideas codified into the notion of a cofibrantly generated model structure. Our exposition follows that historical perspective.

19.3. Localizations with respect to homology theories

We work with the q-model structure on $\mathscr{U}$ as our starting point in this specialization of the general theory above. We consider the class $\mathscr{L}_E$ of E_*-isomorphisms defined in Definition 19.2.2, where E_* is any homology theory. Of course, $\mathscr{L}_E$ is a category of weak equivalences. To simplify notation, write $\mathscr{F}_E = \mathscr{F}_{\mathscr{L}_E}$ for the class of maps that satisfy the RLP with respect to $\mathscr{C}_q \cap \mathscr{L}_E$. To complete the long-promised construction of E_*-localizations of spaces, in particular of localizations and completions at a set of primes T, it suffices to prove that $(\mathscr{L}_E, \mathscr{C}_q, \mathscr{F}_E)$ is a model category. As explained after Theorem 19.2.11, it suffices for that to prove the following factorization theorem.

THEOREM 19.3.1. *Every map* $f \colon X \longrightarrow Y$ *of topological spaces factors as an* $\mathscr{L}_E$*-acyclic cofibration* $i \colon X \longrightarrow Z$ *followed by an* $\mathscr{L}_E$*-fibration* $p \colon Z \longrightarrow Y$.

The rest of this section is devoted to the proof and some remarks on how the conclusion relates to our earlier construction of localizations and completions. Our homology theory E_* is required to satisfy the usual axioms, including the additivity axiom; see, for example, [93, Ch. 14]. Thus it converts disjoint unions of pairs to direct sums, hence its reduced variant $\tilde{E}$ (defined on $\mathscr{T}$) converts wedges to direct sums. Generalizing [93, §14.6], it can be deduced from the axioms that E_* commutes with filtered colimits and, in particular, transfinite composites. We used a more elementary argument to verify this for ordinary homology in Proposition 2.5.4.

Let κ be a regular cardinal greater than or equal to the cardinality of the underlying set of the abelian group $\oplus_{n \in \mathbb{Z}} E_n(*)$. For a CW complex X, let $\kappa(X)$ denote the number of cells of X. By a CW pair (X, A), we understand an inclusion $i \colon A \longrightarrow X$ of a subcomplex A in a CW complex X. Thus i is an E_*-isomorphism if and only if $E_*(X, A) = 0$.

LEMMA 19.3.2. *If* (X, A) *is a CW pair and* $\kappa(X) \leq \kappa$, *then* $E_*(X, A)$ *has at most* κ *elements.*

PROOF. Since $E_*(X, A) \cong \tilde{E}_*(X/A)$, we may assume that A is a point. We start from the case $X = S^n$, we then pass to wedges of copies of S^n by additivity, we next pass to skeleta of X by induction using the long exact sequences associated to cofibrations, and we conclude by passage to colimits. □

NOTATION 19.3.3. Let $\mathscr{K}$ denote the class of CW pairs (B, A) such that

(i) $\kappa(B) \leq \kappa$ and
(ii) $E_*(B, A) = 0$.

Since we have only sets of attaching maps to choose from in constructing such CW pairs, we can choose a subset K of the class $\mathscr{K}$ such that every map in $\mathscr{K}$ is isomorphic to a map in K.

We isolate the following observation, which will allow us to characterize the class $\mathscr{F}_E$ of E-fibrations in terms of K. When $E_*(X, A) = 0$ with X perhaps very large, it shows how to find a subpair in K.

LEMMA 19.3.4. *If (X, A) is a CW pair such that $A \neq X$ and $E_*(X, A) = 0$, then there exists a subcomplex $B \subset X$ such that $B \not\subset A$ and $(B, A \cap B)$ is in K.*

PROOF. We use that X is the colimit of its finite subcomplexes and that E_* commutes with filtered colimits. We take B to be the union of an expanding sequence of subcomplexes $\{B_n\}$ of X such that $B_n \not\subset A$, $\kappa(B_n) \leq \kappa$, and the induced map

$$E_*(B_n, A \cap B_n) \longrightarrow E_*(B_{n+1}, A \cap B_{n+1})$$

is zero for each $n \geq 1$. Using that E_* commutes with sequential colimits, it will follow that $E_*(B, A \cap B) = 0$. To construct the B_n, start with any $B_1 \subset X$ such that $B_1 \not\subset A$ and $\kappa(B_1) \leq \kappa$. Given B_n, for each element $x \in E_*(B_n, A \cap B_n)$, choose a finite subcomplex C_x such that x maps to zero in $E_*(B_n \cup C_x, A \cap (B_n \cup C_x))$. There exists such a C_x since $\operatorname{colim} E_*(C, A \cap C) = E_*(X, A) = 0$, where C runs over the finite subcomplexes of X. We can then take $B_{n+1} = B_n \cup (\cup_x C_x)$. □

PROPOSITION 19.3.5. *A map $p : Z \longrightarrow Y$ has the RLP with respect to K if and only if p is in $\mathscr{F}_E$; that is, $K^{\square} = \mathscr{F}_E$.*

PROOF. Since the maps in K are E-equivalences and q-cofibrations, necessity is obvious. Thus assume that p has the RLP with respect to K. We may as well insist that each pair $(D^n \times I, D^n)$ is in K, and then our hypothesis ensures that p is a q-fibration. It is especially this innocuous-seeming point that makes it convenient to work with the q- rather than the m-model structure. We must show that there is a lift ℓ in each diagram

19.3.6

$$
\begin{array}{ccc}
A & \longrightarrow & Z \\
{\scriptstyle i}\downarrow & {\ell}\nearrow & \downarrow{\scriptstyle p} \\
X & \longrightarrow & Y
\end{array}
$$

in which i is an E-equivalence and a q-cofibration. By Lemmas 15.5.6 and 15.5.7, we may replace i by a cofibrant approximation and, by CW approximation of pairs [93, §10.6], we can arrange that our cofibrant approximation is given by an inclusion of a subcomplex in a CW complex. Thus we may assume without loss of generality that i is the inclusion of a subcomplex A in a CW complex X.

With the notation of Lemma 19.3.4, if we set $C = A \cup B$ then p has the RLP with respect to the pushout $A \longrightarrow C$ of $A \cap B \longrightarrow B$. Thus for each CW pair (X, A) such that $E_*(X, A) = 0$, we have a CW pair $(C, A) \subset (X, A)$ such that $C \neq A$, $E_*(C, A) = 0$, and p has the RLP with respect to $A \longrightarrow C$. Consider the set S of pairs (V, ℓ_V) such that V is a subcomplex of X that contains A, $E_*(V, A) = 0$, and ℓ_V is a lift in the restricted diagram obtained from (19.3.6) by replacing X by V. Order these pairs by $(V, \ell_V) \leq (W, \ell_W)$ if $V \subset W$ and $\ell_W | V = \ell_V$. The colimit of an ordered chain in S is again in S, hence S has a maximal element (W, ℓ_W) by Zorn's lemma. We claim that $W = X$. If not, there is a CW pair $(C, W) \subset (X, W)$ such that $C \neq W$, $E_*(C, W) = 0$, and p has the RLP with respect to $W \longrightarrow C$. We can extend ℓ_W to ℓ_C via the diagram

$$
\begin{array}{ccc}
& \ell_W & \\
W & \longrightarrow & Z \\
\downarrow & {\ell_C}\nearrow & \downarrow{\scriptstyle p} \\
C & \longrightarrow & Y,
\end{array}
$$

and this contradicts the maximality of (W, ℓ_W). $\qquad\square$

The proof of Theorem 19.2.11 is now completed by the small object argument, Proposition 15.1.11, whose smallness hypothesis is verified in Proposition 2.5.4.

THEOREM 19.3.7. *Any map $f : X \longrightarrow Y$ factors as the composite of a relative $\mathcal{K}$-cell complex $i : X \longrightarrow Z$ and a map $p : Z \longrightarrow Y$ that satisfies the RLP with* *respect to $\mathcal{K}$ and is thus in $\mathscr{F}_E$. Therefore the $\mathscr{L}$-localization model structure exists and $\mathscr{C}(\mathcal{K}) = \mathscr{C}_q \cap \mathscr{L}_E$.*

PROOF. Proposition 15.1.11 gives a WFS $(\mathscr{C}(\mathcal{K}), \mathcal{K}^{\boxtimes})$. Together with Proposition 19.3.5, that gives the first statement. As noted in the previous section, the model axioms follow. In particular, $\mathscr{C}(\mathcal{K}) = \mathscr{C}_q \cap \mathscr{L}_E$ since both are $^{\boxtimes}\mathscr{F}_E$. □

The properties of homology theories imply that $\mathscr{C}_q \cap \mathscr{L}_E$ is a saturated class of maps, in the sense of Definition 14.1.7. Therefore a relative $\mathcal{K}$-cell complex is an E_*-isomorphism. This verifies the acyclicity condition of Theorem 15.2.3 for the E-localization model structure $(\mathscr{L}_E, \mathscr{C}_q, \mathscr{F}_E)$. Its compatibility condition states that $\mathcal{I}^{\boxtimes} = \mathcal{K}^{\boxtimes} \cap \mathscr{L}_E$, where $\mathcal{I}$ is the generating set of cofibrations for the q-model structure. Since $\mathcal{K}^{\boxtimes} = \mathscr{F}_E$, this condition holds by Lemma 19.2.12. Therefore Proposition 19.3.5 and Theorem 19.3.7 complete the proof of Theorem 19.2.11. Together with Proposition 19.2.13, they also complete the proof of the following result.

THEOREM 19.3.8. *The classes $(\mathscr{L}_E, \mathscr{C}_q, \mathscr{F}_E)$ give $\mathscr{U}$ a left proper cofibrantly generated model structure with the generating sets $\mathcal{I}$ of q-cofibrations and $\mathcal{K}$ of $\mathscr{L}_E$-acyclic q-cofibrations.*

Proposition 19.2.13 gives the following direct consequence. It will seem familiar to the observant reader: it can be viewed as an analogue of the dual Whitehead theorem; compare Theorems 3.3.8 and 3.3.9. However, the model theoretic proof is altogether different, requiring no use or even mention of cohomology.

COROLLARY 19.3.9. *A map $\xi : X \longrightarrow Y$ is an E-equivalence if and only if*

$$\xi^* : [Y, Z] \longrightarrow [X, Z]$$

is a bijection for all E-local spaces Z.

REMARK 19.3.10. This result generalizes the equivalence of (i) and (ii) in Theorem 6.1.2 from nilpotent spaces to all spaces. Similarly, it generalizes the equivalence of (i) and (ii) in Theorem 11.1.2 from nilpotent spaces to all spaces. Note that Proposition 3.3.11 gives an elementary comparison of homological and cohomological equivalences that does not require nilpotency. In the earlier parts of the book, the focus on nilpotency allowed us to go beyond the characterization of localizations at T in terms of T-local homology to characterizations in terms of homotopy groups or integral homology groups. Similarly, it allowed us to go beyond the characterization of completions at p in terms of mod p homology to obtain a characterization in terms of homotopy groups in the finite-type case, together with a description of the homotopy groups in the general nilpotent case. Little is known about the behavior on homotopy groups of localizations and completions of non-nilpotent spaces. Of course, for generalized homology theories E_*, no such concrete descriptions can be expected, even for nilpotent spaces. The determination of the homotopy groups of particular E-localizations of spectra is a major part of stable homotopy theory.

REMARK 19.3.11. In view of the already excessive length of this book, we shall say nothing about the alternative definitions of localizations and completions of non-nilpotent spaces. The source cited most often is the book [21] of Bousfield and Kan. Sullivan gave a quite different early construction [133], and Morel [109, 110] showed how to compare and apply the Bousfield-Kan and Sullivan constructions. Again, there are many other interesting references. For example, Bousfield [18] showed that, already for $S^1 \vee S^n$, the Bousfield-Kan completion at p does not induce an isomorphism on mod p homology. We reiterate that, with any definition, relatively little is known about the behavior on homotopy groups in general.

19.4. Bousfield localization at sets and classes of maps

We return to a general model category $\mathcal{M}$. As stated at the beginning of §19.2, the theory of Bousfield localization admits an alternative starting point. While less relevant to our work in this book, it gives a more general kind of localization, of which E-localization is a special case. The starting point in this section dualizes to give an alternative starting point for §19.2. However, we warn the reader that these definitions are naive and provisional. They will be superceded in the next section.

DEFINITION 19.4.1. Let $\mathcal{K}$ be any class of maps in $\mathcal{M}$.

(i) An object $Z \in \mathcal{M}$ is $\mathcal{K}$-local if it is fibrant and

$$f^* : [Y, Z] \longrightarrow [X, Z]$$

is a bijection for all maps $f : X \longrightarrow Y$ in $\mathcal{K}$.

(ii) A map $\xi \colon X \longrightarrow Y$ is a $\mathscr{K}$-equivalence if

$$\xi^* \colon [Y, Z] \longrightarrow [X, Z]$$

is a bijection for all $\mathscr{K}$-local objects Z.

(iii) A map $\phi \colon X \longrightarrow X_{\mathscr{K}}$ is a localization of X at $\mathscr{K}$ if $X_{\mathscr{K}}$ is $\mathscr{K}$-local and ϕ is a $\mathscr{K}$-equivalence, so that

$$\phi^* \colon [X_{\mathscr{K}}, Z] \longrightarrow [X, Z]$$

is a bijection for all $\mathscr{K}$-local objects Z.

Define $\mathscr{L}_{\mathscr{K}}$ to be the class of $\mathscr{K}$-equivalences in $\mathscr{M}$. When $\mathscr{K}$ is a set, we change typeface to $\mathcal{K}$; here we have relative $\mathcal{K}$-cell complexes in mind. When $\mathcal{K}$ is a singleton set $\{f\}$, we write f-local, f-equivalence, and f-localization for the notions just defined.

Observe that $\mathscr{K}$-localization, if it exists, is a functor $L_{\mathscr{K}} \colon \mathscr{M} \longrightarrow \mathrm{Ho}\mathscr{M}$ such that ϕ is natural. Observe too that a map in $\mathscr{L}_{\mathscr{K}}$ between $\mathscr{K}$-local objects is a weak equivalence. We have allowed $\mathscr{K}$ to be a class and not just a set of morphisms. However, when $\mathscr{K} = \mathcal{K}$ is a set, there is an evident reduction.

LEMMA 19.4.2. *If $\mathcal{K}$ is a set of maps and f is the coproduct of the maps in $\mathcal{K}$, then the notions of $\mathcal{K}$-local, $\mathcal{K}$-equivalence, and $\mathcal{K}$-localization coincide with the notions of f-local, f-equivalence, and f-localization.*

PROOF. This holds since we have a natural isomorphism $[\amalg X_i, Z] \cong \times_i [X_i, Z]$ for objects X_i indexed on any set $\{i\}$ and any object Z. $\qquad\square$

We give some simple observations that compare our two starting points.

LEMMA 19.4.3. *For any class of maps $\mathscr{K}$, $\mathscr{L}_{\mathscr{K}}$ is a subcategory of weak equivalences that contains $\mathscr{W}$.*

PROOF. Since $[-, Z]$ is a functor, it is easy to check that $\mathscr{L}_{\mathscr{K}}$ is a subcategory of $\mathscr{M}$ that satisfies the two out of three property and is closed under retracts. It contains $\mathscr{W}$ since w^* is a bijection for $w \in \mathscr{W}$ and any object Z. $\qquad\square$

The following two observations are essentially tautologies from the definitions.

LEMMA 19.4.4. *For any class of maps $\mathscr{K}$, the notions of $\mathscr{K}$-local, $\mathscr{K}$-equivalence, and $\mathscr{K}$-localization coincide with the notions of $\mathscr{L}_{\mathscr{K}}$-local, $\mathscr{L}_{\mathscr{K}}$-equivalence, and $\mathscr{L}_{\mathscr{K}}$-localization.*

PROOF. It is immediate from the definitions that Z is $\mathscr{K}$-local if and only if it is $\mathscr{L}_{\mathscr{K}}$-local, and the rest follows. $\qquad\square$

Of course, it might happen that $\mathscr{K}$ itself is a subcategory of weak equivalences, which brings us back to the starting point of §19.2. Here we have the following comparison of contexts, which is just a reformulation of Proposition 19.2.13. In perhaps confusing notation, it tells us that every subcategory $\mathscr{L}$ of weak equivalences for which the $\mathscr{L}$-localization model structure exists is of the form $\mathscr{L}_{\mathscr{K}}$ for some class of maps $\mathscr{K}$, namely $\mathscr{K} = \mathscr{L}$.

PROPOSITION 19.4.5. *If $\mathscr{M}$ is left proper and $\mathscr{K}$ is a subcategory of weak equivalences in $\mathscr{M}$ such that the $\mathscr{K}$-localization model structure exists, then $\mathscr{K} = \mathscr{L}_{\mathscr{K}}$ and the $\mathscr{K}$-localization model structure is left proper.*

Returning to the context of the previous section, we can now reinterpret localization at E as localization at a map. The following result is a corollary of Proposition 19.3.5. We use its notations, and we write $\mathscr{K}_E$ and $\mathcal{K}_E$ for the class $\mathscr{K}$ and set $\mathcal{K}$ defined in Notation 19.3.3.

COROLLARY 19.4.6. *The following conditions on a space Z are equivalent.*

(i) Z is E-local.
(ii) Z is $\mathscr{F}_E$-fibrant.
(iii) Z is $\mathcal{K}_E$-local.
(iv) $Z \longrightarrow *$ satisfies the RLP with respect to $\mathcal{K}_E$.
(v) Z is f_E-local, where f_E is the disjoint union of the maps in $\mathcal{K}_E$.

Therefore the $\mathcal{K}_E$-equivalences are the same as the E-equivalences and localization at E coincides with localization at f_E.

PROOF. (i) and (ii) are equivalent by Proposition 19.2.13, (ii) and (iv) are equivalent by Proposition 19.3.5, and (iv) and (v) are equivalent by Lemma 19.4.2. Analogously to the equivalence of (i) and (ii), we claim that (iii) and (iv) are equivalent, and (iii) implies (iv) by Lemma 15.5.8. For the converse, since the maps $i: A \longrightarrow X$ in $\mathcal{K}_E$ are q-cofibrations between q-cofibrant objects, if we construct $j: \mathrm{Spl}\, i \longrightarrow \mathrm{Cyl}\, X$ using the usual double mapping cylinder construction, then j is in the class $\mathscr{K}_E$ and so has an isomorphic representative in the set $\mathcal{K}_E$. Thus (iv) implies (iii) by Lemma 15.5.3. The last statement follows from Proposition 19.4.5, since it implies that the E-equivalences are the maps that induce an isomorphism on $[-, Z]$ for those spaces satisfying our equivalent conditions. □

19.5. Bousfield localization in enriched model categories

We said that Definition 19.4.1 is naive and provisional. One good reason is that the localization asked for in that definition need not always exist [40, p. 3]. Another reason is that, in practice, the model category $\mathscr{M}$ we start off with is generally a $\mathscr{V}$-model category for some cosmos $\mathscr{V}$ with a monoidal model structure. We write $(\mathscr{W}, \mathscr{C}, \mathscr{F})$ for the model structures on both $\mathscr{M}$ and $\mathscr{V}$. We regard these model structures as fixed and are concerned with localizations of $\mathscr{M}$. In the presence of such enrichment, Definition 19.4.1 is better replaced by the following analogue. Others of our earlier definitions also admit such enriched alternatives.

DEFINITION 19.5.1. Let $\mathscr{K}$ be any class of cofibrations between cofibrant objects of $\mathscr{M}$.

(i) An object $Z \in \mathscr{M}$ is $\mathscr{K}$-local if it is fibrant and
$$f^* : \underline{\mathscr{M}}(Y, Z) \longrightarrow \underline{\mathscr{M}}(X, Z)$$
is a weak equivalence for all maps f in $\mathscr{K}$.

(ii) A map $\xi : X \longrightarrow Y$ is a $\mathscr{K}$-equivalence if
$$(\xi')^* : \underline{\mathscr{M}}(Y', Z) \longrightarrow \underline{\mathscr{M}}(X', Z)$$
is a bijection for all $\mathscr{K}$-local objects Z and some (and hence any) cofibrant approximation $\xi' : X' \longrightarrow Y'$ of ξ.

(iii) A map $\phi : X \longrightarrow X_{\mathscr{K}}$ is a localization of X at $\mathscr{K}$ if $X_{\mathscr{K}}$ is $\mathscr{K}$-local and ϕ is a $\mathscr{K}$-equivalence.

Define $\mathscr{L}_{\mathscr{K}}$ to be the class of $\mathscr{K}$-equivalences in $\mathscr{M}$ and define $\mathscr{F}_{\mathscr{K}}$ to be the class of maps that satisfy the RLP with respect to $\mathscr{C} \cap \mathscr{L}_{\mathscr{K}}$. We say that the $\mathscr{L}_{\mathscr{K}}$-localization model structure exists if $(\mathscr{L}_{\mathscr{K}}, \mathscr{C}, \mathscr{F}_{\mathscr{K}})$ is a model structure on $\mathscr{M}$.

In view of cofibrant replacement of maps, Lemma 15.5.6, our assumption on the maps in $\mathscr{K}$ loses no generality in the context of the previous section;

it ensures that $\underline{\mathscr{M}}(X, Z)$ is homotopically meaningful. As in Lemma 19.4.2, sets $\mathcal{K}$ can be replaced by the disjoint union of the maps in them. Rather than compare our two notions of localizations at sets of maps abstractly, we return to the context of topological spaces, but now we prefer to work in the category $\mathscr{U}_*$ of based topological spaces. Here we quote the following result, which is due to Casacuberta and Rodriguez [26]. As usual, we let $F(X, Y)$ denote the function space of based maps $X \longrightarrow Y$ and let X_+ denote the union of an unbased space X and a disjoint basepoint.

THEOREM 19.5.2. *Let $f : A \longrightarrow B$ be a map between CW complexes and let Z be any based space. Then $f^* : F(B, Z) \longrightarrow F(A, Z)$ is a weak equivalence if and only if the induced functions*

$$[S_+^n, F(B, Z)] \longrightarrow [S_+^n, F(A, Z)],$$

or equivalently

$$[B \wedge S_+^n, Z] \longrightarrow [A \wedge S_+^n, Z],$$

are bijections for all $n \geq 0$.

COROLLARY 19.5.3. *Let $\mathcal{K}$ be a class of relative based CW complexes $A \longrightarrow B$ that is closed under the extended suspension functors $(-) \wedge S_+^n$ for $n \geq 0$. Then the $\mathcal{K}$-local spaces and the $\mathcal{K}$-equivalences in Definitions 19.4.1 and 19.5.1 coincide, hence so do the two notions of $\mathcal{K}$-localization and the two $\mathscr{L}_{\mathcal{K}}$-model structures.*

From here, working in the q-model structure on the category $\mathscr{U}_*$ of based topological spaces, one can elaborate the methods of §19.3 to prove the following existence theorem for the f-model structure. We leave the details to the interested reader, or to the references in the following remarks.

THEOREM 19.5.4. *Let $\mathscr{F}$ denote the set consisting of a relative based CW complex $f : A \longrightarrow B$ and its extended suspensions $f \wedge S_+^n$ for $n \geq 0$. Then the $\mathscr{F}$-localization model structure exists. It is a left proper $\mathscr{U}_*$-model structure.*

REMARK 19.5.5. There is an extensive literature on f-localizations for a map f. The first existence proof is due to Bousfield [17], and the books [53, 65] study the foundations in detail. These sources work with simplicial sets. Dror Fajoun's monograph [40] gives several variant existence proofs, and he explains how either simplicial sets or topological spaces can be used. His monograph analyzes many interesting examples in detail. Bousfield [19] gives a nice overview of this area, with many references. More recently, Jeff Smith (unpublished) has proven that the f-localization model structure exists for any map f in any left proper combinatorial simplicial model category.

REMARK 19.5.6. One can ask whether localizations at classes, rather than sets, of maps always exist. Remarkably, Casacuberta, Scevenels, and Smith [27] prove that this holds if Vopěnka's principle (a certain large cardinal axiom) is valid, but that it cannot be proven using only the usual ZFC axioms of set theory.

We end our discussion of model categories and localization with a philosophical remark that contains a puzzling and interesting open problem.

REMARK 19.5.7. The notion of localizing a category $\mathcal{M}$, or a homotopy category Ho$\mathcal{M}$, at a subcategory $\mathcal{L}$ of weak equivalences makes sense as a general matter of homotopical algebra, independent of model category theory. However, the general theory here depends on the chosen model structure since we have required $\mathcal{L}$-local objects to be fibrant in a given model structure on $\mathcal{M}$. Thus our definition of an $\mathcal{L}$-local object really defines the notion of being $\mathcal{L}$-local relative to a given model structure. For a fixed ambient category of weak equivalences $\mathcal{W}$, there may be several model structures $(\mathcal{W}, \mathcal{C}, \mathcal{F})$ on $\mathcal{M}$ with different good properties.

In particular, in the categories of spaces or chain complexes, we have the q-model structure and the m-model structure with the same weak equivalences and therefore with equivalent homotopy categories. In these cases, all objects are fibrant and so the definition of $\mathcal{L}$-local objects is the same in the two cases. However, the question of the existence of the $\mathcal{L}$-model structure is still model dependent. Clearly, asking whether $(\mathcal{L}, \mathcal{C}, \mathcal{F}_{\mathcal{L}})$ is a model structure is a different question for $\mathcal{C} = \mathcal{C}_q$ and for $\mathcal{C} = \mathcal{C}_m$. Of course, $\mathcal{F}_{\mathcal{L}}$ depends on which choice we make, even though the fibrant objects are the same with both choices.

The known existence proofs for $\mathcal{L}$-local (or $\mathcal{K}$-local) model structures on spaces start from the q-model structure. We actually do not know whether or not $(\mathcal{L}, \mathcal{C}_m, \mathcal{F}_\ell)$ is a model structure even when we do know that $(\mathcal{L}, \mathcal{C}_q, \mathcal{F}_\ell)$ is a model structure. We regard this as a quite unsatisfactory state of affairs, but we leave the existence of model structures $(\mathcal{L}, \mathcal{C}_m, \mathcal{F}_\ell)$ as an open problem.

PART 5

Bialgebras and Hopf algebras

20

BIALGEBRAS AND HOPF ALGEBRAS

We define bialgebras, Hopf algebras, and related algebraic structures, largely following the original paper [104] of Milnor and Moore but incorporating various simplifications and amplifications. The reader is urged to recall our conventions on grading and commutativity from Warning 0.0.6. The theme is the definition of algebraic structures by use of dual commutative diagrams. Thus the familiar concepts of algebra and module dualize to concepts of coalgebra and comodule, and the structures of algebra and coalgebra combine to give the notion of a bialgebra. Incorporating antipodes (sometimes called conjugations), we obtain the notion of a Hopf algebra. In the cocommutative case, bialgebras and Hopf algebras can be viewed as monoids and groups in the symmetric monoidal category of cocommutative coalgebras.

20.1. Preliminaries

We shall work over a commutative ground ring R. The reader may prefer to take R to be a field, since that holds in most applications. Unless otherwise specified, $\otimes = \otimes_R$ and $\mathrm{Hom} = \mathrm{Hom}_R$. Recall that these are defined on graded R-modules by

$$(A \otimes B)_n = \sum_{i+j=n} A_i \otimes B_j \ \text{ and } \ \mathrm{Hom}_n(A, B) = \prod_i \mathrm{Hom}(A_i, B_{i+n}).$$

We think of R, or any other ungraded R-module, as concentrated in degree 0. We define the dual A^* of A by $A^* = \mathrm{Hom}(A, R)$, so that $A^n = \mathrm{Hom}(A_n, R)$; here we have implicitly reversed the grading to superscripts (with a sign change).

Of course, $\otimes$ is associative and unital (with unit R) up to natural isomorphism and has the natural commutativity isomorphism

$$\gamma : A \otimes B \to B \otimes A$$

specified by $\gamma(a \otimes b) = (-1)^{\deg a \deg b} b \otimes a$. We introduce such a sign whenever two entities are permuted. By a harmless standard abuse, we omit the unit and associativity isomorphisms from diagrams and treat them as if they were identifications. Use of the commutativity isomorphism is always made explicit. In categorical language, the category $\mathcal{M}_R$ of graded R-modules is symmetric monoidal, and it is closed in the sense that there is a natural isomorphism

$$\text{Hom}\,(A \otimes B, C) \cong \text{Hom}\,(A, \text{Hom}\,(B, C));$$

it sends f to g, where $g(a)(b) = f(a \otimes b)$. There are further natural maps

$$\nu:\ \text{Hom}\,(A, B) \otimes C \to \text{Hom}\,(A, B \otimes C),$$

$$\rho: A \to A^{**},$$

and

$$\alpha:\ \text{Hom}\,(A, C) \otimes \text{Hom}\,(B, D) \longrightarrow \text{Hom}\,(A \otimes B, C \otimes D),$$

which specializes to

$$\alpha:\ A^* \otimes B^* \to (A \otimes B)^*.$$

These maps are specified by

$$\nu(f \otimes c)(a) = (-1)^{\deg(c)\,\deg(a)} f(a) \otimes c,$$

$$\rho(a)(f) = (-1)^{\deg(a)\,\deg(f)} f(a),$$

and

$$\alpha(f \otimes g)(a \otimes b) = (-1)^{\deg(g)\,\deg(b)} f(a)g(b).$$

We say that A is projective if each A_i is projective (over R), and we say that A is of finite type if each A_i is finitely generated (over R). We say that A is bounded if it is nonzero in only finitely many degrees. Thus A is finitely generated if and only if it is bounded and of finite type. We say that A is bounded below (or above) if $A_i = 0$ for i sufficiently small (or large). Then ν is an isomorphism if A is bounded and either A or C is projective of finite type, ρ is an isomorphism if A is projective of finite type, and the last map α is an isomorphism if A and B are bounded below and A or B is projective of finite type. In these assertions, boundedness hypotheses ensure that the products appearing in our Hom's are finite, so that they become sums, and projective of finite type hypotheses allow us to apply the analogous assertions for ungraded modules. Henceforward, we implicitly restrict attention to nonnegatively graded modules, for which $A_i = 0$ if $i < 0$, since that is the case of greatest interest in algebraic topology.

Virtually all of our substantive results will be proven by use of filtrations and bigraded modules. We usually have $A_{p,q} = 0$ for all $p < 0$ or all $p > 0$. The signs occurring in the study of bigraded modules always refer to the total degree $p + q$. The tensor product of bigraded modules is given by

$$(A \otimes B)_{p,q} = \sum_{i+j=p, k+l=q} A_{i,k} \otimes B_{j,l}.$$

Similarly, the dual A^* is given by $A^{p,q} = \text{Hom}(A_{p,q}, R)$.

A filtration $\{F_p A\}$ of a graded module A is an expanding sequence of submodules $F_p A$. A filtration is said to be complete if

$$A \cong \text{colim} F_p A \quad \text{and} \quad A \cong \lim A/F_p A.$$

In most cases that we consider, we have either $F_p A = A$ for $p \geq 0$ and $\cap_p F_p A = 0$ or $F_p A = 0$ for $p < 0$ and $A = \cup_p F_p A$. In such cases, completeness is clear. We give R the trivial filtration, $F_p R = 0$ for $p < 0$ and $F_p R = R$ for $p \geq 0$. The tensor product of filtered modules is filtered by

$$F_p(A \otimes B) = \text{Im} \left(\sum_{i+j=p} F_i A \otimes F_j B \right) \subset A \otimes B.$$

We say that a filtration of A is flat if each $A/F_p A$ is a flat R-module; we say that a filtration is split if each sequence

$$0 \to F_p A \to A \to A/F_p A \to 0$$

is split exact over R. Of course, these both hold automatically when R is a field.

The associated bigraded module $E^0 A$ of a filtered module A is specified by

$$E^0_{p,q} A = (F_p A/F_{p-1} A)_{p+q}.$$

Of course, E^0 is a functor from filtered modules to bigraded modules.

PROPOSITION 20.1.1. *Let $f : A \to B$ be a map of complete filtered R-modules. If $E^0 f : E^0 A \to E^0 B$ is a monomorphism, or an epimorphism, or an isomorphism, then f and all its restrictions $F_p f$ are also monomorphisms, or epimorphisms, or isomorphisms.*

PROOF. The commutative diagram

$$0 \longrightarrow F_pA/F_{p-1}A \longrightarrow F_qA/F_{p-1}A \longrightarrow F_qA/F_pA \longrightarrow 0$$

$$0 \longrightarrow F_pB/F_{p-1}B \longrightarrow F_qB/F_{p-1}B \longrightarrow F_qB/F_pB \longrightarrow 0$$

implies inductively that f induces a monomorphism or epimorphism or isomorphism $F_qA/F_pA \to F_qB/F_pB$ for all $p < q$. Passing to colimits over q, we find that the same is true for $A/F_pA \to B/F_pB$ for all p. Since lim is left exact and preserves isomorphisms, we obtain the conclusions for the monomorphism and isomorphism cases by passage to limits. Since lim is not right exact, we must work a little harder in the epimorphism case. Here we let C_p be the kernel of the epimorphism $A/F_pA \to B/F_pB$ and let $C = \lim C_p$. A chase of the commutative exact diagram

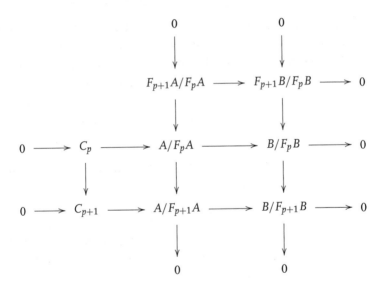

shows that $\{C_p\}$ is an inverse system of epimorphisms. Therefore $\lim^1 C_p = 0$ and each map $C \to C_p$ is an epimorphism. The exact sequence of inverse systems

$$0 \to \{C_p\} \to \{A/F_pA\} \to \{B/F_pB\} \to 0$$

gives rise to an exact sequence $0 \to C \to A \to B \to 0$ and a chase of the commutative exact diagram

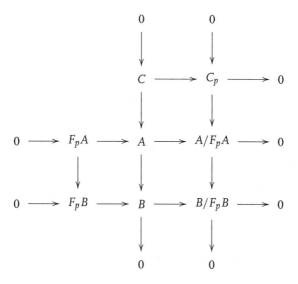

shows that $F_p A \to F_p B$ is an epimorphism. □

Chases of congeries of exact sequences give the following comparison assertion.

PROPOSITION 20.1.2. *Let A and B be filtered R-modules such that A and B are either both split or both flat. Then the natural map*

$$E^0 A \otimes E^0 B \to E^0(A \otimes B)$$

is an isomorphism of bigraded R-modules.

20.2. Algebras, coalgebras, and bialgebras

We give the most basic definitions in this section.

DEFINITION 20.2.1. An R-algebra $A = (A, \phi, \eta)$ is a graded R-module A together with a product $\phi : A \otimes A \to A$ and unit $\eta : R \to A$ such that the following diagrams commute.

A is commutative if the following diagram also commutes.

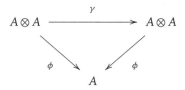

An augmentation of A is a morphism of algebras $\varepsilon : A \to R$. Given ε, $\ker \varepsilon$ is denoted IA and called the augmentation ideal of A; since $\varepsilon\eta = \mathrm{id}$, $A \cong R \oplus IA$. If A and B are algebras, then so is $A \otimes B$; its unit and product are

$$R = R \otimes R \xrightarrow{\eta\otimes\eta} A \otimes B \qquad \text{and} \qquad A \otimes B \otimes A \otimes B \xrightarrow{(\phi\otimes\phi)(\mathrm{id}\otimes\gamma\otimes\mathrm{id})} A \otimes B.$$

An algebra A is commutative if and only if $\phi : A \otimes A \to A$ is a map of algebras.

DEFINITION 20.2.2. An R-coalgebra $C = (C, \psi, \varepsilon)$ is a graded R-module C together with a coproduct $\psi : C \to C \otimes C$ and counit (or augmentation) $\varepsilon : C \to R$ such that the following diagrams commute.

$$
\begin{array}{ccc}
C & \xrightarrow{\psi} & C \otimes C \\
\psi \downarrow & & \downarrow \mathrm{id}\otimes\psi \\
C \otimes C & \xrightarrow{\psi\otimes\mathrm{id}} & C \otimes C \otimes C
\end{array}
\qquad \text{and} \qquad
\begin{array}{ccccc}
 & & C & & \\
 & \swarrow & \downarrow\psi & \searrow & \\
C \otimes R & \xleftarrow{\mathrm{id}\otimes\varepsilon} & C \otimes C & \xrightarrow{\varepsilon\otimes\mathrm{id}} & R \otimes C
\end{array}
$$

C is cocommutative if the following diagram also commutes.

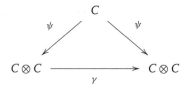

A unit (sometimes called coaugmentation) for C is a morphism of coalgebras $\eta : R \to C$; given η, define $JC = \mathrm{coker}\,\eta$. Since $\varepsilon\eta = \mathrm{id}$, $C \cong R \oplus JC$. If C and D are coalgebras, then so is $C \otimes D$; its augmentation and coproduct are

$$C \otimes B \xrightarrow{\varepsilon\otimes\varepsilon} R \otimes R = R \quad \text{and} \quad C \otimes D \xrightarrow{(\mathrm{id}\otimes\gamma\otimes\mathrm{id})(\psi\otimes\psi)} C \otimes D \otimes C \otimes D.$$

A coalgebra C is cocommutative if and only if ψ is a map of coalgebras.

DEFINITION 20.2.3. Let A be a flat R-module. A bialgebra $(A, \phi, \psi, \eta, \varepsilon)$ is an algebra (A, ϕ, η) with augmentation ε and a coalgebra (A, ψ, ε) with unit η such that the following diagram is commutative.

$$
\begin{array}{ccccc}
A \otimes A & \xrightarrow{\ \phi\ } & A & \xrightarrow{\ \psi\ } & A \otimes A \\
{\scriptstyle \psi \otimes \psi}\downarrow & & & & \uparrow{\scriptstyle \phi \otimes \phi} \\
A \otimes A \otimes A \otimes A & & \xrightarrow[{\mathrm{id} \otimes \gamma \otimes \mathrm{id}}]{} & & A \otimes A \otimes A \otimes A
\end{array}
$$

That is, ϕ is a morphism of coalgebras or, equivalently, ψ is a morphism of algebras. If the associativity of ϕ and coassociativity of ψ are deleted from the definition, then A is said to be a quasi-bialgebra.[1] There result notions of coassociative quasi-bialgebra and of associative quasi-bialgebra.

The flatness of A is usually not assumed but holds in practice; in its absence, the notion of bialgebra is perhaps too esoteric to be worthy of contemplation.

LEMMA 20.2.4. *Let A be projective of finite type.*

(i) *(A, ϕ, η) is an algebra if and only if (A^*, ϕ^*, η^*) is a coalgebra, ε is an augmentation of A if and only if ε^* is a unit of A^*, and A is commutative if and only if A^* is cocommutative.*

(ii) *$(A, \phi, \psi, \eta, \varepsilon)$ is a bialgebra if and only if $(A^*, \psi^*, \phi^*, \varepsilon^*, \eta^*)$ is a bialgebra.*

Similar conclusions hold for quasi-bialgebras.

DEFINITION 20.2.5. We define indecomposable and primitive elements.

(i) Let A be an augmented algebra. Define the R-module QA of indecomposable elements of A by the exact sequence

$$
IA \otimes IA \xrightarrow{\ \phi\ } IA \longrightarrow QA \longrightarrow 0.
$$

Note that QA is well-defined even if A is not associative.

(ii) Let C be a unital coalgebra. Define the R-module PC of primitive elements of C by the exact sequence

$$
0 \longrightarrow PC \longrightarrow JC \xrightarrow{\ \psi\ } JC \otimes JC.
$$

1. This use of "quasi" is due to Milnor and Moore [104]; Drinfeld later gave a more precise meaning to the term "quasi-Hopf algebra" [38].

Let $IC = \ker \varepsilon$. We say that $x \in IC$ is primitive if its image in JC lies in PC. Note that PC is well-defined even if C is not coassociative.

LEMMA 20.2.6. *If C is a unital coalgebra and $x \in IC$, then*

$$\psi(x) = x \otimes 1 + \sum x' \otimes x'' + 1 \otimes x,$$

where $\sum x' \otimes x'' \in IC \otimes IC$. If x is primitive, then

$$\psi(x) = x \otimes 1 + 1 \otimes x.$$

PROOF. $C \otimes C = (R \otimes R) \oplus (IC \otimes R) \oplus (R \otimes IC) \oplus (IC \otimes IC)$, where $R = \operatorname{Im} \eta$, and the natural map $IC \to JC$ is an isomorphism. The first statement holds since

$$(\varepsilon \otimes \operatorname{id})\psi(x) = x = (\operatorname{id} \otimes \varepsilon)\psi(x),$$

and the second statement is immediate from the definition. $\square$

When $x \in IC$, we usually write $\psi(x) = \sum x' \otimes x''$ generically for the coproduct,[2] including the terms $x \otimes 1$ and $1 \otimes x$ and omitting an index of summation.

LEMMA 20.2.7. *If A is an augmented algebra, then $P(A^*) = (QA)^*$. If, further, A is projective of finite type, then*

$$IA \otimes IA \longrightarrow IA \longrightarrow QA \longrightarrow 0$$

is split exact if and only if

$$0 \longrightarrow P(A^*) \longrightarrow I(A^*) \longrightarrow I(A^*) \otimes I(A^*)$$

is split exact; when this holds, $P(A^*)^* = QA$.

DEFINITION 20.2.8. Let A be a quasi-bialgebra. Define $\nu : PA \to QA$ to be the composite

$$PA \longrightarrow JA \cong IA \longrightarrow QA$$

(or, equivalently, the restriction of $IA \to QA$ to PA if PA is regarded as contained in A). A is said to be primitive, or primitively generated, if ν is an epimorphism; A is said to be coprimitive if ν is a monomorphism.

---

2. In the algebraic literature, the more usual convention is to write $\psi(x) = \sum x_{(1)} \otimes x_{(2)}$.

A structure A (algebra, coalgebra, bialgebra, etc.) is filtered if it has a split filtration such that all of the structure maps preserve the filtration. It follows that $E^0 A$ is a structure of the given type. The following definitions give basic tools for the study of (quasi) bialgebras by passage to associated bigraded primitive or coprimitive bialgebras. We warn the reader that the filtrations in the following two definitions are not necessarily complete. In the first case, that is a familiar fact from classical algebra since the intersection of the powers of a (two-sided) ideal in a ring can be nonzero [6, p. 110].

DEFINITION 20.2.9. Let A be an augmented algebra. Define the product filtration $\{F_p A\}$ by $F_p A = A$ if $p \geq 0$ and $F_p A = (IA)^{-p}$ if $p < 0$. Observe that

$$E^0_{p,*} A = 0 \text{ if } p > 0, \quad E^0_{0,*} A = E^0_{0,0} A = R, \quad \text{and} \quad E^0_{-1,*} A = QA.$$

If A is an associative quasi-bialgebra with split product filtration, then $E^0 A$ is a primitive bialgebra since the elements of $E^0_{-1,*} A$ generate $E^0 A$ and are evidently primitive, and this implies coassociativity.

DEFINITION 20.2.10. Let C be a unital coalgebra. Define the coproduct filtration $\{F_p C\}$ by $F_p C = 0$ if $p < 0$, $F_0 C = R$, and $F_p C = \ker \bar{\psi}_p$ if $p > 0$, where $\bar{\psi}_p$ is the composite

$$IC \subset C \xrightarrow{\;\psi_p\;} C \otimes \cdots \otimes C \longrightarrow JC \otimes \cdots \otimes JC, \quad p \text{ factors.}$$

Observe that

$$E^0_{p,*} C = 0 \text{ if } p < 0, \quad E^0_{0,*} C = E^0_{0,0} C = R, \quad \text{and} \quad E^0_{1,*} C = PC.$$

If C is a coassociative quasi-bialgebra with split coproduct filtration, then $E^0 C$ is a coprimitive bialgebra since the elements of $E^0_{-1,*} C$ are evidently indecomposable and include all the primitives; Lemma 21.1.1 below implies that $E^0 C$ is associative.

20.3. Antipodes and Hopf algebras

For a monoid G, the monoid ring $R[G]$ is a bialgebra with product, coproduct, unit, and counit induced by the product, diagonal, identity element, and trivial function $G \longrightarrow \{pt\}$. If G is a group, its inverse function induces an antipode on $R[G]$, in the sense of the following definition.

DEFINITION 20.3.1. An antipode χ on a bialgebra A is a map $\chi : A \to A$ of R-modules such that the following diagrams commute.

A Hopf algebra is a bialgebra with a given antipode.

If A and B have antipodes χ and χ' then $A \otimes B$ has the antipode $\chi \otimes \chi'$.

REMARK 20.3.2. The original definition of an antipode in Milnor and Moore [104] required only one of these two diagrams to commute, since in the cases of interest in algebraic topology, if one of them commutes, then so does the other. Actually, in [104] and most of the topological literature, the term "conjugate" is used instead of "antipode". Historically, the concept of Hopf algebra originated in algebraic topology, where the term "Hopf algebra" was used for what we are calling a bialgebra. The term "bialgebra" was introduced later and is still rarely used in topology. In fact, as we shall see in §21.3 below, the bialgebras that usually appear in algebraic topology automatically have antipodes, so that it is reasonable to ignore the distinction, and we do so where no confusion can arise. We have followed the algebraic literature in using the name antipode and distinguishing between bialgebras and Hopf algebras because of the more recent interest in Hopf algebras of a kind that do not seem to appear in algebraic topology, such as quantum groups.

REMARK 20.3.3. In general, the existence and properties of antipodes is a subtle question. For example, χ can exist but not satisfy $\chi^2 = \mathrm{id}$. The order of an antipode χ is defined to be the minimum n such that $\chi^n = \mathrm{id}$. It can be any even number or can even be infinite [136, p. 89].

In the cocommutative case, the concepts of bialgebra and Hopf algebra can be given a pleasant conceptual form. It is a standard and easy observation that the tensor product is the categorical coproduct in the category of commutative algebras. The units of A and B induce maps of algebras $i : A \to A \otimes B \leftarrow B : j$, and for any algebra maps $f : A \longrightarrow C \longleftarrow B : g$, the composite of $f \otimes g$ and the product on C gives the unique map of algebras $h : A \otimes B \longrightarrow C$ such that $h \circ i = f$ and $h \circ j = g$. We are interested in the dual observation. Recall that, in any category with products, we have the notion of a monoid, namely an object with an associative and unital product, and of a group, namely a monoid with an antipode. The following result is immediate from the definitions.

PROPOSITION 20.3.4. *The tensor product is the categorical product in the category $\mathscr{C}$ of commutative coalgebras. A cocommutative bialgebra is a monoid in $\mathscr{C}$, and a cocommutative Hopf algebra is a group in $\mathscr{C}$.*

There is another conceptual way of thinking about antipodes. It is based on the following construction.

CONSTRUCTION 20.3.5. Let C be a coalgebra and A be an algebra. Then $\mathrm{Hom}\,(C, A)$ is an algebra, called a convolution algebra. Its unit element is the composite $C \xrightarrow{\varepsilon} R \xrightarrow{\eta} A$ and its product is the composite

$$*\colon \mathrm{Hom}\,(C, A) \otimes \mathrm{Hom}\,(C, A) \xrightarrow{\alpha} \mathrm{Hom}\,(C \otimes C, A \otimes A) \xrightarrow{\mathrm{Hom}\,(\psi,\phi)} \mathrm{Hom}\,(C, A).$$

If C is unital with unit η and A is augmented with augmentation ε, then the set $G(C, A)$ of maps of R-modules $f\colon C \longrightarrow A$ such that $f\eta = \eta$ and $\varepsilon f = \varepsilon$ is a submonoid of $\mathrm{Hom}\,(C, A)$ under the convolution product $*$.

REMARK 20.3.6. Visibly, when A is a bialgebra, an antipode is a (two-sided) inverse to the identity map $A \longrightarrow A$ in the monoid $G(A, A)$. Therefore χ is unique if it exists. This remark is one reason to prefer the two-sided rather than the one-sided definition of an antipode.

Clearly, a sensible way to prove that a bialgebra A is a Hopf algebra is to prove more generally that $G(A, A)$ is a group. We return to this point in §21.3, where we give an easy general result of this form that applies to the examples of interest in algebraic topology.

20.4. Modules, comodules, and related concepts

There are many further basic pairs of dual algebraic definitions.

DEFINITION 20.4.1. Let (A, ϕ, η) be an algebra. A left A-module (N, ξ) is an R-module N and action $\xi \colon A \otimes N \to N$ such that the following diagrams commute.

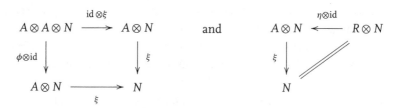

For an R-module N, $(A \otimes N, \phi \otimes \text{id})$ is an A-module and is said to be an extended A-module. For an A-module (N, ξ), ξ is a morphism of A-modules. With kernels and cokernels defined degreewise, the category of left A-modules is abelian. There is an analogous abelian category of right A-modules. For a right A-module (M, λ) and a left A-module (N, ξ), the tensor product $M \otimes_A N$, which of course is just an R-module, can be described as the cokernel of

$$\lambda \otimes \text{id} - \text{id} \otimes \xi : M \otimes A \otimes N \to M \otimes N;$$

$\otimes_A$ is a right exact functor of M and of N.

DEFINITION 20.4.2. Given an augmentation $\varepsilon : A \to R$ of A, regard R as a (left and right) A-module via ε and define

$$Q_A N = R \otimes_A N = N/IA \cdot N;$$

$Q_A N$ is called the module of A-indecomposable elements of N and is abbreviated QN when there is no danger of confusion. Observe that $Q_A(IA) = QA$.

DEFINITION 20.4.3. Let (C, ψ, ε) be a coalgebra. A left C-comodule (N, ν) is an R-module N and coaction $\nu : N \to C \otimes N$ such that the following diagrams commute.

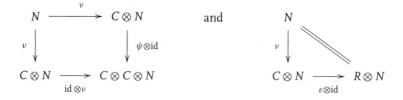

For an R-module N, $(C \otimes N, \psi \otimes \text{id})$ is a C-comodule, said to be a coextended C-comodule. For a C-comodule (N, ν), ν is a morphism of C-comodules. Since $\otimes$ is right but not left exact, the category of left C-comodules does not admit kernels in general; it is abelian if C is a flat R-module. There is an analogous category of right C-comodules. For a right C-comodule (M, μ) and a left C-comodule (N, ν), define the cotensor product $M\square_C N$ to be the kernel of

$$\mu \otimes \text{id} - \text{id} \otimes \nu : M \otimes N \to M \otimes C \otimes N.$$

The functor $\square$ is left exact with respect to sequences of left or right C-comodules that are split exact as sequences of R-modules (in the sense that the kernel at each position is a direct summand).

DEFINITION 20.4.4. Given a unit $\eta \colon R \to C$, regard R as a (left and right) C-comodule via η and define

$$P_C N = R\square_C N = \{n \mid \nu(n) = 1 \otimes n\};$$

$P_C N$ is called the module of C-primitive elements of N and is abbreviated PN when there is no danger of confusion. Observe that $P_C(JC) = PC$.

The following definition is fundamental. For a general algebra A, the tensor product (over R) of A-modules is an $A \otimes A$-module, but for bialgebras we can internalize this structure by pullback along ψ.

DEFINITION 20.4.5. Let $(A, \phi, \psi, \eta, \varepsilon)$ be a bialgebra. For left A-modules (N, ξ) and (N', ξ'), the following composite defines a left A-module structure on $N \otimes N'$.

$$A \otimes N \otimes N' \xrightarrow{(\text{id} \otimes \gamma \otimes \text{id})(\psi \otimes \text{id})} A \otimes N \otimes A \otimes N' \xrightarrow{\xi \otimes \xi'} N \otimes N'$$

An A-structure (module, coalgebra, algebra, bialgebra, Hopf algebra, etc.) is an A-module and a structure of the specified type such that all the maps that define the structure are morphisms of A-modules. Dually, for left A-comodules (N, ν) and (N', ν'), the following composite defines a left A-comodule structure on $N \otimes N'$.

$$N \otimes N' \xrightarrow{\nu \otimes \nu'} A \otimes N \otimes A \otimes N' \xrightarrow{(\phi \otimes \text{id})(\text{id} \otimes \gamma \otimes \text{id})} A \otimes N \otimes N'$$

The dual notion of an A-comodule and a structure whose structural maps are morphisms of A-comodules will be referred to as an A-comodule structure.

LEMMA 20.4.6. *Let A be an algebra and N be an R-module, both projective of finite type.*

(i) (N, ξ) is a left A-module if and only if (N^*, ξ^*) is a left A^*-comodule and then, if QN is also projective of finite type, $(QN)^* = P(N^*)$.

(ii) If A is a bialgebra, then N is a left A-structure if and only if N^* is a left A^*-comodule structure of the dual type.

LEMMA 20.4.7. Let A be a bialgebra and C be a left A-coalgebra. Then $Q_A C$ admits a unique structure of coalgebra such that the natural epimorphism $\pi : C \to Q_A C$ is a morphism of coalgebras.

PROOF. The augmentation of $Q_A C = R \otimes_A C$ is the map

$$\mathrm{id} \otimes \varepsilon : R \otimes_A C \to R \otimes_A R = R$$

and the coproduct is the composite of

$$\mathrm{id} \otimes \psi : R \otimes_A C \to R \otimes_A (C \otimes C)$$

and the natural map $R \otimes_A (C \otimes C) \to (R \otimes_A C) \otimes (R \otimes_A C)$.   □

Note that any bialgebra C that contains A as a subbialgebra is certainly a left A-coalgebra.

LEMMA 20.4.8. Let A be a bialgebra and B be a left A-comodule algebra. Then $P_A B$ admits a unique structure of algebra such that the natural monomorphism $\iota : P_A B \to B$ is a morphism of algebras.

DEFINITION 20.4.9. A morphism $f : A \to B$ of augmented algebras is said to be normal if the images of the composites

$$IA \otimes B \xrightarrow{\; f \otimes \mathrm{id} \;} B \otimes B \xrightarrow{\; \phi \;} B \quad \text{and} \quad B \otimes IA \xrightarrow{\; \mathrm{id} \otimes f \;} B \otimes B \xrightarrow{\; \phi \;} B$$

are equal and if the quotient map $\pi : B \to B//f$ is a split epimorphism, where $B//f$ is defined to be the R-module

$$Q_A B = R \otimes_A B = B/IA \cdot B = B/B \cdot IA = B \otimes_A R.$$

When f is an inclusion, $B//f$ is generally written $B//A$. Clearly $B//f$ admits a unique structure of augmented algebra such that π is a morphism of augmented algebras, and the following is an exact sequence of R-modules.

$$QA \xrightarrow{Qf} QB \xrightarrow{Q\pi} Q(B//f) \longrightarrow 0$$

DEFINITION 20.4.10. A morphism $g : B \to C$ of unital coalgebras is said to be conormal if the kernels of the composites

$$B \xrightarrow{\psi} B \otimes B \xrightarrow{g \otimes id} JC \otimes B \qquad \text{and} \qquad B \xrightarrow{\psi} B \otimes B \xrightarrow{id \otimes g} B \otimes JC$$

are equal and if the inclusion $\iota : B\backslash\backslash g \to B$ is a split monomorphism, where $B\backslash\backslash g$ is defined to be the R-module

$$P_C B = R \square_C B = \ker(g \otimes id)\psi = \ker(id \otimes g)\psi = B \square_C R.$$

When g is an epimorphism, $B\backslash\backslash g$ is generally written $B\backslash\backslash C$. Clearly $B\backslash\backslash g$ admits a unique structure of unital coalgebra such that ι is a morphism of unital coalgebras, and the following is an exact sequence of R-modules

$$0 \longrightarrow P(B\backslash\backslash g) \xrightarrow{P\iota} PB \xrightarrow{Pg} PC.$$

When R is a field, any morphism of commutative augmented algebras is normal and any morphism of cocommutative unital coalgebras is conormal.

REMARK 20.4.11. Let $f : A \to B$ be a morphism of bialgebras. If f is normal, then $B//f$ is a quotient bialgebra of B by Lemma 20.4.7. If f is conormal, then $A\backslash\backslash f$ is a subbialgebra of A by Lemma 20.4.8. The first assertion generalizes. A two-sided ideal $J \subset IB$ is said to be a Hopf ideal if

$$\psi(J) \subset B \otimes J + J \otimes B,$$

and then B/J (if flat) is a quotient bialgebra of B.

We emphasize that the previous few definitions and results work equally well if bialgebras are replaced by Hopf algebras everywhere.

21

CONNECTED AND COMPONENT
HOPF ALGEBRAS

An R-module A such that $A_i = 0$ for $i < 0$ (as we have tacitly assumed through-out) and $A_0 = R$ is said to be connected. Note that a connected algebra admits a unique augmentation and a connected coalgebra admits a unique unit. We shall see in §21.3 that a connected bialgebra always admits a unique antipode. Except in §21.3, we therefore follow the literature of algebraic topology and only use the term "Hopf algebra" in this chapter, since there is no real difference between the notions when A is connected. Connected structures arise ubiquitously in topology and have many special properties. For example, the homology of a connected homotopy associative H-space X is a connected Hopf algebra. The homology of nonconnected but grou-plike ($\pi_0(X)$ is a group) homotopy associative H-spaces leads to the more general notion of a component Hopf algebra. When concentrated in degree zero, these are just the classical group algebras $R[G]$. These too have unique antipodes.

We prove basic theorems on the splitting of connected algebras and coal-gebras over a connected Hopf algebra in §21.2, and we prove the self-duality of free commutative and cocommutative connected Hopf algebras on a single generator in §21.4. To illustrate the power of these beautiful but elementary algebraic results, we show how they can be used to prove Thom's calculation of unoriented cobordism and Bott's periodicity theorem for BU in §21.5 and §21.6.

21.1. Connected algebras, coalgebras, and Hopf algebras

We here prove various special properties that hold in the connected case but do not hold in general. However, they generally do apply to bigraded objects that are connected to the eyes of one of the gradings, and such structures can arise from filtrations of objects that are not connected.

LEMMA 21.1.1. *Let A be a connected coprimitive quasi-Hopf algebra. Then A is associative and commutative. If the characteristic of R is a prime p, then the pth power operation ξ (defined only on even degree elements of A if $p > 2$) is identically zero on IA.*

PROOF. Write $a(x, y, z) = x(yz) - (xy)z$ and $[x, y] = xy - (-1)^{\deg x \deg y} yx$. If x, y, and z are primitive elements of IA, then $a(x, y, z)$, $[x, y]$, and $\xi(x)$ are also primitive by direct calculation from Lemma 20.2.6 and the fact that the coproduct is a map of algebras. Since these elements obviously map to zero in QA, they must be zero. Now proceed by induction on $q = \deg x$, for fixed q by induction on $r = \deg y$, and for fixed q and r by induction on $s = \deg z$. By calculation from the induction hypothesis at each stage, we find that $a(x, y, z)$, $[x, y]$, and $\xi(x)$ are primitive and therefore zero. Here we prove commutativity before handling p^{th} powers so as to ensure that $(x + y)^p = x^p + y^p$. ☐

A Prüfer ring is an integral domain all of whose ideals are flat. A Noetherian Prüfer ring is a Dedekind ring. Recall that we require Hopf algebras to be flat R-modules.

LEMMA 21.1.2. *A connected Hopf algebra A over a Prüfer ring R is the colimit of its Hopf subalgebras of finite type.*

PROOF. Since R is Prüfer, every submodule of the flat R-module A is flat. Any element of A lies in a finitely generated subalgebra B, and B is clearly of finite type. An inductive argument based on the form of $\psi(x)$ given in Lemma 20.2.6 shows that the smallest Hopf subalgebra of A that contains B is also finitely generated. ☐

PROPOSITION 21.1.3. *If $f : A \to B$ is a morphism of augmented algebras, where B is connected, then f is an epimorphism if and only if Qf is an epimorphism.*

PROOF. Certainly Qf is an epimorphism if f is. Suppose that Qf is an epimorphism. By application of the five lemma to the commutative diagram with exact rows

$$
\begin{array}{ccccccc}
IA \otimes IA & \longrightarrow & IA & \longrightarrow & QA & \longrightarrow & 0 \\
\downarrow {\scriptstyle f \otimes f} & & \downarrow {\scriptstyle f} & & \downarrow {\scriptstyle Qf} & & \\
IB \otimes IB & \longrightarrow & IB & \longrightarrow & QB & \longrightarrow & 0
\end{array}
$$

we see by induction on n that f is an epimorphism in degree n for all n since f is trivially an epimorphism in degree 0 by the connectivity of B. □

PROPOSITION 21.1.4. *If $f : A \to B$ is a morphism of R-flat unital coalgebras, where A is connected, then f is a monomorphism if and only if $Pf : PA \to PB$ is a monomorphism.*

PROOF. The argument is dual to that just given. The flatness hypothesis ensures that $f \otimes f : JA \otimes JA \to JB \otimes JB$ is a monomorphism in degree n if f is a monomorphism in degrees less than n. □

The following result is a version of "Nakayama's lemma". It and its dual are used constantly in algebraic topology.

LEMMA 21.1.5. *If A is a connected algebra and N is a left A-module, then $N = 0$ if and only if $QN = 0$.*

PROOF. Clearly $QN = 0$ if and only if $IA \otimes N \to N$ is an epimorphism, and this implies that N is zero by induction on degrees. □

LEMMA 21.1.6. *If A is a connected algebra and $f : N \to N'$ is a morphism of left A-modules, then f is an epimorphism if and only if $Qf : QN \to QN'$ is an epimorphism.*

PROOF. The functor Q is right exact, hence $Q \operatorname{coker} f = 0$ and therefore $\operatorname{coker} f = 0$ if Qf is an epimorphism. □

The duals of the previous two results read as follows.

LEMMA 21.1.7. *If C is a connected coalgebra and N is a left C-comodule, then $N = 0$ if and only if $PN = 0$.*

LEMMA 21.1.8. *If C is an R-flat connected coalgebra and $f : N \to N'$ is a morphism of left C-comodules, then f is a monomorphism if and only if Pf is a monomorphism.*

21.2. Splitting theorems

Here we prove the basic results of Milnor and Moore [104] on tensor product decompositions of connected Hopf algebras. These play a key role in many calculations, for example, in the calculation of the cobordism rings of manifolds.

THEOREM 21.2.1. *Let A be a connected Hopf algebra and B be a connected left A-coalgebra. Write $QB = Q_A B$ and assume that the quotient map $\pi : B \to QB$ is a split epimorphism. Define $\iota : A \to B$ by $\iota(a) = a\eta(1)$ and assume that $\iota \otimes \mathrm{id} : A \otimes QB \to B \otimes QB$ is a monomorphism. Then there is an isomorphism $f : B \to A \otimes QB$ that is a map of both left A-modules and right QB-comodules.*

PROOF. Since π is a split epimorphism, we can choose a map of R-modules $\sigma : QB \to B$ such that $\pi\sigma = \mathrm{id}$. Let $g : A \otimes QB \to B$ be the induced map of left A-modules. Since $Qg : QB = Q(A \otimes QB) \to QB$ is the identity, g is an epimorphism by Lemma 21.1.6. We have the following composite of morphisms of A-modules.

$$h : A \otimes QB \xrightarrow{\ \ g\ \ } B \xrightarrow{\ \ \psi\ \ } B \otimes B \xrightarrow{\ \mathrm{id}\, \otimes\pi\ } B \otimes QB$$

Here A acts through $\varepsilon : A \to R$ on QB and acts diagonally on the tensor products. We claim that h is a monomorphism, so that g is a monomorphism and therefore an isomorphism. Filter $A \otimes QB$ by the degrees of elements of QB,

$$F_p(A \otimes QB) = \sum_{i \le p} A \otimes Q_i B.$$

The associated bigraded module of $A \otimes QB$ satisfies

$$E_{p,q}^0(A \otimes QB) = A_q \otimes Q_p B.$$

Filter $B \otimes QB$ similarly. Since h is a morphism of A-modules and is clearly filtration-preserving when restricted to QB, it is filtration-preserving. Since $\pi(an) = 0$ unless $\deg(a) = 0$, we see that $E^0 h = \iota \otimes \mathrm{id}$ and thus $E^0 h$ is a monomorphism by hypothesis. By Proposition 20.1.1, it follows that h is a monomorphism, as claimed. Now observe that $g(a \otimes \eta(1)) = \iota(a)$ for $a \in A$ and thus $(\mathrm{id} \otimes \varepsilon)g^{-1}\iota = \mathrm{id} : A \to A$, $\varepsilon : QB \to R$. Define f to be the composite

$$B \xrightarrow{\ \ \psi\ \ } B \otimes B \xrightarrow{\ \mathrm{id}\, \otimes\pi\ } B \otimes QB \xrightarrow{\ g^{-1}\otimes\mathrm{id}\ } A \otimes QB \otimes QB \xrightarrow{\ \mathrm{id}\, \otimes\varepsilon\otimes\mathrm{id}\ } A \otimes QB.$$

Clearly f is a morphism of left A-modules and right QB-comodules. Recall the filtration on $A \otimes QB$. Inspection shows that $fg : A \otimes QB \to A \otimes QB$ is filtration-preserving and that

$$E^0(fg) = (\mathrm{id} \otimes \varepsilon)g^{-1}\iota \otimes \pi\sigma = \mathrm{id}.$$

Therefore, by Proposition 20.1.1, fg is an isomorphism, hence so is f. $\qquad\square$

Note that, in the hypotheses, $\iota \otimes \mathrm{id}$ will be a monomorphism if ι is a monomorphism and QB is flat. Since a direct summand of a flat module is flat, the assumption on π implies that QB is flat if B is flat. Of course, when R is a field, as is the case in most applications, the only assumption is that $\iota : A \to B$ be a monomorphism.

The dual result reads as follows.

THEOREM 21.2.2. *Let C be a connected Hopf algebra and B be a connected left C-comodule algebra. Write $PB = P_C B$ and assume that the inclusion $\iota : PB \to B$ is a split monomorphism. Define $\pi : B \to C$ to be the composite of the coaction $\nu : B \to C \otimes B$ and $\mathrm{id} \otimes \varepsilon : C \otimes B \to C$ and assume that $\pi \otimes \mathrm{id} : B \otimes PB \to C \otimes PB$ is an epimorphism. Then there is an isomorphism $g : C \otimes PB \to B$ that is a map of both left C-comodules and right PB-modules.*

When R is a field, the only assumption is that $\pi : B \to C$ be an epimorphism.

These results are frequently applied to morphisms of Hopf algebras. Recall Definitions 20.4.9 and 20.4.10.

THEOREM 21.2.3. *Let $\iota : A \to B$ and $\pi : B \to C$ be morphisms of connected Hopf algebras. The following are equivalent.*

(i) ι is a normal monomorphism, $C = B//A$, and π is the quotient map.

(ii) π is a conormal epimorphism, $A = B\backslash\backslash C$, and ι is the inclusion.

(iii) There is an isomorphism $f : A \otimes C \to B$ of left A-modules and right C-comodules and an isomorphism $g : C \otimes A \to B$ of right A-modules and left C-comodules.

When (i)–(iii) hold,

$$f(\mathrm{id} \otimes \eta) = \iota = g(\eta \otimes \mathrm{id}), \quad (\varepsilon \otimes \mathrm{id})f^{-1} = \pi = (\mathrm{id} \otimes \varepsilon)g^{-1},$$

and the following is a commutative diagram with exact rows.

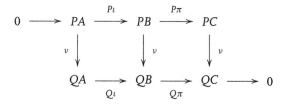

PROOF. Clearly (i) implies (iii) by Theorem 21.2.1 and symmetry; similarly, (ii) implies (iii) by Theorem 21.2.2 and symmetry. When (iii) holds, the descriptions of ι and π in terms of f and g follow from the module and comodule morphism properties of f and g, and (i) and (ii) follow by inspection. The diagram is obvious. □

COROLLARY 21.2.4. *Let $A \to B$ and $B \to C$ be normal monomorphisms of connected Hopf algebras. Then $B \to C$ induces a normal monomorphism of connected Hopf algebras $B//A \to C//A$, and $(C//A)//(B//A)$ is isomorphic to $C//B$.*

PROOF. $C \cong B \otimes C//B$, hence $C//A \cong B//A \otimes C//B$, and the conclusions follow. □

COROLLARY 21.2.5. *Let $A \to B$ and $B \to C$ be conormal epimorphisms of connected Hopf algebras. Then $A \to B$ induces a conormal epimorphism of connected Hopf algebras $A\backslash\backslash C \to B\backslash\backslash C$, and $(A\backslash\backslash C)\backslash\backslash(B\backslash\backslash C)$ is isomorphic to $A\backslash\backslash B$.*

21.3. Component coalgebras and the existence of antipodes

To prove the existence and develop the properties of χ on a bialgebra A, we need to make some hypothesis. However, the usual hypothesis in algebraic topology, connectivity, is too restrictive for many applications. We give a more general hypothesis, but still geared toward the applications in algebraic topology.

DEFINITION 21.3.1. We define grouplike algebras and component coalgebras.

(i) An augmented algebra A is said to be grouplike if the set $\varepsilon^{-1}(1)$ of degree 0 elements is a group under the product of A.

(ii) Let C be a coalgebra such that C_0 is R-free and define

$$\pi C = \{g \mid \psi(g) = g \otimes g \text{ and } g \neq 0\} \subset C_0.$$

For $g \in \pi C$, $g = \varepsilon(g)g$ by the counit property and thus $\varepsilon(g) = 1$ since C_0 is assumed to be R-free. Define the component C_g of g by letting $C_g = Rg \oplus \bar{C}_g$, where the R-module $\bar{C}_g$ of positive degree elements of C_g is

$$\{x \mid \psi(x) = x \otimes g + \sum x' \otimes x'' + g \otimes x, \ \deg x' > 0 \text{ and } \deg x'' > 0\}.$$

(iii) Say that C is a component coalgebra if C_0 is R-free, each C_g is a sub-coalgebra of C, and C is the direct sum of the C_g.

If C is unital then it has a privileged component, namely C_1. Note that primitivity becomes a less general notion in component coalgebras than intuition might suggest: elements x with $\psi(x) = x \otimes g + g \otimes x$, $g \neq 1$, are not primitive according to Definition 20.2.5.

If X is a based space, then $H_*(X; R)$, if R-flat, is a unital component coalgebra. Similarly, $H_*(\Omega X; R)$, if R-flat, is a grouplike component Hopf algebra; it is connected if and only if X is simply connected.

Now recall Construction 20.3.5. We implement the idea at the end of §20.3.

LEMMA 21.3.2. *If C is a unital component coalgebra and A is a grouplike augmented algebra, then $G(C, A)$ is a group under the convolution product $*$.*

PROOF. Let $f \in G(C, A)$. We must construct f^{-1}. Define $f^{-1}(g) = f(g)^{-1}$ for $g \in \pi C$ and extend f^{-1} to all of C_0 by R-linearity. Proceeding by induction on degrees, define $f^{-1}(x)$ for $x \in \bar{C}_g$ by

$$f^{-1}(x) = -f(g)^{-1}f(x)f(g)^{-1} - \sum f(g)^{-1}f(x')f^{-1}(x''),$$

where $\psi(x) = x \otimes g + \sum x' \otimes x'' + g \otimes x$, $\deg x > 0$ and $\deg x'' > 0$. Extend f^{-1} to C by R-linearity. Then $f * f^{-1} = \eta\varepsilon$ by direct inductive calculation. Of course, since every f has a right inverse, $f^{-1} * f = \eta\varepsilon$ follows formally. □

PROPOSITION 21.3.3. *Let A be a grouplike component bialgebra. Then A admits a (unique) antipode χ, so that*

$$\phi(\text{id} \otimes \chi)\psi = \eta\varepsilon = \phi(\chi \otimes \text{id})\psi.$$

Further, the following two diagrams are commutative.

$$
\begin{array}{ccc}
A \xrightarrow{\psi} A \otimes A \xrightarrow{\gamma} A \otimes A & \qquad \qquad & A \otimes A \xrightarrow{\gamma} A \otimes A \xrightarrow{\phi} A \\
\Big\downarrow{\chi} \qquad\qquad \Big\downarrow{\chi \otimes \chi} & and & \Big\downarrow{\chi \otimes \chi} \qquad\qquad\qquad \Big\downarrow{\chi} \\
A \xrightarrow{\psi} A \otimes A & & A \otimes A \xrightarrow{\phi} A.
\end{array}
$$

Moreover, if A is either commutative or cocommutative, then $\chi^2 \equiv \chi \circ \chi = \mathrm{id}$.

PROOF. The first statement is immediate from Lemma 21.3.2. For the first diagram, we claim that both $\psi\chi$ and $(\chi \otimes \chi)\gamma\psi$ are the inverse of $\psi : A \to A \otimes A$ in the group $G(A, A \otimes A)$. Indeed, we have

$$
\psi * \psi\chi = (\phi \otimes \phi)(\mathrm{id} \otimes \gamma \otimes \mathrm{id})(\psi \otimes \psi\chi)\psi = \psi\phi(\mathrm{id} \otimes \chi)\psi = \psi\eta\varepsilon = \eta\varepsilon
$$

by the very definition of a bialgebra. Since χ is natural and $\gamma : A \otimes A \to A \otimes A$ is an automorphism of Hopf algebras, $(\chi \otimes \chi)\gamma = \gamma(\chi \otimes \chi)$. Thus

$$
\begin{aligned}
\psi * (\chi \otimes \chi)\gamma\psi &= (\phi \otimes \phi)(\mathrm{id} \otimes \gamma \otimes \mathrm{id})(\psi \otimes \gamma(\chi \otimes \chi)\psi)\psi \\
&= (\phi \otimes \phi)(\mathrm{id} \otimes \gamma \otimes \mathrm{id})(\mathrm{id} \otimes \mathrm{id} \otimes \gamma)(\mathrm{id} \otimes \mathrm{id} \otimes \chi \otimes \chi)(\psi \otimes \psi)\psi \\
&= (\phi \otimes \mathrm{id})(\mathrm{id} \otimes \gamma)(\mathrm{id} \otimes \phi \otimes \mathrm{id})(\mathrm{id} \otimes \mathrm{id} \otimes \chi \otimes \chi)(\mathrm{id} \otimes \psi \otimes \mathrm{id})(\psi \otimes \mathrm{id})\psi \\
&= (\phi \otimes \mathrm{id})(\mathrm{id} \otimes \gamma)(\mathrm{id} \otimes \eta\varepsilon \otimes \chi)(\psi \otimes \mathrm{id})\psi \\
&= (\phi \otimes \mathrm{id})(\mathrm{id} \otimes \chi \otimes \eta\varepsilon)(\psi \otimes \mathrm{id})\psi \\
&= (\eta\varepsilon \otimes \eta\varepsilon)\psi = \eta\varepsilon.
\end{aligned}
$$

The proof of the second diagram is dual. Finally, to show that $\chi^2 = \mathrm{id}$, it suffices to show that χ^2 is the inverse of χ in the group $G(A, A)$. If A is commutative, the second diagram in the statement gives

$$
\chi^2 * \chi = \phi(\chi^2 * \chi)\psi = \phi(\chi \otimes \chi)(\chi \otimes \mathrm{id})\psi = \chi\phi\gamma(\chi \otimes \mathrm{id})\psi = \chi\eta\varepsilon = \eta\varepsilon.
$$

The proof that $\chi^2 = \mathrm{id}$ when A is cocommutative is dual. $\qquad\square$

Note that the second diagram of the statement asserts that χ is a graded involution. In the connected case, an easy induction gives the following explicit formula for the antipode. If $\deg x > 0$ and $\psi(x) = x \otimes 1 + \sum x' \otimes x'' + 1 \otimes x$,

$\deg x' > 0$ and $\deg x''>0$, then

21.3.4
$$\chi(x) = -x - \sum x' \chi(x'')$$

21.4. Self-dual Hopf algebras

The homology Hopf algebras $H_*(BU; \mathbb{Z})$ and $H_*(BO; \mathbb{F}_2)$ enjoy a very special property: they are self-dual, so that they are isomorphic to the cohomology Hopf algebras $H^*(BU; \mathbb{Z})$ and $H^*(BO; \mathbb{F}_2)$. The proof of this basic result is purely algebraic and explicitly determines the homology Hopf algebras from the cohomology Hopf algebras (or vice versa if one calculates in the opposite order). We assume that the reader knows that the cohomology Hopf algebras are given by

21.4.1
$$H^*(BU; \mathbb{Z}) = P\{c_i \mid i \geq 1\} \quad \text{with} \quad \psi(c_n) = \sum_{i+j=n} c_i \otimes c_j$$

and

21.4.2
$$H^*(BO; \mathbb{F}_2) = P\{w_i \mid i \geq 1\} \quad \text{with} \quad \psi(w_n) = \sum_{i+j=n} w_i \otimes w_j.$$

The calculations of $H^*(BU(n); \mathbb{Z})$ and $H^*(BO(n); \mathbb{F}_2)$ are summarized in [93, pp. 187, 195], and passage to colimits over n gives the stated conclusions. Thus determination of the homology algebras is a purely algebraic problem in dualization.[1]

Recall that the dual coalgebra of a polynomial algebra $P[x]$ over R is written $\Gamma[x]$; when $P[x]$ is regarded as a Hopf algebra with x primitive, $\Gamma[x]$ is called a divided polynomial Hopf algebra.

Clearly $H^*(BU(1); \mathbb{Z}) = P[c_1]$ and $H^*(BO(1); \mathbb{F}_2) = P[w_1]$ are quotient algebras of $H^*(BU; \mathbb{Z})$ and $H^*(BO; \mathbb{F}_2)$. Write $H_*(BU(1); \mathbb{Z}) = \Gamma[\gamma_1]$; it has basis $\{\gamma_i \mid i \geq 0\}$ and coproduct $\psi(\gamma_n) = \sum_{i+j=n} \gamma_i \otimes \gamma_j$, where $\gamma_0 = 1$ and γ_i is dual to c_1^i. Write $H_*(BO(1); \mathbb{F}_2) = \Gamma[\gamma_1]$ similarly. The inclusions $BU(1) \longrightarrow BU$ and $BO(1) \longrightarrow BO$ induce identifications of these homologies with subcoalgebras of $H_*(BU; \mathbb{Z})$ and $H_*(BO; \mathbb{F}_2)$, and we prove that these subcoalgebras freely generate the respective homology algebras.

THEOREM 21.4.3. $H_*(BU; \mathbb{Z}) = P\{\gamma_i \mid i \geq 1\}$, where $\gamma_i \in H_*(BU(1); \mathbb{Z})$ is dual to c_1^i. The basis $\{p_i\}$ for the primitive elements of $H_*(BU; \mathbb{Z})$ such that $\langle c_i, p_i \rangle = 1$

---

1. We thank John Rognes, who texed this section from the first author's notes in 1996.

is specified inductively by

$$p_1 = \gamma_1 \quad \text{and} \quad p_i = (-1)^{i+1} i \gamma_i + \sum_{j=1}^{i-1} (-1)^{j+1} \gamma_j p_{i-j} \quad \text{for } i > 0.$$

This recursion formula is generally ascribed to Newton, of course in a different but related context, although the following explicit evaluation was known even earlier (to Girard, in a 1629 paper).

REMARK 21.4.4. An explicit formula for p_i is given by

$$p_i = \sum_{E} (-1)^{|E|+i} \frac{(|E|-1)! i}{e_1! \cdots e_r!} \gamma^E.$$

Here the sum is taken over all sequences $E = (e_1, \ldots, e_r)$ with $e_q \geq 0$ and $\sum q e_q = i$; $|E| = \sum e_q$ and $\gamma^E = \gamma_1^{e_1} \cdots \gamma_r^{e_r}$.

THEOREM 21.4.5. $H_*(BO; \mathbb{F}_2) = P\{\gamma_i \mid i \geq 1\}$, where $\gamma_i \in H_*(BO(1); \mathbb{F}_2)$ is dual to w_1^i. The nonzero primitive elements of $H_*(BO; \mathbb{F}_2)$ are specified inductively by

$$p_1 = \gamma_1 \quad \text{and} \quad p_i = i \gamma_i + \sum_{j=1}^{i-1} \gamma_j p_{i-j} \quad \text{for } i > 0.$$

Comparison of these theorems to (21.4.1) and (21.4.2) shows that $H^*(BU; \mathbb{Z})$ and $H^*(BO; \mathbb{F}_2)$ are self–dual; that is, they are isomorphic as Hopf algebras to their own duals. Following Moore [108], we shall carry out the proofs by considering self–duality for certain general types of Hopf algebras.

We work in the category of connected free R-modules X of finite type, so that $X_i = 0$ for $i < 0$ and $X_0 = R$. Throughout the discussion, all algebras are to be commutative and all coalgebras are to be cocommutative. Thus all Hopf algebras are to be commutative and cocommutative.

DEFINITION 21.4.6. We define some universal Hopf algebras.

(i) A universal enveloping Hopf algebra of a coalgebra C is a Hopf algebra LC together with a morphism $i \colon C \longrightarrow LC$ of coalgebras that is universal with respect to maps of coalgebras $f \colon C \longrightarrow B$, where B is a Hopf algebra. That is, any such f factors uniquely as $\tilde{f} \circ i$ for a morphism $\tilde{f} \colon LC \longrightarrow B$ of Hopf algebras.

(ii) A universal covering Hopf algebra of an algebra A is a Hopf algebra MA together with a morphism $p\colon MA \longrightarrow A$ of algebras that is universal with respect to maps of algebras $f\colon B \longrightarrow A$, where B is a Hopf algebra. That is, any such f factors uniquely as $p \circ \tilde{f}$ for a morphism $\tilde{f}\colon B \longrightarrow MA$ of Hopf algebras.

LEMMA 21.4.7. *Universal Hopf algebras exist and are unique. That is,*

(i) any coalgebra C admits a universal enveloping Hopf algebra $i\colon C \longrightarrow LC$;

(ii) any algebra A admits a universal covering Hopf algebra $p\colon MA \longrightarrow A$.

PROOF. Of course, uniqueness up to isomorphism follows from universality. For (i), we have $C = R \oplus JC$, where JC is the module of positive degree elements of C. As an algebra, we take $LC = A(JC)$, the free (graded) commutative algebra generated by JC. Let $i\colon C \longrightarrow LC$ be the natural inclusion $JC \longrightarrow LC$ in positive degrees and the identity map id of R in degree zero. If ψ is the coproduct of C, the coproduct of LC is defined to be the unique map of algebras $\psi\colon LC \longrightarrow LC \otimes LC$ that makes the following diagram commute:

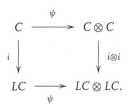

That ψ defines a coalgebra and thus a Hopf algebra structure on LC and that $i\colon C \longrightarrow LC$ is universal follow directly from the universal property of LC as an algebra. For (ii), since all modules are taken to be free of finite type, $p\colon MA \longrightarrow A$ can be specified as $i^*\colon (L(A^*))^* \longrightarrow A^{**} = A$. $\quad\square$

REMARK 21.4.8. Similar constructions may be obtained when we omit some or all of the commutativity hypotheses. We can define universal enveloping commutative Hopf algebras for arbitrary coalgebras and universal covering cocommutative Hopf algebras for arbitrary algebras. These will coincide with our present constructions under our hypotheses. The universal enveloping noncommutative Hopf algebra is of course a quite different construction.

We shall shortly require a pair of dual lemmas, for which we need some notations. For an R-module X, let X^n denote the n-fold tensor product of X

with itself. With the usual sign $(-1)^{\deg x \deg y}$ inserted when x is permuted past y, the symmetric group Σ_n acts on X^n. If X is an algebra or coalgebra, then so is X^n, and Σ_n acts as a group of automorphisms. Let Σ_n act trivially on LC and MA.

LEMMA 21.4.9. *Let C be a coalgebra. For $n > 0$, define $\iota_n \colon C^n \longrightarrow LC$ to be the composite of $i^n \colon C^n \longrightarrow (LC)^n$ and the iterated product $\phi \colon (LC)^n \longrightarrow LC$. Then ι_n is a morphism of both Σ_n-modules and coalgebras. If $C_q = 0$ for $0 < q < m$, then ι_n is an epimorphism in degrees $q \leq mn$.*

PROOF. The first statement is immediate from the definitions and the second statement follows from the fact that the image of ι_n is the span of the monomials in C of length at most n. ☐

LEMMA 21.4.10. *Let A be an algebra. For $n > 0$, define $\pi_n \colon MA \longrightarrow A^n$ to be the composite of the iterated coproduct $\psi \colon MA \longrightarrow (MA)^n$ and $p^n \colon (MA)^n \longrightarrow A^n$. Then π_n is a morphism of both Σ_n-modules and algebras. If $A_q = 0$ for $0 < q < m$, then ι_n is a monomorphism in degrees $q \leq mn$.*

PROOF. This follows by dualizing the previous lemma. ☐

DEFINITION 21.4.11. Let X be a positively graded R-module, so that $X_i = 0$ for $i \leq 0$. Define $LX = L(R \oplus X)$, where $R \oplus X$ is R in degree zero and has the trivial coalgebra structure, in which every element of X is primitive. Define $MX = M(R \oplus X)$, where $R \oplus X$ has the trivial algebra structure, in which the product of any two elements of X is zero. There is a natural morphism of Hopf algebras $\lambda \colon LMX \longrightarrow MLX$, which is defined in two equivalent ways. Indeed, consider the following diagram:

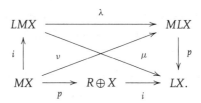

Define μ to be $A(p) \colon A(JMX) \longrightarrow A(X)$, which is the unique morphism of algebras that extends $i \circ p$, and then obtain λ by the universal property of $p \colon MLX \longrightarrow LX$. Define ν to be the dual of $A(i^*) \colon A((JLX)^*) \longrightarrow A(X^*)$, so that

ν is the unique morphism of coalgebras that covers $i \circ p$, and obtain λ by the universal property of $i: MX \longrightarrow LMX$. To see that the two definitions coincide, note that if λ is defined by the first property, then $\lambda \circ i = \nu$ by uniqueness and so λ also satisfies the second property.

Observe that $(R \oplus X)^*$ may be identified with $R \oplus X^*$. Since $MA = (L(A^*))^*$, it follows that

$$MX \equiv M(R \oplus X) = (L(R \oplus X^*))^* \equiv (L(X^*))^*.$$

In turn, with $A = L(X^*)$, this implies

$$ML(X^*) = MA = (L(A^*))^* = (L(L(X^*))^*)^* = (LMX)^*.$$

If X is R-free on a given basis, then the isomorphism $X \cong X^*$ determined by use of the dual basis induces an isomorphism of Hopf algebras

$$\beta: MLX \cong ML(X^*) = (LMX)^*.$$

When $\lambda: LMX \longrightarrow MLX$ is an isomorphism, it follows that LMX is self-dual. While λ is not always an isomorphism, it is so in the cases of greatest topological interest. We now regard $i: C \longrightarrow LC$ as an inclusion, omitting i from the notation. Write $\langle -, - \rangle$ for the usual pairing between a free R-module and its dual.

THEOREM 21.4.12. *Let X be free on one generator x of degree m, where either m is even or R has characteristic two. Then $\lambda: LMX \longrightarrow MLX$ is an isomorphism. Moreover if*

$$c_i = \gamma_i(x) \in \Gamma[x] = MX \quad and \quad \gamma_i = (\beta \circ \lambda)(c_i) \in (LMX)^*,$$

then γ_i is the basis element dual to c_1^i and the basis $\{p_i\}$ for the primitive elements of $(LMX)^*$ such that $\langle c_i, p_i \rangle = 1$ is specified inductively by

$$p_1 = \gamma_1 \quad and \quad p_i = (-1)^{i+1} i \gamma_i + \sum_{j=1}^{i-1} (-1)^{j+1} \gamma_j p_{i-j} \quad for \ i > 0.$$

Here $LMX = P\{c_i \mid i \geq 1\}$ with $\psi(c_n) = \sum_{i+j=n} c_i \otimes c_j$, where $c_0 = 1$. When $R = \mathbb{Z}$ and $m = 2$, LMX may be identified with $H^*(BU; \mathbb{Z})$ and $(LMX)^*$ may be identified with $H_*(BU; \mathbb{Z})$. Thus this result immediately implies

Theorem 21.4.3. Similarly, with $R = \mathbb{F}_2$ and $m = 1$, it implies Theorem 21.4.5. The rest of the section is devoted to the proof.

PROOF. Note that $LX = P[x]$ and write $P[x]^n = P[x_1, \ldots, x_n]$, where $x_i = 1 \otimes \cdots \otimes 1 \otimes x \otimes 1 \otimes \cdots \otimes 1$ with x in the ith position. Let $\sigma_1, \ldots, \sigma_n$ be the elementary symmetric functions in the x_i. Consider $\pi_n \lambda \colon LMX \longrightarrow P[x]^n$, where $\pi_n = p^n \psi \colon MP[x] \longrightarrow P[x]^n$ is as specified in Lemma 21.4.10. From the diagram that defines λ, we see that $p\lambda \colon LMX \longrightarrow P[x]$ is given on generators by

$$p\lambda c_j = ipc_j = \begin{cases} x & \text{if } j = 1 \\ 0 & \text{if } j > 1. \end{cases}$$

Since λ is a morphism of Hopf algebras, it follows that

$$\pi_n \lambda c_j = p^n \psi \lambda c_j = p^n \lambda^n \psi c_j = (p\lambda)^n \left(\sum_{i_1 + \cdots + i_n = j} c_{i_1} \otimes \cdots \otimes c_{i_n} \right) = \begin{cases} \sigma_j & \text{if } j \leq n \\ 0 & \text{if } j > n. \end{cases}$$

Since $LMX = P[c_i]$, the map $\pi_n \lambda \colon P[c_i] \longrightarrow P[\sigma_1, \ldots, \sigma_n]$ is an isomorphism in degrees $q \leq mn$. By Lemma 21.4.10, π_n also takes values in $P[\sigma_1, \ldots, \sigma_n]$ and is a monomorphism in degrees $q \leq mn$. Therefore π_n and λ are both isomorphisms in degrees $q \leq mn$. Since n is arbitrary, this proves that λ is an isomorphism.

To see the duality properties of the γ_i, consider the map $\nu \colon \Gamma[x] \longrightarrow MP[x]$ in the diagram defining λ. Here ν is dual to $A(i^*) \colon A(J\Gamma[x^*]) \longrightarrow P[x^*]$, where x^* is the basis element of X^* dual to x, and i^* maps $\gamma_1(x^*)$ to x^* and annihilates $\gamma_i(x^*)$ for $i > 1$. Since $c_i = \gamma_i(x)$ is dual to $(x^*)^i$, $\nu(c_i)$ is dual to $\gamma_1(x^*)^i$, and thus $\eta\nu(c_i) = \gamma_i$ is dual to c_1^i.

Since the primitive elements of $(LMX)^*$ are dual to the indecomposable elements of LMX, they are free on one generator dual to c_i in each degree mi. We shall prove inductively that this generator is p_i, the case $i = 1$ having been handled above. Consider the term $\gamma_j p_{i-j}$, $1 \leq j \leq i - 1$, in the iterative expression for p_i. Let c^E be a monomial in the c_k, so that $E = (e_1, \ldots, e_r)$ and $c^E = c_1^{e_1} \cdots c_r^{e_r}$. Then

$$\langle c^E, \gamma_j p_{i-j} \rangle = \langle \psi c^E, \gamma_j \otimes p_{i-j} \rangle = \langle \psi c^E, (c_1^j)^* \otimes c_{i-j}^* \rangle$$

by the induction hypothesis and the calculation above. Consideration of the form of ψc^E shows that this is zero unless c^E is either $c_1^j c_{i-j}$ or $c_1^{j-1} c_{i-j+1}$, when it is one in all cases except the case $\langle c_1^i, \gamma_{i-1} p_1 \rangle = i$. It follows that $\langle c^E, p_i \rangle = 0$ except for the case $\langle c_i, p_i \rangle = 1$. An alternative argument is to verify inductively that each p_i is in fact primitive and then check that $\langle c_i, p_i \rangle = 1$. $\qquad \square$

21.5. The homotopy groups of *MO* and other Thom spectra

In [93, Ch. 25], we explained Thom's classical computation of the real cobordism of smooth manifolds. In fact, the exposition there was something of a
cheat. Knowing the splitting theorems of §21.2 and the self-duality theorem of
§21.4, the senior author simply transcribed the first and quoted the second to
give the main points of the calculation. That obscures the conceptual simplicity
of the idea and its implementation. We explain in this section how the general
theory applies. A punch line, explained at the end of the section, is that the
conceptual argument applies to much more sophisticated cobordism theories,
where the actual calculations are far more difficult. We take all homology and
cohomology with coefficients in $\mathbb{F}_2$ in this section.

Recall the description of the Hopf algebra $H^*(BO)$ from (21.4.2). The structure of the dual Hopf algebra $H_*(BO)$ is given in Theorem 21.4.5. To conform
to the notation of [93, Ch. 25], write $\gamma_i = b_i$. It is the image in $H_*(BO)$ of
the nonzero class $x_i \in H_*(\mathbb{R}P^\infty)$. Thus $H_*(BO)$ is the polynomial algebra on
the b_i, and $\psi(b_k) = \sum_{i+j=k} b_i \otimes b_j$.

The Thom prespectrum TO and its associated Thom spectrum MO are
described in [93, pp. 216, 229], but we are not much concerned with
the foundations of stable homotopy theory here. The ring structure on
TO gives its homology an algebra structure, and the Thom isomorphism
$\Phi\colon H_*(TO) \longrightarrow H_*(BO)$ is an isomorphism of algebras [93, p. 221]. Write
$a_i = \Phi^{-1}(b_i)$. The Thom space $TO(1)$ of the universal line bundle is equivalent
to $\mathbb{R}P^\infty$ and, with $a_0 = 1$, a_i is the image of x_{i+1} in $H_*(TO)$.

Let A be the mod 2 Steenrod algebra and A_* be its dual. Then A acts on
the cohomology of spaces, prespectra, and spectra, and the action of A on the
cohomology of a ring prespectrum T dualizes to give $H_*(T)$ a structure of left
A-comodule algebra, as in Theorem 21.2.2. The composite

$$\pi = (\mathrm{id} \otimes \varepsilon)\nu\colon H_*(TO) \longrightarrow A_* \otimes H_*(TO) \longrightarrow A_*$$

is computed on [93, p. 224]. The computation just translates the easy compu-
tation of the action of A on $H^*(\mathbb{R}P^\infty)$ to a formula for the coaction of A_*. As an
algebra, A_* is a polynomial algebra on certain generators ξ_r of degree $2^r - 1$,
and $\pi(a_{2^r-1}) = \xi_r$. Thus π is an epimorphism.

By Theorem 21.2.2, this implies that there is an isomorphism

$$A_* \otimes P_{A_*}(H_*(TO)) \cong H_*(TO)$$

of left A_*-comodules and right $P_{A_*}(H_*(TO))$-modules. Since we know that A_*
and $H_*(TO)$ are polynomial algebras such that the generators of A_* map to

some of the generators of $H_*(TO)$, it is clear that $P_{A_*}(H_*(TO)) \equiv N_*$ must be a polynomial algebra on (abstract) generators u_i of degree i, where $i > 1$ and $i \neq 2^r - 1$. Dually $H^*(TO) = H^*(MO)$ is isomorphic as an A-module to $A \otimes N^*$. As explained informally in [93, §25.7], this implies that MO is a product of suspensions of Eilenberg-Mac Lane spectrum $H\mathbb{F}_2$ and that $\pi_*(MO) \cong N_*$ as an algebra. This gives the now standard way of obtaining Thom's calculation [137] of $\pi_*(MO)$.

The theorem applies to unoriented smooth manifolds, but one might consider less structured manifolds, such as piecewise-linear (PL) or topological manifolds. Focusing on PL manifolds for definiteness, which makes sense since the theory of PL manifolds was designed to get around the lack of obvious transversality in the theory of topological manifolds, one can adapt Thom's theorem to prove geometrically that the *PL*-cobordism groups are isomorphic to the homotopy groups of a Thom prespectrum *TPL*. By neglect of structure, we obtain a map of Thom prespectra $TO \longrightarrow TPL$. We have the same formal structure on *TPL* as we have on *TO*, and we have a commutative diagram

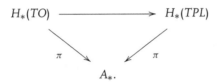

Even without any calculational knowledge of $H_*(BPL)$ and $H_*(TPL)$, we conclude that π on the right must also be an epimorphism.

Therefore, as a matter of algebra, Theorem 21.2.2 gives us an isomorphism

$$A_* \otimes P_{A_*}(H_*(TPL)) \cong H_*(TPL)$$

of left A_*-comodules and right $P_{A_*}H_*(TPL)$-algebras. Here again, the Thom isomorphism $\Phi\colon H_*(TPL) \longrightarrow H_*(BPL)$ is an isomorphism of algebras. Therefore, if we can compute $H_*(BPL)$ as an algebra, then we can read off what $P_{A_*}(H_*(TPL))$ must be as an algebra. The same formal argument as for *MO* shows that *MPL* is a product of suspensions of $H\mathbb{F}_2$ and that $\pi_*(MPL) \cong P_{A_*}(H_*(TPL))$ as algebras. In fact, this argument was understood and explained in [22] well before $H_*(BPL)$ was determined. The calculation of $H_*(BPL; \mathbb{F}_p)$ at all primes p is described in [81, 31], but that is another story.[2]

---

2. It is part of the 1970s story of infinite loop space theory and E_∞ ring spectra; see [90] for a 1970s overview and [96] for a modernized perspective.

In any case, this sketch should give some idea of the algebraic power of the splitting theorems in §21.2.

21.6. A proof of the Bott periodicity theorem

The self-duality of $H^*(BU)$ described in (21.4.1) and Theorem 21.4.3 also plays a central role in a quick proof of (complex) Bott periodicity. We describe how that works in this section. As discussed briefly in [93, §24.2], the essential point is to prove the following result. Homology and cohomology are to be taken with coefficients in $\mathbb{Z}$ in this section.

THEOREM 21.6.1. *There is a map* $\beta\colon BU \longrightarrow \Omega SU$ *of H-spaces that induces an isomorphism on homology.*

It follows from the dual Whitehead theorem that β must be an equivalence.

We begin by defining the Bott map β, following Bott [14]. Write $U(V)$ for the compact Lie group of unitary transformations $V \longrightarrow V$ on a complex vector space V with a given Hermitian product. If V is of countable dimension, let $U(V)$ denote the colimit of the $U(W)$ where W runs through the finite dimensional subspaces of V with their induced Hermitian products. Fixing the standard inclusions $\mathbb{C}^n \longrightarrow \mathbb{C}^\infty$, we specify $BU = U/U \times U$ to be the colimit of the Grassmannians $U(2n)/U(n) \times U(n)$. We let U be the colimit of the $U(2n)$ and SU be its subgroup colim $SU(2n)$ of unitary transformations with determinant one. For convenience, we write $\mathbb{V} = \mathbb{C}^\infty$ and let $\mathbb{V}^n$ denote the direct sum of n copies of $\mathbb{V}$.

It is also convenient to use paths and loops of length π. Taking $0 \le \theta \le \pi$, define $v(\theta) \in U(\mathbb{V}^2)$ by

$$v(\theta)(z', z'') = (e^{i\theta} z', e^{-i\theta} z'').$$

Note that $v(0)$ is multiplication by 1, $v(\pi)$ is multiplication by -1, and $v(\theta)^{-1} = v(-\theta)$. Define

$$\beta\colon U(\mathbb{C}^\infty \oplus \mathbb{C}^\infty) \longrightarrow \Omega SU(\mathbb{C}^\infty \oplus \mathbb{C}^\infty)$$

by letting

$$\beta(T)(\theta) = [T, v(\theta)] = Tv(\theta)T^{-1}v(-\theta),$$

where $T \in U(\mathbb{V}^2)$. Clearly $[T, v(\theta)]$ has determinant one and $\beta(T)$ is a loop at the identity element e of the group $SU(\mathbb{V}^2)$. Moreover, since $v(\theta)$ is just a scalar multiplication on each summand $\mathbb{V}$, if $T = T' \times T'' \in U(\mathbb{V}) \times U(\mathbb{V})$, then $\beta(T)(\theta) = e$. Therefore β passes to orbits to give a well-defined map

$$\beta \colon BU = U/U \times U \longrightarrow \Omega SU.$$

To define the H-space structure on BU, choose a linear isometric isomorphism $\xi \colon V^2 \longrightarrow V$ and let the product $T_1 T_2$ be the composite

$$V^2 \xrightarrow{(\xi^{-1})^2} V^4 \xrightarrow{T_1 \oplus T_2} V^4 \xrightarrow{\mathrm{id}\, \oplus \gamma \oplus \mathrm{id}} V^4 \xrightarrow{\xi^2} V^2,$$

where $\gamma \colon V^2 \longrightarrow V^2$ interchanges the two summands. Up to homotopy, the product is independent of the choice of ξ. The H-space structure we use on ΩSU is the pointwise product, $(\omega_1 \omega_2)(\theta) = \omega_1(\theta)\omega_2(\theta)$. We leave it as an exercise to verify that β is an H-map.[3]

Let $\{e'_i\}$ and $\{e''_i\}$ denote the standard bases of two copies of V and let $\mathbb{C}_1^n$ and $\mathbb{C}_2^n$, respectively, be spanned by the first n vectors in each of these bases. Let

$$j \colon U(\mathbb{C}_1^n \oplus \mathbb{C}_2^1) \longrightarrow U(\mathbb{C}_1^n \oplus \mathbb{C}_2^n)$$

be the inclusion. Restrictions of β give a commutative diagram

$$
\begin{array}{ccc}
\mathbb{C}P^n = U(\mathbb{C}_1^n \oplus \mathbb{C}_2^1)/U(\mathbb{C}_1^n) \times U(\mathbb{C}_2^1) & \xrightarrow{\ \alpha\ } & \Omega SU(\mathbb{C}_1^n \oplus \mathbb{C}_2^1) = \Omega SU(n+1) \\
\downarrow{\scriptstyle j} & & \downarrow{\scriptstyle \Omega j} \\
U(2n)/U(n) \times U(n) = U(\mathbb{C}_1^n \oplus \mathbb{C}_2^n)/U(\mathbb{C}_1^n) \times U(\mathbb{C}_2^n) & \xrightarrow{\ \beta\ } & \Omega SU(\mathbb{C}_1^n \oplus \mathbb{C}_2^n) = \Omega SU(2n).
\end{array}
$$

Passing to colimits over n, we obtain the commutative diagram

$$
\begin{array}{ccc}
\mathbb{C}P^\infty & \xrightarrow{\ \alpha\ } & \Omega SU \\
\downarrow{\scriptstyle j} & {\scriptstyle \simeq}\ \big\downarrow{\scriptstyle \Omega j} \\
BU & \xrightarrow{\ \beta\ } & \Omega SU.
\end{array}
$$

The right arrow is an equivalence, as we see from a quick check of homology or homotopy groups.

We claim that $H_*(\Omega SU)$ is a polynomial algebra on generators δ_i of degree $2i$, $i \geq 1$, and that $\alpha_* \colon H_*(\mathbb{C}P^\infty) \longrightarrow H_*(\Omega SU)$ is a monomorphism onto the free abelian group spanned by suitably chosen polynomial generators δ_i.

---

3. This is also part of the 1970s infinite loop space story; details generalizing these H-space structures and maps to the context of actions by an E_∞ operad may be found in [89, pp. 9–17].

The algebra in §21.4 implies the topological statement that $j_*: H_*(\mathbb{C}P^\infty) \longrightarrow H_*(BU)$ is a monomorphism onto the free abelian group generated by a set $\{\gamma_i\}$ of polynomial generators for $H_*(BU)$, hence the claim will complete the proof of Theorem 21.6.1.

Think of S^1 as the quotient of $[0, \pi]$ obtained by setting $0 = \pi$. Let

$$i: U(\mathbb{C}_1^{n-1} \oplus \mathbb{C}_2^1) \longrightarrow U(\mathbb{C}_1^n \oplus \mathbb{C}_2^1)$$

be the inclusion. It induces a map $i: \mathbb{C}P^{n-1} \longrightarrow \mathbb{C}P^n$ that leads to the left diagram below, and the right diagram is its adjoint.

21.6.2

$$
\begin{array}{ccc}
\mathbb{C}P^{n-1} & \xrightarrow{\alpha} & \Omega SU(n) \\
\downarrow{\scriptstyle i} & & \downarrow{\scriptstyle \Omega i} \\
\mathbb{C}P^n & \xrightarrow{\alpha} & \Omega SU(n+1) \\
\downarrow{\scriptstyle \rho} & & \downarrow{\scriptstyle \Omega\pi} \\
S^{2n} & \xrightarrow{h} & \Omega S^{2n+1}
\end{array}
\qquad
\begin{array}{ccc}
\Sigma\mathbb{C}P^{n-1} & \xrightarrow{\hat{\alpha}} & SU(n) \\
\downarrow{\scriptstyle \Sigma i} & & \downarrow{\scriptstyle i} \\
\Sigma\mathbb{C}P^n & \xrightarrow{\hat{\alpha}} & SU(n+1) \\
\downarrow{\scriptstyle \Sigma\rho} & & \downarrow{\scriptstyle \pi} \\
\Sigma S^{2n} & \xrightarrow{\hat{h}} & S^{2n+1}
\end{array}
$$

Here $\rho: \mathbb{C}P^n \longrightarrow \mathbb{C}P^n/\mathbb{C}P^{n-1} \cong S^{2n}$ is the quotient map and $\pi(T) = T(e'_n)$.

LEMMA 21.6.3. *The composite $\Omega\pi \circ \alpha \circ i$ is trivial, so that $\Omega\pi \circ \alpha$ factors as the composite $h\rho$ for a map h. Moreover, the adjoint $\hat{h}$ of h is a homeomorphism.*

PROOF. Let $T \in U(\mathbb{C}_1^n \oplus \mathbb{C}_2^1)$ represent $\bar{T} \in \mathbb{C}P^n$ and let T_1^{-1} and T_2^{-1} denote the projections of T^{-1} on $\mathbb{C}_1^n$ and on $\mathbb{C}_2^1$. We have

$$
\begin{aligned}
(\Omega\pi)\alpha(T)(\theta) &= Tv(\theta)T^{-1}v(-\theta)(e'_n) \\
&= Tv(\theta)T^{-1}(e^{-i\theta}e'_n) \\
&= T(T_1^{-1}(e'_n), e^{-2i\theta}T_2^{-1}(e'_n)) \\
&= e'_n + (e^{-2i\theta} - 1)TT_2^{-1}(e'_n),
\end{aligned}
$$

as we see by adding and subtracting $TT_2^{-1}(e'_n)$. If $T(e'_n) = e'_n$, so that T is in the image of $U(\mathbb{C}_1^{n-1} \oplus \mathbb{C}_2^1)$ and $\bar{T}$ is in the image of $\mathbb{C}P^{n-1}$, then $T_2^{-1}(e'_n) = 0$ and thus $(\Omega\pi)\alpha(T)(\theta) = e'_n$ for all θ. To prove that $\hat{h}$ is a homeomorphism, it suffices to check that it is injective. Its image will then be open by invariance of domain and closed by the compactness of ΣS^{2n}, hence will be all of S^{2n+1} since S^{2n+1} is connected. Denote points of ΣX as $[x, \theta]$ for $x \in X$ and $\theta \in S^1$. We have

$$\hat{h}(\Sigma\rho)[\bar{T}, \theta] = \pi\hat{\alpha}[\bar{T}, \theta] = (\Omega\pi)\alpha(T)(\theta) = e_n' + (e^{-2i\theta} - 1)TT_2^{-1}(e_n').$$

Since T^{-1} is the conjugate transpose of T, $T_2^{-1}(e_n') = \bar{c}e_1''$, where c is the coefficient of e_n' in $T(e_1'')$. Here $T \notin \mathbb{C}P^{n-1}$ if and only if $c \neq 0$, and then $TT_2^{-1}(e_n') = e_n' + T'(e_n')$, where T' denotes the projection of T on $\mathbb{C}^{n-1} \oplus \mathbb{C}_2^1$. Therefore

$$\hat{h}[\rho(\bar{T}), \theta] = e^{-2i\theta}e_n' + T'(e_n')$$

when $\bar{T} \notin \mathbb{C}P^{n-1}$. The injectivity is clear from this. $\qquad\square$

Armed with this elementary geometry, we return to homology. The rightmost column in the second diagram of (21.6.2) is a fibration, and we use it to compute $H_*(\Omega SU(n+1))$ by induction on n. We have $SU(2) \cong S^3$, and we claim inductively that the cohomology Serre spectral sequence of this fibration satisfies $E_2 = E_\infty$. This leads to a quick proof that

$$H_*(SU(n+1)) = E\{\gamma_{2i+1} | 1 \leq i \leq n\}$$

as a Hopf algebra, where γ_{2i+1} has degree $2i+1$ and $\pi_*(\gamma_{2n+1})$ is a generator of $H_{2n+1}(S^{2n+1})$. Indeed, assume that we know this for $SU(n)$. Then, since the cohomology spectral sequence is multiplicative and the exterior algebra generators of $H^*(SU(n)) = E_2^{0,*}$ have degrees less than $2n$, they must be permanent cycles. Therefore $E_2 = E_\infty$. This implies that $H^*(SU(n+1))$ is an exterior algebra. Moreover, by the edge homomorphisms, i^* is an isomorphism in degrees less than $2n+1$ and the last exterior algebra generator is $\pi^*(i_{2n+1})$. Inductively, the exterior generators in degrees less than $2n$ are primitive. Since i is a map of topological groups, i^* is a map of Hopf algebras. Since $i^*\pi^* = 0$, inspection of the coproduct shows that the generator in degree $2n+1$ must also be primitive.

Using the Serre spectral sequence of the path space fibration over $SU(n+1)$, we conclude that

$$H_*(\Omega SU(n+1)) \cong P\{\delta_i | 1 \leq i \leq n\},$$

where δ_i has degree $2i$. The classical way to see this is to construct a test multiplicative spectral sequence with

$$E_{*,*}^2 = P\{\delta_i | 1 \leq i \leq n\} \otimes E\{\gamma_{2i+1} | 1 \leq i \leq n\}$$

and with differentials specified by requiring y_{2i+1} to transgress to δ_i. This ensures that E_∞ is zero except for $\mathbb{Z} = E_{0,0}^\infty$. We can map the test spectral sequence to the homology Serre spectral sequence of the path space fibration by a map that is the identity on $E_{0,*}^2$ and commutes with the transgression. The conclusion follows by the comparison theorem, Theorem 24.6.1. The argument shows that the polynomial generators transgress to the exterior algebra generators and thus that the exterior algebra generators suspend to the polynomial algebra generators. At the risk of belaboring the obvious, we spell things out explicitly via the following commutative diagram, in which the unlabeled isomorphisms are suspension isomorphisms.

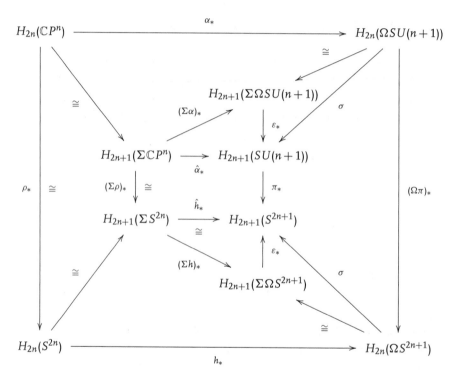

Here ε denotes the evaluation map of the (Σ, Ω) adjunction, and the suspension σ is defined to be the composite of ε_* and the suspension isomorphism. The algebra generator δ_n maps to a fundamental class under $\pi_*\sigma$. By the diagram, so does the basis element $x_{2n} \in H_{2n}(\mathbb{C}P^n)$. Therefore, modulo decomposable elements that are annihilated by σ, $\alpha_*(x_{2i}) = \delta_i$ as claimed.

22

LIE ALGEBRAS AND HOPF ALGEBRAS
IN CHARACTERISTIC ZERO

All of the structure theorems for Hopf algebras in common use in algebraic topology are best derived by filtration techniques from the Poincaré-Birkhoff-Witt (PBW) theorem for graded Lie algebras and restricted Lie algebras. In this chapter, we first introduce Lie algebras and prove the PBW theorem for their universal enveloping algebras. We next show that primitive (= primitively generated) Hopf algebras in characteristic zero are the universal enveloping algebras of their Lie algebras of primitive elements. We then use this fact to study the algebra structure of commutative Hopf algebras in characteristic zero.

While some of these results first appeared in Milnor and Moore [104], the most basic structure theorems go back to earlier work of Hopf, Leray, and Borel.

22.1. Graded Lie algebras

We continue to work over a fixed commutative ring R. The following witty definition is due to John Moore and is used in [104].

DEFINITION 22.1.1. A (graded) Lie algebra over R is a (graded) R-module L together with a morphism of R-modules $L \otimes L \to L$, denoted $[-, -]$ and called the bracket operation, such that there exists an associative R-algebra A and a monomorphism of R-modules $j : L \to A$ such that $j([x, y]) = [jx, jy]$ for $x, y \in L$, where the bracket operation in A is the (graded) commutator,

$$[a, b] = ab - (-1)^{\deg a \deg b} ba.$$

A morphism of Lie algebras is a morphism of R-modules that commutes with the bracket operation.

The following identities are immediate consequences of the definition. It would be more usual to take them as the defining properties of the bracket operation, but we shall see that for particular ground rings R the definition can imply more relations than are listed.

LEMMA 22.1.2. *Let L be a Lie algebra and let $x \in L_p$, $y \in L_q$, and $z \in L_r$. Then the following identities hold.*

(i) $[x, y] = -(-1)^{pq}[y, x]$;

(ii) $[x, x] = 0$ *if either* char $R = 2$ *or p is even*;

(iii) $(-1)^{pr}[x, [y, z]] + (-1)^{pq}[y, [z, x]] + (-1)^{rq}[z, [x, y]] = 0$;

(iv) $[x, [x, x]] = 0$ *if p is odd*.

Formula (iii) is called the Jacobi identity. When p is even, (i) implies $2[x, x] = 0$; when p is odd, (iii) implies $3[x, [x, x]] = 0$. We shall see that, at least if R is a field, any R-module with a bracket operation satisfying these identities can be embedded in a bracket-preserving way in an associative algebra and is therefore a Lie algebra. This is not true for a general R. For instance, $[x, 2x] = 0$ if char $R = 4$ is an identity not implied by those of the lemma (when deg (x) is odd). Of course, for any R, any associative alegbra is a Lie algebra under the commutator operation.

DEFINITION 22.1.3. The universal enveloping algebra of a Lie algebra L is an associative algebra $U(L)$ together with a morphism of Lie algebras $i : L \to U(L)$ such that, for any morphism of Lie algebras $f : L \to A$, where A is an associative algebra, there exists a unique morphism of algebras $\tilde{f} : U(L) \to A$ such that $\tilde{f}i = f$.

Clearly $U(L)$ is unique up to canonical isomorphism, if it exists.

PROPOSITION 22.1.4. *Any Lie algebra L has a universal enveloping algebra $U(L)$, and $i : L \to U(L)$ is a monomorphism whose image generates $U(L)$ as an algebra. Moreover, $U(L)$ is a primitive Hopf algebra.*

PROOF. Let $T(L)$ be the tensor algebra, or free associative algebra, generated by L. Explicitly, $T(L) = \sum_{n \geq 0} T_n(L)$, where $T_0(L) = R$ and $T_n(L) = L \otimes \cdots \otimes L$, n factors of L, if $n > 0$. The product in $T(L)$ is obtained by passage to direct sums from the evident isomorphisms

$T_m(L) \otimes T_n(L) \to T_{m+n}(L)$. Define $i : L \to T(L)$ to be the identification of L with $T_1(L)$. For an associative algebra A, a map of R-modules $f : L \to A$ extends uniquely to a map of algebras $\tilde{f} : T(L) \to A$. Let I be the two-sided ideal in $T(L)$ generated by the elements

$$xy - (-1)^{\deg x \deg y} yx - [x, y], \ x, y \in L,$$

define $U(L) = T(L)/I$, and let $i : L \to U(L)$ be the evident composite. Clearly i has the required universal property. Of course, the injectivity of i is built into our definition of a Lie algebra, and $i(L)$ generates $U(L)$ since $i(L)$ generates $T(L)$. By the universal property, a morphism $f : L \to L'$ of Lie algebras induces a unique morphism $U(f)$ of associative algebras such that the following diagram commutes.

$$
\begin{array}{ccc}
L & \xrightarrow{\ f\ } & L' \\
{\scriptstyle i}\big\downarrow & & \big\downarrow {\scriptstyle i'} \\
U(L) & \xrightarrow[\ U(f)\ ]{} & U(L')
\end{array}
$$

If we take $L' = \{0\}$, then $U(L') = R$ and we obtain an augmentation of $U(L)$. The product $L \times L'$ of Lie algebras inherits a structure of Lie algebra, and the algebra $U(L) \otimes U(L')$ together with the evident morphism of Lie algebras $i : L \times L' \to U(L) \otimes U(L')$,

$$i(x, x') = x \otimes 1 + 1 \otimes x'$$

is easily checked to satisfy the universal property that defines $U(L \times L')$. The diagonal $\triangle : L \to L \times L$ is a map of Lie algebras and therefore induces a morphism of algebras $\psi : U(L) \to U(L) \otimes U(L)$ such that the following diagram commutes.

$$
\begin{array}{ccc}
L & \xrightarrow{\ \triangle\ } & L \times L \\
{\scriptstyle i}\big\downarrow & & \big\downarrow {\scriptstyle i} \\
U(L) & \xrightarrow[\ \psi\ ]{} & U(L) \otimes U(L)
\end{array}
$$

Thus $U(L)$ is a bialgebra, and $i(L) \subset PU(L)$ by the diagram, so that $U(L)$ is primitive. For the antipode, we have the opposite Lie algebra L^{op} with bracket

$[-,-]\gamma$, and $x \longrightarrow -x$ defines a map of Lie algebras $L \longrightarrow L^{op}$. We can identify $U(L^{op})$ as $U(L)^{op}$, and then the universal property gives a map of algebras $\chi : U(L) \longrightarrow U(L)^{op}$, that is, an involution on $U(L)$ itself. Writing the obvious equalities $[x, -x] = 0 = [-x, x]$ as diagrams and passing to the corresponding diagrams induced on the level of universal enveloping algebras, we see that χ is an antipode on $U(L)$. □

22.2. The Poincaré-Birkhoff-Witt theorem

The Poincaré-Birkhoff-Witt theorem gives a complete description of the associated graded Hopf algebra of $U(L)$ with respect to a suitable filtration (under appropriate hypotheses) and therefore gives a complete description of the additive structure of $U(L)$. We require a definition.

DEFINITION 22.2.1. Let L be a Lie algebra. The Lie filtration of $U(L)$ is specified by $F_p U(L) = 0$ if $p < 0$, $F_0 U(L) = R$, and $F_p U(L) = (R \oplus L)^p$ if $p \geq 1$. Clearly $U(L) = \cup_p F_p U(L)$, so the filtration is complete.

Provided that the Lie filtration is split or flat, so that $E^0(U(L) \otimes U(L))$ is isomorphic to $E^0 U(L) \otimes E^0 U(L)$, $E^0 U(L)$ inherits a structure of primitive Hopf algebra from $U(L)$. Since the commutator in $U(L)$ of elements in L agrees with the bracket operation in L and since L generates $U(L)$, we see immediately that $E^0 U(L)$ is commutative. Clearly we have

$$QE^0 U(L) = E^0_{1,*} U(L) = L, \quad \text{where} \quad E^0_{1,q} U(L) = L_{q+1}.$$

Let $L^\sharp$ denote the underlying R-module of L regarded as an abelian Lie algebra and write $A(L) = U(L^\sharp)$. Then $A(L)$ is the free commutative algebra generated by L. Explicitly, $A(L) = T(L)/J$ where J is the commutator ideal.

For a filtered R-module A, write $E^\oplus A$ for the graded R-module that is obtained by regrading the associated bigraded R-module $E^0 A$ by total degree:

$$E^\oplus_n A = \sum_{p+q=n} E^0_{p,q} A.$$

If $E^0 A$ is a bigraded algebra, Hopf algebra, and so on, then $E^\oplus A$ is a graded algebra, Hopf algebra, and so on.

By the universal property of $A(L)$, the evident inclusion of L in $E^\oplus U(L)$ induces a natural map of commutative algebras $f : A(L) \to E^\oplus U(L)$.

NOTATION 22.2.2. If char $R = 2$, let $L^+ = L$ and $L^- = \{0\}$. If char $R \neq 2$, let L^+ and L^- be the R-submodules of L concentrated in even and in odd degrees.

The hypotheses on the characteristic of R in the next result ensure that the identities of Lemma 22.1.2 suffice to characterize our Lie algebras, as we shall see.

THEOREM 22.2.3 (POINCARÉ-BIRKHOFF-WITT). *Let L be an R-free Lie algebra. Assume that* char $R = 2$, *or that 2 is invertible in R, or that $L = L^+$ so that L is concentrated in even degrees. Then $f : A(L) \to E^\oplus U(L)$ is an isomorphism of Hopf algebras.*

PROOF. It will fall out of the proof that the Lie filtration of $U(L)$ is split, so that $E^\oplus U(L)$ is a primitively generated Hopf algebra, and f will preserve coproducts since it is the identity on the R-module L of primitive generators. Of course, f is an epimorphism since L generates $E^\oplus U(L)$. Filter $E^\oplus U(L)$ by filtration degree,

$$F_p E^\oplus U(L) = \sum_{i \leq p} E^0_{i,*} U(L).$$

Obviously $E^0 E^\oplus U(L) = E^0 U(L)$. Give $A(L)$ its Lie filtration. Clearly $E^0 A(L)$ is the free commutative bigraded algebra generated by L regarded as a bigraded R-module via $L_{1,q} = L_{q+1}$. The map f is filtration preserving, and it suffices to prove that $E^0 f$ is a monomorphism.

Give $T(L)$ the evident filtration, $F_p T(L) = \sum_{n \leq p} T_n(L)$, and observe that the quotient maps $\pi : T(L) \to A(L)$ and $\rho : T(L) \to U(L)$ are both filtration-preserving. Let $I = \ker \rho$. We shall construct a filtration-preserving morphism of R-modules $\sigma : T(L) \to A(L)$ such that $\sigma(I) = 0$ and $E^0 \sigma = E^0 \pi$. It will follow that σ factors as $\bar{\sigma} \rho$ for a filtration-preserving R-map $\bar{\sigma} : U(L) \to A(L)$. We will have $E^0 \bar{\sigma} E^0 \rho = E^0 \pi$ and, since $E^0 \rho$ and $E^0 \pi$ are epimorphisms of algebras, $E^0 \bar{\sigma}$ will be a morphism of algebras. The composite

$$E^0 A(L) \xrightarrow{E^0 f} E^0 U(L) \xrightarrow{E^0 \bar{\sigma}} E^0 A(L)$$

will be the identity since it will be a morphism of algebras that restricts to the identity on the R-module L of generators. Thus $E^0 f$ will be a monomorphism and the proof will be complete.

To construct σ, let $\{z_k\}$ be an R-basis for L indexed on a totally ordered set. The set of monomials

22.2.4 $\qquad \{z_{k_1} \cdots z_{k_n} | k_1 \leq \cdots \leq k_n \text{ and } k_i < k_{i+1} \text{ if } z_{k_i} \in L^-\}$

is an R-basis for $A(L)$. Let y_k denote z_k regarded as an element of $T(L)$. Then $y_{k_1} \cdots y_{k_n}$ is a typical basis element of $T_n(L)$. Of course, the sequence $\{k_1, \ldots, k_n\}$ will generally not be ordered as in (22.2.4). When it is so ordered, we require σ to satisfy the formula

22.2.5 $\quad \sigma(y_{k_1} \cdots y_{k_n}) = z_{k_1} \cdots z_{k_n}$ if $k_1 \leq \cdots \leq k_n$ and $k_i < k_{i+1}$ if $z_{k_i} \in L^-$.

For a general sequence $\{k_1, \ldots, k_n\}$ and $1 \leq i < n$, we require

22.2.6 $\quad \sigma(y_{k_1} \cdots y_{k_n}) = (-1)^{\deg z_{k_i} \deg z_{k_{i+1}}} \sigma(y_{k_i} \cdots y_{k_{i-1}} y_{k_{i+1}} y_{k_i} y_{k_{i+2}} \cdots y_{k_n})$

$$+ \sigma(y_{k_1} \cdots y_{k_{i-1}} [y_{k_i}, y_{k_{i+1}}] y_{k_{i+2}} \cdots y_{k_n}).$$

Clearly, if there is a well-defined map $\sigma \colon T(L) \longrightarrow A(L)$ of R-modules that satisfies these formulae, the desired relations $I \subset \ker \sigma$ and $E^0 \sigma = E^0 \pi$ will follow.

We define $\sigma \colon T(L) \to A(L)$ by induction on the filtration degree n, with $\sigma(1) = 1$ and $\sigma(y_k) = z_k$ handling filtration degrees 0 and 1. Assume that σ has been defined on $F_{n-1} T(L)$. Define the index q of a sequence $\{k_1, \ldots, k_n\}$ to be the number of transpositions required to put it in nondecreasing order. We define σ by induction on n and, for fixed n, by induction on the index q. We have defined σ for $n \leq 1$, so we assume that $n > 1$. Define σ by (22.2.5) for sequences of index 0 unless some $k_i = k_{i+1}$ with $z_{k_i} \in L^-$, in which case define σ by the formula

22.2.7 $\qquad \sigma(y_{k_1} \cdots y_{k_n}) = \frac{1}{2}\sigma(y_{k_1} \cdots y_{k_{i-1}} [y_{k_i}, y_{k_i}] y_{k_{i+2}} \cdots y_{k_n}).$

Observe that (22.2.7) is consistent with and in fact forced by (22.2.6). Assuming that σ has been defined on sequences of index less than q and that $\{k_1, \ldots, k_n\}$ has index q, we define $\sigma(y_{k_1} \cdots y_{k_n})$ by (22.2.6) if $k_i > k_{i+1}$ and by (22.2.7) if $k_i = k_{i+1}$ and $z_{k_i} \in L^-$. To complete the proof, we must show that σ is actually well-defined, that is, that our definition of σ by (22.2.6) and (22.2.7) is independent of the choice of i. The argument is tedious, but elementary, and we shall not give full details. There are four cases to be checked, each with two subcases.

Case 1. $k_i = k_{i+1}$ and $k_j = k_{j+1}, j \geq i+1$, with $z_{k_i} \in L^-$ and $z_{k_j} \in L^-$.

Subcase $j > i+1$. We must show that

$$\sigma(\gamma_{k_1} \cdots [\gamma_{k_i}, \gamma_{k_i}] \cdots \gamma_{k_n}) = \sigma(\gamma_{k_1} \cdots [\gamma_{k_j}, \gamma_{k_j}] \cdots \gamma_{k_n}).$$

Here (22.2.6) and induction on n show that both sides are equal to

$$\frac{1}{2}\sigma(\gamma_1 \cdots [\gamma_{k_i}, \gamma_{k_i}] \cdots [\gamma_{k_j}, \gamma_{k_j}] \cdots \gamma_{k_n}).$$

Subcase $j = i+1$. Here (22.2.6), induction on n, and the identity $[x, [x, x]] = 0$ imply the equality

$$\sigma(\gamma_{k_1} \cdots \gamma_{k_i}[\gamma_{k_i}, \gamma_{k_i}] \cdots \gamma_{k_n}) = \sigma(\gamma_{k_1} \cdots [\gamma_{k_i}, \gamma_{k_i}]\gamma_{k_i} \cdots \gamma_{k_n}).$$

Case 2. $k_i = k_{i+1}$ with $z_{k_i} \in L^-$ and $k_j > k_{j+1}, j \geq i+1$.

Subcase $j > i+1$. The argument here is similar to (but has more terms to check than) the argument in the subcase $j > i+1$ of Case 1, the induction hypotheses on both n and q being required.

Subcase $j = i+1$. Let $u = \gamma_{k_i} = \gamma_{k+1}$ and $v = \gamma_{k_{i+2}}$ and let $\deg v = p$; of course, $\deg u$ is odd. We must show that

$$\frac{1}{2}\sigma(\gamma_{k_1} \cdots [u, u]v \cdots \gamma_{k_n}) = (-1)^p \sigma(\gamma_{k_1} \cdots uvu \cdots \gamma_{k_n})$$

$$+ \sigma(\gamma_{k_1} \cdots u[u, v] \cdots \gamma_{k_n}).$$

By (22.2.6) and induction on q and n, we have

$$\sigma(\gamma_{k_1} \cdots uvu \cdots \gamma_{k_n}) = (-1)^p \sigma(\gamma_{k_1} \cdots vuu \cdots \gamma_{k_n}) + \sigma(\gamma_{k_1} \cdots [u, v]u \cdots \gamma_{k_n})$$

$$= \tfrac{1}{2}(-1)^p \sigma(\gamma_{k_1} \cdots v[u, u] \cdots \gamma_{k_n})$$

$$+ \sigma(\gamma_{k_1} \cdots [u, v]u \cdots \gamma_{k_n})$$

$$= \tfrac{1}{2}(-1)^p (\sigma(\gamma_{k_1} \cdots [u, u]v \cdots \gamma_{k_n})$$

$$+ \sigma(\gamma_{k_1} \cdots [v, [u, u]] \cdots \gamma_{k_n}))$$

$$- (-1)^p \sigma(\gamma_{k_1} \cdots u[u, v] \cdots \gamma_{k_n})$$

$$+ \sigma(\gamma_{k_1} \cdots [[u, v], u] \cdots \gamma_{k_n}).$$

The Jacobi and anticommutativity formulas imply

$$\frac{1}{2}[v, [u, u]] + (-1)^p[[u, v], u] = \frac{1}{2}[v, [u, u]] + [u, [u, v]]$$

$$= \frac{1}{2}([v, [u, u]] - (-1)^p[u, [v, u]]$$

$$+ [u, [u, v]]) = 0.$$

Comparing formulas, we obtain the desired equality.

Case 3. $k_i > k_{i+1}$ and $k_j = k_{j+1}$ with $z_k \in L^-, j \geq i + 1$.
The proof in this case is symmetric to that in Case 2.

Case 4. $k_i > k_{i+1}$ and $k_j > k_{j+1}$ with $j \geq i + 1$.
The proof when $j > i + 1$ is straightforward by induction, as in the subcase $j > i + 1$ of Case 1, and the proof when $j = i + 1$ is a calculation similar to that in the subcase $j = i + 1$ of Case 2. □

We retain the hypotheses of the theorem in the following corollary.

COROLLARY 22.2.8. *Let $\{x_i\}$ and $\{y_j\}$ be R-bases for L^- and L^+ indexed on totally ordered sets. Then $U(L)$ is the free R-module on the basis*

$$\{x_{i_1} \cdots x_{i_m} y_{j_1}^{r_1} \cdots y_{j_n}^{r_n} | i_1 < \cdots < i_m, j_1 < \cdots < j_n \text{ and } r_k \geq 1\}.$$

PROOF. Since $E^\oplus U(L)$ is a free R-module, it is isomorphic as an R-module to $U(L)$. The conclusion follows from the evident analogue for $U(L^\sharp) = A(L)$. □

COROLLARY 22.2.9. *Let L be a free R-module together with a bracket operation satisfying the identities listed in Lemma 22.1.2. Assume that char $R = 2$, or 2 is invertible in R, or $L = L^+$. Then L is a Lie algebra.*

PROOF. Construct $U(L)$ as in the proof of Proposition 22.1.4 and give it the Lie filtration of Definition 22.2.1. The proof of Theorem 22.2.3 only used the cited identities and so gives that $A(L) \cong E^\oplus U(L)$. Thus $L \to U(L)$ is a bracket-preserving monomorphism of R-modules. □

22.3. Primitively generated Hopf algebras in characteristic zero

Throughout this section and the next, R is assumed to be a field of characteristic zero. However, all of the results remain valid if R is any ring of characteristic zero in which 2 is invertible and all R-modules in sight are R-free.

A quick calculation shows that the R-module PA of primitive elements of a Hopf algebra A is a Lie subalgebra. The universal property of $U(PA)$ thus gives a natural map of Hopf algebras $g : U(PA) \to A$, and g is clearly an epimorphism if A is primitive. Let $\mathscr{L}$ and $\mathscr{PH}$ denote the categories of Lie algebras and of primitive Hopf algebras over R. We have functors $U : \mathscr{L} \to \mathscr{PH}$ and $P : \mathscr{PH} \to \mathscr{L}$, a natural inclusion $L \subset PU(L)$, and a natural epimorphism $g : U(PA) \to A$, where $L \in \mathscr{L}$ and $A \in \mathscr{PH}$. This much would be true over any commutative ring R, but when R is a field of characteristic zero we have the following result.

THEOREM 22.3.1. *The functors $U : \mathscr{L} \to \mathscr{PH}$ and $P : \mathscr{PH} \to \mathscr{L}$ are inverse equivalences of categories. More explicitly,*

(i) $PU(L) = L$ for any Lie algebra L and

(ii) $g : U(PA) \to A$ is an isomorphism for any primitive Hopf algebra A.

PROOF. We first prove (i). Consider the Lie filtration of $U(L)$. Let $x \in F_p U(L)$, $x \notin F_{p-1} U(L)$, and suppose that $x \in PU(L)$. It suffices to prove that $p = 1$. The image of x in $E^0_{p,*} U(L)$ is primitive. By the PBW-theorem, $E^{\oplus} U(L) \cong A(L)$ as a Hopf algebra. Consider the basis for $A(L)$ given in Corollary 22.2.8. The generators x_i and y_j there are primitive. Using the notation (i, j) for the evident binomial coefficient considered as an element of R, we see that

$$\psi(\gamma^n) = \sum_{i+j=n} (i,j)\gamma^i \otimes \gamma^j \text{ if } \gamma \in L^+.$$

Since char $R = 0$, we check from this that no decomposable basis element is primitive and that no two basis elements have any summands of their coproducts in common, so that no linear combination of decomposable elements is primitive. This implies that $p = 1$ and proves (i).

To prove (ii), define the primitive filtration of a Hopf algebra A by

$$F_p A = 0 \text{ if } p < 0, \quad F_0 A = A, \text{ and } F_p A = (R \oplus PA)^p \text{ if } p > 0.$$

This filtration is complete, $A = \cup_p F_p A$, if and only if A is primitive. By (i), the Lie and primitive filtrations coincide on $U(L)$. For $A \in \mathscr{PH}$ the epimorphism $g : U(PA) \to A$ is filtration-preserving and, since $PU(PA) = PA$, g is an isomorphism on the R-modules of primitive elements. Therefore $E^0 g$ is a monomorphism on the primitive elements of $E^0 U(PA)$, that is, on $E^0_{1,*} U(PA)$. Since $E^0 U(PA)$ is connected with respect to its filtration degree, $E^0 g$ is a monomorphism by Proposition 21.1.4 and g is a monomorphism by Proposition 20.1.1. □

We emphasize that A itself is not assumed to be connected here.

COROLLARY 22.3.2. *If A is a commutative primitive Hopf algebra, then A is isomorphic as a Hopf algebra to the free commutative algebra generated by PA.*

PROOF. $A \cong U(PA) = A(PA)$ since PA is an abelian Lie algebra. $\qquad \square$

Among other things, our next corollary shows that a connected Hopf algebra is primitive if and only if it is cocommutative.

COROLLARY 22.3.3. *Let A be a connected quasi-Hopf algebra.*

(i) $v : PA \to QA$ *is a monomorphism if and only if A is associative and commutative.*

(ii) $v : PA \to QA$ *is an epimorphism if and only if A is coassociative and cocommutative.*

(iii) $v : PA \to QA$ *is an isomorphism if and only if A is a commutative and cocommutative Hopf algebra.*

PROOF. By Lemma 21.1.1, if v is a monomorphism, then A is associative and commutative. Conversely, suppose that A is associative and commutative. Give A its product filtration (see Definition 20.2.9). Then $E^0 A$ is a commutative primitive Hopf algebra, hence $E^0 A \cong A(PE^0 A)$ by the previous corollary. It follows that $PE^0 A = E^0_{-1,*} A$. If $x \in PA$, $x \in F_p A$, and $x \notin F_{p-1} A$, then the image of x in $E^0_{p,*} A$ is primitive and we must have $p = -1$. This implies that v is a monomorphism. When A is of finite type, (ii) follows from (i) by dualization since $A \cong A^{**}$ and (i) holds for A^*. The general case of (i) follows by passage to colimits, using Lemma 21.1.2, since the functors P and Q commute with (directed) colimits. Part (iii) follows from (i) and (ii). $\qquad \square$

COROLLARY 22.3.4. *A Hopf subalgebra of a primitive Hopf algebra is primitive.*

PROOF. Let $A \subset B$, where B is primitive. Since B is cocommutative, so is A. If A is connected, the conclusion follows from (ii) of the previous corollary. For the general case, let $A' = U(PA)$ and let $g : A' \to A$ be the natural map. Give A' and B their primitive filtrations and filter A by $F_p A = A \cap F_p B$. These filtrations are all complete, and g and the inclusion $A \to B$ are filtration-preserving. Clearly $F_1 A' = R \oplus PA = F_1 A$. The induced map $E^0 A \to E^0 B$ is again a monomorphism. Since $E^0 A$ is connected (with respect to its filtration degree) and cocommutative, it is primitively generated. Since $PE^0 B = E^0_{1,*} B$, we find

$$PE^0 A' = E^0_{1,*} A' \cong E^0_{1,*} A = PE^0 A.$$

Thus $E^0 g$ is an isomorphism on primitives and therefore also an epimorphism on indecomposables. By Propositions 21.1.3 and 21.1.4, this implies that $E^0 g$ is an isomorphism and thus g is an isomorphism. □

COROLLARY 22.3.5. *A Hopf subalgebra A of a primitive Hopf algebra B is a normal subalgebra if and only if PA is a Lie ideal of PB. When this holds, $B//A = U(PB/PA)$ and*

$$0 \longrightarrow PA \longrightarrow PB \longrightarrow P(B//A) \longrightarrow 0$$

is an exact sequence of Lie algebras.

PROOF. Assume PA is a Lie ideal in PB. If $x \in PA$ and $y \in PB$, then $[x, y] \in PA$ and, since A is primitive, the equation $xy = [x, y] + (-1)^{\deg x \deg y} yx$ therefore implies that $IA \cdot B = B \cdot IA$. Conversely, assume that A is a normal subalgebra of B and let $C = B//A$. The exact sequence

$$0 \to PA \to PB \to PC$$

implies that PA is a Lie ideal of PB. It is easily checked that C and the inclusion $PB/PA \to C$ satisfy the universal property required of $U(PB/PA)$, and the remaining conclusions follow. □

22.4. Commutative Hopf algebras in characteristic zero

Again, let R be a field of characteristic zero. We prove the classical structure theorems for commutative Hopf algebras in characteristic zero. As before, $A(X)$ denotes the free commutative algebra generated by an R-module X. If we write $E(X)$ for the exterior algebra generated by an R-module X concentrated in odd degrees and $P(X)$ for the polynomial algebra generated by an R-module X concentrated in even degrees, then, for a general R-module X,

$$A(X) = E(X^-) \otimes P(X^+),$$

where X^- and X^+ denote the submodules of X concentrated in odd and even degrees, respectively.

THEOREM 22.4.1 (LERAY). *Let A be a connected, commutative, and associative quasi-Hopf algebra. Let $\sigma : QA \to IA$ be a morphism of R-modules such that $\pi\sigma = \mathrm{id}$, where $\pi : IA \to QA$ is the quotient map. Then the morphism of algebras $f : A(QA) \to A$ induced by σ is an isomorphism.*

PROOF. Give $A(QA)$ and A their product filtrations. These filtrations are complete since A is connected, and f is filtration-preserving. Since $E^0 A$ is a commutative primitive Hopf algebra, $A(PE^0 A) = E^0 A$ by Corollary 22.3.2, and similarly for $A(QA)$. Now $PE^0 A \to QE^0 A$ is just the composite

$$E^0_{-1,*}A(QA) = QA \xrightarrow{\ \sigma\ } IA \xrightarrow{\ \pi\ } QA = E^0_{-1,*}A$$

and is thus the identity. Therefore $E^0 f$ is an isomorphism of Hopf algebras and f is an isomorphism of algebras. □

The following immediate consequence of the previous theorem was the theorem of Hopf that initiated the study of Hopf algebras.

COROLLARY 22.4.2 (HOPF). *Let A be a connected, commutative, and associative quasi-Hopf algebra such that $Q_n A = 0$ if n is even. Then $A \cong E(QA)$ as an algebra. In particular, the conclusion holds if $A_n = 0$ for all sufficiently large n.*

PROOF. For the last statement, note that an even degree indecomposable would give rise to a polynomial subalgebra. □

If the coproduct is coassociative, we can strengthen the conclusion of the preceding corollary.

COROLLARY 22.4.3. *Let A be a connected commutative Hopf algebra such that $Q_n A = 0$ if n is even. Then $A \cong E(PA)$ as a Hopf algebra.*

PROOF. By Corollary 22.3.2, it suffices to prove that $\nu : PA \to QA$ is an epimorphism. By Corollary 22.3.3, ν is a monomorphism and it suffices to prove that A is cocommutative. By Lemma 21.1.2, we may assume that A is of finite type. Then A^* is a primitive Hopf algebra and $P_n A^* = 0$ if n is even. Thus $[x, y] = 0$ if $x, y \in PA^*$ and A^* is commutative. Therefore A is cocommutative. □

We conclude with the following basic result. By Corollary 22.3.3, it is just a restatement of the connected case of Corollary 22.3.2.

THEOREM 22.4.4. *Let A be a connected, commutative, and cocommutative Hopf algebra. Then $A \cong E((PA)^-) \otimes P((PA)^+)$ as a Hopf algebra.*

23

RESTRICTED LIE ALGEBRAS AND HOPF ALGEBRAS IN CHARACTERISTIC p

This chapter is precisely parallel to the previous one. We first introduce restricted Lie algebras and prove the PBW theorem for their universal enveloping algebras. We next show that primitive Hopf algebras in characteristic p are the universal enveloping algebras of their restricted Lie algebras of primitive elements. We then use this fact to study the algebra structure of commutative Hopf algebras in characteristic p.

Most of these results first appeared in Milnor and Moore [104], but with different proofs, and some go back to earlier work of Leray, Samelson, and Borel. §23.4 is a corrected version of results in [84].

23.1. Restricted Lie algebras

In this section and the next, we work over a commutative ring R of prime characteristic p. Of course, either $2 = 0$ or 2 is invertible in R. As before, we let X^+ and X^- denote the R-submodules of even and odd degree elements of an R-module X, with the convention that $X^+ = X$ and $X^- = \{0\}$ if char $R = 2$.

DEFINITION 23.1.1. A restricted Lie algebra over R is a Lie algebra L together with a function $\xi : L^+ \to L^+$ with $\xi(L_n) \subset L_{pn}$, such that there exists an associative algebra A and a monomorphism of Lie algebras $j : L \to A$ such that $j\xi(x) = \xi j(x)$, where $\xi : A^+ \to A^+$ is the pth power operation. A morphism of restricted Lie algebras is a morphism of Lie algebras that commutes with the "restrictions" ξ.

LEMMA 23.1.2. Let L be a restricted Lie algebra. Let $x \in L$, $y \in L_n^+$, $z \in L_n^+$, and $r \in R$. Define $(\mathrm{ad}y)(x) = [x, y]$ and, inductively, $(\mathrm{ad}y)^i(x) = [(\mathrm{ad}y)^{i-1}(x), y]$. Then the following identities hold.

(i) $[x, \xi(y)] = (\text{ad} y)^p(x)$

(ii) $\xi(ry) = r^p \xi(y)$

(iii) $\xi(y + z) = \xi(y) + \xi(z) + \sum_{i=1}^{p-1} s_i(y, z)$, where is$_i(y, z)$ is the coefficient of a^{i-1} in the expression $\text{ad}(ay + z)^{p-1}(y)$; here a is a degree zero indeterminant.

PROOF. Part (ii) is trivial. Consider the polynomial algebra $P[b, c]$ on two indeterminates b and c of the same degree n, where n is even if char $R > 2$. We have the identities

(1) $(b - c)^p = b^p - c^p$ and

(2) $(b - c)^{p-1} = \sum_{i=0}^{p-1} b^i c^{p-1-i}$.

Thus the same identities hold for two commuting elements in any R-algebra. Embed L in an associative algebra A, as in the definition. Left and right multiplication by y are commuting elements in the algebra $\text{Hom}_R(A, A)$, hence (1) implies (i) and (2) implies

(3) $(\text{ad} y)^{p-1}(x) = \sum_{i=0}^{p-1} y^i x y^{p-1-i}$.

To prove (iii), consider the polynomial algebra $A[a]$. Write

(4) $(ay + z)^p = a^p y^p + z^p + \sum_{i=1}^{p-1} s_i(y, z) a^i$.

We must evaluate the coefficients $s_i(y, z)$, which a priori lie in A, as elements of L. Formal differentiation of (4) with respect to a, using $d(a^i) = i a^{i-1}$, gives

(5) $\sum_{i=0}^{p-1} (ay + z)^i y (ay + z)^{p-1-i} = \sum_{i=1}^{p-1} i s_i(y, z) a^{i-1}$.

Replacing x and y by y and $ay + z$, respectively, in (3) and comparing the result to (5), we find that $is_i(y, z)$ admits the description given in (iii). Setting $a = 1$ in (5), we obtain (iii). $\qquad \square$

Observe that (iii) shows that $\xi(y + z) - \xi(y) - \xi(z)$ is in the Lie subalgebra of L generated by y and z.

We shall see that, at least if R is a field, any Lie algebra L with a restriction ξ satisfying these identities can be embedded in a restriction-preserving way as a Lie subalgebra of an associative algebra and is therefore a restricted Lie algebra. Of course, any associative algebra is a restricted Lie algebra under the commutator and p^{th} power operations.

DEFINITION 23.1.3. The universal enveloping algebra of a restricted Lie algebra L is an associative algebra $V(L)$ together with a morphism of restricted Lie algebras $i : L \to V(L)$ such that, for any morphism of restricted Lie algebras

$f : L \to A$, where A is an associative algebra, there exists a unique morphism of algebras $\tilde{f} : V(L) \to A$ such that $\tilde{f} \circ i = f$.

Clearly $V(L)$ is unique up to canonical isomorphism, if it exists.

PROPOSITION 23.1.4. *Any restricted Lie algebra L has a universal enveloping algebra $V(L)$, and $i : L \to V(L)$ is a monomorphism whose image generates $V(L)$ as an algebra. Moreover, $V(L)$ is a primitively generated Hopf algebra.*

PROOF. Let $I \subset U(L)$ be the two-sided ideal generated by all elements of the form $x^p - \xi(x)$, $x \in L^+$. Define $V(L) = U(L)/I$ and let $i : L \to V(L)$ be the composite of $i : L \to U(L)$ and the quotient map $U(L) \to V(L)$. The universal property is easily checked, and it is then clear that i is a monomorphism whose image generates $V(L)$. The proof of the last statement is exactly the same as for $U(L)$, the essential point being that $V(L \times L')$ is isomorphic to $V(L) \otimes V(L')$ for restricted Lie algebras L and L'. $\square$

23.2. The restricted Poincaré-Birkhoff-Witt theorem

We here obtain the Poincaré-Birkhoff-Witt theorem for restricted Lie algebras L. The Lie filtration of $V(L)$ is defined exactly as was the Lie filtration of $U(L)$; see Definition 22.2.1 and the discussion following it. We shall describe the associated graded algebra $E^{\oplus} V(L)$ when L is R-free. In $V(L)$, $x^p = \xi(x)$ for $x \in L^+$. Since $\xi(x)$ has filtration one, $x^p = 0$ in the commutative algebra $E^{\oplus} V(L)$.

Let $L^{\sharp}$ denote the underlying R-module of L regarded as an abelian restricted Lie algebra with restriction zero and write $B(L) = V(L^{\sharp})$. Then $B(L) = A(L)/J$, where J is the ideal generated by $\{x^p | x \in L^+\}$. Clearly the inclusion of L in $E^{\oplus} V(L)$ induces a natural map of algebras $f : B(L) \to E^{\oplus} V(L)$.

THEOREM 23.2.1 (POINCARÉ-BIRKHOFF-WITT). *Let L be an R-free restricted Lie algebra. Then $f : B(L) \to E^{\oplus} V(L)$ is an isomorphism of Hopf algebras.*

PROOF. Give $B(L)$ its Lie filtration and $E^{\oplus} V(L)$ its filtration by filtration degree. Then $E^0 B(L)$ is obtained by application of B to L regarded as a bigraded R-module via $L_{1,q} = L_{q+1}$, and $E^0 E^{\oplus}(L) = E^0 V(L)$. Since f is evidently a filtration-preserving epimorphism, it suffices to prove that $E^0 f$ is a monomorphism. Observe that the quotient maps $\pi : A(L) \to B(L)$ and $\rho : U(L) \to V(L)$ are filtration-preserving. Let $I = \ker \rho$. Recall the map of R-modules $\bar{\sigma} : U(L) \longrightarrow A(L)$ from the proof of Theorem 22.2.3. We shall construct a filtration-preserving morphism of R-modules $\tau : A(L) \to B(L)$ such that $\tau \bar{\sigma}(I) = 0$ and $E^0 \tau = E^0 \pi$. It will follow that $\tau \bar{\sigma} = \bar{\tau} \rho$ for a filtration-preserving R-map $\bar{\tau} : V(L) \to B(L)$ and that $E^0 \bar{\tau}$ is a morphism of algebras. The composite

$$E^0 B(L) \xrightarrow{\quad E^0 f \quad} E^0 V(L) \xrightarrow{\quad E^0 \bar{\tau} \quad} E^0 B(L)$$

will be the identity morphism of algebras, hence $E^0 f$ will be a monomorphism and the proof will be complete.

To construct τ, let $\{y_j\}$ be an R-basis for L^+. Clearly L^- plays a negligible role here, and we let x denote an arbitrary basis element of $A(L^-) = B(L^-)$ and define $\tau(x) = x$. Let z_j denote y_j regarded as an element of $B(L)$. We define τ by induction on the filtration degree. We define τ by the formulas

23.2.2 $\qquad \tau(x y_{j_1}^{r_1} \cdots y_{j_n}^{r_n}) = x z_{j_1}^{r_1} \cdots z_{j_n}^{r_n}$ for each $r_i < p$

and

23.2.3 $\qquad \tau(x y_{j_1}^{r_1} \cdots y_{j_n}^{r_n}) = \tau(x y_{j_1}^{r_1} \cdots y_{j_{i-1}}^{r_{i-1}} \xi(y_{j_1}) y_{j_i}^{r_i - p} y_{j_{i+1}}^{r_{i+1}} \cdots y_{j_n}^{r_n})$ if $r_i \geq p$.

By induction on the filtration degree, these formulas uniquely determine a well-defined filtration-preserving morphism of R-modules $\tau : A(L) \longrightarrow B(L)$ such that $E^0 \tau = E^0 \pi$. It remains to check that $\tau \bar{\sigma}(I) = 0$. By definition, I is the two-sided ideal in $U(L)$ generated by $\{y^p - \xi(y) | y \in L^+\}$. If y is a linear combination $\sum k_i y_{j_i}$, the identities (ii) and (iii) of Lemma 23.1.2 and the agreement of commutators and Lie brackets of elements of L in $U(L)$ imply that

$$y^p - \xi(y) = \sum_i k_i^p (y_{j_i}^p - \xi(y_{j_i})).$$

Thus I is the two-sided ideal in $U(L)$ generated by $\{y_j^p - \xi(y_j)\}$. Now a calculation from the identity (i) of Lemma 23.1.2 and the inductive definitions of $\bar{\sigma}$ and τ gives the conclusion. $\qquad \square$

The following corollaries are deduced precisely as in the case of Lie algebras.

COROLLARY 23.2.4. *Let $\{x_i\}$ and $\{y_j\}$ be R-bases for L^- and L^+ indexed on totally ordered sets. Then $V(L)$ is the free R-module on the basis*

$$\{x_{i_1} \cdots x_{i_m} y_{j_1}^{r_1} \cdots y_{j_n}^{r_n} | i_1 < \cdots < i_m, \, j_1 < \cdots < j_n \text{ and } 1 \leq r_k < p\}.$$

COROLLARY 23.2.5. *Let L be an R-free Lie algebra together with a restriction operation satisfying the identities listed in Lemma 23.1.2. Then L is a restricted Lie algebra.*

23.3. Primitively generated Hopf algebras in characteristic p

In this section, R is assumed to be a field of characteristic p. Again, all of the results remain valid if R is any ring of characteristic p and all R-modules in sight are R-free.

The R-module PA of primitive elements of a Hopf algebra A is a restricted Lie subalgebra. The universal property of $V(PA)$ thus gives a natural map of Hopf algebras $g : V(PA) \to A$, and g is an epimorphism if A is primitive. Let $\mathscr{RL}$ and $\mathscr{PH}$ denote the categories of restricted Lie algebras and of primitive Hopf algebras over R. We have functors $V : \mathscr{RL} \to \mathscr{PH}$ and $P : \mathscr{PH} \to \mathscr{RL}$, a natural inclusion $L \subset PV(L)$, and a natural epimorphism $g : V(PA) \to A$, where $L \in \mathscr{RL}$ and $A \in \mathscr{PH}$.

THEOREM 23.3.1. *The functors $V : \mathscr{RL} \to \mathscr{PH}$ and $P : \mathscr{PH} \to \mathscr{RL}$ are inverse equivalences of categories. More explicitly,*

(i) $PV(L) = L$ for any restricted Lie algebra L and

(ii) $g : V(PA) \to A$ is an isomorphism for any primitive Hopf algebra A.

PROOF. To prove (i), we consider the Lie filtration of $V(L)$. By the PBW theorem, $E^{\oplus} V(L) \cong B(L)$ as a Hopf algebra. Arguing precisely as in the characteristic zero case, we find that $PE^0 V(L) = E^0_{1,*} V(L)$ and conclude that $PV(L) \subset F_1 V(L)$. This proves (i). To prove (ii), consider the primitive filtration of A, as specified in the proof of Theorem 22.3.1. The Lie and primitive filtrations on $V(L)$ coincide and g is filtration-preserving. It follows just as in the characteristic zero case that $E^0 g$ is a monomorphism and that g is therefore an isomorphism. $\square$

COROLLARY 23.3.2. *If A is a commutative primitive Hopf algebra such that $x^p = 0$ if $x \in (IA)^+$, then A is isomorphic as a Hopf algebra to $B(PA)$.*

PROOF. $A \cong V(PA) \cong B(PA)$ since PA is an abelian restricted Lie algebra with restriction zero. $\square$

Unlike its characteristic zero analogue, Corollary 23.3.2 fails to describe arbitrary commutative primitive Hopf algebras A over R. We have $A \cong V(PA)$, and we shall study $V(PA)$ in more detail in the next section. For similar reasons, the characteristic p analogue of Corollary 22.3.3 takes the following weaker form. We again emphasize that A was not assumed to be connected in the results above.

COROLLARY 23.3.3. *Let A be a connected quasi-Hopf algebra.*

(i) $\nu : PA \to QA$ *is a monomorphism if and only if A is associative and commutative and satisfies* $x^p = 0$ *for* $x \in (IA)^+$.

(ii) If A is commutative and associative and if $\xi(A)$ *is the quasi-Hopf subalgebra of A whose positive degree elements are spanned by* $\{x^p | x \in (IA)^+\}$, *then the following is an exact sequence of R-modules.*

$$0 \longrightarrow P\xi(A) \longrightarrow PA \overset{\nu}{\longrightarrow} QA$$

(iii) If A is a commutative and cocommutative Hopf algebra, then the following is an exact sequence of R-modules.

$$0 \longrightarrow P\xi(A) \longrightarrow PA \overset{\nu}{\longrightarrow} QA \longrightarrow Q\lambda(A) \longrightarrow 0$$

Here $\lambda(A)$ *is the quotient Hopf algebra* $\xi(A^*)^*$ *of A if A is of finite type and, in general,* $\lambda(A)$ *is the colimit of the* $\lambda(B)$, *where B runs over the Hopf subalgebras of A that are of finite type.*

PROOF. If ν is a monomorphism, then A is associative and commutative and $x^p = 0$ for $x \in (IA)^+$ by Lemma 21.1.1. Conversely, give A its product filtration, which is complete since A is connected. The previous corollary applies to give $E^0 A \cong B(PE^0 A)$. It follows as in the proof of Corollary 22.3.3 that ν is a monomorphism. To prove (ii), let $B = A//\xi A$. Then B satisfies the hypotheses of (i). By Theorem 21.2.3, we have the commutative diagram with exact rows

$$
\begin{array}{ccccccc}
0 & \longrightarrow & P\xi(A) & \longrightarrow & PA & \longrightarrow & PB \\
 & & \downarrow{\scriptstyle \nu} & & \downarrow{\scriptstyle \nu} & & \downarrow{\scriptstyle \nu} \\
 & & Q\xi(A) & \longrightarrow & QA & \longrightarrow & QB & \longrightarrow & 0
\end{array}
$$

Here $\nu : PB \to QB$ is a monomorphism and $Q\xi(A) \to QA$ is zero. Now (ii) follows by a simple diagram chase. When A is of finite type, (iii) follows from (ii) by dualization, and the general case then results by passage to colimits. $\square$

COROLLARY 23.3.4. *A Hopf subalgebra of a primitive Hopf algebra is itself primitive.*

PROOF. Let $A \subset B$, where B is primitive. By precisely the same argument as in the proof of Corollary 22.3.4, it suffices to prove the result when A

and B are connected. By Lemma 21.1.2, we may assume that A is of finite type. The proof of Lemma 21.1.2 applies to show that B is the colimit of its primitive Hopf subalgebras of finite type, and A will necessarily be contained in one of them. Thus we may assume that B is also of finite type. Then A^* is a quotient hopf algebra of B^*. Since $v : PB^* \to QB^*$ is a monomorphism, part (i) of the previous corollary applies to show that B^* is associative and commutative with zero p^{th} powers. Therefore A^* also has these properties and $v : PA \to QA$ is a monomorphism. Dualizing back, we have that $v : PA \to QA$ is an epimorphism. □

COROLLARY 23.3.5. *A Hopf subalgebra A of a primitive Hopf algebra B is a normal subalgebra if and only if PA is a restricted Lie ideal of PB. When this holds, $B//A = V(PB/PA)$ and*

$$0 \to PA \to PB \to P(B//A) \to 0$$

is an exact sequence of restricted Lie algebras.

PROOF. The argument is the same as for Lemma 20.4.7, but with the observation that, since A is a subalgebra of B, PA is automatically closed under the restriction in B and is thus a restricted Lie ideal if and only if it is a Lie ideal. □

23.4. Commutative Hopf algebras in characteristic p

In this section, R is assumed to be a perfect field of characteristic p. We need R to be perfect since relations of the form $x^{p^q} = ry^p$, where r has no p^{th} root, would lead to counterexamples to the main results.

THEOREM 23.4.1. *Let A be a connected, commutative, and associative quasi-Hopf algebra. For a morphism of R-modules $\sigma : QA \to IA$ such that $\pi\sigma = \text{id}$, where $\pi : IA \to QA$ is the quotient map, let $R(A;\sigma)$ be the abelian restricted Lie subalgebra of A generated by the image of σ. For a suitable choice of σ, the morphism of algebras $f : V(R(A;\sigma)) \to A$ induced by the inclusion of $R(A;\sigma)$ in A is an isomorphism.*

PROOF. Clearly f is an epimorphism for any choice of σ. Let $\mathscr{F}$ be the family of pairs (B, σ), where B is a quasi-Hopf subalgebra of A and $\sigma : QB \to B$ is a R-splitting of $\pi : B \to QB$ such that the following properties hold.

(1) The map of algebras $f : V(R(B;\sigma)) \to B$ associated to σ is an isomorphism.

(2) The map $QB \to QA$ induced by the inclusion of B in A is a monomorphism, and $(QB)_q = 0$ for $q > n$ if $(QB)_n \to (QA)_n$ is not an isomorphism.

Partial order $\mathscr{F}$ by $(C, \tau) < (B, \sigma)$ if $C \subset B$ and σ extends τ. Note that $QC \to QB$ is then a monomorphism such that $(QC)_q = 0$ for $q > n$ if $(QC)_n \to (QB)_n$ is not an isomorphism. The family $\mathscr{F}$ is nonempty since it contains $(R, 0)$, and the union of a chain in $\mathscr{F}$ is an element of $\mathscr{F}$. Therefore $\mathscr{F}$ has a maximal element (C, τ). Assume for a contradiction that $C \neq A$. Let n be minimal such that $(QC)_n \neq (QA)_n$. Then $(QC)_q = 0$ for $q > n$. Choose $y \in A_n$ such that $\pi(y)$ is not in QC and let B be the subalgebra of A generated by C and y; B is necessarily a quasi-Hopf subalgebra. The quotient $B//C$ is a monogenic Hopf algebra with primitive generator the image z of y.

A check of coproducts shows that the minimal m such that $y^m = 0$ must be a power of p. Define the height of y by $\mathrm{ht}(y) = t$ if $y^{p^t} = 0$ but $y^{p^{t-1}} \neq 0$, or $\mathrm{ht}(y) = \infty$ if there exists no such t. It is possible that $y \in B$ has greater height than $z \in B//C$, but we claim that there exists $x \in B$ such that x also has image z in $B//C$ and $\mathrm{ht}(x) = \mathrm{ht}(z)$. Granting the claim, we complete the proof as follows. By Theorem 21.2.3, the composite map of algebras

$$C \otimes B//C \xrightarrow{\ i \otimes \sigma\ } B \otimes B \xrightarrow{\ \phi\ } B$$

is an isomorphism, where $i : C \to B$ is the inclusion and $\sigma(z) = x$. If we extend $\tau : QC \to C$ to $\sigma : QB \to B$ by setting $\sigma \pi(y) = x$, then the associated map of algebras $V(R(B; \sigma)) \to B$ is an isomorphism and $(C, \tau) < (B, \sigma)$.

Thus it remains to prove the claim. There is nothing to prove if $p > 2$ and n is odd or if z has infinite height. Thus let $\mathrm{ht}(z) = s$. Let $C' = \xi^s(C)$. Since R is perfect, C' is a quasi-Hopf subalgebra of C. Consider the commutative diagram

$$
\begin{array}{ccccc}
C & \longrightarrow & B & \longrightarrow & B//C \\
\downarrow{\scriptstyle \gamma} & & \downarrow{\scriptstyle \beta} & & \downarrow{\scriptstyle \alpha} \\
C//C' & \longrightarrow & B//C' & \longrightarrow & (B//C')//(C//C')
\end{array}
$$

where the vertical arrows are quotient maps. By Corollary 21.2.4, α is an isomorphism, and we regard it as an identification. Let $x' \in B//C'$ map to $z \in B//C$. Then x' has height p^s. Indeed, $\xi^s(x')$ is primitive since

$\psi(x') = x' \otimes 1 + u + 1 \otimes x'$, where $u \in I(C//C') \otimes I(C//C')$ and thus $\xi^s(u) = 0$. However, $B//C'$ has no nonzero primitive elements of degree $p^s n$ in view of the exact sequence

$$0 \longrightarrow P(C//C') \longrightarrow P(B//C') \longrightarrow P(B//C)$$

and the fact that all indecomposable elements of $C//C'$ have degree $\leq n$ and all $(p^s)^{th}$ powers of elements of $C//C'$ are zero (and similarly for $B//C$). Now choose $w \in B$ such that $\beta(w) = x'$. Then $\psi(w) = w \otimes 1 + v + 1 \otimes w$, where $v \in IC \otimes IC$. Since $(\mathrm{id} \otimes \beta)(v) \in IC \otimes I(C//C')$, $\xi^s(\mathrm{id} \otimes \beta)(v) = 0$. It follows that $(\mathrm{id} \otimes \beta)\psi \xi^s(w) = \xi^s(w) \otimes 1$. By Theorem 21.2.3, this implies that $\xi^s(w)$ is in C'. Let $\xi^s(w) = \xi^s(w')$, where $w' \in C$. If $x = w - w'$, then $\xi^s(x) = 0$ and x projects to z in $B//C$. $\square$

EXAMPLE 23.4.2. The theorem fails if σ is not chosen properly. For a counterexample, let $p = 2$ and take $A = P\{x\} \otimes E\{y\}$, where x and y are primitive elements of degrees one and three. If one foolishly defines $\sigma : QA \to A$ by $\sigma\pi(x) = x$ and $\sigma\pi(y) = y + x^3$, then $V(R(A; \sigma))$ is a polynomial algebra on two generators.[1]

We have the following immediate corollary for quasi-Hopf algebras having only odd-degree generators. We say that a commutative algebra is strictly commutative if $x^2 = 0$ for all odd-degree elements x; of course, this always holds if char $R \neq 2$.

COROLLARY 23.4.3. *Let A be a connected, strictly commutative, and associative quasi-Hopf algebra such that $Q_n A = 0$ if n is even. Then $A \cong E(QA)$ as an algebra.*

Again, we obtain a stronger conclusion when the coproduct is coassociative.

COROLLARY 23.4.4 (LERAY-SAMELSON). *Let A be a connected strictly commutative Hopf algebra such that $Q_n A = 0$ if n is even. Then $A \cong E(PA)$ as a Hopf algebra.*

PROOF. This follows from Corollaries 23.3.2 and 23.3.3 by the same arguments used to prove Corollary 22.4.3. $\square$

---

1. This example is due to Paul Goerss [private communication].

Using Corollary 23.3.5, we can obtain an analogue of Theorem 22.4.4, but this result gives considerably less complete information than was obtainable in the characteristic zero case.

COROLLARY 23.4.5. *Let A be a connected, commutative, and cocommutative Hopf algebra over R, where char $R > 2$. Let $B = E(PA^-)$ and $C = A//B$. Then $A \cong B \otimes C$ as a Hopf algebra.*

PROOF. By (iii) of Corollary 23.3.3, $\nu : PA \to QA$ is an isomorphism in odd degrees. We may assume that A is of finite type. Dualizing, we have that $\nu : PA^* \to QA^*$ is also an isomorphism in odd degrees, and there results a map of Hopf algebras $\pi^* : B^* \to A^*$ such that the evident composite $B \to A \to B$ is the identity. Let $\rho : A \to C$ be the natural epimorphism and define $\omega : A \to B \otimes C$ to be the composite

$$A \xrightarrow{\ \psi\ } A \otimes A \xrightarrow{\ \pi \otimes \rho\ } B \otimes C.$$

Since A is cocommutative, ψ is a morphism of Hopf algebras, hence so is ω. Since $(\epsilon \otimes \mathrm{id})\omega = \rho$ and $(\mathrm{id} \otimes \epsilon)\omega = \pi$, ω is clearly an epimorphism. Using the exact sequence of primitives in Corollary 23.3.5 and the fact that $P(B \otimes C) = PB \oplus PC$, we see that ω is an isomorphism on primitives and therefore a monomorphism. $\quad\square$

To complete our results, we must still determine the structure of $V(L)$, where L is an abelian restricted Lie algebra. Clearly, it suffices to study L itself.

THEOREM 23.4.6. *Let L be an abelian restricted Lie algebra such that L_0 is finitely generated as a restricted Lie algebra. Then L is isomorphic to a direct sum of monogenic abelian restricted Lie algebras.*

PROOF. Clearly $L = L^- \times L^+$ as an abelian restricted Lie algebra. Since L^- is just a vector space, with no additional structure, we may as well assume that $L = L^+$. Let $L(n)$ be the abelian restricted Lie subalgebra of L generated by the L_i for $i \leq n$. Since L is the union of the $L(n)$, it suffices to prove the result when $L = L(n)$. We proceed by induction on n.

We first consider the case $L = L(0) = L_0$, which is exceptional. Let $P[t]$ denote the noncommutative polynomial algebra in one indeterminate t with $tr = r^p t$ for $r \in R$. If $R = \mathbb{F}_p$, $P[t]$ is the ordinary polynomial algebra. The relation $\xi(rx) = r^p \xi(x)$ shows that L is a $P[t]$-module via $tx = \xi(x)$. Since R is perfect, $r \to r^p$ is an automorphism of R, and $P[t]$ is a principal ideal domain by Jacobson [72, p. 30]. Therefore, by [72, pp. 43–44], every finitely generated $P[t]$-module is a direct sum of cyclic modules. This says that L is a finite direct sum of monogenic abelian restricted Lie algebras.

In general, L is the direct sum of $L(0)$ and its restricted Lie subalgebra of positive-degree elements, so we may now assume that $L_0 = 0$. Consider the case $L = L(n)$, where $n > 0$ and the conclusion holds for $L(n-1)$. Choose a splitting σ of the epimorphism $L(n)_n \longrightarrow L(n)_n/L(n-1)_n$ and choose a basis for $L(n)_n/L(n-1)_n$. The image under σ of the chosen basis gives a set of generators of degree n of $L(n)$. Since, in contrast with the case $L(0)$, there is no further structure in sight in degree n, we may apply a passage to colimits argument to see that the conclusion holds in general if it holds when there are only finitely many generators, q say, of degree n. We proceed by induction on q, there being nothing to prove if there are no such generators. Thus assume first that L has q generators of degree n and let L' be the abelian restricted Lie subalgebra of L generated by $L(n-1)$ together with $q-1$ of these generators. Let $L'' = L/L'$. By the induction hypothesis, L' is a sum of monogenic abelian restricted Lie algebras. By construction, L'' is an abelian restricted Lie algebra generated by a single element, x say, of degree n. It suffices to prove that L is isomorphic to $L' \oplus L''$. To show this, it suffices to construct a morphism $f : L'' \to L$ of abelian restricted Lie algebras such that $\pi f = \mathrm{id}$, where $\pi : L \to L''$ is the quotient map.

Define the height of an element $z \in L$ by $\mathrm{ht}(z) = s$ if $\xi^s(z) = 0$ but $\xi^{s-1}(z) \neq 0$ and $\mathrm{ht}(z) = \infty$ if $\xi^m(z) \neq 0$ for all m. Of course, if $\pi(y) = x$, then $\mathrm{ht}(y) \geq \mathrm{ht}(x)$. To construct f, it suffices to find $y \in L$ such that $\pi(y) = x$ and $\mathrm{ht}(y) = \mathrm{ht}(x)$ since $f(x) = y$ then determines f. If $\mathrm{ht}(x) = \infty$, any y such that $\pi(y) = x$ will do. Thus assume that x has finite height s. Since R is perfect, $\xi^s(L')$ is a abelian restricted Lie subalgebra of L'. Let $M' = L'/\xi^s(L')$ and $M = L/\xi^s(L')$. We may identify L'' with M/M'. Choose $z \in M$, which projects to $x \in L''$, and $w \in L$, which projects to z. We have $M_t = 0$ for $t \geq p^s n$ by construction and $L''_t = 0$ for $t \geq p^s n$ since $\mathrm{ht}(x) = s$. Thus $M_t = 0$ for $t \geq p^s n$ and $\xi^s(z) = 0$. Therefore $\xi^s(w) = \xi^s(w')$ for some $w' \in L'$. Let $y = w - w'$. Then $\pi(y) = x$ and $\mathrm{ht}(y) = s$. $\qquad\square$

Observe that $V(\oplus_i L_i) \cong \otimes_i V(L_i)$. One way to see this formally is to ignore the coproduct and observe that, as a left adjoint, the functor V from abelian restricted Lie algebras to commutative algebras commutes with categorical coproducts, which are direct sums on the Lie algebra level and tensor products on the algebra level. The following two theorems are therefore direct consequences of Theorems 23.3.1 and 23.4.1. Note that a connected mono-genic Hopf algebra is of the form $E[x]$, where $x \in (IA)^-$, or $P[x]/(x^{p^q})$ or $P[x]$, where $x \in (IA)^+$.

THEOREM 23.4.7. *If A is a primitive commutative Hopf algebra and A_0 is finitely generated as an algebra, then A is isomorphic as a Hopf algebra to a tensor product of monogenic Hopf algebras.*

THEOREM 23.4.8 (BOREL). *If A is a connected, commutative, and associative quasi-Hopf algebra, then A is isomorphic as an algebra to a tensor product of monogenic Hopf algebras.*

24

A PRIMER ON SPECTRAL SEQUENCES

This chapter contains those results about spectral sequences that we used earlier in the book, incorporated into a brief background compendium of the very minimum that anybody interested in algebraic topology needs to know about spectral sequences. Introductory books on algebraic topology usually focus on the different kinds of chain and cochain complexes that can be used to define ordinary homology and cohomology. It is a well-kept secret that the further one goes into the subject, the less one uses such complexes for actual calculation. Rather, one starts with a few spaces whose homology and cohomology groups can be computed by hand, using explicit chain complexes. One then bootstraps up such calculations to the vast array of currently known computations using a variety of spectral sequences. McCleary's book [98] is a good encyclopedic reference for the various spectral sequences in current use. Other introductions can be found in many texts in algebraic topology and homological algebra [79, 123, 142]. However, the truth is that the only way to master the use of spectral sequences is to work out many examples in detail.

All modules are over a commutative ring R and understood to be graded, whether or not the grading is mentioned explicitly or denoted. In general, we leave the gradings implicit for readability. The preliminaries on tensor product and Hom functors of §20.1 remain in force in this chapter.

24.1. Definitions

While spectral sequences arise with different patterns of gradings, the most commonly encountered homologically and cohomologically graded spectral sequences fit into the patterns given in the following pair of definitions.

DEFINITION 24.1.1. A homologically graded spectral sequence $E = \{E^r\}$ consists of a sequence of $\mathbb{Z}$-bigraded R modules $E^r = \{E^r_{p,q}\}_{r \geq 1}$ together with differentials

$$d^r : E^r_{p,q} \to E^r_{p-r,q+r-1}$$

such that $E^{r+1} \cong H_*(E^r)$. A morphism $f : E \to E'$ of spectral sequences is a family of morphisms of complexes $f^r : E^r \to E'^r$ such that f^{r+1} is the morphism $H_*(f^r)$ induced by f^r.

DEFINITION 24.1.2. A cohomologically graded spectral sequence $E = \{E_r\}$ consists of $\mathbb{Z}$-bigraded R-modules $E_r = \{E_r^{p,q}\}_{r \geq 1}$ together with differentials

$$d_r : E_r^{p,q} \to E_r^{p+r,q-r+1}$$

such that $E_{r+1} \cong H_*(E_r)$. We can regrade E_r homologically by setting $E_r^{p,q} = E^r_{-p,-q}$, so in principle the two grading conventions define the same concept.

Let $E = \{E^r\}$ be a spectral sequence. Let Z^1, the cycles, be the kernel of d^1 and B^1, the boundaries, be the image of d^1. Then, under the identification of $H_*(E^1)$ with E^2, d^2 is a map

$$Z^1/B^1 \to Z^1/B^1.$$

Continuing this identification, E^r is identified with Z^{r-1}/B^{r-1} and the map

$$d^r : Z^{r-1}/B^{r-1} \to Z^{r-1}/B^{r-1}$$

has kernel Z^r/B^{r-1} and image B^r/B^{r-1}. These identifications give a sequence of submodules

$$0 = B^0 \subset B^1 \subset \cdots \subset Z^2 \subset Z^1 \subset Z^0 = E^1.$$

Define $Z^\infty = \cap_{r=1}^\infty Z^r$, $B^\infty = \cup_{r=1}^\infty B^r$, and $E^\infty_{p,q} = Z^\infty_{p,q}/B^\infty_{p,q}$, writing $E^\infty = \{E^\infty_{p,q}\}$. We say that E is a first-quadrant spectral sequence if $E^r_{p,q} = 0$ for $p < 0$ or $q < 0$. In a first-quadrant spectral sequence the terms $\{E^r_{p,0}\}$ are called the base terms and the terms $\{E^r_{0,q}\}$ are called the fiber terms. Note that elements of $E^r_{p,0}$ cannot be boundaries for $r \geq 2$ since the differential

$$d^r : E^r_{p+r,-r+1} \longrightarrow E^r_{p,0}$$

has domain the 0 group. Thus

$$E^{r+1}_{p,0} = \mathrm{Ker}\,(d^r : E^r_{p,0} \to E^r_{p-r,r-1})$$

and there is a sequence of monomorphisms

$$e_B \colon E^{\infty}_{p,0} = E^{p+1}_{p,0} \to E^p_{p,0} \to \cdots \to E^3_{p,0} \to E^2_{p,0}.$$

Similarly, for $r \geq 1$, $E^r_{0,q}$ consists only of cycles and so there are epimorphisms

$$e_F \colon E^2_{0,q} \to E^3_{0,q} \to \cdots \to E^{q+2}_{0,q} = E^{\infty}_{0,q}.$$

The maps e_B and e_F are called edge homomorphisms. From these maps we define a "map" $\tau = e_F^{-1} d^p e_B^{-1} \colon E^2_{p,0} \to E^2_{0,p-1}$, as in the following diagram.

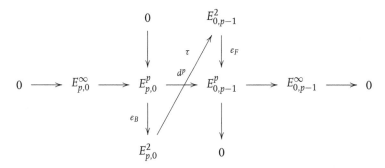

This map is called the transgression. It is an additive relation [79, II.6] from a submodule of $E^2_{p,0}$ to a quotient module of $E^2_{0,p-1}$.

A cohomologically graded first-quadrant spectral sequence E is also defined to have $E^{p,q}_r = 0$ for $p < 0$ or $q < 0$. However, when regraded homologically it becomes a third-quadrant spectral sequence. Again, its base terms have $q = 0$ and its fiber terms have $p = 0$. It has edge homomorphisms

$$e_B \colon E^{p,0}_2 \to E^{p,0}_3 \to \cdots \to E^{p,0}_p \to E^{p,0}_{p+1} = E^{p,0}_{\infty}$$

(which are epimorphisms) and

$$e_F \colon E^{0,q}_{\infty} = E^{0,q}_{q+2} \longrightarrow E^{0,q}_{q+1} \longrightarrow \cdots \longrightarrow E^{0,q}_3 \to E^{0,q}_2$$

(which are monomorphisms). Its transgression $\tau = e_B^{-1} d_p e_F^{-1}$ is induced by the differential $d_p \colon E^{0,p-1}_p \to E^{p,0}_p$. It is an additive relation from a submodule of $E^{0,p-1}_2$ to a quotient module of $E^{p,0}_2$.

24.2. Exact couples

Exact couples provide an especially useful and general source of spectral sequences. We first define them in general, with unspecified gradings. This leads to the most elementary example, called the Bockstein spectral sequence. We then describe the gradings that usually appear in practice.

DEFINITION 24.2.1. Let D and E be modules. An exact couple $\mathscr{C} = \langle D, E; i, j, k \rangle$ is a diagram

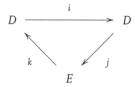

in which $\operatorname{Ker} j = \operatorname{Im} i$, $\operatorname{Ker} k = \operatorname{Im} j$, and $\operatorname{Ker} i = \operatorname{Im} k$.

If $d = jk \colon E \to E$, then $d \circ d = jkjk = 0$. Construct $\mathscr{C}' = \langle D', E'; i', j', k' \rangle$ by letting

$$D' = i(D) \quad \text{and} \quad E' = H_*(E; d),$$

and, writing overlines to denote passage to homology classes,

$$i' = i|_{i(D)}, \quad j'(i(x)) = \overline{j(x)} = j(x) + jk(E), \quad \text{and} \quad k'(\bar{y}) = k'(y + jk(E)) = k(y).$$

That is, i' is a restriction of i and j' and k' are induced from j and k by passing to homology on targets and sources. We easily check that j' and k' are well-defined and the following result holds.

LEMMA 24.2.2. $\mathscr{C}'$ is an exact couple.

Starting with $\mathscr{C} = \mathscr{C}^{(1)}$, we can iterate the construction to form $\mathscr{C}^{(r)} = \langle D^r, E^r, i^r, j^r, k^r \rangle$. (The notation might be confusing since the maps i^r, j^r, and k^r given by the construction are not iterated composites). Then, with $d^r = j^r k^r$, $\{E^r\}$ is a spectral sequence. It can be graded differently than in the previous section since we have not specified conditions on the grading of D and E. The Bockstein spectral sequence in the proof of Lemma 4.3.4 comes from a particularly simple exact couple and is singly graded rather than bigraded.

EXAMPLE 24.2.3. Let C be a torsion-free chain complex over $\mathbb{Z}$. From the short exact sequence of groups

$$0 \longrightarrow \mathbb{Z} \overset{p}{\longrightarrow} \mathbb{Z} \longrightarrow \mathbb{Z}/p\mathbb{Z} \longrightarrow 0$$

we obtain a short exact sequence of chain complexes

$$0 \longrightarrow C \longrightarrow C \longrightarrow C \otimes \mathbb{Z}/p\mathbb{Z} \longrightarrow 0.$$

The induced long exact homology sequence is an exact couple

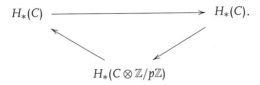

The resulting spectral sequence is called the mod p Bockstein spectral sequence. Here $d^r : E^r_n \to E^r_{n-1}$ for all $r \geq 1$ and all n, and we have short exact sequences

$$0 \longrightarrow (p^{r-1} H_n(C)) \otimes \mathbb{Z}/p\mathbb{Z} \longrightarrow E^r_n \longrightarrow \mathrm{Tor}\,(p^{r-1} H_{n-1}(C), \mathbb{Z}/p\mathbb{Z}) \longrightarrow 0.$$

When $r = 1$, this is the universal coefficient exact sequence for calculating $H_n(C; \mathbb{F}_p)$, and we may view it as a higher universal coefficient exact sequence in general.

We can describe this spectral sequence in very elementary terms. Let Σ^n be the functor on graded abelian groups given by $(\Sigma^n A)_{q+n} = A_q$. For a cyclic abelian group π, we have a $\mathbb{Z}$-free resolution $C(\pi)$ given by $\mathbb{Z}$ in degree 0 if $\pi = \mathbb{Z}$ and by copies of $\mathbb{Z}$ in degrees 0 and 1 with differential q^s if $\pi = \mathbb{Z}/q^s$. Assume that $H_*(C)$ is of finite type and write $H_n(C)$ as a direct sum of cyclic groups. For each cyclic summand, choose a representative cycle x and, if $\pi = \mathbb{Z}/q^s$, a chain y such that $d(y) = q^s x$. For each cyclic summand π, these choices determine a chain map $\Sigma^n C(\pi) \longrightarrow C$. Summing over the cyclic summands and over n, we obtain a chain complex C' and a chain map $C' \longrightarrow C$ that induces an isomorphism on homology and on Bockstein spectral sequences.

The Bockstein spectral sequences $\{E^r\}$ of the $\Sigma^n C(\pi)$ are trivial to compute. When $\pi = \mathbb{Z}$, $E^r_n = \mathbb{Z}$ and $E^r_m = 0$ for $m \neq n$ for all r. When $\pi = \mathbb{Z}/q^s$ for $q \neq p$, $E^r_n = 0$ for all n and r. When $\pi = \mathbb{Z}/p^s$, $E^1 = E^s$ is $\mathbb{F}_p$ in degrees n and $n+1$, $d^s : E^s_{n+1} \longrightarrow E^s_n$ is an isomorphism, and $E^r = 0$ for $r > s$. Returning to C, we see that $E^\infty \cong (H_*(C)/TH_*(C)) \otimes \mathbb{F}_p$, where $T\pi$ denotes the torsion subgroup of a finitely generated abelian group π. Moreover, there is one summand $\mathbb{Z}/p^s$ in $H_*(C)$ for each summand $\mathbb{F}_p$ in the vector space $d^s E^s$. The higher universal coefficient exact sequences are easy to see from this perspective.

We conclude that complete knowledge of the Bockstein spectral sequences of C for all primes p allows a complete description of $H_*(C)$ as a graded abelian group.

The previous example shows that if X is a space whose homology is of finite type and if one can compute $H_*(X;\mathbb{Q})$ and $H_*(X;\mathbb{F}_p)$ together with the mod p Bockstein spectral sequences for all primes p, then one can read off $H_*(X;\mathbb{Z})$. For this reason, among others, algebraic topologists rarely concern themselves with integral homology but rather focus on homology with field coefficients. This is one explanation for the focus of this book on rationalization and completion at primes.

This is just one particularly elementary example of an exact couple. More typically, D and E are $\mathbb{Z}$-bigraded and, with homological grading, we have

$$\deg i = (1,-1), \quad \deg j = (0,0), \quad \text{and} \quad \deg k = (-1,0).$$

This implies that

$$\deg i^r = (1,-1), \quad \deg j^r = (-(r-1), r-1), \quad \text{and} \quad \deg k^r = (-1,0).$$

Since $d^r = j^r k^r$, we then have

$$d^r : E^r_{p,q} \to E^r_{p-r,q+r-1},$$

as in our original definition of a spectral sequence.

24.3. Filtered complexes

Filtered chain complexes give rise to exact couples and therefore to spectral sequences. This is one of the most basic sources of spectral sequences. The Serre spectral sequence, which we describe in §24.5 below, could be obtained as an example, although we shall construct it differently.

Let A be a $\mathbb{Z}$-graded complex of modules. An (increasing) filtration of A is a sequence of subcomplexes

$$\cdots \subset F_{p-1}A \subset F_pA \subset F_{p+1}A \subset \cdots$$

of A. The associated graded complex E^0A is the bigraded complex defined by

$$E^0_{p,q}A = (F_pA/F_{p-1}A)_{p+q},$$

with differential d^0 induced by that of A. The homology $H_*(A)$ is filtered by

$$F_pH_*(A) = \mathrm{Im}(H_*(F_pA) \to H_*(A)),$$

and thus $E^0H_*(A)$ is defined.

Let $A_{p,q} = (F_pA)_{p+q}$. The inclusion $F_{p-1}A \subset F_pA$ restricts to inclusions $i: A_{p-1,q+1} \to A_{p,q}$ and induces quotient maps $j: A_{p,q} \to E^0_{p,q}$. The short exact sequence

24.3.1
$$0 \longrightarrow F_{p-1}A \overset{i}{\longrightarrow} F_p A \overset{j}{\longrightarrow} E_p^0 A \longrightarrow 0$$

of chain complexes induces a long exact sequence

$$\cdots \longrightarrow H_n(F_{p-1}A) \overset{i_*}{\longrightarrow} H_n(F_p A) \overset{j_*}{\longrightarrow} H_n(E_p^0 A) \overset{k_*}{\longrightarrow} H_{n-1}(F_{p-1}A) \longrightarrow \cdots.$$

Let $D_{p,q}^1 = H_{p+q}(F_p A)$ and $E_{p,q}^1 = H_{p+q}(E_p^0)$. Then

$$\langle D^1, E^1; i_*, j_*, k_* \rangle$$

is an exact couple. It gives rise to a spectral sequence $\{E^r A\}$, which is functorial on the category of filtered complexes.

THEOREM 24.3.2. *If $A = \cup_p F_p A$ and for each n there exists $s(n)$ such that $F_{s(n)}A_n = 0$, then $E_{p,q}^\infty A = E_{p,q}^0 H_*(A)$.*

The proof is tedious, but elementary. We give it in the last section of the chapter for illustrative purposes. The conclusion of the theorem, $E_{p,q}^\infty A \cong E_{p,q}^0 H_*(A)$, is often written

$$E_{p,q}^2 A \Rightarrow H_{p+q}(A),$$

and E^r is said to converge to $H_*(A)$.

The filtration of A is said to be canonically bounded if $F_{-1}A = 0$ and $F_n A_n = A_n$ for all n, and in this case E^r certainly converges to $H_*(A)$.

Dually, cohomology spectral sequences arise naturally from decreasing filtrations of complexes. Regrading complexes cohomologically, so that the differentials are maps $\delta \colon A^n \longrightarrow A^{n+1}$, a decreasing filtration is a sequence

$$\cdots \supset F^p A \supset F^{p+1} A \supset \cdots.$$

If we rewrite A as a complex, $A_n = A^{-n}$, and define $F_p A = F^{-p} A$, then our construction of homology spectral sequences immediately gives a cohomology spectral sequence $\{E_r A\}$. With evident changes of notation, Theorem 24.3.2 takes the following cohomological form.

THEOREM 24.3.3. *If $A = \cup_p F^p A$ and for each n there exists $s(n)$ such that $F^{s(n)}A^n = 0$, then $E_\infty^{p,q} A = E_0^{p,q} H^*(A)$.*

A decreasing filtration is canonically bounded if $F^0 A = A$ and $F^{n+1}A^n = 0$ for all n, and in this case E_r certainly converges to $H^*(A)$.

In practice, we often start with a homological filtered complex and dualize it to obtain a cohomological one, setting $A^* = \mathrm{Hom}\,(A, R)$ and filtering it by

$$F^p A^* = \mathrm{Hom}\,(A/F_{p-1}A, R).$$

At least when R is a field, the resulting cohomology spectral sequence is dual to the homology spectral sequence.

24.4. Products

Recall that a differential graded algebra (DGA) A over R is a graded algebra with a product that is a map of chain complexes, so that the Leibnitz formula

$$d(xy) = d(x)y + (-1)^{\deg x} x d(y)$$

is satisfied. When suitably filtered, A often gives rise to a spectral sequence of DGA's, meaning that each term E^r is a DGA. It is no exaggeration to say that the calculational utility of spectral sequences largely stems from such multiplicative structure. We give a brief description of how such structure arises in this section. We work more generally with exact couples rather than filtered chain complexes, since our preferred construction of the Serre spectral sequence largely avoids the use of chains and cochains.

Let $\mathscr{C}_1 = \langle D_1, E_1; i_1, j_1, k_1 \rangle$, $\mathscr{C}_2 = \langle D_2, E_2; i_2, j_2, k_2 \rangle$, and $\mathscr{C} = \langle D, E, i, j, k \rangle$ be exact couples. A pairing

$$\phi\colon E_1 \otimes E_2 \to E$$

is said to satisfy the condition μ_n if for any $x \in E_1$, $y \in E_2$, $a \in D_1$ and $b \in D_2$ such that $k_1(x) = i_1^n(a)$ and $k_2(y) = i_2^n(b)$ there exists $c \in D$ such that

$$k(xy) = i^n(c)$$

and

$$j(c) = j_1(a)y + (-1)^{\deg x} x j_2(b).$$

We write the pairing by concatenation rather than using ϕ to minimize notation. By convention, we set $i_1^0 = \mathrm{id}$ and $i_2^0 = \mathrm{id}$. Then the only possible choices are $a = k_1(x)$, $b = k_2(y)$, and $c = k(xy)$, so that μ_0 is the assertion that

$$jk(xy) = j_1 k_1(x)y + (-1)^{\deg x} x j_2 k_2(y).$$

Since the differential on E is jk, and similarly for E_1 and E_2, μ_0 is precisely the assertion that ϕ is a map of chain complexes, and it then induces

$$\phi'\colon E_1' \otimes E_2' \longrightarrow E'.$$

We say that ϕ satisfies the condition μ if ϕ satisfies μ_n for all $n \geq 0$.

PROPOSITION 24.4.1. *Assume that ϕ satisfies μ_0. Then ϕ satisfies μ_n if and only if ϕ': $E_1' \otimes E_2' \to E'$ satisfies μ_{n-1}.*

PROOF. Suppose that ϕ satisfies μ_n. Let $x' \in E_1'$, $y' \in E_2'$, $a' \in D_1'$, and $b' \in D_2'$ satisfy $k_1'(x') = i_1'^{n-1}(a')$ and $k_2'(y') = i_2'^{n-1}(b')$. If $x' = \bar{x}$, $y' = \bar{y}$, $a' = i_1(a)$, and $b' = i_2(b)$, we find that

$$k_1(x) = i_1^n(a) \quad \text{and} \quad k_2(y) = i_2^n(b).$$

It follows that there exits $c \in D$ such that

$$k(xy) = i^n(c) \quad \text{and} \quad j(c) = j_1(a)y + (-1)^{\deg x} x j_2(b).$$

Taking $c' = i(c)$, we find that

$$k'(\overline{xy}) = k'(\bar{x}\bar{y}) = i'^{n-1}(c')$$

and

$$j'(c') = j_1'(a')y' + (-1)^{\deg x'} x' j_2'(b').$$

The converse is proven similarly. $\square$

COROLLARY 24.4.2. *If ϕ satisfies μ, then so does ϕ', and therefore so do all successive*

$$\phi^r: E_1^r \otimes E_2^r \to E^r,$$

$r \geq 1$, where ϕ^{r+1} is the composite

$$H_*(E_1^r) \otimes H_*(E_2^r) \to H_*(E_1^r \otimes E_2^r) \longrightarrow H_*(E^r)$$

of the Künneth map and $H_*(\phi^r)$. Thus each ϕ^r is a map of chain complexes.

The point is that it is usually quite easy to see explicitly that ϕ satisfies μ, and we are entitled to conclude that the induced pairing of E^r terms satisfies the Leibnitz formula for each $r \geq 1$.

EXAMPLE 24.4.3. The cup product in the singular cochains $C^*(X)$ gives rise to the product

$$\phi: H^*(X; \mathbb{F}_p) \otimes H^*(X; \mathbb{F}_p) \to H^*(X; \mathbb{F}_p).$$

Regarding $H^*(X; \mathbb{F}_p)$ as the E_1 term of the Bockstein spectral sequence of the cochain complex $C^*(X)$, we find that ϕ satisfies μ. Therefore each E^r in the mod p cohomology Bockstein spectral sequence of X is a DGA, and $E^{r+1} = H^*(E^r)$ as an algebra.

Now let A, B, and C be filtered complexes. Filter $A \otimes B$ by

$$F_p(A \otimes B) = \sum_{i+j=p} F_i A \otimes F_j B.$$

Suppose $\phi \colon A \otimes B \to C$ is a morphism of filtered complexes, so that

$$F_p A \cdot F_q B \subset F_{p+q} C.$$

Then ϕ induces a morphism of spectral sequences

$$E^r(A \otimes B) \to E^r(C).$$

Since $E^r A \otimes E^r B$ is a complex, we have a Künneth map

$$E^r A \otimes E^r B \to E^r(A \otimes B),$$

and its composite with $H_*(\phi)$ defines a pairing

$$E^r A \otimes E^r B \to E^r C.$$

This is a morphism of complexes since an easy verification shows that

$$\phi_* \colon E^1 A \otimes E^1 B \to E^1 C$$

satisfies the condition μ. If R is a field, or more generally if our Künneth map is an isomorphism, then $\{E^r A \otimes E^r B\}$ is a spectral sequence isomorphic to $\{E^r(A \otimes B)\}$ and the product is actually a morphism of spectral sequences. In general, however, we are concluding the Leibnitz formula even when the Künneth map of E^r terms does not induce an isomorphism on homology.

If, further, each of the filtered complexes A, B, and C satisfies the hypothesis of the convergence theorem, Theorem 24.3.2, then inspection of its proof shows that the product

$$E^\infty A \otimes E^\infty B \to E^\infty C$$

agrees with the product

$$E^0 H_*(A) \otimes E^0 H_*(B) \longrightarrow E^0 H_*(C)$$

induced by passage to quotients from the induced pairing

$$H_*(A) \otimes H_*(B) \to H_*(C).$$

24.5. The Serre spectral sequence

We give what we feel is perhaps the quickest construction of the Serre spectral sequence, but, since we do not want to go into details of local coefficients, we leave full verifications of its properties, in particular the identification of the E_2 term, to the reader. In applications, the important thing is to understand what the properties say. Their proofs generally play no role. In fact, this is true of most spectral sequences in algebraic topology. It is usual to construct the Serre spectral using the singular (or, in Serre's original work [124], cubical) chains of all spaces in sight. We give a more direct homotopical construction that has the advantage that it generalizes effortlessly to a construction of the Serre spectral sequence in generalized homology and cohomology theories.

For definiteness, we take $R = \mathbb{Z}$ here, and we fix an abelian group π of coefficients. We could just as well replace $\mathbb{Z}$ by any commutative ring R and π by any R-module. Let $p \colon E \to B$ be a Serre fibration with fiber F and connected base space B. It is usual to assume that F too is connected, but that is not really necessary. Fixing a basepoint $b \in B$, we may take $F = p^{-1}(b)$, and that fixes an inclusion $i \colon F \longrightarrow E$. Using [93, p. 48], we may as well replace p by a Hurewicz fibration. This is convenient since it allows us to exploit a relationship between cofibrations and fibrations that does not hold for Serre fibrations. Using [93, p. 75], we may choose a based weak equivalence f from a CW complex with a single vertex to B. Pulling back p along f, we may as well replace p by a Hurewicz fibration whose base space is a CW complex B with a single vertex b. Having a CW base space gives a geometric filtration with which to work, and having a single vertex fixes a canonical basepoint and thus a canonical fiber.

Give B its skeletal filtration, $F_p B = B^p$, and define $F_p E = p^{-1}(F_p B)$. Observe that $F_0 E = F$. By Lemma 1.3.1, the inclusions $F_{p-1} E \subset F_p E$ are cofibrations. They give long exact sequences of pairs on homology with coefficients in any fixed abelian group π. We set

$$D^1_{p,q} = H_{p+q}(F_p E; \pi) \quad \text{and} \quad E^1_{p,q} = H_{p+q}(F_p E, F_{p-1} E; \pi),$$

and we may identify $E^1_{p,q}$ with $\tilde{H}_{p+q}(F_p E / F_{p-1} E; \pi)$. The cited long exact sequences are given by maps

$$i^1 \colon H_{p+q}(F_{p-1} E; \pi) \longrightarrow H_{p+q}(F_p E; \pi)$$

and

$$j^1 \colon H_{p+q}(F_p E; \pi) \longrightarrow H_{p+q}(F_p E, F_{p-1} E; \pi)$$

induced by the inclusions $i \colon F_{p-1} E \subset F_p E$ and $j \colon (F_p E, \emptyset) \subset (F_p E, F_{p-1} E)$ and by connecting homomorphisms

$$k^1 \colon H_{p+q}(F_p E, F_{p-1} E; \pi) \longrightarrow H_{p+q-1}(F_{p-1} E; \pi).$$

We have an exact couple and therefore a spectral sequence. Let $C_*(B)$ denote the *cellular* chains of the CW complex B. Filter $H_*(E; \pi)$ by the images of the $H_*(F_p E; \pi)$.

THEOREM 24.5.1 (HOMOLOGY SERRE SPECTRAL SEQUENCE). *There is a first-quadrant homological spectral sequence* $\{E^r, d^r\}$, *with*

$$E^1_{p,q} \cong C_p(B; \mathcal{H}_q(F; \pi)) \quad \text{and} \quad E^2_{p,q} \cong H_p(B; \mathcal{H}_q(F; \pi))$$

that converges to $H_*(E; \pi)$. *It is natural with respect to maps*

$$
\begin{array}{ccc}
D & \xrightarrow{\ g\ } & E \\
{\scriptstyle q}\downarrow & & \downarrow{\scriptstyle p} \\
A & \xrightarrow[\ f\ ]{} & B
\end{array}
$$

of fibrations. Assuming that F is connected, the composite

$$H_p(E; \pi) = F_p H_p(E; \pi) \longrightarrow F_p H_p(E; \pi)/F_{p-1} H_p(E; \pi) = E^\infty_{p,0} \xrightarrow{\ e_B\ } E^2_{p,0}$$

$$= H_p(B; \pi)$$

is the map induced by $p\colon E \to B$. The composite

$$H_q(F; \pi) = H_0(B; H_q(F; \pi)) = E^2_{0,q} \xrightarrow{\ e_F\ } E^\infty_{0,q} = F_0 H_q(E; \pi) \subset H_q(E; \pi)$$

is the map induced by $i\colon F \subset E$. The transgression $\tau\colon H_p(B; \pi) \to H_{p-1}(F; \pi)$ is the inverse additive relation to the suspension $\sigma_*\colon H_{p-1}(F; \pi) \longrightarrow H_p(B; \pi)$.

SKETCH PROOF. Consider the set of p-cells

$$e\colon (D^p, S^{p-1}) \longrightarrow (B^p, B^{p-1}).$$

When we pull the fibration p back along e, we obtain a trivial fibration since D^p is contractible. That is, $p^{-1}(D^p) \simeq D^p \times F$. Implicitly, since $F = p^{-1}(b)$ is fixed, we are using a path from b to a basepoint in D^p when specifying this equivalence, and it is here that the local coefficient systems $\mathcal{H}_q(F)$ enter into the picture. These groups depend on the action of $\pi_1(B, b)$ on F. We prefer not to go into the details of this since, in most of the usual applications, $\pi_1(B, b)$ acts trivially on F and $\mathcal{H}_q(F)$ is just the ordinary homology group $H_q(F)$, so that

$$E^2_{p,q} = H_p(B; H_q(F)).$$

For $p = 0$, the local coefficients have no effect and we may use ordinary homology, as we have done when describing the fiber edge homomorphism. Of course, $F_p B / F_{p-1} B$ is the wedge over the maps e of the spheres $D^p / S^{p-1} \cong S^p$. We conclude that $F_p E / F_{p-1} E$ is homotopy equivalent to the wedge over e of copies of $S^p \wedge F_+$. Therefore, as an abelian group, $H_{p+q}(F_p E, F_{p-1} E; \pi)$ is the direct sum over e of copies of $H_q(F)$ and can be identified with $C_p(B) \otimes H_q(F)$. Using the precise description of cellular chains in terms of cofiber sequences given in [93, pp. 96–97], we can compare the cofiber sequences of the filtration of E with those of the filtration of B to check that $E^1_{*,q}$ is isomorphic as a chain complex to $C_*(B; \mathcal{H}_q(F))$. This is straightforward when $\pi_1(B)$ acts trivially on F, and only requires more definitional details in general. The identification of E^2 follows. We shall return to the proof of convergence in §24.7. The naturality is clear. The statements about the edge homomorphisms can be seen by applying naturality to the maps of fibrations

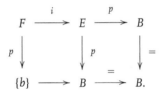

The additive relation $\sigma_*: H_{p-1}(F; \pi) \longrightarrow H_p(B; \pi)$, $p \geq 1$, admits several equivalent descriptions. The most convenient one here is in terms of the following diagram.

$$H_p(E; \pi) \xrightarrow{\ j_*\ } H_p(E, F; \pi) \xrightarrow{\ \partial\ } H_{p-1}(F; \pi) \xrightarrow{\ i_*\ } H_{p-1}(E; \pi)$$

$$\downarrow{p_*} \quad\quad \nearrow{\sigma_*}$$

$$H_p(B, b; \pi)$$

The additive relation σ_* is defined on $\operatorname{Ker} i_*$ and takes values in $\operatorname{Coker} p_* j_*$. If $i_*(x) = 0$, there exists y such that $\partial(y) = x$, and $\sigma_*(x) = p_*(y)$. Thinking in terms of a relative spectral sequence or using $(F_p E, F_0 E) \subset (E, F)$, we see that $d^r(p_*(y)) = 0$ for $r < p$, so that the transgression $\tau(p_*(y)) = d^p(p_*(y))$ is defined. Since $i_*(x) = 0$, x cannot survive the spectral sequence. A check from the definition of the differentials in terms of our exact couple shows that $d^p(p_*(y)) = x$. $\qquad\square$

There is also a cohomological Serre spectral sequence. When $\pi = R$ is a commutative ring, this is a spectral sequence of DGA's by an application of Corollary 24.4.2. To construct this variant, we use the cohomological exact couple obtained from the long exact sequences in cohomology of the pairs $(F_p E, F_{p-1} E)$. The diagonal map gives a map of fibrations

$$
\begin{array}{ccc}
E & \xrightarrow{\Delta} & E \times E \\
{\scriptstyle p}\downarrow & & \downarrow{\scriptstyle p \times p} \\
B & \xrightarrow{\Delta} & B \times B
\end{array}
$$

and therefore gives a map of cohomological spectral sequences.

THEOREM 24.5.2 (COHOMOLOGY SERRE SPECTRAL SEQUENCE). *There is a first-quadrant cohomological spectral sequence* $\{E_r, d_r\}$, *with*

$$E_1^{p,q} \cong C^p(B; \mathscr{H}^q(F; \pi)) \quad and \quad E_2^{p,q} \cong H^p(B; \mathscr{H}^q(F; \pi))$$

that converges to $H^*(E; \pi)$. *It is natural with respect to maps of fibrations. Assuming that F is connected, the composite*

$$H^p(B; \pi) = H^p(B; H^0(F; \pi)) = E_2^{p,0} \xrightarrow{e_B} E_\infty^{p,0} \to H^p(E; \pi)$$

is the map induced by $p \colon E \to B$. *The composite*

$$H^q(E; \pi) \longrightarrow E_\infty^{0,q} \xrightarrow{e_F} E_2^{0,q} = H^0(B; H^q(F; \pi)) = H^q(F; \pi)$$

is the map induced by $i \colon F \subset E$. *The transgression* $\tau \colon H^{p-1}(F; \pi) \to H^p(B; \pi)$ *is the inverse additive relation to the suspension* $\sigma^* \colon H^p(B; \pi) \longrightarrow H^{p-1}(F; \pi)$. *If* $\pi = R$ *is a commutative ring, then* $\{E_r\}$ *is a spectral sequence of DGA's such that* $E_2 = H^*(B; \mathscr{H}^*(F; R))$ *as an R-algebra and* $E_\infty = E^0 H^*(E; R)$ *as R-algebras.*

SKETCH PROOF. Up to the last statement, the proof is the same as in homology. For the products, we already have the map of spectral sequences induced by Δ, so it suffices to work externally, in the spectral sequence of $E \times E$. Since we are using cellular chains, we have a canonical isomorphism of chain complexes $C_*(B) \otimes C_*(B) \cong C_*(B \times B)$ [93, p. 99]. Using this, it is not difficult to define a pairing of E_1 terms

$$\phi \colon C^*(E; \mathscr{H}^*(F)) \otimes C^*(E; \mathscr{H}^*(F)) \longrightarrow C^*(E; \mathscr{H}^*(F))$$

that satisfies μ. Then the last statement follows from Corollary 24.4.2. $\qquad\Box$

A short exact sequence

$$1 \longrightarrow G' \longrightarrow G \longrightarrow G'' \longrightarrow 1$$

of (discrete) groups gives a fibration sequence

$$K(G',1) \longrightarrow K(G,1) \longrightarrow K(G'',1),$$

and there result Serre spectral sequences in homology and cohomology. Focusing on cohomology for definiteness, it takes the following form. This spectral sequence can also be constructed purely algebraically, and it is then sometimes called the Lyndon spectral sequence. It is an example where local coefficients are essential.

PROPOSITION 24.5.3. (LYNDON-HOCHSCHILD-SERRE SPECTRAL SEQUENCE). *Let G' be a normal subgroup of a group G with quotient group G'' and let π be a G-module. Then there is a spectral sequence with*

$$E_2^{p,q} \cong H^p(G''; H^q(G';\pi))$$

that converges to $H^*(G;A)$.

PROOF. The point that needs verification in a topological proof is that the E_2 term of the Serre spectral sequence agrees with the displayed algebraic E_2 term. The latter is shortened notation for

$$\mathrm{Ext}^p_{\mathbb{Z}[G'']}(\mathbb{Z}, \mathrm{Ext}^q_{\mathbb{Z}[G']}(\mathbb{Z},\pi)),$$

where the group actions on $\mathbb{Z}$ are trivial. The algebraic action of G'' on G' coming from the short exact sequence agrees with the topologically defined action of the fundamental group of $\pi_1(K(G'',1))$ on $\pi_1(K(G',1))$. We can take account of the G''-action on π when defining the local cohomology groups $\mathscr{H}^*(K(G',1);\pi)$ and identifying E_2, and then the point is to identify the displayed Ext groups with

$$H^p(K(G'',1); \mathscr{H}^*(K(G',1);\pi)).$$

The details are elaborations of those needed to work [93, Ex. 2, pp. 127, 141]. □

24.6. Comparison theorems

We have had several occasions to use the following standard result. We state it in homological terms, but it has an evident cohomological analogue.

THEOREM 24.6.1. (COMPARISON THEOREM, [79, THM. XI.11.1]). *Let* $f: E \to {}'E$ *be a homomorphism of first-quadrant spectral sequences of modules over a commutative ring. Assume that E_2 and ${}'E_2$ admit universal coefficient exact sequences as displayed in the following diagram, and that, on the E_2 level, f is given by a map of short exact sequences as displayed.*

$$
\begin{array}{ccccccccc}
0 & \longrightarrow & E^2_{p,0} \otimes E^2_{0,q} & \longrightarrow & E^2_{p,q} & \longrightarrow & \mathrm{Tor}_1\,(E^2_{p-1,0}, E^2_{0,q}) & \longrightarrow & 0 \\
 & & \downarrow{\scriptstyle f \otimes f} & & \downarrow{\scriptstyle f} & & \downarrow{\scriptstyle \mathrm{Tor}\,(f,f)} & & \\
0 & \longrightarrow & {}'E^2_{p,0} \otimes {}'E^2_{0,q} & \longrightarrow & {}'E^2_{p,q} & \longrightarrow & \mathrm{Tor}_1\,({}'E^2_{p-1,0}, {}'E^2_{0,q}) & \longrightarrow & 0
\end{array}
$$

Write $f^r_{p,q}: E^r_{p,q} \longrightarrow {}'E^r_{p,q}$. *Then any two of the following imply the third.*

(i) $f^2_{p,0}: E^2_{p,0} \longrightarrow {}'E^2_{p,0}$ *is an isomorphism for all $p \geq 0$.*

(ii) $f^2_{0,q}: E^2_{0,q} \longrightarrow {}'E^2_{0,q}$ *is an isomorphism for all $q \geq 0$.*

(iii) $f^\infty_{p,q}: E^\infty_{p,q} \longrightarrow {}'E^\infty_{p,q}$ *is an isomorphism for all p and q.*

Details can be found in [79, XI.11]. They amount to well-arranged induction arguments. The comparison theorem is particularly useful for the Serre spectral sequence when the base and fiber are connected and the fundamental group of the base acts trivially on the homology of the fiber. The required conditions on the E^2 terms are then always satisfied. In §13.3, we made use of a refinement due to Hilton and Roitberg [64, Thms. 3.1, 3.2] that applies when we allow the base spaces to be nilpotent and to act nilpotently rather than trivially on the homology of the fibers.

THEOREM 24.6.2. *Consider a map of fibrations*

$$
\begin{array}{ccccc}
F & \longrightarrow & E & \longrightarrow & B \\
\downarrow{\scriptstyle f_1} & & \downarrow{\scriptstyle f} & & \downarrow{\scriptstyle f_2} \\
F' & \longrightarrow & E' & \longrightarrow & B',
\end{array}
$$

in which B and B' are nilpotent and act nilpotently on $H_*(F)$ and $H_*(F')$, respectively, where all homology is taken with coefficients in some fixed abelian group. The following conclusions hold.

(i) Let $P \geq 2$ and $Q \geq 0$ be fixed integers. Suppose that

(a) $H_q(f_1)$ is injective for $q < Q$, $H_Q(f_1)$ is surjective,

(b) $H_p(f_2)$ is injective for $p < P$, and $H_P(f_2)$ is surjective.

Then $H_n(f)$ is injective for $n < N \equiv \min(P, Q)$ and $H_N(f)$ is surjective.

(ii) Let $P \geq 2$ and $N \geq 0$ be fixed integers. Suppose that
(a) $H_n(f)$ is injective for $n < N$, $H_N(f)$ is surjective,
(b) $H_p(f_2)$ is injective for $p < P$, and $H_P(f_2)$ is surjective.
Then $H_q(f_1)$ is injective for $q < Q \equiv \min(N, P-1)$ and $H_Q(f_1)$ is surjective.

24.7. Convergence proofs

To give a little more insight into the inner workings of spectral sequences, we give the proof of Theorem 24.3.2 in detail. In fact, the convergence proof for filtered complexes will give us an alternative description of the entire spectral sequence that avoids explicit use of exact couples. If we had given a chain-level construction of the Serre spectral sequence, its convergence would be a special case. The proof of convergence with the more topological construction that we have given is parallel, but simpler, as we explain at the end of the section.

We begin with a description of the E^∞-term of the spectral sequence of an arbitrary exact couple $\langle D, E; i, j, k \rangle$. Recall that, for any (homological) spectral sequence, we obtain a sequence of inclusions

$$0 = B^0 \subset B^1 \subset \cdots \subset Z^2 \subset Z^1 \subset Z^0 = E^1$$

such that $E^{r+1} \cong Z^r / B^r$ for $r \geq 1$ by setting $Z^r = \text{Ker}(d^r)$ and $B^r = \text{Im}(d^r)$.

When $\{E^r\}$ arises from an exact couple, $d^r = j^r k^r$. Here $Z^r = k^{-1}(\text{Im } i^r)$. Indeed, $j^r k^r(z) = 0$ if and only if $k^r(z) \in \text{Ker } j^r = \text{Im } i^r$. Since k^r is the map induced on homology by k, $z \in k^{-1}(\text{Im } i^r)$. Similarly, $B^r = j(\text{Ker } i^r)$. Indeed, $b = j^r k^r(c)$ for some $c \in C^{r-1}$ if and only if $b \in j^r(\text{Im } k^r) = j^r(\text{Ker } i^r)$. Since j^r is induced from j acting on D, $b \in j(\text{Ker } i^r)$. Applying this to the calculation of E^r rather than E^{r+1}, we obtain

24.7.1 $$E^r = Z^{r-1}/B^{r-1} = k^{-1}(\text{Im } i^{r-1})/j(\text{Ker } i^{r-1})$$

and therefore

24.7.2 $$E^\infty = k^{-1}D^\infty/jD^0,$$

where $D^\infty = \cap_{r \geq 1}\text{Im } i^r$ and $D^0 = \cup_{r \geq 1} \text{Ker } i^r$.

Now let A be a filtered complex. Define a (shifted) analogue C^r of Z^r by

$$C^r_{p,q} = \{a | a \in F_p A_{p+q} \text{ and } d(a) \in F_{p-r}A_{p+q-1}\}.$$

These are the cycles up to filtration r. We shall prove shortly that

24.7.3 $$E^r_{p,q}A = (C^r_{p,q} + F_{p-1}A_{p+q})/(d(C^{r-1}_{p+r-1,q-r+2}) + F_{p-1}A_{p+q})$$

for $r \geq 1$ and therefore

24.7.4 $E_{p,q}^\infty A = (C_{p,q}^\infty + F_{p-1}A_{p+q})/(d(C^\infty)_{p,q} + F_{p-1}A_{p+q}),$

where $C_{p,q}^\infty = \cap_{r \geq 1} C_{p,q}^r$ and $d(C^\infty)_{p,q} = \cup_{r \geq 1} d(C_{p+r-1,q-r+2}^{r-1})$.

Recall that j denotes the quotient map $F_p A \longrightarrow F_p A / F_{p-1} A = E_p^0 A$, which fits into the exact sequence (24.3.1). Formula (24.7.3) rigorizes the intuition that an element $x \in E_{p,q}^r$ can be represented as $j(a)$ for some cycle up to filtration r, say $a \in C_{p,q}^r$, and that if $d(a) = b \in F_{p-r}A$, then $j(b)$ represents $d^r(x)$ in $E_{p-r,q+r-1}^r$. The formula can be turned around to give a construction of $\{E^r A\}$ that avoids the use of exact couples. Historically, the alternative construction came first. Assuming this formula for the moment, we complete the proof of Theorem 24.3.2 as follows.

PROOF OF THEOREM 24.3.2. We are assuming that $A = \cup F_p A$ and, for each n, there exists $s(n)$ such that $F_{s(n)}A_n = 0$. Give the cycles and boundaries of A the induced filtrations

$$F_p Z_{p+q} = Z^{p+q}(A) \cap F_p A \text{ and } F_p B_{p+q} = B^{p+q}(A) \cap F_p A.$$

Then $F_p B \subset F_p Z$ and $H(F_p A) = F_p Z / F_p B$. Since $F_p H(A)$ is the image of $H(F_p A)$ in $H(A)$, we have

$$F_p H(A) = (F_p Z + B)/B \text{ and } E_{p,*}^0 H(A) = F_p H_*(A)/F_{p-1} H_*(A).$$

With a little check for the third equality, this implies

$$E_{p,*}^0 H_*(A) = (F_p Z + B)/(F_{p-1}Z + B)$$

$$= (F_p Z)/(F_p Z \cap (F_{p-1}Z + B))$$

$$= (F_p Z)/(F_p Z \cap (F_{p-1}A + F_p B))$$

$$= (F_p Z + F_{p-1}A)/(F_p B + F_{p-1}A).$$

For each q and for sufficiently large r, namely $r \geq p - s(p+q-1)$, we have

$$F_p Z_{p+q} + F_{p-1}A_{p+q} = C_{p,q}^r + F_{p-1}A_{p+q} = C_{p,q}^\infty + F_{p-1}A_{p+q}.$$

Therefore

$$F_p Z + F_{p-1}A = C_{p,*}^\infty + F_{p-1}A.$$

If $b \in F_p B_{p+q}$, then $b = d(a)$ for some $a \in A_{p+q+1}$. By assumption, $a \in F_t A$ for some t, and then, by definition, $a \in C_{t,p+q+1-t}^{t-p} = C_{p+r-1,q-r+2}^{r-1}$, where

$r = t + 1 - p$. Therefore

$$F_p B + F_{p-1} A = d(C^\infty_{p,*}) + F_{p-1} A.$$

By (24.7.4), we conclude that $E^0 H(A) = E^\infty A$. □

PROOF OF (24.7.3). To see the starting point, observe that $j\colon F_p A \longrightarrow E^0_p A$ carries $C^1_{p,q}$ onto the cycles of $E^0_{p,q} A$ and carries $d(C^0_{p,q+1})$ onto the boundaries of $E^0_{p,q} A$. The proof of (24.7.3) has four steps. We show first that j induces a map

$$\bar{j}\colon C^r_{p,q} + F_{p-1} A_{p+q} \longrightarrow Z^{r-1}_{p,q}.$$

We show next that $\bar{j}$ is surjective. We then observe that

$$\bar{j}(d(C^{r-1}_{p+r,p-q+1}) + F_{p-1} A_{p+q}) \subset B^{r-1}_{p,q}.$$

Finally, we show that the inverse image of $B^{r-1}_{p,q}$ is exactly $d(C^{r-1}_{p+r,p-q+1}) + F_{p-1} A_{p+q}$. These statements directly imply (24.7.3).

Let $x \in C^r_{p,q}$ and let $y = d(x) \in F_{p-r} A$. Note that y is a cycle, but not generally a boundary, in the chain complex $F_{p-r} A$ and continue to write y for its homology class. Note too that $y \in C^\infty_{p-r,q+r-1}$ since $d(y) = 0$. Write $\bar{x}$ for the element of $E^1_{p,q} A$ represented by $j(x)$. The connecting homomorphism

$$k_*\colon E^1_{p,q} = H_{p+q}(E^0_p A) \longrightarrow H_{p+q-1}(F_{p-1} A) = D^1_{p-1,q}$$

takes $\bar{x}$ to $i^{r-1}_*(y)$. Therefore $\bar{x} \in Z^{r-1}_{p,*}$ and we can set $\bar{j}(x) = \bar{x}$.

To see the surjectivity, consider an element $w \in Z^{r-1}_{p,q} \subset E^1_{p,q}$. We have $k_*(w) = i^{r-1}_*(y)$ for some $y \in H_{p+q-1}(F_{p-r} A)$, and we again also write y for a representative cycle. Let w be represented by $j(x')$, where $x' \in F_p A$. Then $k_*(w)$ is represented by $d(x') \in F_{p-1} A$, and $d(x')$ must be homologous to y in $F_{p-1} A$, say $d(x'') = d(x') - y$. Let $x = x' - x''$. Then $d(x) = y$ and $j(x) = j(x')$ since $x'' \in F_{p-1} A$. Therefore $\bar{j}(x) = w$ and $\bar{j}$ is surjective.

Now let $v \in d(C^{r-1}_{p+r-1,q-r+2}) \subset C^r_{p,q}$, say $v = d(u)$, where $u \in F_{p+r-1} A$. Again, v is a cycle but not necessarily a boundary in $F_p A$, and we continue to write v for its homology class. Since v becomes a boundary when included into $F_{p+r-1} A$, $i^{r-1}_*(v) = 0$. Thus the class $\bar{v}$ represented by $j(v)$ is in $j_*(\operatorname{Ker} i^{r-1}_*) = B^{r-1}$.

Conversely, suppose that $\bar{j}(x) \in B^{r-1}_{p,q}$, where $x \in C^r_{p,q}$. This means that $\bar{j}(x) = j_*(v)$ for some $v \in \operatorname{Ker} i^{r-1}_*$. Then $j(x)$ is a chain, also denoted v, such that $v = d(u)$ for some chain $u \in F_{p+r-1} A$. Since $j(x - d(u)) = 0$, $x - d(u) \in F_{p-1} A$. Thus $x = d(u) + (x - d(u))$ is an element of $d(C^{r-1}_{p+r,p-q+1}) + F_{p-1} A_{p+q}$. □

PROOF OF CONVERGENCE OF THE SERRE SPECTRAL SEQUENCE. For a large enough r, we have $E_{p,q}^{\infty} = E_{p,q}^{r}$. Precisely, $E_{p,q}^{r}$ consists of permanent cycles when $r > p$ and it consists of non-bounding elements when $r > q + 1$, since the relevant differentials land in or come from zero groups. Fix $r > \max(p, q + 1)$ and consider the description of $E_{p,q}^{r}$ given in (24.7.1). Omitting the coefficient group π from the notation, we have the exact sequence

$$\cdots \longrightarrow H_{p+q}(F_{p-1}E) \xrightarrow{i_*} H_{p+q}(F_p E) \xrightarrow{j_*}$$

$$H_{p+q}(F_p E, F_{p-1}E) \xrightarrow{k_*} H_{p+q-1}(F_{p-1}E) \longrightarrow \cdots.$$

With $D_{p,q}^1 = H_{p+q}(F_p E)$ and $E_{p,q}^1 = H_{p+q}(F_p E, F_{p-1}E)$, this displays our exact couple. Consider $Z_{p,q}^{r-1}$, which is $k_*^{-1}(\operatorname{Im} i_*^{r-1})$. The domain of i_*^{r-1} is zero with our choice of r, so that

$$Z_{p,q}^{r-1} = \operatorname{Im} j_*.$$

Similarly, consider $B_{p,q}^{r-1}$, which is $j_*(\operatorname{Ker} i_*^{r-1})$. With our choice of r, $\operatorname{Ker} i_*^{r-1}$ is the kernel of the map

$$i_{p,*}^{\infty} : H_{p+q}(F_p E) \longrightarrow H_{p+q}(E),$$

so that

$$B_{p,q}^{r-1} = j_*(\operatorname{Im}(\partial : H_{p+q+1}(E, F_p E) \longrightarrow H_{p+q}(F_p E))).$$

Recall that $F_p H_{p+q}(E) = \operatorname{Im} i_*^{\infty}$ and define

$$\bar{j} : F_p H_{p+q}(E) \longrightarrow Z_{p,q}^{r-1}/B_{p,q}^{r-1} = E_{p,q}^{\infty}$$

by

$$\bar{j}(i_*^{\infty}(x)) = j_*(x).$$

This is well-defined since $\operatorname{Ker} i_*^{\infty} = \operatorname{Im} \partial$, and it is clearly surjective. Its kernel is $F_{p-1} H_{p+q}^*(E) = \operatorname{Im} i_{p-1,*}^{\infty}$ since $i_{p-1,*}^{\infty} = i_{p,*}^{\infty} \circ i_*$ and $\operatorname{Ker} j_* = \operatorname{Im} i_*$. $\quad\square$

Bibliography

[1]   J. F. Adams. On the non-existence of elements of Hopf invariant one. *Ann. of Math.* (2), 72:20–104, 1960.

[2]   J. F. Adams. The sphere, considered as an *H*-space mod *p*. *Quart. J. Math. Oxford. Ser.* (2), 12:52–60, 1961.

[3]   M. Aguilar, S. Gitler, and C. Prieto. *Algebraic topology from a homotopical viewpoint.* Universitext. Springer-Verlag, New York, N. Y., 2002. Translated from the Spanish by Stephen Bruce Sontz.

[4]   K. K. S. Andersen and J. Grodal. The classification of 2-compact groups. *J. Amer. Math. Soc.*, 22(2):387–436, 2009.

[5]   K. K. S. Andersen, J. Grodal, J. M. Møller, and A. Viruel. The classification of *p*-compact groups for *p* odd. *Ann. of Math.* (2), 167(1):95–210, 2008.

[6]   M. F. Atiyah and I. G. Macdonald. *Introduction to commutative algebra.* Addison-Wesley, Reading, M. A.–London–Don Mills, Ont., 1969.

[7]   L. Auslander and G. Baumslag. Automorphism groups of finitely generated nilpotent groups. *Bull. Amer. Math. Soc.*, 73:716–717, 1967.

[8]   G. Baumslag. Automorphism groups of nilpotent groups. *Amer. J. Math.*, 91:1003–1011, 1969.

[9]   T. Beke. Sheafifiable homotopy model categories. *Math. Proc. Cambridge Philos. Soc.*, 129(3):447–475, 2000.

[10]   V. Belfi and C. Wilkerson. Some examples in the theory of *P*-completions. *Indiana Univ. Math. J.*, 25(6):565–576, 1976.

[11]   F. Borceux. *Handbook of categorical algebra. 1*, volume 50 of *Encyclopedia of mathematics and its applications.* Cambridge University Press, Cambridge, 1994. Basic category theory.

[12]   F. Borceux. *Handbook of categorical algebra. 2*, volume 51 of *Encyclopedia of mathematics and its applications.* Cambridge University Press, Cambridge, 1994. Categories and structures.

[13]   A. Borel. Some finiteness properties of adele groups over number fields. *Inst. Hautes Études Sci. Publ. Math.*, (16):5–30, 1963.

[14]   R. Bott. Quelques remarques sur les théorèmes de périodicité. *Bull. Soc. Math. France*, 87:293–310, 1959.

[15]   N. Bourbaki. *Algebra. I. Chapters 1–3.* Elements of Mathematics (Berlin). Springer-Verlag, Berlin, 1989. Translated from the French, reprint of the 1974 edition.

[16] A. K. Bousfield. The localization of spaces with respect to homology. *Topology*, 14:133–150, 1975.

[17] A. K. Bousfield. Constructions of factorization systems in categories. *J. Pure Appl. Algebra*, 9(2):207–220, 1976/77.

[18] A. K. Bousfield. On the *p*-adic completions of nonnilpotent spaces. *Trans. Amer. Math. Soc.*, 331(1):335–359, 1992.

[19] A. K. Bousfield. Homotopical localizations of spaces. *Amer. J. Math.*, 119(6):1321–1354, 1997.

[20] A. K. Bousfield and V. K. A. M. Gugenheim. On PL de Rham theory and rational homotopy type. *Mem. Amer. Math. Soc.*, 8(179):ix+94, 1976.

[21] A. K. Bousfield and D. M. Kan. *Homotopy limits, completions and localizations*, volume 304 of *Lecture notes in mathematics*. Springer-Verlag, Berlin, 1972.

[22] W. Browder, A. Liulevicius, and F. P. Peterson. Cobordism theories. *Ann. of Math. (2)*, 84:91–101, 1966.

[23] E. H. Brown, Jr. and R. H. Szczarba. Real and rational homotopy theory. In *Handbook of algebraic topology*, pages 867–915. North-Holland, Amsterdam, 1995.

[24] K. S. Brown. *Cohomology of groups*, volume 87 of *Graduate texts in mathematics*. Springer-Verlag, New York, N. Y., 1994. Corrected reprint of the 1982 original.

[25] H. Cartan and S. Eilenberg. *Homological algebra*. Princeton University Press, Princeton, N. J., 1956.

[26] C. Casacuberta and J. L. Rodríguez. On towers approximating homological localizations. *J. London Math. Soc. (2)*, 56(3):645–656, 1997.

[27] C. Casacuberta, D. Scevenels, and J. H. Smith. Implications of large-cardinal principles in homotopical localization. *Adv. Math.*, 197(1):120–139, 2005.

[28] W. Chachólski. On the functors CW_A and P_A. *Duke Math. J.*, 84(3):599–631, 1996.

[29] J. D. Christensen and M. Hovey. Quillen model structures for relative homological algebra. *Math. Proc. Cambridge Philos. Soc.*, 133(2):261–293, 2002.

[30] F. R. Cohen. A course in some aspects of classical homotopy theory. In *Algebraic topology (Seattle, Wash., 1985)*, volume 1286 of *Lecture notes in mathematics*, pages 1–92. Springer, Berlin, 1987.

[31] F. R. Cohen, T. J. Lada, and J. P. May. *The homology of iterated loop spaces*, volume 533 of *Lecture notes in mathematics*. Springer-Verlag, Berlin, 1976. Also available at math.uchicago.edu/~may.

[32] M. Cole. The homotopy category of chain complexes is a homotopy category. Preprint.

[33] M. Cole. Mixing model structures. *Topology Appl.*, 153(7):1016–1032, 2006.

[34] J. F. Davis and P. Kirk. *Lecture notes in algebraic topology*, volume 35 of *Graduate studies in mathematics*. American Mathematical Society, Providence, R. I., 2001.

[35] P. Deligne, P. Griffiths, J. Morgan, and D. Sullivan. Real homotopy theory of Kähler manifolds. *Invent. Math.*, 29(3):245–274, 1975.

[36] T. tom Dieck. *Algebraic topology*. EMS textbooks in mathematics. European Mathematical Society (EMS), Zürich, 2008.

[37] A. Dold and R. Thom. Quasifaserungen und unendliche symmetrische Produkte. *Ann. of Math. (2)*, 67:239–281, 1958.

[38] V. G. Drinfel'd. Quasi-Hopf algebras. *Algebra i Analiz*, 1(6):114–148, 1989. English translation in *Leningrad Math. J.* (1), 1419–1457, 1990.

[39] E. Dror, W. G. Dwyer, and D. M. Kan. An arithmetic square for virtually nilpotent spaces. *Illinois J. Math.*, 21(2):242–254, 1977.

[40]   E. Dror Farjoun. *Cellular spaces, null spaces and homotopy localization*, volume 1622 of *Lecture notes in mathematics*. Springer-Verlag, Berlin, 1996.

[41]   W. G. Dwyer. Strong convergence of the Eilenberg-Moore spectral sequence. *Topology*, 13:255–265, 1974.

[42]   W. G. Dwyer, P. S. Hirschhorn, D. M. Kan, and J. H. Smith. *Homotopy limit functors on model categories and homotopical categories*, volume 113 of *Mathematical surveys and monographs*. American Mathematical Society, Providence, R. I., 2004.

[43]   W. G. Dwyer and J. Spaliński. Homotopy theories and model categories. In *Handbook of algebraic topology*, pages 73–126. North-Holland, Amsterdam, 1995.

[44]   W. G. Dwyer and C. W. Wilkerson. Homotopy fixed-point methods for Lie groups and finite loop spaces. *Ann. of Math. (2)*, 139(2):395–442, 1994.

[45]   S. Eilenberg and N. Steenrod. *Foundations of algebraic topology*. Princeton University Press, Princeton, N. J., 1952.

[46]   A. D. Elmendorf, I. Kriz, M. A. Mandell, and J. P. May. *Rings, modules, and algebras in stable homotopy theory*, volume 47 of *Mathematical surveys and monographs*. American Mathematical Society, Providence, R. I., 1997. With an appendix by M. Cole.

[47]   L. Fajstrup and J. Rosický. A convenient category for directed homotopy. *Theory Appl. Categ.*, 21(1):7–20, 2008.

[48]   Y. Félix, S. Halperin, and J.-C. Thomas. *Rational homotopy theory*, volume 205 of *Graduate texts in mathematics*. Springer-Verlag, New York, N. Y., 2001.

[49]   R. Fritsch and R. A. Piccinini. *Cellular structures in topology*, volume 19 of *Cambridge studies in advanced mathematics*. Cambridge University Press, Cambridge, 1990.

[50]   P. Gabriel and M. Zisman. *Calculus of fractions and homotopy theory*. Ergebnisse der Mathematik und ihrer Grenzgebiete, Band 35. Springer-Verlag, New York, N. Y., 1967.

[51]   J. Gillespie. The flat model structure on Ch(R). *Trans. Amer. Math. Soc.*, 356(8):3369–3390 (electronic), 2004.

[52]   J. Gillespie. The flat model structure on complexes of sheaves. *Trans. Amer. Math. Soc.*, 358(7):2855–2874 (electronic), 2006.

[53]   P. Goerss and J. Jardine. *Simplicial homotopy theory*, volume 174 of *Progress in mathematics*. Birkhäuser Verlag, Basel, 1999.

[54]   P. Goerss and K. Schemmerhorn. Model categories and simplicial methods. In *Interactions between homotopy theory and algebra*, volume 436 of *Contemporary mathematics*, pages 3–49. American Mathematical Society, Providence, R. I., 2007.

[55]   B. I. Gray. Spaces of the same n-type, for all n. *Topology*, 5:241–243, 1966.

[56]   M. Hall, Jr. *The theory of groups*. Macmillan, New York, N. Y., 1959.

[57]   P. Hall. Finiteness conditions for soluble groups. *Proc. London Math. Soc. (3)*, 4:419–436, 1954.

[58]   D. K. Harrison. Infinite abelian groups and homological methods. *Ann. of Math. (2)*, 69:366–391, 1959.

[59]   A. Hatcher. *Algebraic topology*. Cambridge University Press, Cambridge, 2002.

[60]   P. Hilton. Localization and cohomology of nilpotent groups. *Math. Z.*, 132:263–286, 1973.

[61]   P. Hilton and G. Mislin. Bicartesian squares of nilpotent groups. *Comment. Math. Helv.*, 50(4):477–491, 1975.

[62]   P. Hilton, G. Mislin, and J. Roitberg. *Localization of nilpotent groups and spaces*. Volume 15 in *North-Holland mathematics studies*. North-Holland Publishing, Amsterdam, 1975.

[63]   P. Hilton and J. Roitberg. On principal S^3-bundles over spheres. *Ann. of Math. (2)*, 90:91–107, 1969.

[64]   P. Hilton and J. Roitberg. On the Zeeman comparison theorem for the homology of quasi-nilpotent fibrations. *Quart. J. Math. Oxford Ser. (2)*, 27(108):433–444, 1976.

[65]   P. S. Hirschhorn. *Model categories and their localizations*, volume 99 of *Mathematical surveys and monographs*. American Mathematical Society, Providence, R. I., 2003.

[66]   M. Hovey. *Model categories*, volume 63 of *Mathematical surveys and monographs*. American Mathematical Society, Providence, R. I., 1999.

[67]   M. Hovey. Model category structures on chain complexes of sheaves. *Trans. Amer. Math. Soc.*, 353(6):2441–2457 (electronic), 2001.

[68]   M. Hovey. Cotorsion pairs and model categories. In *Interactions between homotopy theory and algebra*, volume 436 of *Contemporary mathematics*, pages 277–296. American Mathematical Society, Providence, R. I., 2007.

[69]   J. R. Hubbuck. On homotopy commutative H-spaces. *Topology*, 8:119–126, 1969.

[70]   W. Hurewicz. On the concept of fiber space. *Proc. Nat. Acad. Sci. U. S. A.*, 41:956–961, 1955.

[71]   D. C. Isaksen. A model structure on the category of pro-simplicial sets. *Trans. Amer. Math. Soc.*, 353(7):2805–2841 (electronic), 2001.

[72]   N. Jacobson. *Lectures in abstract algebra. Vol. I. Basic concepts.* D. Van Nostrand, Toronto–New York–London, 1951.

[73]   C. U. Jensen. On the vanishing of $\varprojlim^{(i)}$. *J. Algebra*, 15:151–166, 1970.

[74]   D. M. Kan. On c. s. s. complexes. *Amer. J. Math.*, 79:449–476, 1957.

[75]   G. M. Kelly. *Basic concepts of enriched category theory*, volume 64 of *London mathematical society lecture note series*. Cambridge University Press, Cambridge, 1982.

[76]   G. M. Kelly and R. Street. Review of the elements of 2-categories. In *Category Seminar (Proc. Sem., Sydney, 1972/1973)*, volume 420 of *Lecture notes in mathematics*, pages 75–103. Springer, Berlin, 1974.

[77]   I. Kříž and J. P. May. Operads, algebras, modules and motives. *Astérisque*, (233): iv+145pp, 1995. Also available at math.uchicago.edu/∼may.

[78]   A. G. Kurosh. *The theory of groups.* Chelsea Publishing, New York, N. Y., 1960. Translated from the Russian and edited by K. A. Hirsch. 2nd English ed. 2 volumes.

[79]   S. Mac Lane. *Homology.* Classics in mathematics. Springer-Verlag, Berlin, 1995. Reprint of the 1975 edition.

[80]   S. Mac Lane. *Categories for the working mathematician*, volume 5 of *Graduate texts in mathematics*. Springer-Verlag, New York, N. Y., 2nd edition, 1998.

[81]   I. Madsen and R. J. Milgram. *The classifying spaces for surgery and cobordism of manifolds*, volume 92 of *Annals of mathematics studies*. Princeton University Press, Princeton, N. J., 1979.

[82]   M. A. Mandell and J. P. May. Equivariant orthogonal spectra and S-modules. *Mem. Amer. Math. Soc.*, 159(755):x+108, 2002. Also available at math.uchicago.edu/∼may.

[83]   M. A. Mandell, J. P. May, S. Schwede, and B. Shipley. Model categories of diagram spectra. *Proc. London Math. Soc. (3)*, 82(2):441–512, 2001. Also available at math .uchicago.edu/∼may.

[84]   J. P. May. The cohomology of restricted Lie algebras and of Hopf algebras. *J. Algebra*, 3:123–146, 1966. Also available at math.uchicago.edu/∼may.

[85]   J. P. May. Some remarks on the structure of Hopf algebras. *Proc. Amer. Math. Soc.*, 23:708–713, 1969. Also available at math.uchicago.edu/∼may.

[86] J. P. May. A general algebraic approach to Steenrod operations. In *The Steenrod algebra and its applications (Proc. Conf. to Celebrate N. E. Steenrod's Sixtieth Birthday, Battelle Memorial Inst., Columbus, Ohio, 1970)*, volume 168 of *Lecture notes in mathematics*, pages 153–231. Springer, Berlin, 1970. Also available at math.uchicago.edu/~may.

[87] J. P. May. *The geometry of iterated loop spaces*, volume 271 of *Lectures notes in mathematics*. Springer-Verlag, Berlin, 1972. Also available at math.uchicago.edu/~may.

[88] J. P. May. Classifying spaces and fibrations. *Mem. Amer. Math. Soc.*, 1(155):xiii+98, 1975. Also available at math.uchicago.edu/~may.

[89] J. P. May. E_∞ *ring spaces and E_∞ ring spectra*, volume 577 of *Lecture notes in mathematics*. Springer-Verlag, Berlin, 1977. With contributions by Frank Quinn, Nigel Ray, and Jørgen Tornehave.

[90] J. P. May. Infinite loop space theory. *Bull. Amer. Math. Soc.*, 83(4):456–494, 1977. Also available at math.uchicago.edu/~may.

[91] J. P. May. The dual Whitehead theorems. In *Topological topics*, volume 86 of *London mathematical society lecture note series*, pages 46–54. Cambridge University Press, Cambridge, 1983. Also available at math.uchicago.edu/~may.

[92] J. P. May. *Simplicial objects in algebraic topology*. Chicago lectures in mathematics. University of Chicago Press, Chicago, Ill. 1992. Reprint of the 1967 original.

[93] J. P. May. *A concise course in algebraic topology*. Chicago lectures in mathematics. University of Chicago Press, Chicago, Ill. 1999.

[94] J. P. May. The additivity of traces in triangulated categories. *Adv. Math.*, 163(1):34–73, 2001. Also available at math.uchicago.edu/~may.

[95] J. P. May. Picard groups, Grothendieck rings, and Burnside rings of categories. *Adv. Math.*, 163(1):1–16, 2001. Also available at math.uchicago.edu/~may.

[96] J. P. May. What precisely are E_∞ ring spaces and E_∞ ring spectra? In *New topological contexts for Galois theory and algebraic geometry (BIRS 2008)*, volume 16 of *Geometry and topology monographs*, pages 215–282. Geom. Topol. Publ., Coventry, 2009. Also available at math.uchicago.edu/~may.

[97] J. P. May and J. Sigurdsson. *Parametrized homotopy theory*, volume 132 of *Mathematical surveys and monographs*. American Mathematical Society, Providence, R. I., 2006. Also available at math.uchicago.edu/~may.

[98] J. McCleary. *A user's guide to spectral sequences*, volume 58 of *Cambridge studies in advanced mathematics*. Cambridge University Press, Cambridge, 2nd edition, 2001.

[99] C. A. McGibbon. Self-maps of projective spaces. *Trans. Amer. Math. Soc.*, 271(1):325–346, 1982.

[100] C. A. McGibbon. The Mislin genus of a space. In *The Hilton Symposium 1993 (Montreal, PQ)*, volume 6 of *CRM proceedings and lecture notes*, pages 75–102. American Mathematical Society, Providence, R. I., 1994.

[101] C. A. McGibbon. On the localization genus of a space. In *Algebraic topology: New trends in localization and periodicity (Sant Feliu de Guíxols, 1994)*, volume 136 of *Progress in mathematics*, pages 285–306. Birkhäuser, Basel, 1996.

[102] J. Milnor. The geometric realization of a semi-simplicial complex. *Ann. of Math. (2)*, 65:357–362, 1957.

[103] J. Milnor. On spaces having the homotopy type of CW-complex. *Trans. Amer. Math. Soc.*, 90:272–280, 1959.

[104] J. Milnor and J. C. Moore. On the structure of Hopf algebras. *Ann. of Math. (2)*, 81:211–264, 1965.

[105] G. Mislin. The genus of an H-space. In *Symposium on algebraic topology (Battelle Seattle Res. Center, Seattle, Wash., 1971)*, volume 249 of *Lecture notes in mathematics*, pages 75–83. Springer, Berlin, 1971.

[106] G. Mislin. Wall's obstruction for nilpotent spaces. *Topology*, 14(4):311–317, 1975.

[107] G. Mislin. Finitely dominated nilpotent spaces. *Ann. of Math. (2)*, 103(3):547–556, 1976.

[108] J. C. Moore. Algèbres de Hopf universelles. In *Périodicité des groupes d'homotopie stables des groupes classiques, d'après Bott*, volume 12 of *Séminaire Henri Cartan*. École Normale Supérieure. Secrétariat mathématique, Paris, 1959–1960.

[109] F. Morel. Quelques remarques sur la cohomologie modulo p continue des pro-p-espaces et les résultats de J. Lannes concernant les espaces fonctionnels hom(BV, X). *Ann. Sci. École Norm. Sup. (4)*, 26(3):309–360, 1993.

[110] F. Morel. Ensembles profinis simpliciaux et interprétation géométrique du foncteur T. *Bull. Soc. Math. France*, 124(2):347–373, 1996.

[111] A. Neeman. *Triangulated categories*, volume 148 of *Annals of mathematics studies*. Princeton University Press, Princeton, N. J., 2001.

[112] P. F. Pickel. Finitely generated nilpotent groups with isomorphic finite quotients. *Trans. Amer. Math. Soc.*, 160:327–341, 1971.

[113] D. Quillen. *Homotopical algebra*. Volume 43 in *Lecture notes in mathematics*. Springer-Verlag, Berlin, 1967.

[114] D. Quillen. The geometric realization of a Kan fibration is a Serre fibration. *Proc. Amer. Math. Soc.*, 19:1499–1500, 1968.

[115] D. Quillen. Rational homotopy theory. *Ann. of Math. (2)*, 90:205–295, 1969.

[116] D. Quillen. On the cohomology and K-theory of the general linear groups over a finite field. *Ann. of Math. (2)*, 96:552–586, 1972.

[117] D. L. Rector. Loop structures on the homotopy type of S^3. In *Symposium on algebraic topology (Battelle Seattle Res. Center, Seattle, Wash., 1971)*, volume 249 of *Lecture notes in mathematics*, pages 99–105. Springer, Berlin, 1971.

[118] E. Riehl. Algebraic model structures. arXiv:0910.2733.

[119] J.-E. Roos. Sur les foncteurs dérivés de $\varprojlim$. Applications. *C. R. Acad. Sci. Paris*, 252: 3702–3704, 1961.

[120] J.-E. Roos. Derived functors of inverse limits revisited. *J. London Math. Soc. (2)*, 73(1): 65–83, 2006.

[121] R. Schön. Fibrations over a CWh-base. *Proc. Amer. Math. Soc.*, 62(1):165–166, 1977.

[122] R. Schwänzl and R. M. Vogt. Strong cofibrations and fibrations in enriched categories. *Arch. Math. (Basel)*, 79(6):449–462, 2002.

[123] P. Selick. *Introduction to homotopy theory*, volume 9 of *Fields Institute monographs*. American Mathematical Society, Providence, R. I., 1997.

[124] J.-P. Serre. Homologie singulière des espaces fibrés. Applications. *Ann. of Math. (2)*, 54:425–505, 1951.

[125] J.-P. Serre. Groupes d'homotopie et classes de groupes abéliens. *Ann. of Math. (2)*, 58:258–294, 1953.

[126] J.-P. Serre. *Cohomologie galoisienne*, volume 5 of *Lecture notes in mathematics*. Springer-Verlag, Berlin, 5th edition, 1994.

[127] M. Shulman. Comparing composites of left and right derived functors. arXiv: 0706.2868v1.

[128] M. Shulman. Homotopy limits and colimits and enriched homotopy theory. arXiv: math/0610194v3.

[129] J. Stasheff. A classification theorem for fibre spaces. *Topology*, 2:239–246, 1963.

[130] N. E. Steenrod. *Cohomology operations*. Study 50 in *Annals of mathematics studies*. Princeton University Press, Princeton, N. J., 1962. Lectures by N. E. Steenrod written and revised by D. B. A. Epstein.

[131] A. Strøm. Note on cofibrations. *Math. Scand.*, 19:11–14, 1966.

[132] A. Strøm. The homotopy category is a homotopy category. *Arch. Math. (Basel)*, 23:435–441, 1972.

[133] D. Sullivan. Genetics of homotopy theory and the Adams conjecture. *Ann. of Math. (2)*, 100:1–79, 1974.

[134] D. Sullivan. Infinitesimal computations in topology. *Inst. Hautes Études Sci. Publ. Math.*, (47):269–331 (1978), 1977.

[135] D. Sullivan. *Geometric topology: localization, periodicity and Galois symmetry*, volume 8 of *K-Monographs in mathematics*. Springer, Dordrecht, 2005. The 1970 MIT notes, edited and with a preface by Andrew Ranicki.

[136] M. E. Sweedler. *Hopf algebras*. Mathematics lecture note series. W. A. Benjamin, New York, N. Y., 1969.

[137] R. Thom. Quelques propriétés globales des variétés différentiables. *Comment. Math. Helv.*, 28:17–86, 1954.

[138] J. Verdier. Catégories dérivées. In P. Deligne, editor, *Cohomologie étale*, volume 569 of *Lecture notes in mathematics*. Springer-Verlag, Berlin–New York, 1977. Séminaire de Géométrie Algébrique du Bois-Marie SGA 4½.

[139] R. M. Vogt. The help-lemma and its converse in quillen model categories. arXiv: 1004.5249.

[140] C. T. C. Wall. Finiteness conditions for CW-complexes. *Ann. of Math. (2)*, 81:56–69, 1965.

[141] R. B. Warfield, Jr. *Nilpotent groups*, volume 513 of *Lecture notes in mathematics*. Springer-Verlag, Berlin, 1976.

[142] C. A. Weibel. *An introduction to homological algebra*, volume 38 of *Cambridge studies in advanced mathematics*. Cambridge University Press, Cambridge, 1994.

[143] G. W. Whitehead. *Elements of homotopy theory*, volume 61 of *Graduate texts in mathematics*. Springer-Verlag, New York, N. Y., 1978.

[144] C. W. Wilkerson. Applications of minimal simplicial groups. *Topology*, 15(2):111–130, 1976.

[145] A. Zabrodsky. Homotopy associativity and finite CW complexes. *Topology*, 9:121–128, 1970.

[146] A. Zabrodsky. On the genus of finite CW-*H*-spaces. *Comment. Math. Helv.*, 49:48–64, 1974.

[147] A. Zabrodsky. *Hopf spaces*, volume 22 of *North-Holland mathematics studies*. North-Holland Publishing, Amsterdam, 1976.

[148] H. Zassenhaus. *The Theory of Groups*. Chelsea Publishing, New York, N. Y., 1949. Translated from the German by Saul Kravetz.

Index